Earth and Planetary Science

Earth and Planetary Science

Edited by **John Wayne**

Callisto Reference
New York

Published by Callisto Reference,
106 Park Avenue, Suite 200,
New York, NY 10016, USA
www.callistoreference.com

Earth and Planetary Science
Edited by John Wayne

International Standard Book Number: 978-1-63239-723-2 (Hardback)

Printed in China.

Contents

Preface

This book aims to highlight the current researches and provides a platform to further the scope of innovations in this area. This book is a product of the combined efforts of many researchers and scientists from different parts of the world. The objective of this book is to provide the readers with the latest information in the field.

Earth and planetary science refers to that field of science which deals with the scientific study of the planets, moons and other celestial bodies, particularly in our solar system. Its objective is to understand their structure, dynamics, evolution, and their history. Scientists use physics, chemistry, biology, chronology and mathematics to analyse these planets, especially earth. This book provides comprehensive insights into this field and the various studies that are constantly contributing towards advancing technologies and evolution of this area are examined in detail. Those with an interest in the subject of earth and planetary science will find this text very resourceful. Coherent flow of topics, reader-friendly language and extensive material provided in this book will make it a beneficial source of knowledge.

I would like to express my sincere thanks to the authors for their dedicated efforts in the completion of this book. I acknowledge the efforts of the publisher for providing constant support. Lastly, I would like to thank my family for their support in all academic endeavors.

Editor

Comparing the visual perception and aesthetic evaluation of natural landscapes in Russia and Japan: cultural and environmental factors

Elena G Petrova[1*], Yury V Mironov[2], Yoji Aoki[8], Hajime Matsushima[3], Satoshi Ebine[3], Katsunori Furuya[4], Anastasia Petrova[5], Norimasa Takayama[6] and Hirofumi Ueda[7]

Abstract

Japan and Russia have deeply rooted cultural traditions regarding natural landscape appreciation, share a common border, and have areas with similar natural environments. They differ, however, in cultural, historical, and economic aspects. The purpose of this study was to reveal the similarities and differences between Russian and Japanese respondents regarding the visual and emotional evaluation of landscapes based on ethno-cultural and regional differences. We asked respondents at universities in Russia (Moscow, Irkutsk, and Kamchatka) and Japan (Hokkaido, Chiba, and Miyazaki) to group and rate 70 landscape images. Unlike theoretical concepts that explain landscape preferences within an evolutionary framework or according to individual and cultural differences, we found that these factors interact in more complicated ways. Cultural traditions and features of the natural environment that were familiar to respondents influenced their visual perception and aesthetic evaluation of landscape. Russian respondents seemed more emotional while Japanese respondents tended to be more restrained in their assessments. However, there was a strong correlation between their estimates of landscape attractiveness, which might confirm the existence of universal human concepts of landscape aesthetics. The most attractive for both Russian and Japanese respondents were waterfalls, mountains, and lakes; waterless plains were the least attractive. At the same time, we found cross-cultural differences in assessing seacoasts, rivers, forests, and swampy plains. There was practically no correlation between Russian and Japanese respondents in their appreciation of exotic/familiar landscapes. For the Russian respondents, the most exotic landscapes were also the most attractive, although we did not observe such a tendency for the Japanese. All respondents appreciated certain familiar landscapes that were symbols of native nature as very attractive. Unlike 'geoscientific' landscape classifications, in the visual and emotional grouping of landscapes, the most important feature appeared to be the presence/absence of water and the type of water basin (river, lake, and sea); for Russian respondents (especially Muscovites), topography was also important, while the Japanese respondents mostly used visual and seasonal characteristics in their classifications. All Japanese respondents assessed the attractiveness and exoticism of landscapes almost identically, while there were some differences among Russian respondents from different regions.

Keywords: Visual landscape classification; Aesthetic evaluation; Attractive landscape; Exotic landscape; Traditional landscape appreciation; Influence of natural environment

* Correspondence: epgeo@mail.ru
[1]Faculty of Geography, Lomonosov Moscow State University, GSP-1, Leninskie Gory, Moscow 119991, Russia
Full list of author information is available at the end of the article

Background

Current urbanization and economic development that lead to increases in industrial, transport, agricultural, recreational, and other man-made loadings on our natural environment threaten natural landscapes at different levels, from local to global. When selecting landscapes for special care and conservation, we should consider not only the objective assessment of their qualities, significance, and usefulness - that is, their ecological, economic, and social value - but also their aesthetic originality since it is very important to preserve aesthetically valuable landscapes for future generations. For example, the scenic beauty and high aesthetic value of the landscape is one of the key prerequisites for assigning National Park status to a given area (IUCN - International Union for Conservation of Nature 1969).

In recent decades, many studies devoted to the psychological and aesthetic evaluation of landscapes have been published (Aoki 1999, 2007). Nevertheless, the concept of landscape appreciation has yet to be clearly defined and there is a lack of agreement on evaluation methods, the factors that determine landscape preferences, the steps in appreciation, and so forth.

There is no consensus on what the phenomenon of 'landscape' truly is (Jones 1991; Brabyn 2009), how people physiologically and mentally experience it, which human attributes affect the perception of it, and why it is apprehended the way it is (Aoki 2007). Some researchers agree that people from various distinctive cultures prefer natural environments to built or otherwise human-influenced ones (e.g., Kaplan and Kaplan 1989; Yang and Kaplan 1990; Ulrich 1993; van den Berg et al. 2003). They explain these preferences in terms of 'attention restoration theory': natural environments offer support for the recovery of attention, which is a very limited resource that is easily exhausted in daily life, including work. Several studies conclude that the similarities in the visual evaluation of landscapes surpass the differences across cultures as well as smaller groups (Purcell et al. 1994; Herzog et al. 2003; Staats and Hartig 2004). Some authors interpret these similarities as supporting an evolutionary theory of human landscape preferences (Appleton 1975; Orians 1986; Kaplan 2001). However, other studies identify individual and intergroup distinctions in environmental preferences within an evolutionary framework, in addition to the similarities (Hull and Reveli 1989; Dramstad et al. 2006). Bourassa (1991) suggests that the difficulties in understanding landscape derive from human evolution and are affected by both ontogenesis and phylogenesis. He proposes a need for three steps of clarification that consider 'biological and cultural constraints and personal idiosyncrasies' (Bourassa, 1991, p. 110). Lothian (1999) emphasizes that many contemporary theories of landscape quality have adopted Kant's subjectivist view of aesthetics, which identifies beauty as 'in the eye of the beholder' rather than in the actual object. Backhaus (2011) concludes that there is no such thing as an absolute perspective on landscapes; people can perceive the same landscape in different ways.

Two kinds of respondent attributes could influence landscape preferences. The first involves the identities of a human group (e.g., nationality, ethnicity, living environment, gender, and age) (Tips and Savasdisara 1986; Takayama et al. 2012). The second includes individual characteristics (e.g., social status, personal history, educational level, and professional interests, hobbies) (Aoki 2007). Kaltenborn and Bjerke (2002) argue that both cultural and genetic factors may contribute to evaluative responses to environmental types. According to Appleton (1975), original landscape appreciation derives from respondents' on-site experiences; it can evolve with physiological, and even sociological, age (Aoki and Kitamura 2001). Thus, aesthetic perception is a complex process that combines a person's sensory abilities, which are both inherent and acquired through training, education, and social development (Nikolaev 2005).

A natural environment that is native, or the most familiar, to respondents can determine landscape appreciation as well. According to Leo Gumilyov (1993), the famous Russian ethnologist and founder of ethnogeny theory, there is an indissoluble union between ethnos and its surrounding landscape; thus, this relationship can exert influence upon people's perceptions of the world. Therefore, when studying human landscape preferences, it is important to compare them with landscapes that respondents consider familiar and unfamiliar (exotic).

The 'outside views' of the representatives of other cultures allow us to see the unusual in the usual, take a fresh look at familiar landscapes, descry their originality and beauty, and experience new aesthetic features. A good example is the influence of Russian literature on forest landscape appreciation in Japan. After Japan was opened to foreigners during the Meiji Era (in the late 19th century), the Japanese writers Doppo Kunikida and Roka Tokutomi became interested in Russian literature. Poetic descriptions of forests by Russian writers such as Ivan Turgenev and Leo Tolstoy stimulated these Japanese men of letters to recognize the beauty of deciduous forests as well, to praise them in their own works (Kunikida 1901), and to use deciduous trees (*zokibayashi* in Japanese) in their formalized gardens. In earlier times, people in Japan did not consider such trees aesthetically valuable and only used them as firewood. In this sense, the Russian appreciation of forests facilitated the first step toward the widespread preservation of forest landscapes in Japan (Okajima et al. 2010).

With regard to the authors mentioned above, in our conception of natural landscape appreciation, we assume that people belonging to different cultures differ in their landscape preferences as a result of a number of ethnocultural features, as well as historical, social, and environmental factors. It is very important to reveal and consider these differences. The purpose of this study is to compare human landscape preferences, visual and emotional perception, and the aesthetic evaluation of natural landscapes by Japanese and Russians. Both Japan and Russia have deeply rooted cultural traditions regarding natural landscape appreciation. Our countries share a common border and also have areas with similar natural environments; they differ, however, in cultural, social, historical, and economic aspects. We believe that such a cross-cultural analysis is not only interesting from a theoretical perspective but also very useful in practical and humane aspects, as it should promote a better understanding for solving many intercultural issues, including international cooperation in landscape protection, management, and planning.

Methods

Visualization of landscapes

All of the senses are involved in the aesthetic perception of landscape. However, we receive almost 90% of aesthetic information through contemplation (Nikolaev 2005). Thus, in contrast to the geoscientific analysis of landscape, which investigates its substantive properties, aesthetic perception mainly deals with landscape appearance - views or scenery that are contemplated by a perceptive and communicative observer. The aesthetic evaluation of landscape is primarily one of visual assessment. Therefore, a very important aspect of our research is visual landscape presentation. In the early stages of landscape studies in Russia, Japan, and other countries, the on-site approach prevailed, wherein respondents visited the survey areas and evaluated them directly (Sevenant and Antrop 2009). However, this approach limited the number of respondents, visits, and sites (Aoki 1999). Thus, we have chosen another approach that focuses on color photos.

First, we created a digital photo database of natural landscapes both similar and unique for Russia and Japan using the principles for taking pictures described by Petrova et al. (2008a). Almost 900 pictures were taken in different geographical, climatic, and landscape zones of both countries. Our goal was to show the greatest variety of Russian and Japanese landscapes. All the pictures are documentary images of landscapes taken with equally sharp focus from front to back under normal lighting at eye level using wide-angle lenses with a focal distance of 28 to 35 mm (equivalent to a 35 mm film camera), which corresponds to the angular span of the human eye. We did not apply any of the special effects used in artistic photography, such as concentrating on certain foreground objects with a blurred background, beautiful sunsets and sunrises, spectacular beams of light that illuminate specific objects, spectacular mists, 'silky' water, and increased contrast that focuses on particular elements. Nor did we use any editing programs that divert attention from the perception of landscapes as a whole and emphasize particular objects.

Selection of photo images

For the survey procedure, we selected 70 photographs from the database. They represented in equal numbers the main types of Russian and Japanese landscapes in seven categories: forests, treeless plains (including tundras, wetlands, and steppes), mountains, river valleys, waterfalls, lakes, and seacoasts. Based on the opinion of six Russian and six Japanese professionals, we defined the number of pictures and the range of categories, which together offer an idea of the diversity of natural landscapes in both countries. We selected the specific images for each category, which included five pictures of Russia and five of Japan, using soft preferential voting by the same experts (Aoki and Petrova 2010). When choosing this selection of 70 images, we were limited by the actual duration of respondent interviews at university centers (two academic hours). In our experience, respondents generally require more time to properly evaluate a larger number of photographs.

During the interviews, each respondent received the same set of 70 photos, which were printed in postcard format (13 cm × 18 cm). The order of the photos was random but the same for all participants. To avoid influencing respondents' opinions, we did not specify geographical characteristics such as landscape category, object, or survey area.

Respondents and survey procedures

Aiming to discover possible landscape preferences based on ethno-cultural and regional differences, we invited respondents with similar social and demographic characteristics (age, education, social status, etc.) from different regions of Japan and Russia. We conducted surveys at universities in Russia (Moscow State University in Central Russia, Irkutsk State University in East Siberia, and Kamchatka State University in Far Eastern Russia) and Japan (Hokkaido University, Chiba University, and Miyazaki University). When selecting these university centers, we were guided by the same principles: the metropolitan areas of both countries and two outlying regions, which differ most strongly in their natural environments, were represented. Students, young teachers, and researchers, mainly in geosciences as well as the humanities, ranging from 17 to 30 years of age participated in the survey. There was a roughly equal number of men and women. We interviewed 248 respondents (about 40 to 50 respondents from each region); this group size is generally considered sufficient for polls within homogenous groups (Ermolaev 2003).

We asked each respondent

- to fill out a form requesting personal data (age, gender, education, etc.);
- to group all 70 landscape images according to personal perceptions of their visual similarities and name each group;
- to rate the attractiveness and exoticism of each landscape image; and
- to characterize the natural landscapes of Russia and Japan, respectively, using just three words.

The groupings of landscape images allowed us to discover which components were more important for respondents in their visual and emotional landscape classifications and how they interpreted the landscapes in the photos.

To evaluate the attractiveness of landscapes, we used a five-point scale where '5' indicated the most attractive landscape while '1' denoted an extremely unattractive landscape. A simpler three-point scale was chosen to assess the exoticism of landscapes, with exotic landscapes marked as '+1' and familiar ones '−1.' When respondents could not decide between these categories, they chose '0.' Comparing the attractiveness evaluation with the 'exotic versus familiar' assessment of the same landscapes, we tried to discover whether 'exotic' or 'familiar' landscapes were more preferable to the respondents and whether these attributes might influence the assessment of landscape attractiveness.

In the Japanese study areas, respondents were interviewed in Japanese; in Russia they were interviewed in Russian; therefore, there were no linguistic difficulties.

Statistical analysis

Data obtained from the interviews were analyzed using elementary and multivariate statistical methods.

We summarized the results of the photo groupings by comparing the seven landscape classifications, thus accumulating the 'collective opinion' of respondents from every region and in all study areas of Russia and Japan as a whole. For this, we calculated how often each image co-occurred with any of the other 69 images within the same landscape group as reported by each respondent. Then, we processed the resulting matrices of co-occurrence frequencies using modified factor analysis in the Statistical Package for the Social Sciences (SPSS 16.00 FULL). This method is based on hierarchical agglomerative clustering in which two images with the highest correlation in a multidimensional (in this case, 70-dimensional) space are merged into a cluster. These clusters consequentially form larger groups with lower correlations. It is important to note that a number of agglomeration steps and a number of clusters were not defined *a priori* but automatically produced by the algorithm. Thus, our initial set of landscape categories, which was unknown to the respondents, could not influence their classifications.

To reveal the similarities and differences in the assessment of landscape attractiveness and landscape exoticism between different groups of Russian and Japanese respondents, we calculated their mean scores for each landscape image. We used Student's t-test to assess the statistical significance of the differences between the estimates of landscape attractiveness by the Russian and Japanese respondents. When the Student's t-test values exceeded *2.58*, the differences between their estimates were considered significant (error probability: $p < 0.01$); for values less than 1.96 ($p > 0.05$), they were taken as insignificant, while for values of $1.96 < t < 2.58$ they were taken as undefined.

We also calculated the words Russian and Japanese respondents used most often to characterize the natural landscapes of both countries.

Statistical analysis allowed us to reveal some of the common features and main points of difference between Russian and Japanese respondents, as well as between groups of Russian respondents from different regions (Moscow, Irkutsk, and Kamchatka), in the visual and emotional evaluation of natural landscapes. Our analysis also discovered similarities between all groups of Japanese respondents in their assessments. Below, we discuss the main findings.

Results and discussion

Visual and emotional grouping of landscapes

Using hierarchical agglomerative clustering of the photographs, we obtained (1) six visual and emotional landscape classifications composed by each group of respondents and (2) a general visual and emotional landscape classification that comprises the 'collective' opinion of the Russian and Japanese respondents. In all cases, the revealed landscape groups generally corresponded to the set of landscape categories we used in our initial classification: forests, mountains, river valleys, waterfalls, lakes, and seacoasts. The only exceptions were 'treeless plains' that initially included images of tundras, wetlands, and steppes. The majority of respondents grouped them by the presence or absence of water in the picture and interpreted some images with water as 'swampy plains' or 'lowland rivers' and those without visible water as 'waterless plains.' This new set of landscapes identified by respondents was used as a legend for the resulting graphs showing the evaluation of attractiveness and exoticism (see pp. 7-8).

At a later step in the hierarchical agglomerative clustering of the photographs, forests, 'waterless plains,' and 'dry' mountains (without visible water in the photo) formed a larger group (super-group) of 'waterless' landscapes; 'swampy plains,' rivers, and waterfalls comprised a group of landscapes associated with river systems, while lakes and seacoasts formed a group of landscapes having large bodies of water (Table 1).

Thus, unlike 'geoscientific' landscape classifications - where physical properties, genetic unity, and homogeneity in both zonal (bioclimatic) and azonal (geology and relief) conditions play the leading roles (Isachenko 1991) - the most important feature in the visual and emotional grouping of landscapes appears to be the presence/absence of water and the type of water basin (river, lake, sea). These results were common among all groups of Russian and Japanese respondents and support an idea regarding similarities in human perceptions and visual evaluations of natural scenes. Our findings are consistent with those of other studies focused on visual landscape characteristics that use water as the key landscape element (Yang and Brown 1992; Hull and Stewart 1995; Ode et al. 2008).

At the same time, we ascertained some differences between Russian and Japanese respondents in grouping landscapes. For example, the division of some landscapes into subgroups of mountains and plains was more important for Russian - especially Muscovite - respondents, who are the least familiar with mountains but most impressed by them. In this regard, our findings only partially correspond to those of other studies that also use topography as the main indicator of landscape character (Brush 1981; Arriaza et al. 2004; Nasar and Li 2004; Ode et al. 2008).

Only Russian respondents used topography as a key parameter when grouping lakes. Unlike the Japanese, the Russians did not yet recognize lakes as a separate group. They combined lowland lakes with lowland rivers (although both are very important natural features in Central Russia, sometimes it is difficult to distinguish them in the photos), characterized some big lakes as seacoasts, and merged mountain lakes with mountains. However, when grouping mountains, Russian respondents separated mountain lakes from 'dry mountains' and folded mountains from volcanoes. In comparison, Japanese respondents did not separate volcanic lakes from lakes in folded mountains; visual characteristics were more important to them: they divided lakes into three subgroups not by their geological origin or topography but by the visible size of the water surface in the picture. Moreover, Japanese (unlike the Russians) did not separate volcanoes from folded mountains when grouping mountain landscapes. They considered mountains as a single group. This is probably because many mountains in Japan are of volcanic origin, while in Russia folded mountains and active volcanoes are usually in different locales. It is also noteworthy that Japanese poets, who often praised the beauty of volcanoes, did not refer to them as 'volcanoes' but as simply 'mountains' (Meshcheryakov 2012).

Another interesting result is that our analysis of forest groupings revealed evident differences not only between Russian and Japanese respondents but also between different groups of Russian respondents. Thus, the Moscow respondents clearly divided forest landscapes into two subgroups of mountains and plain forests. Japanese respondents used, again, purely visual characteristics, separating distant views of forests from close-up views within forests and distinguishing winter forests (seasonal characteristics). However, the Irkutsk and Kamchatka respondents did not actually identify any subgroups of forest landscapes.

Ueda et al. (2012) obtained similar results when analyzing sketches of forest landscape images made by Russian and Japanese respondents. In the Japanese research sites, 'close-up view' was the predominant category followed by 'bird's-eye view.' However, the results for the Russian research sites showed important differences: the 'sideways view' in Moscow, the 'close-up view' in Moscow and Irkutsk, and the 'bird's-eye view' and 'distant view' in

Table 1 Landscape categories in the initial classification and landscape groups revealed by respondents

Landscape groups (revealed by respondents)	Landscape categories (in the initial classification)						
	Forests	Waterfalls	Rivers	Treeless plains	Lakes	Seacoasts	Mountains
Waterless landscapes							
Forests	10			1			
Waterless plains				4			
Waterless mountains				1			7
Landscapes associated with river systems							
Waterfalls		10					
Swampy plains			1	4			
Rivers			8				
Landscapes having large bodies of water							
Seacoasts					2	10	
Lakes					9		3

The numbers indicate the number of photos belonging to this category or group of landscapes.

Figure 1 Landscape of Oze National Park in Japan.

Irkutsk and Kamchatka. We can explain these variations in terms of topography. The flat, vast terrains around Moscow do not provide 'distant' or 'bird's-eye' views of forests there, in contrast to the mountainous areas around Irkutsk, in Kamchatka, and in Japan. These differences between the survey regions explain why topography was the most important factor for the Moscow respondents' groupings of forest images, while the Japanese, Irkutsk, and Kamchatka respondents - who are more familiar with views from the outside of forests, which they usually see on mountain slopes in the background - did not consider this factor in their groupings.

In some cases, respondents interpreted the same landscape images differently. For example, the majority of Japanese respondents characterized the picture of Oze National Park in Japan (Figure 1) as a plain forest, Irkutsk and Kamchatka respondents saw it as a mountain forest, and the Moscow respondents considered it a swamp. We believe these interpretations also vary as a result of the differences in terrains between the study regions. For example, swampy landscapes are more familiar to the Moscow respondents than to the Japanese, who interpreted the stunted trees in the swamp as a forest; however, the Irkutsk and Kamchatka respondents focused on the mountains in the background rather than the swamp in the foreground.

Evaluation of landscape attractiveness and exoticism

During the interviews, we learned that the respondents considered attractive landscapes beautiful and comfortable, not only for long-term stays but also for permanent living. Exotic landscapes were considered unusual and foreign to the respondents. According to the respondents, even if such landscapes were unsightly, they would be interesting to visit at least once. Therefore, when assessing the attractiveness of familiar landscapes, respondents focused on both their aesthetic feelings and a sense of belonging - a sense of connection to the native landscape. However, in their assessments of the attractiveness of exotic landscapes, their cognitive interest in the unknown played an important role.

In general, there was a rather strong correlation between the estimates of landscape attractiveness by the Russian and Japanese respondents (correlation coefficient: $R = 0.73$), which might indicate the existence of universal human concepts regarding landscape aesthetics. However, Russian respondents appeared more emotional while the Japanese tended to be more restrained in their assessments. As Table 2 shows, Russian respondents assessed almost all the landscape types significantly higher than the Japanese did. The only exceptions were treeless (both swampy and waterless) plains; all respondents assessed them lowly and almost identically. The most attractive types for both Russian and Japanese respondents were waterfalls, mountains, and lakes; the least attractive were

Table 2 Statistical significance of differences between estimates of landscapes attractiveness by Russian and Japanese respondents

Group of landscapes	Russian respondents			Japanese respondents			Student's *t*-test
	n	*x*	*s*	*n*	*x*	*s*	
Waterfalls	1,270	4.26	0.97	1,230	3.78	1.03	**12**
Mountains	1,016	4.2	0.99	984	3.75	1.15	**9.4**
Lakes	1,524	4.15	0.99	1,476	3.73	1.09	**11.1**
Seacoasts	1,524	3.92	1.06	1,476	3.34	1.14	**14.4**
Rivers	1,016	3.86	1.05	984	3.69	1.03	**3.7**
Forests	1,397	3.72	1.08	1,353	3.32	1.16	**9.4**
Swampy plains	635	3.56	1.11	615	3.4	1.16	***2.5***
Waterless plains	508	3.08	1.21	492	3.07	1.18	*0.1*
All	8,890	3.93	1.08	8,610	3.54	1.13	**23.3**

n, number of estimates; *x*, average score; *s*, standard deviation.
Significant values ($t > 2.58$; $p < 0.01$) are marked in bold, undefined values ($1.96 < t < 2.58$; $0.01 > p < 0.05$) in bold italics, and insignificant values ($t < 1.96$; $p > 0.05$) in italics; *p*, error probability.

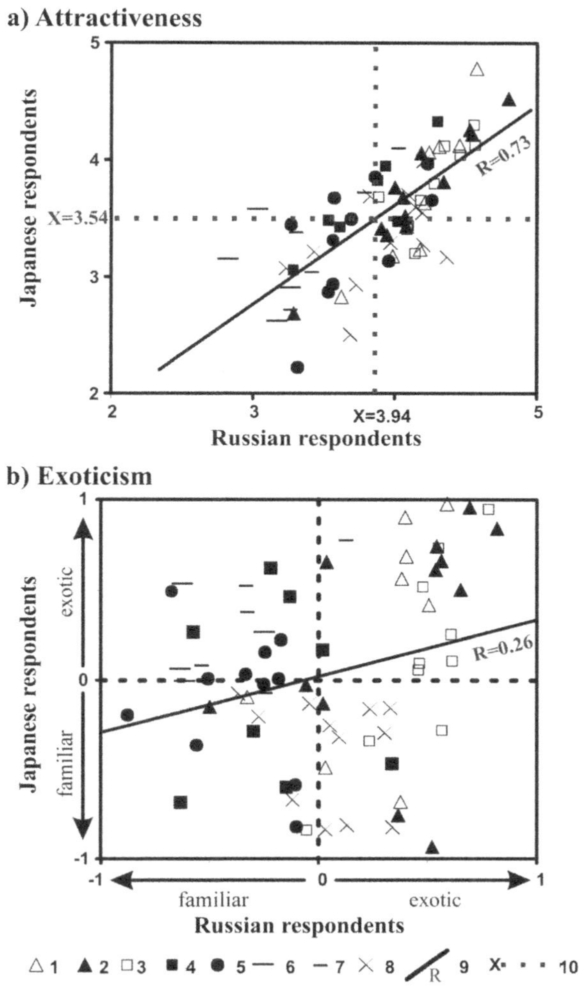

Figure 2 Correlation between the estimates of Russian and Japanese respondents. Correlation between the estimates of Russian and Japanese respondents regarding the **(a)** attractiveness and **(b)** exoticism of different landscape categories: 1 - mountains, 2 - lakes, 3 - waterfalls, 4 - rivers, 5 - forests, 6 - 'waterless' plains, 7 - swampy plains, 8 - sea coasts, 9 - trend lines and correlation coefficients, 10 - mean score.

waterless plains (Figure 2a, Tables 3 and 4). The special importance of water and topography in the human experience of landscape is also evident in earlier research on preference (e.g., Brush 1981; Wherrett 2000; Arriaza et al. 2004; Nasar and Li 2004). However, the Russians and Japanese evaluated some other types of landscapes, including waterscapes, differently. Such differences apply to seacoasts, rivers, forests, and swampy plains, which tended to be ranked intermediately in their assessments. Russian respondents assessed the attractiveness of both seacoasts and rivers significantly higher than forests and swampy plains and substantially higher than waterless plains. For Japanese respondents, rivers were almost as attractive as waterfalls, mountains, and lakes, but seacoasts were noticeably less attractive.

There was practically no correlation between Russian and Japanese respondents in evaluating exotic/familiar landscapes ($R = 0.26$). The majority of Russian respondents considered mountain landscapes (including mountain lakes and waterfalls) and seacoasts as the most exotic, while forests, rivers, and treeless plains were the most familiar. For Japanese respondents, seacoasts were familiar and treeless plains exotic; all other landscapes varied considerably in their assessments (Figure 2b).

All groups of Japanese respondents were virtually identical in their assessments of the attractiveness and exoticism of landscapes (correlation coefficients between their scores: $R = 0.90$ to 0.96). Among the Russian respondents, in comparison, participants from different regions evaluated them differently. This is probably because Japan is a more culturally homogeneous nation (Meshcheryakov 2010). However, the natural environment can also play a role here: the regions of Russia in which we performed our investigations differ significantly in their natural characteristics; meanwhile the Japanese regions vary mainly in their climatic conditions but are geomorphologically similar.

To discuss the ethno-cultural aspects of the perception and aesthetic evaluation of landscapes, it is appropriate to compare the survey data from Kamchatka and

Table 3 Statistical significance of differences (Student's *t*-test) between the attractiveness estimates of different landscape types by Russian respondents

X	Group of landscapes	Waterfalls	Mountains	Lakes	Seacoasts	Rivers	Forests	Swampy plains	Waterless plains
4.26	**Waterfalls**	0	*1.5*	**2.9**	**8.8**	**9.4**	**13.5**	**14.1**	**22.1**
4.2	**Mountains**		0	*1.2*	**6.7**	**8.5**	**11.1**	**12.2**	**19.9**
4.15	**Lakes**			0	**7.7**	**7.1**	**11.2**	**12.2**	**20.4**
3.92	**Seacoasts**				0	*1.4*	**5**	**7.1**	**15.3**
3.86	**Rivers**					0	**3.2**	**5.5**	**13.4**
3.72	**Forests**						0	**3.1**	**11.3**
3.56	**Swampy plains**							0	**7.2**
3.08	**Waterless plains**								0

The legend for Table 2 applies here as well.

Table 4 Statistical significance of differences (Student's *t*-test) between the attractiveness estimates of different landscape types by Japanese respondents

X	Group of landscapes	Waterfalls	Mountains	Lakes	Rivers	Swampy plains	Seacoasts	Forests	Waterless plains
3.78	**Waterfalls**	0	*0.6*	*1.2*	**2**	7.2	10.4	10.6	12.4
3.75	**Mountains**		0	*0.4*	*1.5*	5.9	8.7	8.9	10.6
3.73	**Lakes**			0	*0.9*	6.2	9.8	9.7	11.4
3.69	**Rivers**				0	5.2	7.7	8	10.4
3.4	**Swampy plains**					0	*1.1*	*1.4*	4.7
3.34	**Seacoasts**						*0*	*0.5*	4.5
3.32	**Forests**							0	4.1
3.07	**Waterless plains**								0

The legend for Table 2 applies here as well.

Hokkaido. Hokkaido is Japan's northernmost island and the closest to Kamchatka in terms of natural conditions. Our analysis revealed that the Kamchatka respondents (along with other Russians) were slightly more attracted to seacoasts (Figure 3a). In all other respects, Kamchatka and Hokkaido respondents evaluated the attractiveness of different landscape types similarly ($R = 0.72$) and about the same way as Russian and Japanese respondents as a whole ($R = 0.73$).

A more complex pattern can be observed when comparing estimates of landscape exoticism (Figure 3b). In general, the correlation coefficient between the estimates of Kamchatka and Hokkaido respondents was low and only slightly higher ($R = 0.37$) than that between Russian and Japanese respondents ($R = 0.26$). This means that the residents of Kamchatka and Hokkaido, despite living in similar environments, did not always evaluate familiar/exotic landscapes similarly. In some assessments, they were closer to the representatives of their culture, living under fundamentally different natural conditions. Cross-cultural differences were most evident in the evaluations of treeless plains, forests, and rivers. Hokkaido respondents (like other Japanese) considered treeless plains exotic, but Kamchatka residents considered them familiar (like other Russians). All Russian respondents, including respondents from Kamchatka, considered forests and rivers familiar, while Japanese respondents, including respondents from Hokkaido, rated them very differently - from the most familiar to the most exotic.

At the same time, Kamchatka respondents' evaluations of some landscapes were virtually identical to those of Japanese respondents (including Hokkaido) and very different from those of Russian respondents from other regions. This applies to seacoasts and mountains, which are both the most remarkable and most similar landscape types in Kamchatka and Japan. Seacoasts were familiar to residents of Japan and Kamchatka but exotic to other Russians. The exoticism of mountain landscapes varied among all groups, but the ratio of 'exotic' to 'familiar' mountains was the same for Kamchatka, Hokkaido,

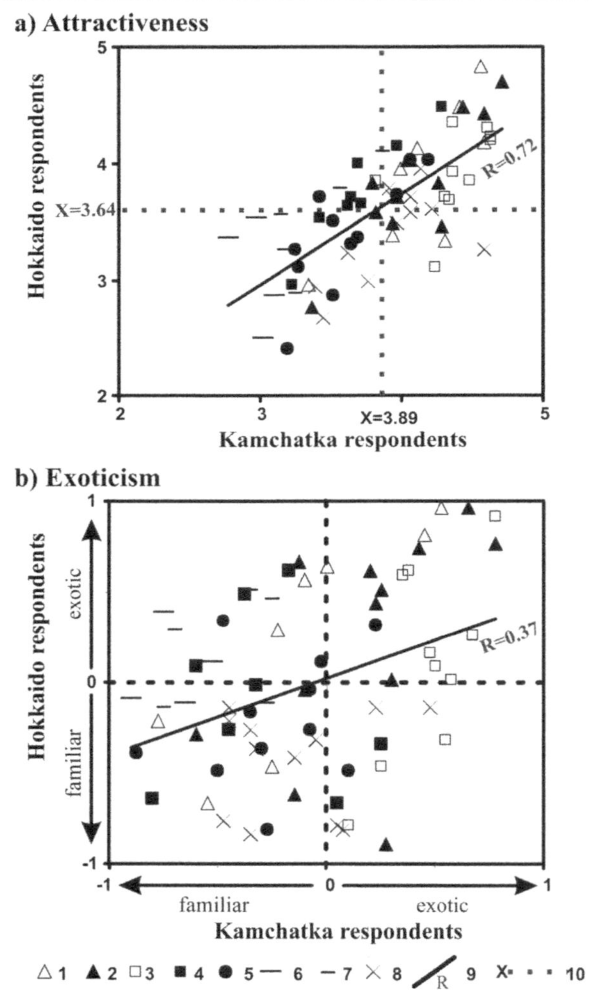

Figure 3 Correlation between the estimates of Kamchatka and Hokkaido respondents. Correlation between the estimates of Kamchatka and Hokkaido respondents regarding the **(a)** attractiveness and **(b)** exoticism of landscapes. Explanations in Figure 2.

and Japan and was equal to two; for Irkutsk it was five, and for Moscow respondents it was 14.

Russian respondents were apparently more attracted to the most exotic landscapes. The correlation coefficient between their estimates of attractiveness and exoticism was 0.80; it was lower ($R = 0.52$) only for the Moscow respondents. For all Russian respondents, mountain lakes, mountains, and waterfalls were the most attractive and simultaneously the most exotic, followed by seacoasts and rivers, at that, rocky coasts were more attractive than sandy beaches, and mountain streams were more attractive than lowland rivers. The majority of Russian respondents considered forests and treeless plains the most familiar and least attractive. However, forests were more attractive than treeless plains, especially for residents of Irkutsk and Kamchatka.

Every group of Russian respondents also appreciated a small group of familiar but nevertheless attractive natural sites. Thus, for most Russians, Lake Baikal (Figure 4) was not only one of the most beautiful but also easily recognizable natural symbols of Russia; for the Irkutsk respondents, it was also a closely located natural object. Similarly, for the Moscow respondents, Moskva River (Figure 5), as well as the forests and lakes of Central Russia, were familiar and attractive sites. Respondents from Petropavlovsk-Kamchatsky appreciated the Koryak, Avachinsky, and Karymsky volcanoes, as well as the 'Three Brothers' cliffs, which are the most beautiful natural objects close to the city. Thus, not only exotic landscapes but also the symbols of native nature were attractive for Russian respondents. Similarly, only one among many volcanoes was familiar and attractive for the Japanese respondents: Mount Fuji (Figure 6).

For both Japanese and Russians, mountains, mountain lakes, and waterfalls were not only the most attractive but also exotic. However, in contrast to the Russian respondents, there was no correlation between the evaluations of exotic and attractive landscapes in any of the Japanese groups ($R = 0.26$ to 0.35). Japanese cultural traditions - a settled way of life and traditional appreciation of Japan's natural environment as the most beautiful and comfortable for living, unlike that of foreign countries (Meshcheryakov 2010) - might play a role here. For example, treeless plains were among of the most exotic but unattractive landscape types in Japan. The attractiveness and exoticism of forests and rivers varied considerably in their estimates (Figure 2).

Seacoasts were familiar for both Japanese and Kamchatka respondents (Figure 3b). However, if the coasts were very attractive for Kamchatka residents, they were less so for the Japanese. This is probably due to traditional perceptions as well. Despite the fact that Japan is surrounded by the sea, for most of its history, the state has positioned itself as not a marine but rather agricultural and mainland nation. In traditional Japanese culture, the sea was more often represented in terms of 'negative elements': Japanese writers and poets did not praise it in their works, educated people never made sea voyages, and fishermen were perceived as rude,

Figure 5 The most attractive of familiar landscapes according to Moscow respondents: Moskva River.

Figure 4 The most attractive of familiar landscapes according to Russian respondents: Lake Baikal.

Figure 6 The most attractive of familiar landscapes according to Japanese respondents: Mount Fuji.

uncouth people with a significantly lower social status than that of peasants (Meshcheryakov 2012).

Accordingly, we can assume that differences in the assessment of attractiveness and exoticism are associated with both cultural differences (traditional perceptions) and the features of the natural environments the respondents live in or have had experiences with. Thus, all Russian respondents regarded as familiar landscapes relevant to not only their own region but also their historical homeland - that is, Central European Russia, the center of Russian civilization.

Prevailing images and the real diversity of landscapes in Russia and Japan

Russian respondents characterized nature in Japan in terms of 'mountain/volcano,' 'island/sea/ocean/coast,' 'beauty,' and '*sakura*' (Japanese cherry tree). This is close to the opinions of Japanese people ('mountains,' 'sea,' 'rivers,' and 'four seasons'). Even the first Russian travelers to the country in the 19th century, such as the writer Ivan Goncharov, admired the exotic and diverse beauty of Japan's natural features. According to Goncharov, 'All is so harmonious, so picturesque, and so different from reality here' (as cited in Petrova et al. 2010, p. 195). The Japanese respondents, however, imagined nature in Russia as a set of quite unattractive landscapes ('coniferous forest/taiga,' 'boundless plains,' 'tundras') and additionally characterized it using words such as 'snow' and 'cold.' This image only partly coincides with that of the Russian respondents: 'forest/taiga,' 'space/plain,' 'river,' and 'mountain' (the last one was added by the Irkutsk and Kamchatka respondents).

Neither image reflects the real natural diversity of both countries (Petrova et al. 2008b). This conclusion is also confirmed by the fact that the Russian respondents, who had never been to Japan, perceived many Japanese landscapes as 'familiar' Russian landscapes (Figure 7a). Moreover, many landscapes in Russia were no less exotic to them than the 'typical' landscapes of Japan. Similarly, the Japanese respondents, who had never been to Russia, often did not distinguish between Russian and Japanese landscapes (Figure 7b). It is also paradoxical that many mountain landscapes, which are characteristic components of nature in both Japan and Russia according to the respondents, were evaluated as not only the most attractive but also the most exotic (Figure 2b). Moreover, a set of six photos of the most attractive exotic landscapes was identical for the Russian and Japanese respondents. All of these pictures were taken in Russia: the Altai, the Caucasus, the Khibiny Mountains (north of European Russia), and the unique mountain plateau Putorana in Central Siberia.

Thus, the prevailing views regarding the 'typical' natural landscapes in Russia and Japan, and the fundamental differences between them, do not adequately reflect the diverse natural features of both countries. Such views could be attributable to the fact that mass media and literature tend to pay special attention to the flat landscapes of the Central European part of Russia and the limited number of famous mountainous landscapes in central and southern Japan; meanwhile, other landscapes in both countries remain practically unknown to the larger population. We hope our study will help to remedy this gap.

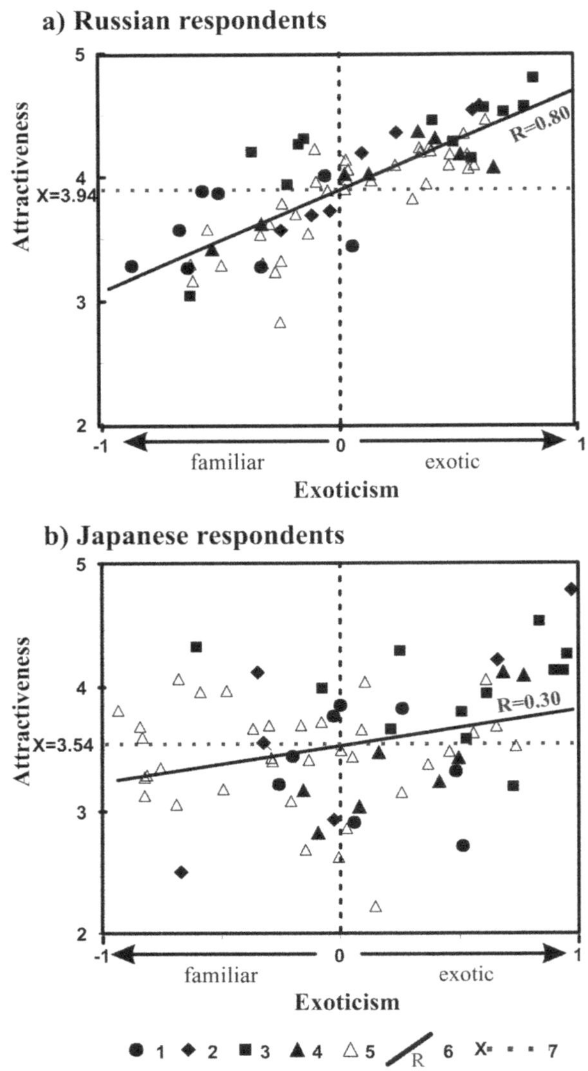

Figure 7 Correlation between the evaluation of attractiveness and exoticism according to Russian and Japanese respondents. Correlation between the evaluation of attractiveness and exoticism according to **(a)** Russian and **(b)** Japanese respondents of different regions: 1 - plains of European Russia, 2 - Khibiny and Caucasus mountains, 3 - Central and East Siberia, 4 - Kamchatka, 5 - Japan, 6 - trend lines and correlation coefficients, 7 - mean score.

Conclusions

Previous investigations of landscape preferences tended to explain the matter through theoretical concepts that focused on either individual and cultural differences or evolutionary processes. Our analysis has shown that these factors interact in a much more complicated way. It allowed us to reveal some common features of and main differences in the perception and visual evaluation of natural landscapes between Russian and Japanese respondents, as well as between Russian respondents from different regions (Moscow, Irkutsk, and Kamchatka).

The survey process involved grouping landscapes according to respondents' personal preferences as well as rating the attractiveness and exoticism of landscapes. The results can be summarized as follows.

The most important feature for grouping landscapes appeared to be the presence or absence of water as well as the type of water basin. Topography was also very important for the Russians (especially the Muscovite respondents), while both visual and seasonal characteristics were more significant for the Japanese.

Japanese respondents tended to be more restrained and the Russians more emotional in their assessments of landscape attractiveness; however, almost all assessed waterfalls, mountains, and lakes as the most attractive and waterless plains as the least. These common features might confirm the existence of universal human concepts regarding landscape aesthetics.

On the other hand, Russian and Japanese respondents evaluated some other types of landscapes differently (e.g., seacoasts, rivers, forests, swampy plains).

While Russian respondents considered the most exotic landscapes the most attractive, there was no such tendency for the Japanese. Respondents from both countries equally regarded the most famous symbols of native nature as very attractive.

It is interesting that all Japanese respondents assessed the attractiveness and exoticism of landscapes almost identically, while Russian respondents from different regions expressed different opinions. These results support the hypothesis that landscape preferences are influenced by not only the cultural factors determined by a country's traditions but also the features of familiar natural environments.

We hope our findings will not only be useful in the theoretical aspect but will also promote a better understanding between both cultures regarding issues of landscape protection, management, and planning. First of all, this refers to natural objects that can be used for recreational purposes because they are aesthetically valuable or exotic. We also hope to continue our research, engaging with more respondents of different ages and from different regions.

Competing interests
The authors have no competing interests to declare.

Authors' contributions
YA proposed the topic and conceived and designed the study. HM and NT helped develop the methodology for interviewing the respondents. YM, EP, YA, and HM created the photo database. EP, SE, KF, HU, and AP conducted the experimental study. YM analyzed and helped interpret the data. HM and AP collaborated with the corresponding author in constructing the manuscript. All authors read and approved the final manuscript.

Authors' information
EP holds a PhD in geography. She is a Senior Researcher with the Faculty of Geography, Lomonosov Moscow State University. She is the Russian side leader of this research project. She is associated with Russian Geographical Society, European Geosciences Union, and Japan Geosciences Union.
YM holds a PhD in geology and mineralogy. He is a Senior Researcher with the Vernadsky State Geological Museum of the Russian Academy of Sciences. His current research interests include aesthetic evaluation and appreciation of natural landscapes. He is associated with Russian Geographical Society.
YA holds a PhD in engineering. Until 2009, he was a special senior researcher in the Social and Environmental Systems Division of the National Institute for Environmental Studies in Tsukuba, Japan. He was the initiator and Japanese-side leader of this research project until 2008. He is now retired. In 2006, he received a prize from the head of the Academic Committee at the Center of Environmental Information Science. He is affiliated with the National Parks Association, Tourism Research Association of Japan, International Haiku Association, Japan IFLA, and others.
HM holds a PhD in agriculture. He is a lecturer at Hokkaido University and has been the Japanese-side leader of this project since 2009. His research interest is the conservation of neighborhood nature and its sustainable recreational use, especially for coastal areas.
SE was previously a student at Hokkaido University and conducted the questionnaire and analysis mainly in 2009. He now works as an officer in Nasushiobara.
KF holds a PhD and is an associate professor at Chiba University. His research interest is the conservation of nature and its sustainable recreational use, especially in national parks.
AP holds a PhD in history. She is a researcher in the Institute for Oriental Studies at the Russian Academy of Sciences. Her research interests include the history of the Orient and Orientalism. She is affiliated with the European Geosciences Union and the Japan Geosciences Union.
NT holds a PhD in agriculture. He is currently a researcher at the Forestry and Forest Products Research Institute in Tsukuba, Japan. He contributed to the research design and joint investigation of this project. His research interest is the effect of forest environments on psychological restorativeness.
HU holds a PhD and is a lecturer at Sapporo City University. His research interest is the establishment of environmental design methods that involve the place attachments of citizens.

Acknowledgements
The Japan Society for the Promotion of Science and Russian Foundation for Basic Research supported this work under grant number 08-05-91204.
We express our gratitude to the following for their help in organizing interviews with respondents: Olga Leontyeva, Yaroslav Muravyev, Alexey Naumov, Naokata Okajima, Lidiya Roman, Nikolay Seliverstov, Yury Semenov, and Alexander Cherkashin. We give special thanks to Richard B. Cathcart for his kind help in improving our English text.

Author details
[1]Faculty of Geography, Lomonosov Moscow State University, GSP-1, Leninskie Gory, Moscow 119991, Russia. [2]Vernadsky State Geological Museum of the Russian Academy of Sciences, Mokhovaya 11, Bldg.11, Moscow 125009, Russia. [3]Research Faculty of Agriculture, Hokkaido University, Kita 9, Nishi 9, Kita-ku, Sapporo 060-8589, Japan. [4]Graduate School of Horticulture, Chiba University, 648, Matsudo, Matsudo-shi, Chiba 271-8510, Japan. [5]Institute for Oriental Studies of the Russian Academy of Sciences, Rozhdestvenka 12, Moscow 107031, Russia. [6]Forestry and Forest Products Research Institute, 1 Matsunosato, Tsukuba, Ibaraki 305-8687, Japan. [7]School of Design, Sapporo City University, Geijutsu-no-mori 1, Minami-ku, Sapporo 005-0864, Japan. [8]Open University of Japan, Nishikamata 5-12-1, Ota-ku, Tokyo 144-0051, Japan.

References

Aoki Y (1999) Trends in the study of the psychological evaluation of landscape. Landsc Res 24:85–94, doi:10.1080/01426399908706552

Aoki Y (2007) Recent trends of English paper on the psychological evaluation of landscape. J Environ Inf Sci 35(5):181–188

Aoki Y, Kitamura S (2001) Ontogenic and phylogenic evolution of the human appreciation of the landscape. Proceedings of the 38th IFLA world congress, Singapore, pp 114–122, 26–29 June 2001

Aoki Y, Petrova E (2010) Bilateral project of JSPS and RFBR on comparison of natural landscape appreciation between Russia and Japan: first stage. J Environ Inf Sci 38(5):105–110

Appleton J (1975) The experience of landscape. John Wiley & Sons, London

Arriaza M, Canas-Ortega JF, Canas-Madueno JA, Ruiz-Aviles P (2004) Assessing the visual quality of rural landscapes. Landsc Urban Plan 69:115–125

Backhaus N (2011) Landscapes, spatial totalities or special regions? Procedia Soc Behav Sci 14:193–202, doi:10.1016/j.sbspro.2011.03.036

Bourassa SC (1991) The aesthetic of landscape. Belhaven Press, London

Brabyn L (2009) Classifying landscape character. Landsc Res 34(3):299–321, doi:10.1080/01426390802371202

Brush RO (1981) Landform and scenic preferences: a research note. Landsc Plan 8:301–306

Dramstad WE, Sundli Tveit M, Fjellstad WJ, Fry GLA (2006) Relationships between visual landscape preferences and map-based indicators of landscape structure. Landsc Urban Plan 78(4):465–474, doi:10.1016/j.landurbplan.2005.12.006

Ermolaev O (2003) Mathematical statistics for psychologists: a textbook. Moscow Psychological and Social Institute. Flint, Moscow (in Russian)

Gumilyov LN (1993) Ethnosphere: history of humans and history of nature. Ecopros, Moscow (in Russian)

Herzog TR, Maguire CP, Nebel MB (2003) Assessing the restorative components of environments. J Environ Psy 23(2):159–170

Hull RB IV, Reveli GRB (1989) Cross-cultural comparison of landscape scenic beauty evaluations: a case study in Bali. J Environ Psy 9(3):177–191, doi:10.1016/S0272-4944(89)80033-7

Hull RB, Stewart WP (1995) The landscape encountered and experienced while hiking. Environ Behav 27:404-426

Isachenko AG (1991) Landscape science and physical–geographical regionalization: a textbook for universities. Vysshaya shkola, Moscow (in Russian)

IUCN - International Union for Conservation of Nature (1969) National parks. http://www.iucn.org. Accessed 1 Jul 2014

Jones M (1991) The elusive reality of landscape. Concepts and approaches in landscape research. Norsk Geografisk Tidsskrift Nor J Geogr 45(4):229–244

Kaltenborn BP, Bjerke T (2002) Associations between environmental value orientations and landscape preferences. Landsc Urban Plan 59:1–11, doi:10.1016/S0169-2046(01)00243-2

Kaplan R (2001) The nature of the view from home: psychological benefits. Environ Behav 33(4):507–542

Kaplan R, Kaplan S (1989) The experience of nature: a psychological perspective. Cambridge University Press, New York

Kunikida D (1901) Musashino [The Musashi Plain]. Minyusha, Tokyo (in Japanese)

Lothian A (1999) Landscape and the philosophy of aesthetics: Is landscape quality inherent in the landscape or in the eye of the beholder? Landsc Urban Plan 44(4):177–198, doi:10.1016/S0169-2046(99)00019-5

Meshcheryakov AN (2010) Japan in the arms of space and time. Natalis, Moscow (in Russian)

Meshcheryakov AN (2012) The most beautiful: the nature of Japan in interpretation of Shiga Shigetaka. Hist Mod 1(15):111–137 (in Russian)

Nasar JL, Li M (2004) Landscape mirror: the attractiveness of reflecting water. Landsc Urban Plan 66:233–238

Nikolaev VA (2005) Landscape: aesthetics and design (a textbook). Aspect Press, Moscow (in Russian)

Ode Å, Tveit MS, Fry G (2008) Capturing landscape visual character using indicators: touching base with landscape aesthetic theory. Landsc Res 33 (1):89–117, doi:10.1080/01426390701773854

Okajima N, Petrova E, Petrova A (2010) The influence of Russian literature on two Japanese literary figures and the Japanese sense of scenic beauty. Proceedings of the 8th annual Hawaii international conference on arts & humanities, Honolulu, pp 3135–3150, 13–16 Jan 2010

Orians GH (1986) An ecological and evolutionary approach to landscape aesthetics. In: Penning-Roswell EC, Lowenthal D (eds) Landscape meanings and values. Allen & Unwin, London, pp 3–25

Petrova E, Aoki Y, Mironov Y, Petrova A, Furuya K, Matsushima H, Takayama N (2008a) Comparison of natural landscapes appreciation between Russia and Japan: methods of investigation. In: Raschi A, Trampetti S (eds) Management for protection and sustainable development: the fourth international conference on monitoring and management of visitor flows in recreational and protected areas. Ospedaletto, Pisa, pp 198–202

Petrova E, Aoki Y, Petrova A (2010) Landscape descriptions by Russian visitors to Japan until 1900. In: Kinda A, Komeie T, Minamide S, Mizoguchi S, Uesugi K (eds) Proceedings of the 14th international conference of historical geographers. Kyoto University, Kyoto University Press, Kyoto, pp 194–195

Petrova EG, Mironov YV, Petrova AA (2008b) National parks of Japan from the point of view of the Russians. Priroda (Nature) 11:28–39 (in Russian)

Purcell AT, Lamb RJ, Mainardi Peron E, Falchero S (1994) Preference or preferences for landscape? J Environ Psy 14(3):195–209, doi:10.1016/S0272-4944(94)80056-1

Sevenant M, Antrop M (2009) Cognitive attributes and aesthetic preferences in assessment and differentiation of landscapes. J Environ Manag 90(9):2889–2899, doi:10.1016/j.jenvman.2007.10.016

Staats H, Hartig T (2004) Alone or with a friend: a social context for psychological restoration and environmental preferences. J Environ Psy 24(2):199–211

Takayama N, Matsushima H, Petrova E, Ueda H, Nakajima T, Furuya K, Aoki Y (2012) Differences in environmental attitudes between Russia and Japan. In: Outdoor recreation in change: current knowledge and future challenges. Proceedings of the 6th international conference on monitoring and management of visitors in recreational and protected areas, Stockholm, pp 404–405, 21–24 Aug 2012

Tips WEJ, Savasdisara T (1986) The influence of the socio-economic background of subjects on their landscape preference evaluation. Landsc Urban Plan 13:225–230, doi:10.1016/0169-2046(86)90036-8

Ueda H, Nakajima T, Takayama N, Petrova E, Matsushima H, Furuya K, Aoki Y (2012) Landscape image sketches of forests in Japan and Russia. For Policy Econ 19:20–30, doi:10.1016/j.forpol.2012.01.002

Ulrich RS (1993) Biophilia, biophobia, and natural landscapes. In: Kellert SR, Wilson EO (eds) The biophilia hypothesis. Island Press, Washington, pp 73–137

Van den Berg AE, Koole SL, Van der Wulp NY (2003) Environmental preference and restoration: (How) are they related? J Environ Psy 23(2):135–146

Wherrett JR (2000) Creating landscape preference models using internet survey techniques. Landsc Res 25:79–96

Yang B, Brown TJ (1992) A cross-cultural comparison of preferences for landscape styles and landscape elements. Environ Behav 24(4):471–507

Yang BE, Kaplan R (1990) The perception of landscape style: a cross-cultural comparison. Landscape and Urban Planning 19:251–262, doi:10.1016/0169-2046(90)90024-V

2

Upwelling: a unit of disturbance in equatorial spread *F*

Roland T Tsunoda

Abstract

Plasma structure in the nighttime equatorial *F* layer, often referred to as equatorial spread *F* (ESF), is not uniformly distributed, either in time or in space. Observations indicate that ESF in the bottomside *F* layer takes the form of patches; plasma structure within the *F* layer takes the form of localized plasma depletions, called equatorial plasma bubbles (EPBs), which tend to occur in clusters. Another observed feature is an upwelling, which has been described as a localized, upward modulation of isodensity contours in the bottomside *F* layer. Interestingly, zonal widths of ESF patches, EPB clusters, and upwellings are similar. Moreover, all display an east-west asymmetry. The objective of this paper is to show, for the first time, that an ESF patch is the bottomside counterpart of an EPB cluster, and that both are products of the electrodynamical process that takes place within an upwelling. The process can be described as having three phases: (1) amplification of upwelling amplitude during the post-sunset rise of the *F* layer, (2) launching of the first EPB of the evening, from crest of the upwelling, and (3) structuring of plasma within the upwelling. Hence, an upwelling, whose presence is responsible for the formation of ESF patches and EPB clusters, can be envisioned as a unit of disturbance that occurs in the nighttime equatorial ionosphere.

Keywords: Equatorial spread F; F-region plasma structure; Equatorial ionosphere; Equatorial plasma bubbles; Deep convective activity in troposphere; Atmospheric gravity waves; Neutral-ion coupling; Swoopers; Transequatorial propagation; Interchange instability

Background

Overview

Plasma structure, found in the nighttime equatorial *F* layer, is often referred to as equatorial spread *F* (ESF). Although the term was coined originally to describe the spreading of the *F* trace (in range) in ionograms (Cohen and Bowles, 1961), it is now used generically to refer to both the structure and the conditions that are conducive for its development. Our knowledge and understanding of ESF have evolved from views of structure in the bottomside of the *F* layer with ground-based sensors, and from measurements made *in situ* within the *F* layer with space-borne instruments. Useful and unifying glimpses, though only occasional, have been provided by rockets and incoherent-scatter (IS) radars.

A better understanding of the physical processes is slowly emerging as theoretical and modeling studies are conducted to interpret observations. Clearly, theories and model outputs cannot be properly evaluated without observations that are both comprehensive and accurate. In this regard, we believe that the measurements needed to answer the outstanding questions on ESF are not being made (e.g., see Tsunoda 2005).

In the meantime, we believe that much can be learned by considering whether the 'patchiness' of ESF and the 'clustering' of equatorial plasma bubbles (EPBs) (e.g., McClure et al. 1977) are related, through some kind of physical process. Clearly, if this kind of relationship can be shown to exist, a unified and more comprehensive description of ESF would emerge. Patches have been detected in the bottomside *F* layer, over several decades, with ground-based instruments such as (1) ionosondes (Calvert and Cohen 1961; Cohen and Bowles 1961), (2) radio experiments (Calvert et al. 1963; Davies and Chang 1968; Röttger 1973), and (3) radars (Clemesha 1964; Kelleher and Skinner 1971). Impressions are that the patches, observed by these different methods, are produced by related plasma structures (e.g., Kelleher and Röttger 1973; Aarons et al. 1978). Within the *F* layer, ESF takes the form of EPBs

Correspondence: tsunoda@sri.com
Center for Geospace Studies, SRI International, 333 Ravenswood Ave, Menlo Park, CA 94025, USA

(e.g., Tsunoda 1981; Tsunoda et al. 1982), which are spaced in longitude. A closer examination reveals that EPBs occur more often in clusters (e.g., Singh et al. 1997), than they do individually. Interestingly, other observations have shown that an upwelling (i.e., localized altitude modulation) often appears in the bottomside *F* layer, prior to the development of ESF or EPBs (e.g., Tsunoda and White 1981; Tsunoda 2005). Moreover, the width of an upwelling is very similar to those of ESF patches and EPB clusters. And, there is further evidence that an upwelling can control when, where, and how an ESF patch or EPB cluster develops (Tsunoda and White 1981; Tsunoda 2005).

Methods

Objectives

The objectives of this paper are to unify the observations of ESF and EPBs, and to validate the following hypothesis:

- The formation of ESF patches and EPB clusters is controlled by the electrodynamics that sets up within an upwelling in bottomside of the equatorial *F* layer

An upwelling is defined as a localized, upward displacement of isodensity contours in the bottomside of the *F* layer, as shown in Figure 1 (an upwelling can appear in apparent isolation, at some longitude, or several can appear as a wave train; the latter is referred to as large-scale wave structure (LSWS)). The contours, shown between 300 and 450 km in altitude, describe the distribution of plasma density (*N*). Here, upwelling amplitude is about 50 km and zonal width is about 300 km (the increased jaggedness in contours associated with lower *N* values is produced by low signal-to-noise ratios, and should be ignored for our purposes here). The smattering of small pink islands that appear in the vicinity of 500 km altitude is associated with the peak of the *F* layer. Perhaps the earliest measurement of an upwelling was made with a 630 nm all-sky imager by Weber et al. (1978, 1980, 1982). A collocated ionosonde showed that local uplifting of isodensity contours was associated with the north-south-aligned channel of nightglow depletion. Unambiguous descriptions of this kind are few and far between because they are available only from ALTAIR, a fully steerable incoherent scatter (IS) radar (Tsunoda et al. 1979; Tsunoda and White 1981; Tsunoda 1983). Less-direct results have been obtained from variations in total electron content (TEC) as a function of longitude (e.g., Thampi et al. 2009; Tsunoda et al. 2010); these more routine measurements have shown that occurrences of upwellings are fairly common (Tulasi Ram et al. 2014).

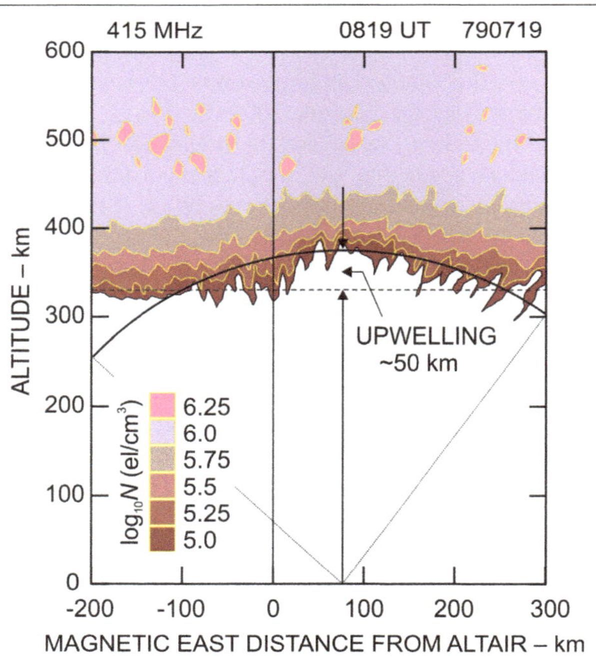

Figure 1 An upwelling in the bottomside *F* layer. Extracted from ALTAIR IS measurements made at 0819 UT on 19 July 1979.

In the remainder of this paper, we show how an upwelling provides the framework for the development of (1) each patch that is detected by a ground-based sensor, and (2) each cluster of EPBs that is encountered by a satellite-borne instrument. We conclude that an upwelling can be viewed as a unit of disturbance in ESF. To place this kind of description in context, we note that usage of the term 'convective ionospheric storm' to refer to ESF has increased (e.g., Makela et al. 2005, 2006; Kelley et al. 2005, 2009, 2011). If the research community chooses to accept this terminology, it may be reasonable to refer to the process, which takes place within an upwelling, as a convective ionospheric substorm.

ESF patches in bottomside *F* layer

The properties of ESF patches found in the bottomside of the equatorial *F* layer, presented below, have been compiled from (1) bistatic radio propagation experiments, in which both the Doppler shift and spread in a continuous-wave (CW) signal were measured; (2) transequatorial propagation (TEP) experiments, in which the delay and azimuthal angle-of-arrival of pulsed transmissions were measured; and (3) ionograms, which were recorded at a station near the magnetic dip equator. We show that, regardless of the method of measurement, ESF patches appear sequentially, starting around the end of the post-sunset rise (PSSR) of the *F* layer, and each patch displays an east-west (EW) asymmetry.

Results

Radio propagation: CW-DOPPLER results

Indications that ESF consists of drifting patches of plasma structure first appeared in radio propagation experiments, which were conducted in Africa during the early 1960s. A radio link was set up from Tripoli, Lybia (32.5° N, 13.1° E) to Accra, Ghana (5.3° N, 0.2° W) (Calvert et al. 1963); later, another link was set up from Monrovia, Liberia (6.2° N, 10.5° W) to Accra (Davies and Chang 1968). In both experiments, a CW signal was transmitted from one location, and Doppler properties of the received signal were monitored at the other. The baseline of the Monrovia-Accra path was about 1,175 km long and aligned east-southeastward from the magnetic dip equator (Davies and Barghausen 1966). The baseline of the Tripoli-Accra path was 3,315 km long and aligned south-southwestward from the magnetic meridian (these baselines can be compared with those used for TEP experiments, in 'Transequatorial propagation results' section).

The patchy nature of signals received at Accra from Monrovia, on 26 to 27 October 1963, can be seen in Figure 2, which is reproduced from Davies and Chang (1968) with added notation. The signal of interest can be seen to have had essentially a zero Doppler shift at 1600 universal time (UT). We interpret this behavior to be associated with a propagation path that underwent a one-hop reflection from the bottomside of a stationary, and horizontally stratified, *F* layer. The Doppler shift of this trace became increasingly more negative during the PSSR of the *F* layer, and reached a maximum value around 1830 UT (indicated by a downward-pointing red arrow), which is close to *E*-region sunset (SS_E). The spreading of the trace indicates the presence of ESF. After 1940 UT, three somewhat spread traces appeared in sequence; all displayed a large positive Doppler shift, which decreased with time. Each of these traces was referred to as a 'swooper' (Davies and Chang 1968). The Doppler shifts of the first two swoopers changed sign, but their magnitudes did not approach those, when the sign was positive. The Doppler shift of the third swooper approached zero and remained small in magnitude. This Doppler signature is a good example of the EW asymmetry, which is a distinctive property of ESF patches (e.g., Röttger 1973; Maruyama and Kawamura 2006) and EPB clusters.

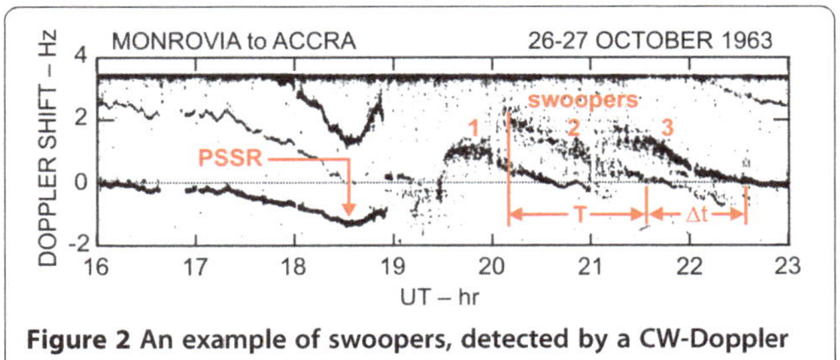

Figure 2 An example of swoopers, detected by a CW-Doppler technique, along the Monrovia-Accra path on 26 to 27 October 1963. Reproduced from Davies and Chang (1968) with added notation.

The finding that two swoopers can coexist at any given time, separated by different Doppler shifts, is also direct evidence that these structures were spatially separated in longitude (that is, different Doppler shifts would be measured, if signals arrive from different longitudes at different elevation angles). In this regard, Davies and Chang (1968) found that only one swooper was usually detected on a given night, although there were a few nights when up to five swoopers were detected. The duration of a swooper varied between 30 min and 2 h; typical duration was 1 h. Similarly, the spacing between two swoopers was also about an hour. We should keep in mind, however, that these properties are colored by the transport velocity of patches and the reflection geometry of the experiments.

Transequatorial propagation results

A more global view of patches is available from TEP experiments, in which both transmitter and receiver are located at substantial distances from the dip equator (e.g., see review of TEP by Nielson and Crochet 1974). For example, the lengths of great-circle TEP paths (1) from Tsumeb (19.3° S, 17.7° E) to Lindau (51.4° N, 10.1° E) (Röttger 1973), and (2) from Shepparton (36.2° S, 135.3° E) to Oarai (36.3° N, 140.6° E) (Maruyama and Kawamura 2006; Saito et al. 2008), were both about 8,000 km. With this viewing geometry, signals that arrive at azimuths ±60° away from the great-circle path have been detected.

Patches were found to first appear around the end of the PSSR (Maruyama and Kawamura 2006), and become distributed, sometimes over 10,000 km in EW extent (Röttger 1973). These propagation modes have been interpreted in terms of side reflections from tilted plasma structures, near the magnetic dip equator, that are aligned with geomagnetic field (*B*) lines. One to six (and as many as ten) patches have been observed simultaneously during an interval of 15 min. The lifetime of a patch could vary between 15 and 120 min, which is essentially the same as that of a swooper. The spacing between patches could be less than 100 km or more than 1,500 km; the median value was 380 km (Röttger 1973). Allowing zonal transport, the typical spacing between swoopers is also about 400 km, which is in excellent agreement with patch separation (for example, nine patches spaced 400 km in longitude, would span 3,200 km, or about 2 h in local time). The median eastward drift speed was found to be 110 m/s (Röttger 1973).

Results from TEP also contained an EW asymmetry, which is reminiscent of that found in swoopers. That is, patches from the Tsumeb-Lindau experiment were observed more often in the sector to the west of the great-circle path, than observed in the sector to the east; the ratio of west-to-east occurrences was a factor of two in the results reported by Röttger (1973). Patches from the Shepparton-Oarai experiment were predominantly from the west sector for the month of March, 2003, according to Maruyama and Kawamura (2006).

Digisonde results

Features found in equatorial ionograms are also consistent with those found in swoopers and TEP patches; that is, range spread *F* (RSF) is also patchy, approaches from the west and recedes to the east, while displaying an EW asymmetry. To illustrate, we have scaled ionograms obtained with a Digisonde at Jicamarca, Peru (11.95° S, 76.87° W), and constructed a family of virtual height (*h*') curves versus UT, which is presented in Figure 3. Each curve is associated with an integer value of the plasma frequency, from 3 to 8 MHz (as labeled near the left ordinate). We have also plotted the minimum virtual height of the bottomside *F* layer (*h*'*F*), which is shown as a gray curve. The times for SS_E and sunset in the *F* region (SS_F) are marked with vertical gray lines. From these curves, we can recognize the PSSR of the equatorial *F* layer, which occurred between 2300 and 0030 UT, and its descent, between 0130 and 0400 UT.

Of interest is the altitude modulation found in the *h*'(3 MHz) curve, and the development and distribution of RSF. A maximum in *h*'(3 MHz) occurred around 0130 UT, accompanied by some RSF at 3 MHz. Thereafter, the *h*'(3 MHz) trace decreased rapidly and reached a minimum value around 0220 UT; the continued presence of RSF (3 MHz) suggests that it occurred more or less from the overhead direction. The *h*'(3 MHz) trace appeared to rise slightly, reaching a relative maximum around 0325 UT. This maximum was accompanied by oblique RSF (5 MHz) traces, which were first seen at earlier times, to the west of Jicamarca (from angle-of-arrival determination). When *h*'(3 MHz) reached its maximum, the RSF appeared to be coming from overhead, together with stronger RSF (5 MHz). As the *h*'(3 MHz) curve started to decrease again, we see that RSF (5 MHz) became stronger, and there was a component of RSF (3 MHz) that came from the east.

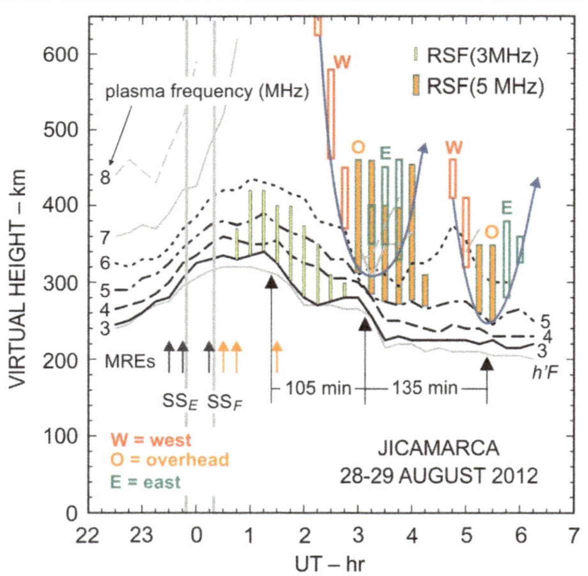

Figure 3 A plot of the virtual height of reflection. At selected ionosonde frequencies, presented as a function of UT on 28 to 29 August 2012. Also included are occurrence times for SS_E and SS_F, and MREs. Vertical bars indicate the times and virtual heights over which RSF was observed.

The EW asymmetry is evident from the fact that RSF (3 MHz) did not reach virtual heights as 'high' to the east, as they did when RSF was first detected to the west. We suggest that the source of this asymmetry is closely related to those seen in swoopers and TEP measurements. A similar pattern in RSF can be seen in Figure 3, between 0500 and 0600 UT, but there is almost no evidence of modulation in the *h*'(3 MHz) curve. The reason for this behavior is likely that the bottomside of *F* layer had reached altitudes, where chemical recombination, not plasma transport, controls the altitude profile of *N*. Moreover, the amplitude of upwellings decreases during the descent of the *F* layer, for the same reason that it increases during the PSSR. In any case, the presence of RSF with this familiar distribution suggests that another patch had passed over the Digisonde at Jicamarca. Hence, the evidence suggests that there were three upwellings that developed on this night, which became 'filled' with ESF.

EPB clusters in *F* layer

Our hypothesis, that EPBs develop within the confines of an upwelling, was actually formulated from the results obtained with ALTAIR, a fully-steerable IS radar (Tsunoda et al. 1979; Tsunoda and White 1981; Tsunoda 1983). Similar conclusions can be drawn from *in situ* measurements made from the Atmospheric Explorer *E* (AE-E) satellite (Singh et al. 1997). A clear example of LSWS with associated upwellings (first pass), which become filled with localized depletions (second pass), is presented in Figure 4; this plot is from Singh et al. (1997) with added notation. This event occurred on 30 March 1977, when AE-E was in a circular orbit at an altitude of 275 km. The ion number density (N_i), measured with an ion-drift meter onboard AE-E during two consecutive orbits, 7192 and 7193, is shown with two traces. Three upwellings (marked by downward-pointing arrows) are clearly evident in the upper trace, from orbit 7192. In comparison, the N_i values in this example are between 10^5 and 10^6 ions/cm^3, which are similar to those associated with the upwelling in

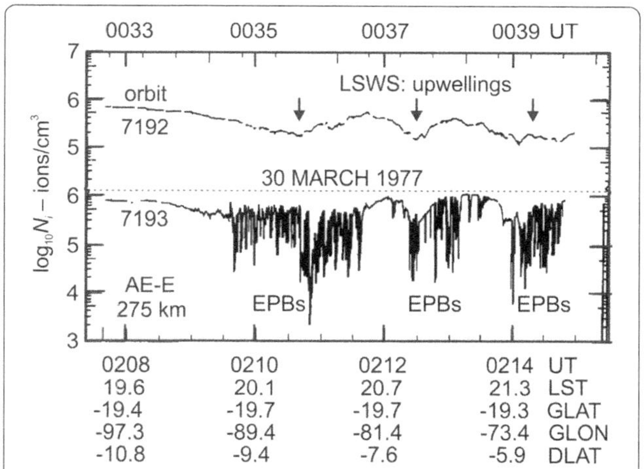

Figure 4 Upwellings measured by the AE-E satellite during one pass, and clusters of EPBs during the next pass. This figure, without annotations, is from Singh et al. (1997).

Figure 1. The UT scale associated with that orbit is shown along the top of the plot. During orbit 7193, considerable structure was present, as seen in the lower trace. The UT scale, together with those for local solar time (LST), geographic latitude (GLAT), geographic longitude (GLON), and magnetic dip latitude (DLAT), are shown along the bottom of the plot. The dotted line indicates where there is a break in the scale for N_i, along the ordinate. Although we have listed GLAT or GLON for orbit 7192, the trajectories were separated by no more than 2° in GLAT near the west end of the LSWS, so that orbit 7193 was essentially collocated with orbit 7192 near the east end of the LSWS. Hence, there is no doubt that the structure developed within upwellings, during the time between the two orbits.

There are several features of interest in Figure 4. The first is the encounter of LSWS, which occurred around 0035 UT, where GLON was 90° W. At this longitude, LST lags UT by 6 h, which means local time was approximately 1835 LST, when the solar zenith angle was about 98.5°, which corresponds with a time just before SS_E. This finding is consistent with earlier results, which showed that LSWS can be detected as early as in the late afternoon (Thampi et al. 2009; Tsunoda et al. 2011, 2013), and is amplified during the PSSR of the *F* layer (Tsunoda and White 1981; Tsunoda et al. 2013). Note that the upper envelope of N_i, which tracks the quiet regions and the maxima of the structure, resembles the modulation of the trace from the first pass. Hence, each cluster of plasma structure can be interpreted as localized depletions in N_i that developed within an upwelling. Stated another way, each upwelling appears to become filled with EPBs. Other examples from AE-E results can be seen in Hysell and Kelley (1997), their Figure 1; Kil and Heelis (1998a), their Figures 2 and 6; Kil and Heelis (1998b), their Figures 2 and 9; and Hei et al. (2005), their Figures 1 and 2.

Upwelling control: three phases

The development of ESF and EPBs can be described in terms of three phases: growth, onset, and structure. During the growth phase ('Growth phase: amplification of upwellings' section), which takes place during the PSSR, upwelling amplitude is amplified. During the onset phase ('Onset phase: ESF patches and primary EPBs' section), which takes place around the end of the PSSR, the first EPB of the evening is observed. And, during structure cluster phase ('Structure phase: clustering and EW asymmetry' section), which takes place during the descent of the *F* layer, plasma structure develops within each upwelling, in the form of ESF patches and EPB clusters. Evidence of this upwelling control is described below. Discussion and interpretation of key findings are presented in 'A working model' section.

Growth phase: amplification of upwellings

Amplification

ALTAIR IS measurements have provided direct evidence that upwellings are amplified during PSSR (Tsunoda and White 1981). Less-direct results have been obtained by examining the variations in TEC as a function of longitude. That is, signals from the CERTO radio beacon (Bernhardt and Siefring 2006) on the Communication/Navigation Outage Forecasting System (C/NOFS) satellite (De la Beaujardiere et al. 2004), in a low-inclination orbit, were recorded at ground stations; TEC was estimated using the differential-phase technique (e.g., Thampi et al. 2009; Tsunoda et al. 2010; Tulasi Ram et al. 2012). Variations in TEC with longitude were interpreted in terms of LSWS in the bottomside *F* layer (Tsunoda et al. 2013), and the results from successive satellite passes have shown that upwelling amplitude does grow during the PSSR (Tsunoda et al. 2010, 2013).

Ionogram signatures

Additional evidence regarding upwelling presence comes from two ionogram signatures, namely multi-reflected echoes (MREs) (Tsunoda 2009; Thampi et al. 2012; Tsunoda et al. 2013) and satellite traces (STs) (Abdu et al. 1981; Tsunoda 2008; Thampi et al. 2012). The finding that MREs appear most often around the time of SS_E (Tsunoda et al. 2015) is consistent with the finding that upwellings are amplified during the PSSR; in fact, the maximum in PSSR rate is centered more or less about SS_E (Tsunoda 2009; Tsunoda et al. 2015). Occurrences of MREs, in Figure 3, are marked with short, upward-pointing arrows. Most were detected during the PSSR (when isodensity contours are sloped upward, from west to the east). Typically, MREs appear first, especially around SS_E, followed by STs (e.g., see Tsunoda et al. (2013), their Figure 1). Both have been referred to as precursors because they precede the development of ESF.

The MREs are produced by multiple ionospheric reflections, which are indicative of an enhancement in reflection coefficient. Enhancement is possible, when reflection occurs from concave isodensity contours, such as those associated with an upwelling (Tsunoda 2009; Thampi et al. 2012). Enhancement should be strongest, when the radius of curvature of the reflecting surface is approximately equal to the height of reflection. We can see that this is the case for the upwelling shown in Figure 1, where we have scribed a circular arc for comparison. Hence, it should be apparent that the zonal wavelength of an LSWS would be about 300 km, when MREs are observed. And, indeed, the median patch size, from TEP experiments, is 380 km (Röttger 1973). Estimates of patch size or spacing, by other means, are also about 400 km (Aarons et al. 1978; Tsunoda and White 1981).

An ST appears to involve an oblique propagation path with reflection from a tilted isodensity surface (e.g., a wall of an upwelling). If the tilt of the reflection surface is slight, the propagation path back to the ionosonde could involve two ionospheric reflections and one ground reflection. In this case, the ST would appear offset from the two-hop (2 *F*) trace (Tsunoda 2008), which is assumed to be produced by two overhead reflections from a horizontally stratified *F* layer. In fact, an ST associated with the 2 *F* trace often precedes an ST associated with the main one-hop (1 *F*) *F* trace; either or both typically precede the appearance of ESF. The ST associated with the 1 *F* trace involves direct reflection from a more steeply tilted surface. If the shape of the isodensity contours in Figure 1 is typical, an ST could be associated with the 2 *F* trace, if an ionospheric reflection that supports a returning signal occurs from the east wall. Or, an ST could be associated with the 1 *F* trace, if a returning signal occurs from a more steeply inclined west wall.

Small-scale irregularities along west wall

A remaining feature, found to occur during growth phase, is radar backscatter from the west wall of an upwelling (Tsunoda 1981, his Figure 3). To illustrate, we present two maps in Figure 5; they were constructed from ALTAIR measurements made on 24 July 1979 (they are replicas of the west upwelling in Tsunoda and White (1981), with added notation). The nominal shape of the upwelling is sketched as a dashed red curve in Figure 5a,b (the purple curve represents the 10^5 el/cm^3 isodensity contour). A tilted echoing region is more or less aligned with the west wall of the upwelling. Hence, this tilted structure can be considered to be a radar signature indicating the presence of an upwelling. An enhancement in amplitude scintillations (produced by km-scale irregularities in *N*) has also been observed along the west wall of an upwelling (Tsunoda et al. 2011; Tulasi Ram et al. 2012). If this plot is typical, we can claim that small-scale irregularities (<1 km) are excited along the west wall during the growth phase, prior to the development of any EPBs. We will return to the difference between west-wall echoes and EPBs in 'Onset phase: ESF patches and primary EPBs' and 'Structure phase: clustering and EW asymmetry' sections.

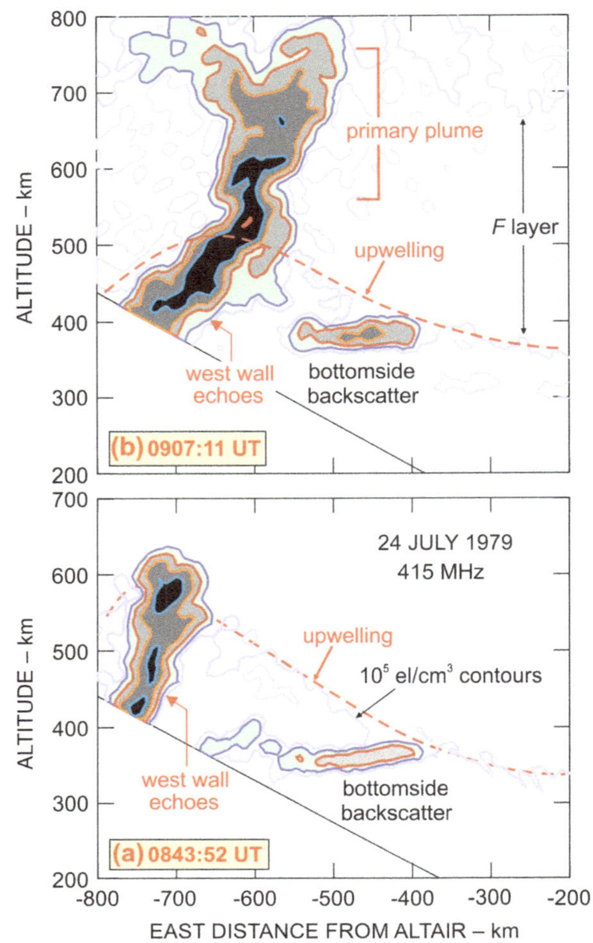

Figure 5 ALTAIR maps showing the radar backscatter. (a) ALTAIR map, which shows that radar backscatter from west wall was present prior to appearance of primary and secondary EPBs. **(b)** ALTAIR map showing the launching of the primary EPB from the crest of the upwelling.

Onset phase: ESF patches and primary EPBs

The onset phase takes place around the end of the PSSR. During this phase, observations in the bottomside *F* layer indicate the onset of RSF is preceded by appearances of MREs and STs (Tsunoda et al. 2013), as can be seen in Figure 3. First appearance of swoopers also occurs during this phase, as can be seen in Figure 2. Both of these bottomside signatures of ESF patches appear at the end of the PSSR (notice that MREs and STs, which appear during the growth phase, have relatively discrete traces and are considered to be precursors of RSF and EPBs). A sketch, that shows the nominal onset time for ESF to be around the end of the PSSR, can be found in Clemesha and Wright (1966).

And, observations in the topside *F* layer, during this phase, include the launching of the first (primary) EPB of the evening (Tsunoda 1981; Yokoyama et al. 2004). The appearance of the primary EPB is noteworthy because EPBs are generally absent during the growth phase, which precedes the onset phase. If EPBs are suppressed during the growth phase, an implication is that the process for EPB generation becomes active during the onset phase. We should note that the launching of the primary EPB occurs from near the crest of the upwelling, that is closest to the end of the PSSR (e.g., Tsunoda and White 1981), rather than envisioning the timing of the launch to be associated with the end of the PSSR.

From our perspective, the concavity in isodensity contours plays a central role in EPB production. Concavity is associated with the PSSR and descent of the *F* layer, and also with LSWS amplitude and zonal wavelength. For example, assume that the PSSR rate is 50 m/s and its duration is an hour; if we allow a symmetrical descent of the *F* layer, we will have concave isodensity contours that have a radius of curvature of about 15,000 km. Although curvature appears small, the contours are higher by about 100 km at the end of the PSSR, than they were before the start of the PSSR. Hence, the ion-neutral collision frequency (ν_{in}) is smallest during onset phase, and the linear growth rate of the interchange instability (driven by gravity) (Dungey 1956) reaches a maximum. The conditions are even more favorable for the launching of an EPB, if we include LSWS. Indeed, whereas the PSSR may raise the *F* layer by 100 km, the local altitude modulation associated with an upwelling could easily add another 50 km (see Figure 1). The finding that ESF is associated with a high *F* layer (e.g., Farley et al. 1970) is consistent with the launching of primary EPBs during the onset phase.

Structure phase: clustering and EW asymmetry

Patches and clusters develop during the descent of the *F* layer. During descent, *E* is westward which means the bottomside of a horizontally stratified *F* layer is stable to the interchange process, unless it is high enough in altitude that gravity is the dominant driver. In either case, patches or clusters would not form. On the other hand, the west wall of an upwelling would be unstable to the interchange process, driven by an eastward *U* (Tsunoda 1983). The elements of this process have already been outlined in several papers; we revisit the more important details.

We have already shown how small-scale irregularities appear along the west wall (Tsunoda 1981; Tsunoda et al. 2011) during the growth phase ('Growth phase: amplification of upwellings' section). We also know that the west wall continues to structure, leading to the appearance of secondary EPBs along that wall (Tsunoda 1983). We have not yet paid much attention, however, to the order in which west-wall echoes and EPBs appear. It turns out that this order provides some clues regarding the underlying physics. We found, from Figure 5a, that west-wall echoes precede EPB development (other maps, from this night, can be found in Tsunoda and White (1981). Indeed, if we now refer to Figure 5b, we can see that a primary EPB (plume) developed from the crest of the upwelling, shortly after appearance of the west-wall echoes. We further note that the shape of the west-wall echoes is relatively smooth, indicating that secondary EPBs have not yet formed.

Next, we revisit the ALTAIR backscatter maps in Tsunoda (1983) to examine in more detail how EPB clusters form. To do so, we have constructed a montage of backscatter maps from 24 July 1979, to highlight the temporal development and evolution of the upwelling and the EPBs. The montage is presented in Figure 6 (details of the ALTAIR scans used to construct the montage are given in Table 1). Two backscatter maps are shown in the center panel, positioned according to the magnetic east distance from ALTAIR (at the times the maps were taken). The one on the left, taken at 0907 UT, is essentially that presented in Figure 5b. The approximate shape and location of the upwelling is sketched with a thin red curve. Although the sketch is based simply on the notion that the backscatter must have occurred from a region of steep westward gradient in *N*, and is highly subjective, we believe it is adequate for showing the location of the primary plume (EPB) relative to the upwelling. Referring back to Figure 6, we have shown the nominal location and shape of the primary plume with a thick red curve, which extends upward from the crest of the upwelling. The plot on the right, taken at 1027 UT, 1 h and 20 min after the first, is much more complex than the first.

The temporal evolution of the primary plume and upwelling was tracked by examining four other backscatter maps that were taken between these two. We have included two of the four backscatter maps, one taken at 0927 UT (top panel) and the other at 0958 UT (bottom panel). The map in the top panel shows how structuring along the west wall has grown, from a simple, tilted echo region to a broader distribution with small protrusions. We are also left with the impression that the primary plume has shifted a bit to the east of center. The map in the bottom panel shows how the protrusions have grown along the west wall, and seemingly, the primary plume

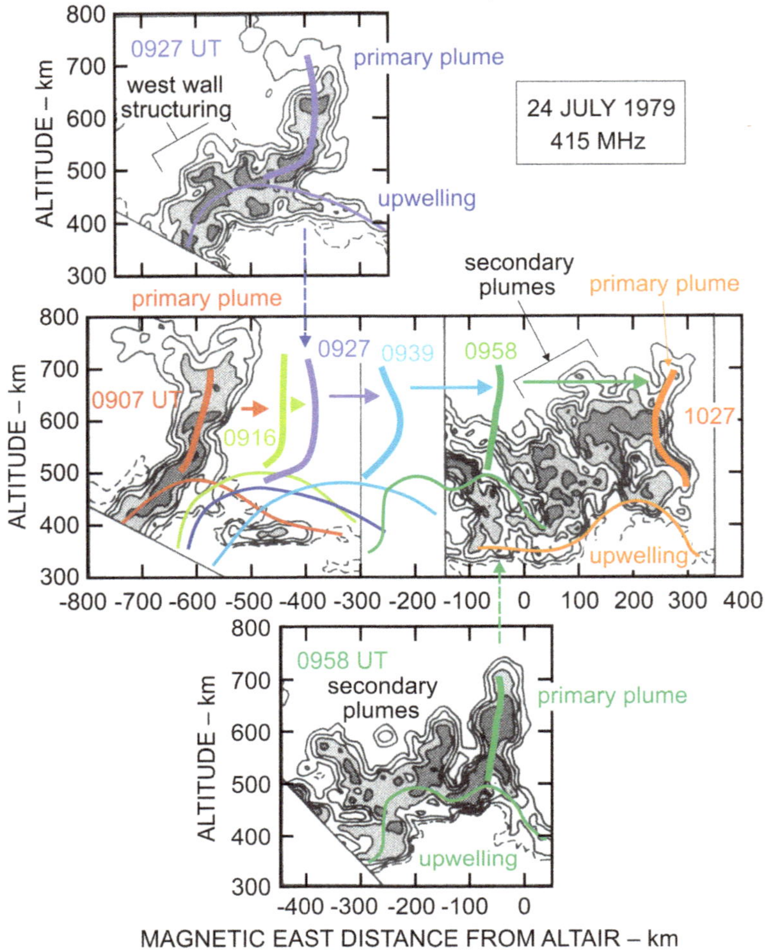

Figure 6 Montage illustration. A montage to illustrate the appearance of primary plume at crest of upwelling (0907 UT), the structuring of the west wall of upwelling (0927 UT), the development of secondary plumes (0958, 1027 UT), and the relative displacement of the primary plume to the eastern edge of the upwelling (1027 UT). Colored curves in the center panel represent the nominal locations of the primary plume and the upwelling, at different times on the night of 24 July 1979. All six of the backscatter maps are shown in Tsunoda (1983).

has moved even further to the east relative to the crest of the upwelling. All positions of the primary plume and upwelling are shown by colored curve segments, which are labeled by the times of maps. The interpretation that the plasma structures along the west wall are secondary plumes, should be evident from the map taken at 1035 UT, which is presented in Figure 7.

Table 1 ALTAIR scans on 24 July 1979

Scan number	Start and stop times (UT)	Direction (rate)
1	0907:11 - 0915:25	W → E(M*)
2	0915:59 - 0924:11	E → W(M)
3	0926:46 - 0931:36	W → E(M)
4	0939:16 - 0944:06	W → E(M)
5	0957:33 - 0959:25	W → E(F*)
6	1026:41 - 1030:55	E → W(F)
7	1035:53 - 1040:09	E → W(F)

*Scan rates: $M = 0.25°/s$; $F = 0.5°/s$.

It is important to note that both the primary and secondary EPBs appear to be transported eastward relative to the position of the upwelling. The EPBs eventually encroach into the region that was occupied originally by the east wall of the upwelling. A consequence of this encroachment appears to be a steepening of the tilt of the east wall, as the primary plume approaches the eastern boundary of the upwelling. This near alignment of the primary plume with the east wall appears to be a common feature, as seen in other ALTAIR maps (e.g., Tsunoda et al. 1982). This appears to be the process by which the entire upwelling becomes filled with plasma structure. This filling of the upwelling region is consistent with the EPB distribution in the AE-E measurements presented in Figure 4. A discussion of why this

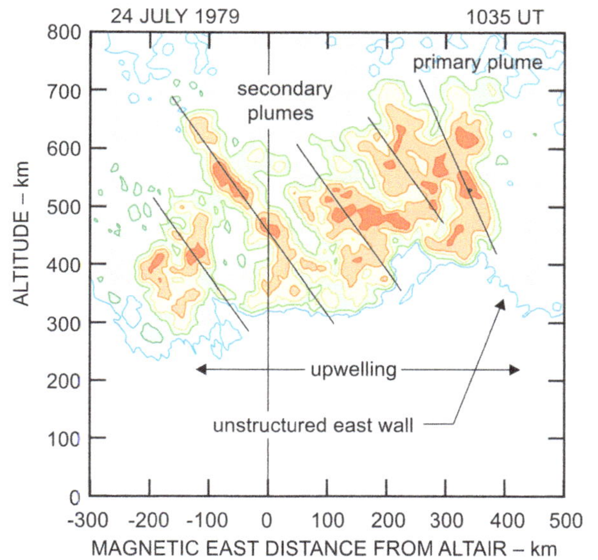

Figure 7 Backscatter distribution obtained with a spatial scan taken at 1035 UT, which clearly shows the cluster of EPBs.

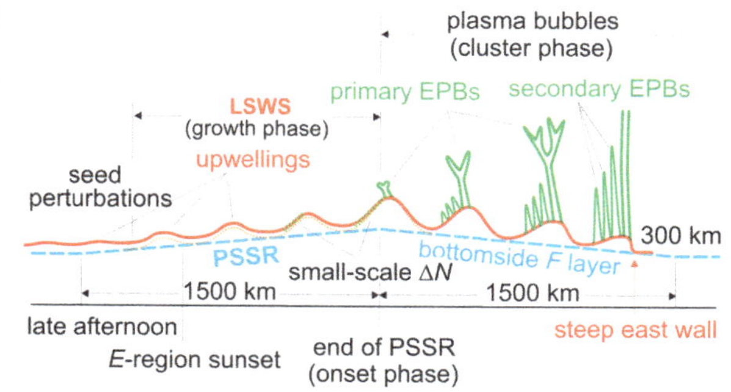

Figure 8 Working hypothesis for the temporal evolution of upwellings and EPBs, from late afternoon to the post-midnight sector.

should happen is presented in 'A working model' section. The process, by which structure develops on the west wall and is transported to the region of the east wall, is clearly susceptible to the development of EW asymmetries.

Discussion

A working model

We believe the evidence supporting the upwelling hypothesis ('Objectives' section) is strong enough to be used as a working model for the development of ESF. Theory, observations, and outputs from computer models can be used to test the robustness of the hypothesis. A sketch that captures the basic elements of this working model is presented in Figure 8. It can be viewed as a snapshot in UT, in which the ordinate is altitude and the abscissa is longitude; the vertical plane is aligned with the magnetic dip equator. The blue dashed line represents an isodensity contour in the bottomside of the F layer (the sketch is centered on the end of the PSSR). At the west end of the sketch, LT is assumed to be the late afternoon sector, where small-amplitude LSWS can be present prior to the growth phase ('Growth phase: amplification of upwellings' section). The wavy red curve represents an isodensity contour, when LSWS is present. Although LSWS is shown in Figure 8 to extend without interruption from late sunset to midnight (or later), upwellings can appear in apparent isolation, or in a group of up to ten upwellings (Röttger 1973). In this sketch, the PSSR is shown to take place in 1 h, which is equivalent to about 1,500 km in zonal distance. The distance between upwellings (i.e., wavelength of LSWS) is about 400 km (e.g., Röttger 1973; Tsunoda and White 1981). Sizes of observed patches are also about 400 km (e.g., Aarons et al. 1978).

Convective activity

The source of LSWS remains to be identified. However, the finding that LSWS can appear in the late afternoon suggests that velocity shear, driven by the F-region dynamo (Guzdar et al. 1982, 1983), is not the primary source. Instead, we suggest, for our working model, that a more viable source may exist in the lower atmosphere. A popular notion is that a localized region of deep convective activity in the troposphere, referred to as a mesoscale convective complex (MCC), will generate atmospheric gravity waves (AGWs). Conceptually, AGWs could reach the lower thermosphere and produce a polarization E in the E region, which could map along B lines to the base of the F layer (the small-amplitude LSWS, found in the late afternoon sector, could be produced in this manner). The smallness of LSWS amplitude could be a matter of smaller-amplitude AGWs in the lower thermosphere. In other words, fluid amplification is less, up to the lower thermosphere, than it is up to the thermosphere. If AGWs do reach the thermosphere, they could produce LSWS, around or after SS_E, through a neutral-ion coupling process that involves the F-region Pedersen current (Klostermeyer 1978).

Studies seeking correlation between convective activity and ESF, however, have produced mixed results (McClure et al. 1998; Su et al. 2008; Tsunoda 2010a,b,c). Correlation was best in the South American sector (Tsunoda 2010c). It turns out that the ESF patches over Jicamarca on the night of 28 to 29 August 2012 were, in fact, associated with deep convective activity in the troposphere. To illustrate, we present four maps of outgoing longwave radiation (OLR), for the Peruvian sector, in Figure 9. Following Waliser et al. (1993), we can interpret a localized blue-colored region as an MCC. The OLR maps were taken every 30 min. Other than that one region, where there may be two MCCs next to each other, the entire area

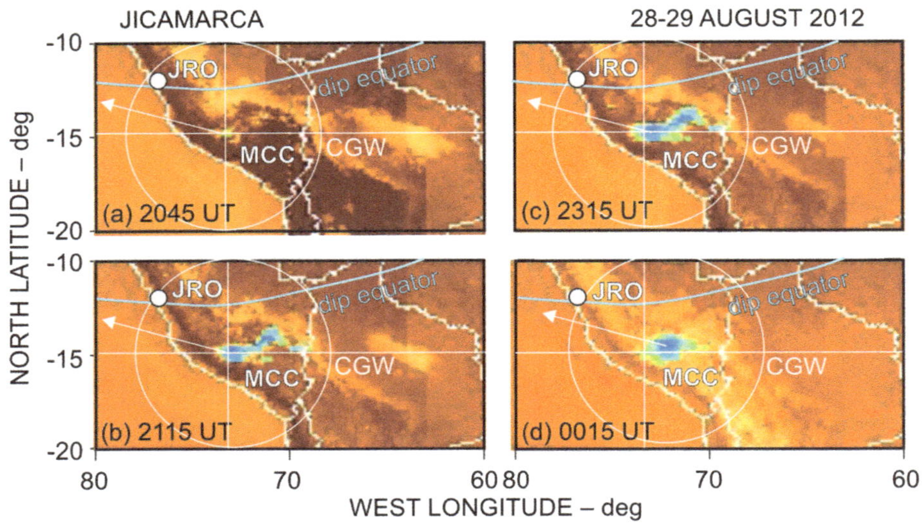

Figure 9 OLR maps showing the temporal development of a MCC (blue colored) region near JRO on 28-29 August 2012. **(a)** 2045 UT: very weak MCC. **(b)** 2115 UT: appearance of an extended region of deep convection. **(c)** 2315 UT: sustained convective activity. **(d)** 0015 UT: change in shape of MCC.

was devoid of convective activity. This event took place during the southern winter period, when both convective activity and strong PSSR rates are rare (Tsunoda 2010c). Prior to 2045 UT on August 28, MCCs were not detected. Activity intensified around 2115 UT and continued through 0015 UT on August 29.

The geometry appears to be ideal for neutral-ion coupling; that is, the MCC was located about 600 km (white circle) to the east-southeast of the Jicamarca Radio Observatory (JRO). Given the localized nature of the MCC, we envision that circular AGWs (CGWs) were excited (Tsunoda 2010b). Westward-propagating CGWs are most likely to penetrate to the thermosphere in the presence of an eastward *U* (e.g., Vadas 2007; Fritts and Vadas 2008). Moreover, neutral-ion coupling is expected to be most efficient, when the CGW phase front is aligned with *B*. Hence, we anticipate that the AGW reached *F*-region altitude in the direction of the white arrow, to the west of JRO. And, according to Waldock and Jones (1987), the source region in the troposphere is likely located between 250 and 1,500 km from the observed ionospheric perturbations. The ESF patches that arrived from the west of JRO are consistent with this scenario.

It is important to keep in mind that an AGW propagates relative to the background neutral gas. If the neutral gas moves relative to the ground reference frame (i.e., *U*), the phase velocity of a westward-propagating AGW will be Doppler-shifted to a value that is the difference of the two velocities. An extended interaction time for the neutral-ion coupling process, often referred to as 'spatial resonance' (Whitehead 1971), is possible when this difference velocity is matched with the zonal transport of the bottomside *F* layer. It turns out that LSWS is either nearly stationary, or drifts slowly westward, during the early stages of the growth phase (Tsunoda and White 1981; Kudeki et al. 1981; Tsunoda 2005, 2007; Tsunoda et al. 1981, 2013), when *E*-region dynamo still exerts control. The drift direction of upwellings becomes eastward during the structure phase (e.g., see Figure 6), when the *F*-region dynamo prevails. More details regarding this kind of geometry for spatial resonance can be found in Tsunoda (2007).

Growth phase

There are three features of interest that appear during the growth phase: (1) amplification of upwellings, (2) appearance of small-scale irregularities along the west wall of upwellings, and (3) the absence of EPBs. The first two, taken together, appear consistent with the notion that they are produced by the interchange instability, driven by orthogonal forces. That is, upwelling amplification could result from the interchange instability driven by the eastward electric field (*E*) that is responsible for the PSSR (gravity is likely a lesser player because the altitude of the bottomside *F* layer is, at this time, still relatively low). Backscatter along the west wall could be produced by the interchange instability driven by an eastward *U*.

Absence of EPBs in the presence of the other two features, however, calls for some discussion. A reasonable explanation might be velocity-shear effects. Without velocity shear, the growth rate of the interchange instability should be larger for smaller-scale structures ($kL >> 1$) than for larger-scale structures ($kL \leq 1$); for example, see Guzdar et al. (1982, 1983), their Figure 1 (here, k is the wavenumber for LSWS, and L is the scale length for the velocity shear). However, in the presence of velocity shear, the growth rate of the interchange instability, for small-scale structures, should be strongly damped (Perkins and Doles 1975). Although the growth rate of larger-scale structures is not enhanced by velocity shear, it is less damped than that for the smaller-scale irregularities (Guzdar et al. 1982, 1983; Huba et al. 1983; Satyanarayana et al. 1984). This filtering process produces a relative maximum at a larger wavelength. Indeed, if we use 40 km as a reasonable estimate for the width of EPBs, and 400 km as that of upwellings, we would have a difference is spatial scale that is an order of magnitude.

A question that remains, is why do we observe small-scale irregularities (< few km) along the west wall of an upwelling, at a time when EPBs appear to be suppressed? One possibility is that the vector direction of the Pedersen current may have been close to tangent to isodensity contours. In this case, there is no velocity shear and the largest growth rate would be associated with small-scale irregularities. But, then, why don't we see secondary EPBs? There are other possible interpretations, which are beyond the scope of this paper. For example, (1) the spatial spectrum of seed plasma perturbations may have a peculiar distribution, (2) there is the question of which AGWs are capable of reaching F-region altitudes, and (3) there is the question of how the properties of LSWS will be colored by neutral-ion coupling. For our purposes here, we are mainly concerned with growth of upwellings, and how their presence controls the properties of ESF patches and clusters of EPBs. Amplification of upwellings during PSSR is a key process because the larger-amplitude upwellings profoundly distort the background conditions that must prevail in the absence of upwellings. In other words, upwelling growth transforms the background into a two-dimensional problem.

Onset phase

The appearance of the primary EPB from the crest of an upwelling argues in favor of curvature acting as an enhancement of the 'seed' amplitude. The implication is that the growth rate of the interchange instability is not large enough to produce EPBs from small-amplitude LSWS. Keeping in mind that upwelling amplitudes can reach 50 km without any appearance of an EPB, we can infer that either seed amplitude for EPBs is smaller than that for an upwelling, or that the growth rate of EPBs is smaller than that for an upwelling during PSSR. Furthermore, the confinement of the primary EPB to the vicinity of the crest of the upwelling, at the end of the PSSR, points to a likelihood that the conditions at the crest, created by upwelling growth, are responsible for the appearance of the primary EPB (Tsunoda 2005). It seems evident that curvature in isodensity contours in the vicinity of the crest will determine the seed amplitude for the primary EPB; ν_{in} should also be lowest at the crest. Both play roles in the interchange instability process.

Structure phase

In 'Structure phase: clustering and EW asymmetry' section, we showed how EPB clusters form, from EPB generation along the west wall, followed by the filling in of the entire upwelling region. Interpretation of west-wall structuring via the interchange instability, driven by an eastward U, appears straightforward. We also showed how EPBs move eastward relative to the upwelling, which leads to the filling of the entire upwelling region with EPBs, as can be seen in Figures 4 and 7. This latter behavior can be explained by the presence of velocity shear near the base of the F layer, where LSWS forms. As mentioned above, zonal drift, around SS_E, is likely controlled by the E-region dynamo at lower altitudes, and by the F-region dynamo at higher altitudes. In this case, plasma drift would be westward (eastward) at lower (higher) altitudes. It turns out that the isodensity contours, on which LSWS is observed, have been found to be either stationary or to drift slowly westward (Tsunoda and White 1981; Tsunoda et al. 2013). After SS_E, we expect control by the E-region dynamo to shift to lower altitudes, which means LSWS should begin to drift eastward as it falls under the control of the F-region dynamo. Nonetheless, the zonal drift speed of LSWS, which is often identified by the 10^5 el/cm^3 isodensity contour (see Figures 1 and 5a), will continue to move more slowly than plasma near the F-region peak, where N should be closer to 10^6 el/cm^3 (e.g., Zalesak et al. 1982). Hence, when EPBs penetrate upward into the F layer, they will drift with the background plasma because vertically elongated structures cannot develop a strong vertical polarization E. Hence, filling of upwellings with EPBs through relative zonal transport is expected.

EW asymmetry

It is not difficult to show that the EW asymmetry that develops within an upwelling is likely responsible for the asymmetries found in ESF patches and swoopers. To illustrate, we present a sketch in Figure 10 that illustrates what likely happens to paths of radio signals, when an upwelling with plasma structure is situated to the west of an ionosonde, and when the same structure is situated to the east of the station. To the west, rays can find surfaces of reflection along the west wall of the

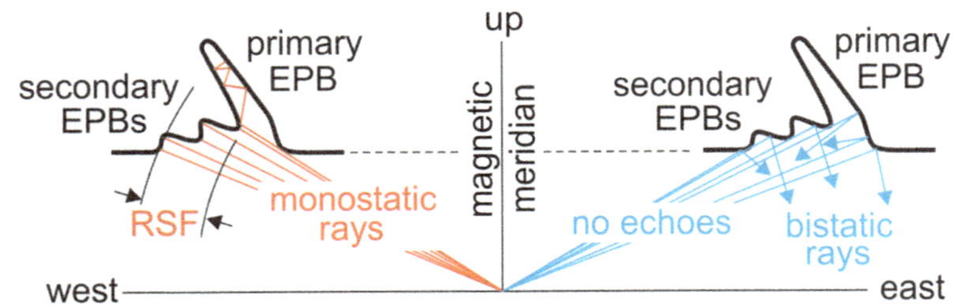

Figure 10 Sketch to illustrate the source of EW asymmetry in patches. Monostatic (red) rays return to ionosonde; bistatic (blue) rays do not return to ionosonde.

upwelling (where the bottomside shape of secondary EPBs are encountered) that will allow the rays to return to the ionosonde (see Thampi et al. (2012) for further discussion). We refer to these rays as monostatic rays; they are drawn with red lines. The extent of RSF would be related to the distribution of structure along the west wall. There is evidence that a ray could find its way into the primary EPB (or even secondary EPBs), as sketched (e.g., Tsunoda et al. 1995; Sales et al. 1996). Because the width of the EPB wedge (Tsunoda et al. 1982) decreases with altitude, up to the peak of the F layer, there is a good chance that the ray can retrace its way back to the ionosonde. This reflection mode is essentially identical to that described in Tsunoda (2012) to explain one form of MRE. To the east, the geometry is such that the rays can only skim the isodensity contours along the west wall (i.e., secondary EPBs), and when they do, the reflected paths are generally away from the ionosonde. We refer to rays that do not return to the ionosonde as bistatic rays, shown here with blue lines. This interpretation is consistent with the observations presented in Figure 3.

The sketch in Figure 10 can also be used to visualize how EW asymmetry manifests itself in swoopers and off-great-circle paths in TEP experiments. In the case of a short baseline (e.g., Davies and Chang 1968), we can treat the monostatic rays as those transmitted from one location, that encounter side reflections, but still reach the receiver at the other end of the bistatic path. This conversion is relatively simple because plasma structures are all expected to be elongated along B lines. In the case of long baselines, which are aligned more or less with the magnetic meridian, the simplest interpretation would be in terms of rays that follow a great-circle path in a direction away from the magnetic meridian, up to the vicinity of the dip equator. The one-hop path that is envisioned to straddle the dip equator is essentially the one just described for a short baseline. After side reflections have occurred, the remaining distance is envisioned to follow a great-circle path to the receiver (this is the geometry used by Röttger (1973) to interpret his TEP results).

Conclusions

From our perspective, the notion that an ESF patch or EPB cluster can be envisioned as a unit of disturbance is appealing. With hindsight, use of the term 'convective ionospheric substorm' does not seem to be so far-fetched. The spatial scale of the substorm is set by the zonal width of the upwelling, which is, perhaps typically, 400 km (e.g., Röttger 1973). The substorm process, which consists of three phases, is completed in one to two (or a few) hours. All of the pertinent electrodynamics take place in a plane transverse to B, with allowance for the mapping of E along B. With this framework for understanding, we are forced to keep in mind that even in the simplest of descriptions, ESF must be considered in terms of a two-dimensional (2D) structure (in plane transverse to B) that evolves with time. For example, it should be evident, from Figure 6, that an altitude-time-intensity description of ESF obtained with a fixed-beam radar (e.g., Woodman and La Hoz 1976) does not capture a true representation of the process.

The importance of a 2D description should be abundantly clear. The envisioned ESF process is still closely associated with the PSSR, which is consistent with most observations (e.g., Fejer et al. 1999). The primary EPBs, for example, are likely direct products of a strong PSSR and a high F layer, which is conducive to the gravitational interchange instability. The PSSR plays a less-direct role in the development of ESF patches and EPB clusters. That is, the PSSR amplifies an upwelling, and it is the zonal gradient along the west wall of the upwelling, together with an eastward U, that leads to the development of a cluster of secondary EPBs. Moreover, structuring of the west wall (but not east wall) is responsible for the EW asymmetry in numerous observations.

The finding that the ESF patches and EPB clusters tend to have spatial scales around 400 km is also appealing because most of the proposed physics appear to favor this spatial scale. The amplification of LSWS, but not EPBs, suggests that velocity shear may be responsible for a preferred larger wavelength (Guzdar et al. 1982, 1983). On the other hand, the observations (although still few) of LSWS prior to SS_E suggests that the source is not the velocity shear in the bottomside *F* layer. If true, the source may be an AGW that is excited in the troposphere. It turns out that AGWs with larger spatial scales have higher phase velocities, which are more likely to survive the journey from troposphere to ionosphere (e.g., Vadas 2007). Once the AGW reaches the *F*-region, LSWS must be produced by AGWs through some form of neutral-ion coupling, which likely involves a polarization *E* (e.g., Klostermeyer 1978). It also turns out that most of the geometries that allow avoidance of polarization-shorting effects, such as the azimuth of propagation for AGWs (Tsunoda 2010a,c), AGWs have circular phase fronts (Tsunoda 2010b), and the presence of a conducting *E* layer on *B* lines that thread the phase fronts of AGWs, favor larger spatial scales.

In closing, our paradigm appears to be a reasonable first step toward constructing a unified working model for ESF. The model, as described, is likely more relevant to conditions of high solar activity during ESF season. Much remains to be done. There are already a significant number of observations that may not appear to fit the paradigm described in this paper. We suggest that care be taken in how one-dimensional descriptions are extrapolated to produce a 2D description, before drawing conclusions. That aside, it is likely that other interpretations will be needed to modify this working model to accommodate new observations, and to extend our understanding of the underlying physics.

Abbreviations

2D: two dimensional; AE-E: atmospheric explorer-E; AGW: atmospheric gravity wave; *B*: geomagnetic field; C/NOFS: communication/navigation outage forecasting system; CGW: circular AGW; CW: continuous wave; DLAT: dip latitude; *E*: electric field; EPB: equatorial plasma bubble; ESF: equatorial spread F; EW: east-west; GLAT: geographic latitude; GLON: geographic longitude; h'F: minimum virtual height of F layer; IS: incoherent scatter; JRO: Jicamarca radio observatory; LST: local solar time; LSWS: large-scale wave structure; MCC: mesoscale convective complex; MRE: multi-reflected echoes; N: plasma density; N_i: ion density; OLR: outgoing longwave radiation; PSSR: post-sunset rise; RSF: range spread F; SS_E: sunset in E-region; SS_F: sunset in F-region; ST: satellite trace; TEC: total electron content; TEP: transequatorial propagation; *U*: neutral wind; UT: universal time; ν_{in}: ion-neutral collision frequency.

Competing interests

The authors' declared that they have no competing interests.

Acknowledgements

Research was supported by the National Science Foundation under grant ATM-1242815, and the Air Force Office of Scientific Research, under Contract FA9550-15-C-0018. Ionograms from the Jicamarca Digisonde are available from website, umlcar.uml.edu/DIDBase/, operated by the University of Massachusetts, Lowell, Mass. The Digisonde is operated by the Jicamarca Radio Observatory. OLR maps are available from website, aviationweather.gov/obs/sat/intl/. The JpGU classification for this paper is space and planetary sciences.

References

Aarons J, Buchau J, Basu S, McClure JP (1978) The localized origin of equatorial *F* region irregularity patches. J Geophys Res 83:1659–1664
Abdu MA, Batista IS, Bittencourt JA (1981) Some characteristics of spread *F* at the magnetic equatorial station Fortaleza. J Geophys Res 86:6836–6842
Bernhardt PA, Siefring CL (2006) New satellite-based systems for ionospheric tomography and scintillation region imaging. Radio Sci 41:RS5S23, doi:10.1029/2005RS003360
Calvert W, Cohen R (1961) The interpretation and synthesis of certain spread *F* configurations appearing on equatorial ionograms. J Geophys Res 66:3125–3140
Calvert W, Davies K, Stiltner E, Brown JT (1963) Equatorial spread-*F* motions. In: Proc Intl Conf Ionosphere. Inst. Phys. Physical Soc, London, pp 316–322
Clemesha BR (1964) An investigation of the irregularities in the *F* region associated with equatorial type spread *F*. J Atmos Terr Phys 26:91–112
Clemesha BR, Wright RWH (1966) A survey of equatorial spread *F*. In: Newman P (ed) Spread F and Its Effects Upon Radio Wave Propagation and Communications. Agardograph 95, 3, W. and J. MacKay and Co., Ltd, London, England
Cohen R, Bowles KL (1961) On the nature of equatorial spread *F*. J Geophys Res 66:1081–1106
Davies K, Barghausen AF (1966) The effect of spread *F* on the propagation of radiowaves near the equator. In: Spread F and its Effects upon Radiowave Propagation and Communication, Chapter 2-13. Agardograph 95, Technivision, Maidenhead, England, pp 437–466
Davies K, Chang NJF (1968) Radio-Doppler observations of the ionosphere near the magnetic equator. In Scatter Propagation of Radio Waves, Part II, AGARD Conf. Proc. No. 37, Paper 52.
De la Beaujardiere O, Jeong L, Basu B, Basu S, Beach T, Bernhardt P, Burke W, Groves K, Heelis R, Holzworth R, Huang C, Hunton D, Kelley M, Pfaff R, Retterer J, Rich F, Starks M, Straus P, Valladares C (2004) C/NOFS: A mission to forecast scintillations. J Atmos Solar-Terr Phys 66:1573–1591
Dungey JW (1956) Convective diffusion in the equatorial *F* region. J Atmos Terr Phys 9:304–310
Farley DT, Balsley BB, Woodman RF, McClure JP (1970) Equatorial spread *F*: Implications of VHF radar observations. J Geophys Res 75:7199–7216
Fejer BG, Scherliess L, de Paula ER (1999) Effects of the vertical plasma drift velocity on the generation and evolution of equatorial spread *F*. J Geophys Res 104:19,859–19,869
Fritts DC, Vadas SL (2008) Gravity wave penetration into the thermosphere: Sensitivity to solar cycle variations and mean winds. Ann Geophys 26:3841–3861
Guzdar PN, Satyanarayana P, Huba JD, Ossakow SL (1982) Influence of velocity shear on the Rayleigh-Taylor instability. Geophys Res Lett 9:547–550
Guzdar PN, Satyanarayana P, Huba JD, Ossakow SL (1983) Correction. Geophys Res Lett 10:492
Hei MA, Heelis RA, McClure JP (2005) Seasonal and longitudinal variation of large-scale topside equatorial plasma depletions. J Geophys Res 110:A12315, doi:10.1029/2005JA011153
Huba JD, Ossakow SL, Satyanarayana P, Guzdar PN (1983) Linear theory of the E x B instability with an inhomogeneous electric field. J Geophys Res 88:425–434
Hysell DL, Kelley MC (1997) Decaying equatorial *F* region plasma depletions. J Geophys Res 102:20,007–20,017
Kelleher RF, Röttger J (1973) Equatorial spread *F* irregularities observed at Nairobi and on the transequatorial path Lindau-Tsumeb. J Atmos Terr Phys 35:1207–1211
Kelleher RF, Skinner NJ (1971) Studies of *F* region irregularities at Nairobi, II. By direct backscatter at 27.8 MHz. Ann Geophys 27:195–200
Kelley MC, de la Beaujardiere O, Retterer J, Makela JJ (2005) Introduction to special section on Communication/Navigation Forecasting System: A next step in Space Weather. Space Wea 3:S12C01, doi:10.1029/2005SW000189
Kelley MC, Rodrigues FS, Makela JJ, Tsunoda R, Roddy PA, Hunton DE, Retterer JM, de La Beaujardiere O, de Paula ER, Ilma RR (2009) C/NOFS and radar observations during a convective ionospheric storm event over South America. Geophys Res Lett 36:L00C07, doi:10.1029/2009GL039378
Kelley MC, Makela JJ, de La Beaujardiere O, Retterer J (2011) Convective ionospheric storms: A review. Rev Geophys 49:RG2003, doi:10.1029/2010RG000340
Kil H, Heelis RA (1998a) Global distribution of density irregularities in the equatorial ionosphere. J Geophys Res 103:407–417
Kil H, Heelis RA (1998b) Equatorial density irregularity structures at intermediate scales and their temporal evolution. J Geophys Res 103:3969–3981
Klostermeyer J (1978) Nonlinear investigation of the spatial resonance effect in the nighttime equatorial *F* region. J Geophys Res 83:3753–3760
Kudeki E, Fejer BG, Farley DT, Ierkic HM (1981) Interferometer studies of equatorial *F*-region irregularities and drifts. Geophys Res Lett 8:377–380
Makela JJ, Kelley MC, Su S-Y (2005) Simultaneous observations of convective ionospheric storms: ROCSAT-1 and ground-based imagers. Space Wea 3: S12C02

Makela JJ, Kelley MC, De La Beaujardiere O (2006) Convective ionospheric storms: A major space weather problem. Space Wea 4. doi:10.1029/2005SW000145.

Maruyama T, Kawamura M (2006) Equatorial ionospheric disturbance observed through a transequatorial HF propagation experiment. Ann Geophys 24:1401–1409

McClure JP, Hanson WB, Hoffman JH (1977) Plasma bubbles and irregularities in the equatorial ionosphere. J Geophys Res 82:2650–2656

McClure JP, Singh S, Bamgboye DK, Johnson FS, Kil H (1998) Occurrence of equatorial *F* region irregularities: Evidence for tropospheric seeding. J Geophys Res 103:29,119–29,135

Nielson DL, Crochet M (1974) Ionospheric propagation of HF and VHF radio waves across the geomagnetic equator. Rev Geophys Space Phys 12:688–702

Perkins FW (1975) Doles JH III. Velocity shear and the E x B instability J Geophys Res 80:211–214

Röttger J (1973) Wave-like structures of large-scale equatorial spread-*F* irregularities. J Atmos Terr Phys 35:1195–1206

Saito S, Maruyama T, Ishii M, Kubota M, Ma G, Chen Y, Li J, Duyen CH, Truong TL (2008) Observations of small- to large-scale ionospheric irregularities associated with plasma bubbles with a transequatorial HF propagation experiment and spaced GPS receivers. J Geophys Res 113:A12313, doi:10.1029/2008JA013149

Sales GS, Reinisch BW, Scali JL, Dozois C, Bullett TW, Weber EJ, Ning P (1996) Spread *F* and the structure of equatorial ionization depletions in the southern anomaly region. J Geophys Res 101:26,819–26,827

Satyanarayana P, Guzdar PN, Huba JD, Ossakow SL (1984) Rayleigh-Taylor instability in the presence of a stratified shear layer. J Geophys Res 89:2945–2954

Singh S, Johnson FS, Power RA (1997) Gravity wave seeding of equatorial plasma bubbles. J Geophys Res 102:7399–7410

Su SY, Chao CK, Liu CH (2008) On monthly/seasonal/longitudinal variations of equatorial irregularity occurrences and their relationship with the postsunset vertical drift velocities. J Geophys Res 113:A05307, doi:10.1029/2007JA012809

Thampi SV, Yamamoto M, Tsunoda RT, Otsuka Y, Tsugawa T, Uemoto J, Ishii M (2009) First observations of large-scale wave structure and equatorial spread *F* using CERTO radio beacon on the C/NOFS satellite. Geophys Res Lett 36: L18111, doi:10.1029/2009GL039887

Thampi SV, Tsunoda R, Jose L, Pant TK (2012) Ionogram signatures of large-scale wave structure and their relation to equatorial spread *F*. J Geophys Res 117: A08314, doi:10.1029/2012JA017592

Tsunoda RT (1981) Time evolution and dynamics of equatorial backscatter plumes, 1, Growth phase. J Geophys Res 86:139–149

Tsunoda RT (1983) On the generation and growth of equatorial backscatter plumes. 2. Structuring of the west walls of upwellings. J Geophys Res 88:4869–4874

Tsunoda RT (2005) On the enigma of day-to-day variability in equatorial spread *F*. Geophys Res Lett 32:L08103, doi:10.1029/2005GL022512

Tsunoda RT (2007) Seeding of equatorial plasma bubbles with electric fields from an E_s-layer instability. J Geophys Res 112:A06304, doi:10.1029/2006JA012103

Tsunoda RT (2008) Satellite traces: An ionogram signature for large-scale wave structure and a precursor for equatorial spread *F*. Geophys Res Lett 35: L20110, doi:10.1029/2008GL035706

Tsunoda RT (2009) Multi-reflected echoes: Another ionogram signature of large-scale wave structure. Geophys Res Lett 36:L01102, doi:10.1029/2008GL036221

Tsunoda RT (2010a) On seeding equatorial spread *F* during solstices. Geophys Res Lett 37:L05102, doi:10.1029/2010GL042576

Tsunoda RT (2010b) On seeding equatorial spread *F*: Circular gravity waves. Geophys Res Lett 37:L10104, doi:10.1029/2010GL043422

Tsunoda RT (2010c) On equatorial spread *F*: Establishing a seeding hypothesis. J Geophys Res 115:A12303, doi:10.1029/2010JA015564

Tsunoda RT (2012) A simple model to relate ionogram signatures to large-scale wave structure. Geophys Res Lett 39:L18107, doi:10.1029/2012GL053179

Tsunoda RT, White BR (1981) On the generation and growth of equatorial backscatter plumes—1. Wave structure in the bottomside F layer J Geophys Res 86:3610–3616

Tsunoda RT, Baron MJ, Owen J, Towle DM (1979) ALTAIR: An incoherent scatter radar for equatorial spread *F* studies. Radio Sci 14:1111–1119

Tsunoda RT, Livingston RC, McClure JP, Hanson WB (1982) Equatorial plasma bubbles: Vertically elongated wedges from the bottomside *F* layer. J Geophys Res 87:9171–9180

Tsunoda RT, Livingston RC, Buonocore JJ, McKinley AV (1995) The frequency-agile radar: A multifunctional approach to remote sensing of the ionosphere. Radio Sci 30:1623–1643

Tsunoda RT, Bubenik DM, Thampi SV, Yamamoto M (2010) On large-scale wave structure and equatorial spread *F* without a post-sunset rise of the *F* layer. Geophys Res Lett 37:L07105, doi:10.1029/2009GL042357

Tsunoda RT, Yamamoto M, Tsugawa T, Hoang TL, Tulasi Ram S, Thampi SV, Chau HD, Nagatsuma T (2011) On seeding, large-scale wave structure, equatorial spread *F*, and scintillations over Vietnam. Geophys Res Lett 38:L20102, doi:10.1029/2011GL049173

Tsunoda RT, Thampi SV, Nguyen TT, Yamamoto M (2013) On validating the relationship of ionogram signatures to large-scale wave structure. J Atmos Solar-Terr Phys 103:30–35

Tsunoda RT, Nguyen TT, Le MH (2015) Effects of tidal forcing, conductivity gradient, and active seeding, on the climatology of equatorial spread *F* over Kwajalein. J Geophys Res 120. doi:10.1002/2014JA020762.

Tulasi Ram S, Yamamoto M, Tsunoda RT, Thampi SV (2012) On the application of differential phase measurements to study the zonal large scale wave structure (LSWS) in the ionospheric electron content. Radio Sci 47:RS2001, doi:10.1029/2011RS004870

Tulasi Ram S, Yamamoto M, Tsunoda RT, Chau HD, Hoang TL, Damtie B, Wassaie M, Yatini CY, Manik T, Tsugawa T (2014) Characteristics of large-scale wave structure observed from African and Southeast Asian longitude sectors. J Geophys Res 119. doi:10.1002/2013JA019712.

Vadas SL (2007) Horizontal and vertical propagation and dissipation of gravity waves in the thermosphere from lower atmospheric and thermospheric sources. J Geophys Res 112:A06305, doi:10.1029/2006JA011845

Waldock JA, Jones TB (1987) Source regions of medium scale travelling ionospheric disturbances observed at mid-latitudes. J Atmos Terr Phys 49:105–114

Waliser DE, Graham NE, Gautier C (1993) Comparison of the highly reflective cloud and outgoing longwave radiation datasets for use in estimating tropical deep convection. J Climate 6:331–353

Weber EJ, Buchau J, Eather RH, Mende SB (1978) North-south aligned equatorial airglow depletions. J Geophys Res 83:712–716

Weber EJ, Buchau J, Moore JG (1980) Airborne studies of equatorial *F* layer ionospheric irregularities. J Geophys Res 85:4631–4641

Weber EJ, Brinton HC, Buchau J, Moore JG (1982) Coordinated airborne and satellite measurements of equatorial plasma depletions. J Geophys Res 87:10,503–10,513

Whitehead JD (1971) Ionization disturbances caused by gravity waves in the presence of an electrostatic field and a background wind. J Geophys Res 76:238–241

Woodman RF, La Hoz C (1976) Radar observations of *F* region equatorial irregularities. J Geophys Res 81:5447–5466

Yokoyama T, Fukao S, Yamamoto M (2004) Relationship of the onset of equatorial *F* region irregularities with the sunset terminator observed with the Equatorial Atmosphere Radar. Geophys Res Lett 31:L24804, doi:10.1029/2004GL021529

Zalesak ST, Ossakow SL, Chaturvedi PK (1982) Nonlinear equatorial spread *F*: The effect of neutral winds and background Pedersen conductivity. J Geophys Res 87:151–166

3

Investigating the particle precipitation of a moving cusp aurora using simultaneous observations from the ground and space

Satoshi Taguchi[1,2,3*], Keisuke Hosokawa[1,2] and Yasunobu Ogawa[4]

Abstract

Using observations of a moving cusp aurora from a high-sensitivity all-sky imager at Longyearbyen, Svalbard, and *in situ* observations of the precipitating particles from a spacecraft that flew over the aurora, we examined the particle precipitation features in the early and final stages of the moving cusp aurora. We focused on two auroral structures created near noon, separated by approximately 3 min, during a southwestward interplanetary magnetic field (IMF) condition on 17 December 2012. The second auroral structure occurred when the IMF turned further southward. Immediately after the appearance of the latter structure, the two auroral structures were adjacently situated, and the DMSP F18 spacecraft passed through these regions. A detailed comparison of the data from particle spectrometers onboard the spacecraft and the 630 nm aurora image data demonstrates that the ion precipitation in the young cusp aurora (i.e., second auroral structure) had a high energy flux, whereas that in the old cusp aurora (i.e., first auroral structure) had a very low energy flux. For the electron precipitation, the features in both regions were found to be very similar; the energy flux at approximately 100 eV often exceeded 1×10^9 eV cm^{-2} s^{-1} sr^{-1} eV^{-1} in both regions. This indicates that the electron precipitation in the moving cusp aurora is maintained at a high flux level over a certain interval from its starting time. Thus, we suggest that the electron precipitation flux in the moving cusp aurora is controlled by a mechanism independent of the ion precipitation.

Keywords: High-latitude ionosphere; Aurora; Particle precipitation; Cusp

Background

A moving mesoscale auroral structure, which is characterized by high red line intensities, is a typical phenomenon in the dayside cusp of the high-latitude ionosphere. Its motion generally has a poleward component and very often also has a strong azimuthal component. As has been shown by many researchers (e.g., Sandholt et al. 1986; Lockwood et al. 1989; Fasel 1995; Sandholt and Farrugia 2003; Oksavik et al. 2004; Taguchi et al. 2012), the azimuthal motion is eastward or westward, which is consistent with the direction of the interplanetary magnetic field (IMF) B_Y-related tension force arising from the curvature of the open magnetic field line after reconnection on the dayside magnetopause.

A moving auroral structure that is observed well away from the main red line auroral band in the daytime sector is recognized as being the ionospheric signature of open magnetic flux motion driven by patchy or intermittent reconnection, known as a flux transfer event (FTE) (Haerendel et al. 1978; Russell and Elphic 1978). For a moving auroral structure that appears to be embedded in the red line auroral band, however, it is difficult to determine whether such a structure is the signature of an FTE. However, this auroral structure should also be located in the open magnetic field line region because the open/closed field line boundary generally coincides with the equatorward edge of the red line auroral band (Milan et al. 1999).

Immediately after the open magnetic field line is created by reconnection, magnetosheath plasmas (both electrons and ions) stream into the flux tube. Since the flux tube is moving, the injected particles with higher (or lower) velocities reach the ionosphere after they travel a shorter (or

* Correspondence: taguchi@kugi.kyoto-u.ac.jp
[1]Department of Communication Engineering and Informatics, University of Electro-Communications, Chofu, Tokyo 182-8585, Japan
[2]Center for Space Science and Radio Engineering, University of Electro-Communications, Chofu, Tokyo 182-8585, Japan
Full list of author information is available at the end of the article

longer) distance. For ions, this behavior is observed over a period of several hundred seconds (Lockwood and Smith 1994) because ion velocities are typically comparable to the velocity of the flux tube. The ion energy flux also tends to decrease with increasing distance from the origin of the open flux tube, as is inferred from the results of the modeling of the particle precipitation for steady-state reconnection by Onsager et al. (1993).

For the electron precipitation in the moving cusp aurora, it is still not understood how the electron precipitation changes with increasing distance from the origin of the open flux tube. Satellite observations immediately inside the magnetopause have demonstrated the motion of flux tubes containing enhanced magnetosheath-like electrons (Owen et al. 2001). Since electrons have high speeds along magnetic field lines, they may come down to the ionosphere in time scales on the order of seconds, suggesting that the electron features of the ionospheric signature of the open flux tube reflect the spatial change in the magnetosheath density as a function of the downstream distance from the reconnection point.

Another possibility is that the precipitating electrons are affected by the ions that are injected with them, rather than simply reflecting the spatial change in the magnetosheath. Burch (1985) suggested that the cusp electrons follow the ions to maintain charge neutrality. If this behavior is operative in the flux tube of the moving cusp aurora and the mechanisms for maintaining the behavior do not change significantly with increasing distance from the origin of the flux tube, a fixed relationship may be seen between the features of the precipitating electrons and those of the precipitating ions, irrespective of the distance from the origin of the open flux tube.

At ionospheric heights, it is difficult to discern the evolution of the moving particle precipitation region using *in situ* data from a single spacecraft because the speed of low-altitude spacecraft is considerably greater than the speed of the precipitation region. The data obtained by a spacecraft would simply reflect the spatial structure inside the region without providing any information on its evolution. One powerful tool that does provide information on the evolution of the precipitation region is the all-sky imager. An all-sky imager can be used to understand when and where auroral structures are created and how they subsequently move. If an all-sky imager observes multiple moving auroral structures whose elapsed times from their appearance differ and a spacecraft flies over these structures, then the spacecraft data can provide information on how the precipitation region changes with elapsed time from the appearance of the structures or with increasing distance from their points of origin.

For a single moving auroral structure, Oksavik et al. (2005) presented a simultaneous observation event of this type. Data from the ion drift meter on the Defense Meteorological Satellite Program (DMSP) F13 spacecraft for this event indicated the presence of mesoscale convection for a moving cusp aurora, whereas the precipitating particle data from this satellite did not show any prominent signatures. This latter result was presumably due to the fact that the satellite observation was made after the aurora disappeared (10:12:00 UT in Figure four of Oksavik et al. 2005). A moving cusp aurora has a short lifetime, typically 3 to 4 min (Fasel 1995). This short lifetime also makes it rare for their presence to coincide with the measurements of *in situ* spacecraft.

In this paper, we report results obtained on 17 December 2012 from the simultaneous observation of two cusp auroral structures by an all-sky imager and a spacecraft. We show the motion of the two auroral structures near local noon using the 630 nm auroral image captured by the all-sky imager. One auroral structure arrived at the satellite position approximately 230 s after its initial appearance (although the height of the aurora was much lower than the altitude of the satellite), whereas the other arrived at a similar position only 60 s after its initial appearance. We present observations of the precipitating particles made by the DMSP F18 spacecraft that flew over these auroral structures and show the precipitation features of the ions and electrons in the young and old flux tubes (i.e., the flux tubes of the auroral structures that appeared second and first, respectively).

Methods

Instrumentation

The instrumentation used in this study consists of a ground-based all-sky imager and particle spectrometers on the DMSP F18 spacecraft, which flew at an altitude of approximately 840 km. The all-sky imager uses an electron multiplier charge-coupled device (EMCCD) camera (Hamamatsu, C9100-13, Hamamatsu, Japan) with an imaging resolution of 512 × 512 pixels and measures emission at two wavelengths, 557.7 and 630.0 nm, using narrow passband interference filters. This ground-based imager has been operating in Longyearbyen, Norway, (geographical latitude 78.1° N and longitude 16.0° E) since October 2011 (Taguchi et al. 2012). As shown in Taguchi et al. (2012), the 630.0 nm (red) line data from this EMCCD camera can provide detailed information about the dynamic features of moving cusp auroral structures. The high sensitivity of this camera is also very effective for identifying the structure of polar cap patches (Hosokawa et al. 2013a, 2013b, 2014; Sakai et al. 2014). Thus, in the current study, we used 630.0 nm wavelength images.

In the first observation season, from the end of October 2011 to the end of February 2012, we obtained 630.0 nm data with an exposure time of 4 s per image. For the second year of observation, which includes the event studied in this paper, we modified the observation mode slightly so that the 630.0 nm images could be obtained with an exposure time of 1 s in addition to the exposure time of 4 s. For this study, we used the imaging data with an exposure time of 1 s. As is demonstrated later, images with this short exposure time were more helpful for identifying the location of auroral brightening, even though the data were not obtained at regular intervals, because of the constraints of the software controlling the automated observation system.

The particle spectrometers on the DMSP F18 satellite measure precipitating electrons and ions with energies between 30 eV and 30 keV once per second (Hardy et al. 1984). Since electron precipitation with an energy of approximately 100 eV strongly affects the 630.0 nm emission rate (e.g., Roble and Rees 1977), out of the full available energy range of the spectrometer, we concentrated on the electron data from three energy intervals near 100 eV, specifically at approximately 70, 100, and 150 eV.

Observations of the cusp aurora on 17 December 2012

Figure 1 shows the 630.0 nm all-sky image taken at 08:26:52 UT. The all-sky image is projected onto an Earth-centered sphere at a height of 250 km, which is the typical altitude for a cusp aurora (e.g., Sandholt et al. 1990). A longitudinally elongated emission band (bluish color in Figure 1) was present across the entire field of view and spanned 09:30 to 13:30 magnetic local time (MLT). The postnoon portion of this auroral oval exhibited enhanced emission (exceeding 4 kR, greenish color), three distinct areas of which are indicated by black arrows in Figure 1. The largest region, which occurred at approximately 13:00 MLT, is hereafter referred to as Aurora *A*. The smaller auroral structure located adjacent to Aurora *A* is called Aurora *B*. The auroral structure located equatorward of Aurora *B* is referred to as Aurora *C*. This study focuses on Auroras *B* and *C*.

Figure 2 shows the solar wind conditions from 07:30 to 09:30 UT. Solar wind data obtained by the Advanced Composition Explorer (ACE) spacecraft positioned approximately 230 R_E upstream in the solar wind are represented by solid black lines, and the 15 s averaged IMF data from the GEOTAIL satellite, which was located at (X_{GSE}, Y_{GSE}, Z_{GSE}) = (23.7 to 23.3, 17.4 to 18.3, 5.1 to 5.5) R_E in

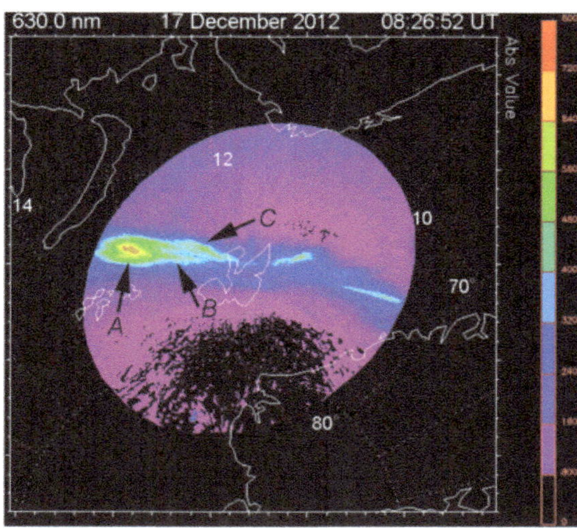

Figure 1 A 630.0 nm dayside all-sky image taken at 08:26:52 UT on 17 December 2012. The all-sky image is projected onto an Earth-centered sphere at a height of 250 km with the geomagnetic coordinate system shown in the background. Black arrows indicate three auroral structures: *A*, *B*, and *C*.

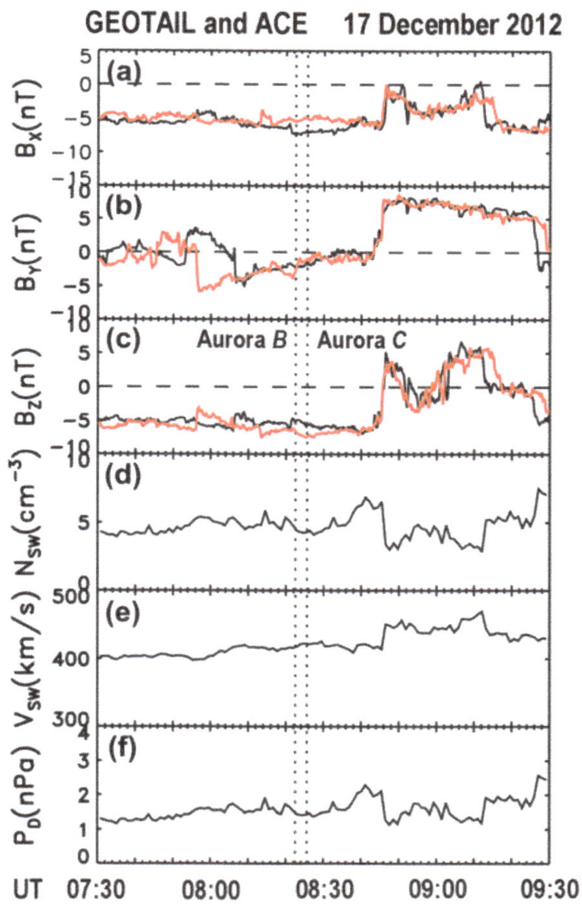

Figure 2 GEOTAIL (red) and ACE (black) solar wind conditions. The data from GEOTAIL in the near-Earth solar wind are advanced by 7 min, whereas those from ACE, located approximately 230 R_E upstream in the solar wind, are delayed by 47 min. The **(a)** *X*-, **(b)** *Y*-, and **(c)** *Z*-components of the IMF in the geocentric solar magnetospheric coordinate, **(d)** the solar wind proton density, **(e)** the flow speed, and **(f)** the dynamic pressure are plotted. The two dotted lines indicate the time when Auroras *B* and *C* appeared in the dayside ionosphere.

the geocentric solar ecliptic (GSE) coordinate system, are plotted in the top three panels (Figure 2a,b,c) as red lines. Figure 2a,b,c shows the *X*-, *Y*-, and *Z*-components, respectively, of the IMF in the geocentric solar magnetospheric (GSM) coordinate system. In these figures, the GEOTAIL data are advanced by 7 min.

To estimate this time lag, we used (Lockwood et al. 1989)

$$T_{\mathrm{delay}} = \left\{ X_{SC} - 1.33X_{MP} - Y_{SC}\left(\frac{B_X}{B_Y}\right) + 2.64X_{MP} \right\} V_{SW}^{-1} + 2\ (\mathrm{min}), \quad (1)$$

where X_{SC} and Y_{SC} represent the spacecraft position in the GSE coordinate system, X_{MP} is the magnetopause location, B_X and B_Y are the *X*- and *Y*-components of the IMF in the GSE coordinate system, and V_{SW} is the solar wind speed. The second term on the right-hand side, 1.33 X_{MP}, represents an approximation of the subsolar standoff distance of the bow shock (Fairfield 1971). The factor of 2.64 in the fourth term comes from the multiplication of 0.33 and 8, where 8 is the slowdown factor of the solar wind in the subsolar magnetosheath (Spreiter and Stahara 1980; Khan and Cowley 1999). The last term, 2 min, represents the Alfven wave travel time from the magnetopause to the ionosphere. We assumed X_{MP} to be 10 R_E and obtained $T_{\mathrm{delay}} = -7$ min by using (X_{SC}, Y_{SC}) = (23.5, 17.8) R_E and (B_X, B_Y) = (−5.2, −1.3) nT as average values for 08:20 to 08:30 UT and taking V_{SW} to be 410 km s^{-1} from the ACE data.

The ACE data (black line) plotted in Figure 2 are 16 and 64 s averages of IMF and plasma data, respectively. A delay of 47 min was introduced in the ACE data to align the sharp increase in the ACE B_Z with a similar increase in the GEOTAIL B_Z at approximately 08:45 UT. The two dotted lines indicate the time at which the aurora images of Auroras *B* and *C* (Figure 1) were first captured, specifically 08:22:28 and 08:25:33 UT, respectively. The aurora images will be shown later. As is shown in both spacecraft data, the IMF B_Z was very stable at approximately −5 nT (Figure 2c) during the interval of approximately 1 h before the appearance of the auroral structures.

The GEOTAIL IMF data (red line) indicate that Aurora *B* appeared when (B_Y, B_Z) was approximately (−4, −6) nT. Immediately following this, the IMF turned further southward, and when (B_Y, B_Z) was approximately (−2, −7) nT, Aurora *C* appeared. As is shown later, the motion of Aurora *C* had a smaller eastward component than that of Aurora *B*, which is consistent with these IMF values.

The *X*-component B_X of the GEOTAIL IMF was relatively constant during the interval of approximately 1 h before the appearance of the auroral structures of interest (Figure 2a). A negative B_X when $B_Z < 0$ may be a relatively favorable condition for stronger auroral emission on the dayside (Yang et al. 2013). The solar wind plasma density N_{sw} was approximately 5 cm^{-3} during that interval (Figure 2d). The solar wind speed V_{sw} and the dynamic pressure P_D were also stable at approximately 410 km s^{-1} (Figure 2e) and in the range of approximately 1 to 2 nPa (Figure 2f), respectively.

Particle observations from the DMSP spacecraft

Figure 3 shows the data from the particle spectrometers on the DMSP F18 satellite from 08:18 to 08:28 UT on 17 December 2012, during which DMSP F18 flew from the dusk to the prenoon sector. The energy fluxes of the electrons and ions are plotted with color codes in the bottom two panels, together with the integral energy flux and average energy of the electrons and ions. Note that the ion energy axis is inverted in the fourth panel. The DMSP satellite initially measured regions on the dusk side from 08:19:15 to 08:20:00 UT, during which the energy flux of the electron precipitation (third panel) was relatively low and the ion precipitation (fourth panel) had a relatively high energy (the so-called central plasma sheet). The spacecraft entered the boundary plasma sheet at approximately 08:20:00 UT; here, the energy of the peak electron energy flux exceeded 1 keV, indicating that the precipitating electrons had accelerated fluxes. The high-latitude portion of this boundary plasma sheet appears to be smoothly connected to the region where the electron energy flux was very low (after approximately 08:22:30 UT).

At approximately 08:25:15 UT, the spacecraft entered the ion precipitation region that is typical for the cusp. Specifically, precipitating ions with energies between 100 eV and 1 keV became evident at 08:25:18 UT, i.e., 75.50° magnetic latitude (MLAT) and 13:33 MLT. The total energy flux and average energy of the ions generally increased with decreasing latitude (first and second panels, red dots) until 08:27 UT. After that, the energy flux became somewhat low (top panel, red dots). After 08:27:14 UT, the ion energy flux in the range of 1 to 10 keV (fourth panel) was lower than 1×10^6 eV cm^{-2} s^{-1} sr^{-1} eV^{-1} (becoming a greenish color). We determined this latitude (72.54° MLAT) to be the equatorward boundary of the cusp. The cusp range is represented by the red horizontal bar between the two spectrograms. Presumably, the region equatorward of this range is on a closed field line because the energy flux of the precipitating electrons at approximately 100 eV (i.e., electrons of magnetosheath origin) is very low. As will be shown in the following section, the most notable result is that the cusp includes the region containing precipitating electrons with a higher energy flux (reddish color in the third panel).

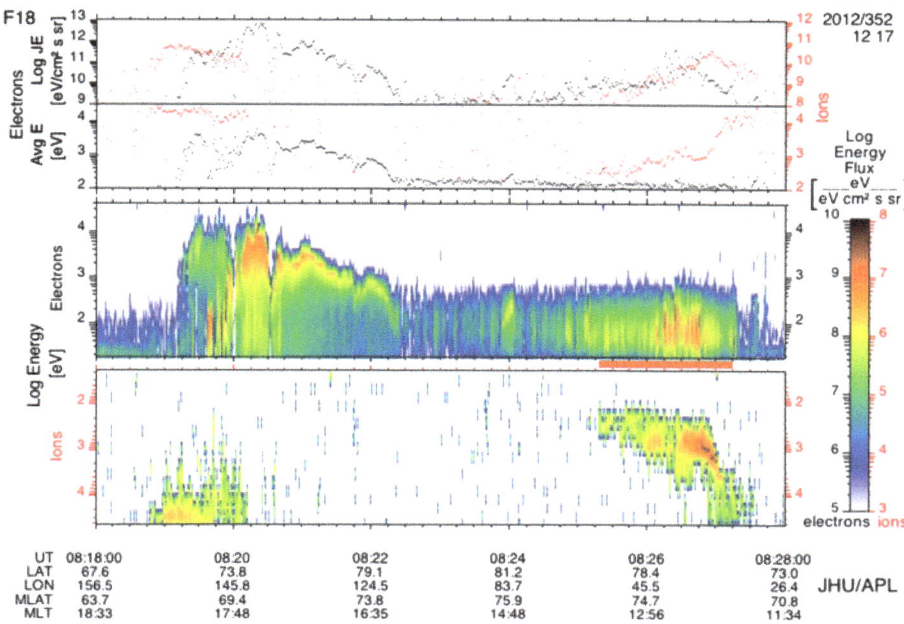

Figure 3 DMSP F18 particle observations along the passage from the dusk to the prenoon sector. The top panel indicates the electron and ion integral energy fluxes, and the second panel shows the average energy. The bottom two panels present the electron and ion differential energy flux spectrograms. The red horizontal bar, which extends from 75.50° to 72.54° MLAT, indicates the region we defined as the cusp.

Results and discussion

Multiple boundary crossings inside the cusp

Figure 4 presents an expanded plot of the DMSP F18 particle data (shown in Figure 3) near the electron precipitation with very high flux. Figure 4a shows the ratio of the integral ion number flux to the integral electron number flux in a logarithmic scale. To obtain this ratio, we first calculated the integral number flux (J_{TOT}) by dividing the integral energy flux (*JE*, shown in the top panel of Figure 3) by the average energy (E_{Avg}, shown in

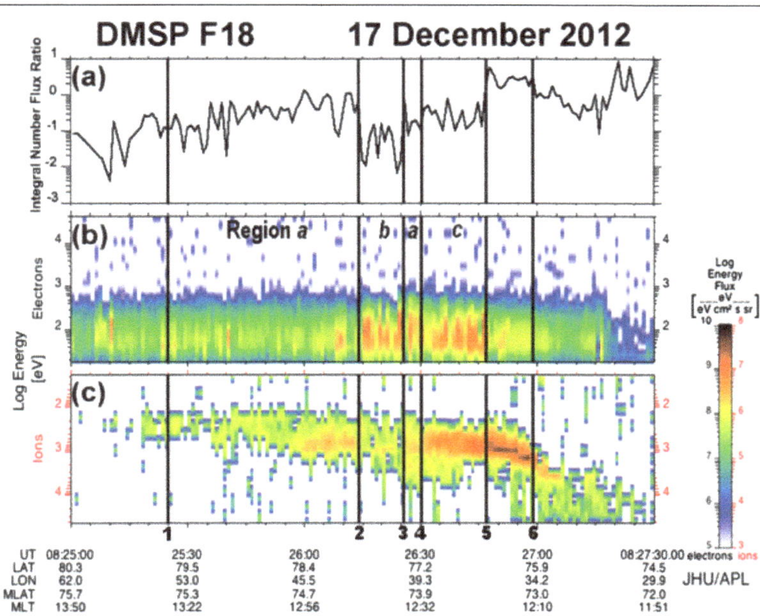

Figure 4 Particle data from DMSP F18 near the electron precipitation with a very high flux. The top panel **(a)** represents the normalized ratio of the integral ion number flux to the integral electron number flux in a logarithmic scale. The format of the bottom two panels **(b,c)** is the same as that of the bottom two panels of Figure 3. The six vertical lines in each panel indicate the boundaries between different particle features.

the second panel of Figure 3) for the electrons and ions (Hardy et al. 1985). We then defined the ratio as

$$R = \left(\frac{1}{k}\right)\frac{J_{TOT(ion)}}{J_{TOT(elec)}}, \quad (2)$$

where k is constant and was determined to be 0.057. This is a normalization factor so that R can be roughly between 10^{-2} and 10. We obtained 0.057 from the median of the ratio $J_{\mathrm{TOT(ion)}}/J_{\mathrm{TOT(elec)}}$ in the region 08:26:48 to 08:27:16 UT, which is on the equatorward side of the high-flux region, i.e., outside of the region of focus in this paper. This means that R is normalized by the features in the steady cusp. Note that the choice of k would not affect the main results of this paper.

Figure 4 shows the boundaries between different precipitating plasma features. Boundary 1, which is the leftmost line of the six lines on the plot, is the position over which the DMSP satellite flew at 08:25:25 UT. In the electron spectrogram (Figure 4b), this boundary represents the point where the maximum energy flux for the three energy steps around 100 eV first exceeded 1×10^8 eV cm^{-2} s^{-1} sr^{-1} eV^{-1} (yellow color) for two consecutive seconds. This point occurred at (MLAT, MLT) = (75.4°, 13:26). These two consecutive data bins, i.e., the observations over the given 2 s, corresponded to a horizontal scale of approximately 15 km along the satellite path. An aurora of this horizontal size would be detected over an area larger than 2 × 2 pixels (out of 512 × 512 pixels) wherever the aurora is situated inside the imager's field of view. The 2 × 2 pixel area is the smallest that can be reasonably resolved. Boundary 1 was very close to the poleward boundary of the cusp ion precipitation (08:25:18 UT), which was defined using the ion precipitation in Figure 3.

Boundary 2 is located at the position the DMSP satellite crossed at 08:26:14 UT; at this position, (MLAT, MLT) = (74.37°, 12:44). This boundary was determined from the sharp drop in the integral number flux ratio (Figure 4a). This can also be seen as a sharp drop in the ion energy, as at this time, the energy of the peak ion energy flux dropped to approximately 200 eV with a decrease in the ion energy flux (Figure 4c). After this time, low-energy ion precipitation appeared intermittently. Just 1 s later, i.e., at 08:26:15 UT, the satellite entered the region where the energy flux of the electron precipitation was very high; the maximum of the energy flux for the three energy bins around 100 eV exceeded 1×10^9 eV cm^{-2} s^{-1} sr^{-1} eV^{-1}. We refer to the region before Boundary 2 and after Boundary 1 as Region *a*.

The next boundary, Boundary 3, was chosen as the position the DMSP satellite crossed at 08:26:26 UT; at Boundary 3, (MLAT, MLT) = (74.05°, 12:35). At this time, there was a significant jump in the integral number flux ratio (Figure 4a). The maximum electron energy flux from the three energy bins around 100 eV decreased to values below 1×10^8 eV cm^{-2} s^{-1} sr^{-1} eV^{-1}, similar to the values often observed prior to Boundary 2. In addition, at this time, the energy flux of the ion precipitation sharply increased and approached that observed before Boundary 2. We refer to the region between Boundaries 2 and 3 as Region *b*. Region *b* is characterized as the region where the integral ion number flux relative to the integral electron number flux is very low compared with Region *a*.

After 08:26:26 UT (Boundary 3), ion precipitation with features similar to those seen in Region *a* appeared and continued for a brief period of time. At 08:26:30 UT, the energy of the peak ion energy flux dropped to well below 1 keV. We designated this position Boundary 4. The electron features in the region between Boundaries 3 and 4 are similar to those near the equatorward edge of Region *a*, i.e., the region observed during 08:26:08 to 08:26:10 UT. As was mentioned above, the electron energy flux dropped sharply at Boundary 3, and that low value is also similar to the values that were often observed prior to Boundary 2. The integral number flux ratios between Boundaries 3 and 4 are also similar to the ratios prior to Boundary 2. We thus regarded the region between Boundaries 3 and 4 as a reappearance of Region *a*.

For a while after 08:26:30 UT (Boundary 4), the satellite continued to observe high electron energy flux. These electron features are very similar to those seen in Region *b*. However, there is a clear difference in the ion precipitation features. The energy flux of the ions was higher after Boundary 4 than in Region *b*. The integral number flux ratio was also higher after Boundary 4 than in Region *b*. These observations are crucial results of this study and will be used later for comparison with the auroral image. At 08:26:47 UT, the integral number flux ratio increased sharply, and the maximum energy flux of the electron precipitation for the three energy bins around 100 eV no longer reached 1×10^9 eV cm^{-2} s^{-1} sr^{-1} eV^{-1}. We designated this position Boundary 5, and we refer to the region between Boundaries 4 and 5 as Region *c*.

Although the energy flux of the electron precipitation shows a sharp drop at Boundary 5, this flux remained relatively high for a while longer. Boundary 6 (at 08:26:58 UT) was determined to be the last point of that relatively high-flux period. After this boundary, there were no further instances of the maximum energy flux for the three energy bins with two consecutive values exceeding 1×10^8 eV cm^{-2} s^{-1} sr^{-1} eV^{-1}. At this boundary, (MLAT, MLT) = (73.08°, 12:12); this latitude was higher than the equatorward boundary of the cusp (72.54°) by approximately 0.5°.

Electron precipitation for each auroral structure

Figures 5 and 6 show how these six boundaries are situated relative to the moving cusp aurora. Figure 5a shows the position of Boundary 1 (red square) on the 630.0 nm all-sky image taken at 08:25:33 UT, which was 8 s after

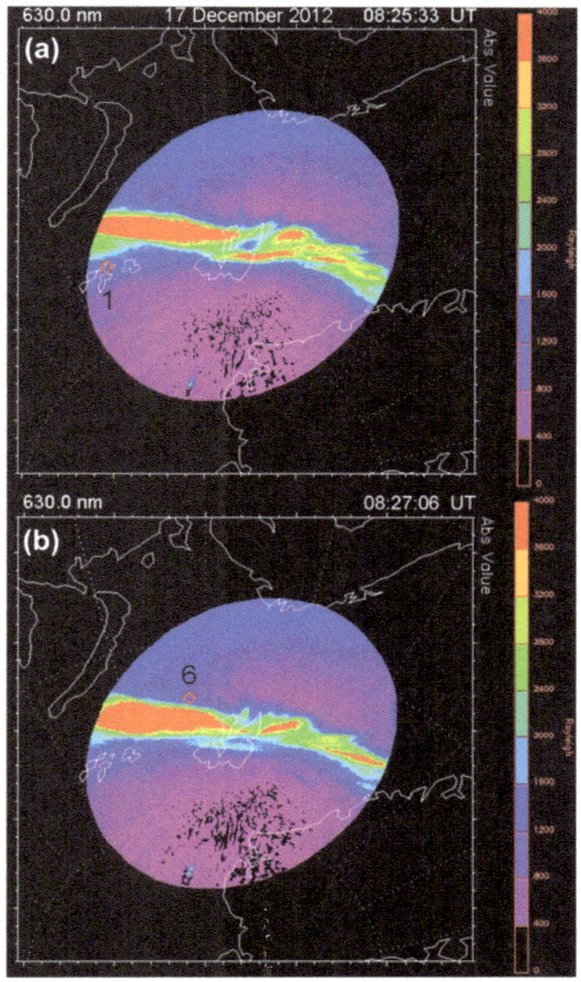

Figure 5 Positions of **(a)** Boundary 1 and **(b)** Boundary 6 relative to the 630.C nm oval. The positions are indicated by red diamonds.

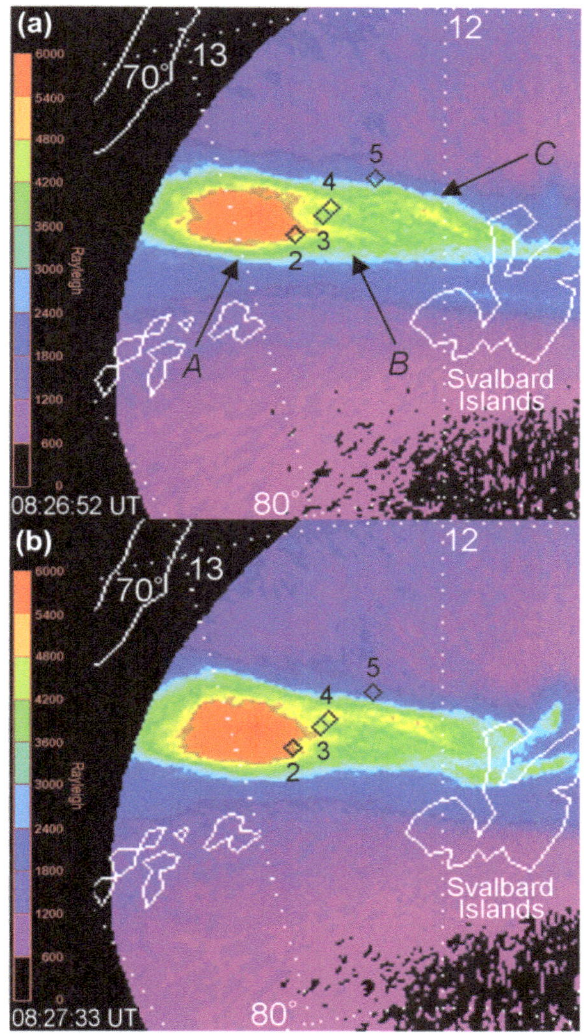

Figure 6 Positions of four boundaries (black diamonds) relative to Auroras *A*, *B*, and *C*. The four boundaries are shown on the 630.0 nm images taken at **(a)** 08:26:52 and **(b)** 08:27:33 UT.

the satellite passed through Boundary 1. Boundary 1 is located near the poleward boundary of the dayside 630 nm auroral oval, whose emission ranged over 0.8 to 1.6 kR (dark blue in Figure 5a). The 630 nm emission is produced by the deexcitation of the ^1D state of atomic oxygen, a state with a theoretical lifetime of 110 s. Considering this lifetime, we checked all auroral images obtained between 08:25:25 and 08:27:15 UT (=08:25:25 + 110 s). The position of Boundary 1 relative to the poleward boundary of the 630 nm auroral oval was very similar throughout this interval.

Figure 5b shows the position of Boundary 6 (red square) on the 630.0 nm all-sky image taken at 08:27:06 UT, which was 8 s after the satellite passed through Boundary 6 at 08:26:58 UT. Boundary 6 was located near the equatorward boundary of the 630 nm dayside auroral oval. This provides clear evidence that the cusp aurora, including its equatorward part, is located in the open field line region, though this feature has already been suggested with HF radar and a meridian-scanning photometer by Milan et al. (1999). We checked all auroral images obtained between 08:26:58 and 08:28:48 UT (=08:26:58 + 110 s). The position of Boundary 6 relative to the equatorward boundary of the 630 nm auroral oval was very similar throughout this interval.

Figure 6a shows the positions of Boundaries 2, 3, 4, and 5 on the 630.0 nm all-sky image taken at 08:26:52 UT, which was also shown in Figure 1. This time is 38, 26, 22, and 5 s after the detection of Boundaries 2, 3, 4, and 5, respectively. As indicated in Figure 6a, the region between Boundaries 2 and 3 (Region *b* in Figure 4) overlapped Aurora *B*, and the region between Boundaries 4 and 5 (Region *c* in Figure 4) overlapped the eastward part of Aurora *C*. As will be shown later, Aurora *B* was moving eastward, whereas Aurora *C* did not move substantially at that time. Subsequently obtained images also show that the poleward and equatorward boundaries of Aurora *C* were Boundaries 4 and 5, respectively. Figure 6b, taken at 08:27:33 UT, is an example of such an image. This image was taken 46 s after the DMSP observation of Boundary 5. At this time, Aurora *C* was still situated between Boundaries 4 and 5. Note that because the excited oxygen atom has a relatively long lifetime (theoretically 110 s, as mentioned above), obtaining a more precise relationship between the extent of the moving auroral structure and the electron precipitation region would require information about how the neutral wind, with which the excited oxygen atom moves, was distributed during that period of time.

Figure 4 shows that the particle precipitation features between Boundaries 3 and 4 were similar to those in regions at latitudes higher than Boundary 2. We know from Figure 6a that both regions were parts of Aurora *A*. These spatial relationships, i.e., Aurora *A*-Region *a*, Aurora *B*-Region *b*, and Aurora *C*-Region *c* are shown schematically in Figure 7, together with Boundaries 2, 3, 4, and 5. Note that the satellite just skimmed Aurora *A* without crossing through its main part. This is why the DMSP observation obtained before Boundary 2 did not include regions where the energy flux of the electron precipitation was very high (Figure 4). Figure 4 also shows that the energy flux of the electron precipitation was very low both immediately before Boundary 2 and just after Boundary 3. Figure 6 implies that those regions may be connected as the outer edge of Aurora *A*.

Change in the positions of Auroras *B* and *C*

From Figure 4, we have shown that Regions *b* and *c* had significantly different ion precipitation features, whereas their electron precipitation features were very similar. In this section, we will show that this difference in ion precipitation is reasonable when considering the motion of Regions *b* (i.e., Aurora *B*) and *c* (i.e., Aurora *C*). Figure 8 shows when and where Auroras *B* and *C* appeared and

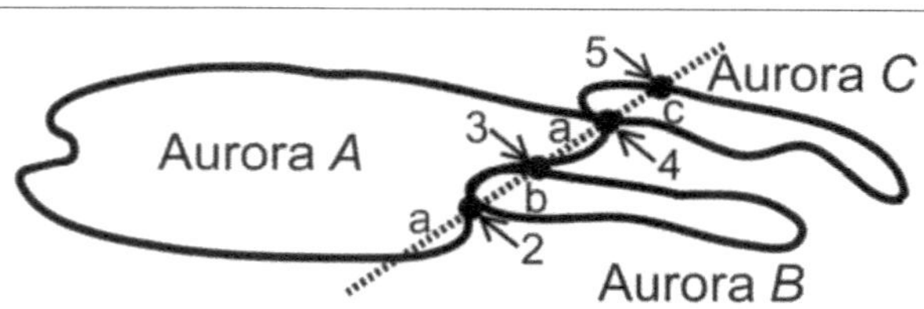

Figure 7 Schematics of the spatial relation between the plasma boundaries and the three auroral structures. The dotted line represents the DMSP F18 pass. Boundaries 2, 3, 4, and 5, which were determined from the DMSP data (Figure 4), are plotted as black dots on the line. Regions *a*, *b*, and *c*, which are indicated within the auroral region, are also based on the DMSP data (Figure 4).

how they moved. Figure 8a (top left) shows the auroral image at 08:22:15 UT; the three white dotted lines indicate 11:00, 12:00, and 13:00 MLT. When the image at 08:22:28 UT (Figure 8b) is compared with this image, a faint intensification, indicated by a black arrow, can be seen in the image at 08:22:28 UT. The location of the peak emission is (MLAT, MLT) = (74.3°, 11:56). The aurora was not conspicuous at this time but became clear in the image shown in Figure 8c, which was taken 47 s after the image in Figure 8b.

As shown by the black arrow in each of the subsequent images, this auroral structure moved slightly poleward until 08:25:53 UT (Figure 8i) and then accelerated strongly eastward, which is consistent with the motion of the reconnected field line for a negative IMF B_Y (Figure 2). The image shown in Figure 8k is from the same data source as those shown in Figures 1 and 6a. The location of Aurora *B* at 08:27:19 UT (Figure 8l) is (MLAT, MLT) = (74.5°, 12:24). We determined this location from the trailing edge of the aurora whose emission is more than 4 kR. The DMSP observation of Region *b* (i.e., 08:26:14 UT for Boundary 2) was obtained in the middle of the eastward motion of the aurora and approximately 230 s after the appearance of Aurora *B*. After 08:27:19 UT (Figure 8l), Aurora *B* was no longer evident, and thus the observation by the spacecraft at 08:26:14 UT was obtained just before Aurora *B* disappeared, i.e., in the final stage of the aurora.

The motion of Aurora *B* suggests that it takes a few minutes (from 08:22:28 to 08:25:53 UT) for the result of the tension exerted by the reconnected field line to become prominent in the ionosphere after the appearance of the cusp aurora. A similar tendency can be identified in a larger eastward-moving auroral structure, which was seen over the mainland of the Svalbard Islands (Figure 8f,g,h,i,j,k). During the interval shown in Figure 8f,g,h,i,j,k, the auroral structure was moving eastward, but prior to that (Figure 8d,e), the auroral structure had been moving poleward, roughly along the 1100 MLT meridian. A recent study based on simultaneous observations from the Super Dual Auroral Radar Network (SuperDARN), a far-ultraviolet instrument on the IMAGE spacecraft, and a ground magnetometer also reported similar flow features for the cusp proton aurora event (Taguchi et al. 2010).

The red arrow in Figure 8h (08:25:33 UT) indicates the initial image of Aurora *C*. The location of the peak emission is (MLAT, MLT) = (74.2°, 12:08). This aurora did not move substantially until 08:27:19 UT (Figure 8l) but later moved poleward. The all-sky images obtained 3 min later (i.e., at 08:30:19 UT) are shown in Figure 9. The location of Aurora *C* is approximately (MLAT, MLT) = (75.3°, 12:15). As is expected from a negative value of the IMF component B_Y, the motion had an eastward component as well as a poleward component. After this time, Aurora *C* continued to move further poleward and

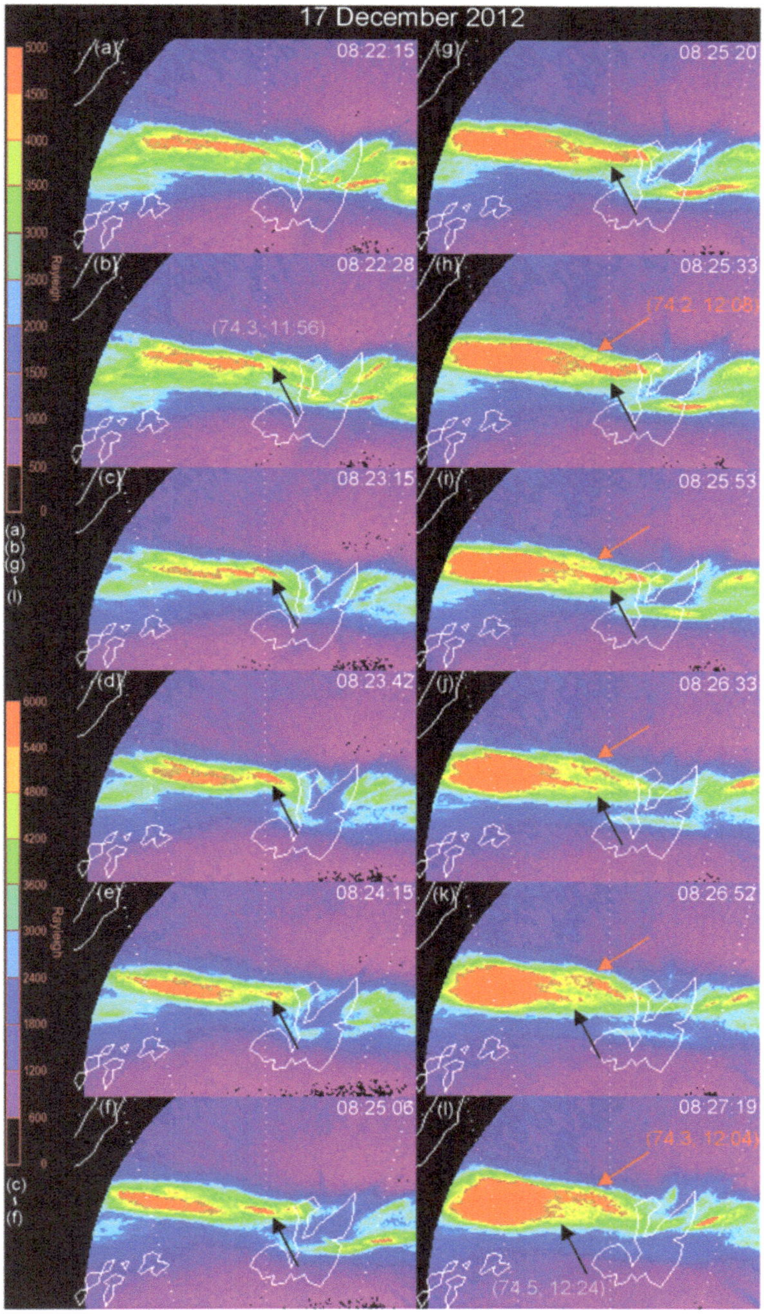

Figure 8 Changes in the positions of Auroras *B* and *C*. The image in **(a)** represents the situation before the appearance of Aurora *B* and Aurora *C*. The black arrow in **(b)**-**(l)** indicates the position of Aurora *B*, and the red arrow, shown in **(h)**-**(l)**, represents the position of Aurora *C*. In **(b)**, **(h)**, and **(l)**, the location of the aurora is given as coordinates (MLAT, MLT). The image in **(k)** has the same source as the images in Figures 1 and 6, but the color coding is different.

disappeared after 08:33:18 UT (not shown). The lifetime exceeded 465 s. The DMSP observation of Region *c* at 08:26:30 UT (i.e., Boundary 4) was obtained approximately 60 s after its appearance, i.e., in the early stage of the aurora.

The interval between the appearance of Aurora *C* and the DMSP observation (60 s) is short when compared with the interval for Aurora *B*, which was approximately 230 s. We have shown that the energy flux of the ion precipitation in Aurora *C* was much higher than that in Aurora *B*. We know that the energy flux of the precipitating ions in the young open flux tube (i.e., Aurora *C*) was much higher than in the old open flux tube (i.e., Aurora *B*). This is consistent with the signatures expected from

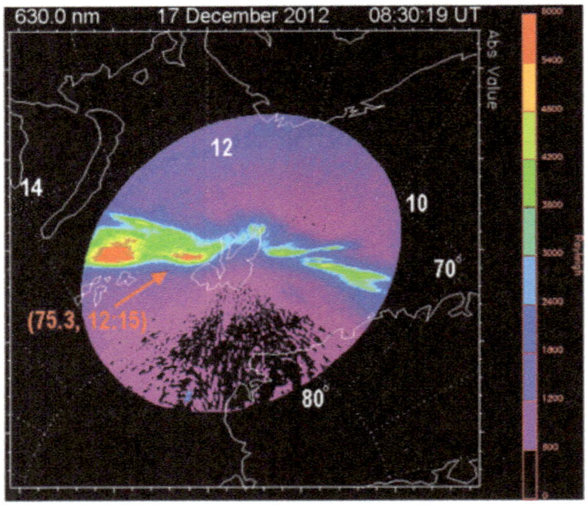

Figure 9 A 630.0 nm dayside all-sky image taken at 08:30:19 UT on 17 December 2012. The format is the same as that of Figure 1. The red arrow indicates Aurora *C*.

reconnection, indicating that the energy flux of the ion precipitation tends to decrease with increasing distance from the footprint of the reconnection point (e.g., Onsager et al. 1993).

High-flux electron precipitation

The high fluxes of the electron precipitation in both Regions *b* and *c* suggest that the energy fluxes of the precipitating electrons in the young and old open flux tubes do not differ significantly. For the present event, the maximum differential energy flux at approximately 100 eV similarly exceeded 1×10^9 eV cm^{-2} s^{-1} sr^{-1} eV^{-1} in both regions. In contrast, as mentioned above, the energy fluxes of the precipitating ions in those regions differed significantly. Thus, it appears that the electron precipitation flux level is controlled by a mechanism independent of the ion precipitation flux.

Further studies are needed to understand the process that directly produces the high electron precipitation flux inside an open flux tube; however, it is estimated that an upward field-aligned current whose carrier is the precipitating electrons and a parallel potential drop, which is presumably produced in conjunction with the upward field-aligned current, may be important factors. A mesoscale convection system consisting of upward and downward field-aligned currents was proposed for the ionospheric signatures of the FTE by Southwood (1987); subsequently, Oksavik et al. (2004) showed that such a flow pattern actually exists. Collocation of the mesoscale upward field-aligned current with intense electron precipitation can be seen in an event reported by Taguchi et al. (1993) (e.g., 02:51:32 to 02:51:40 UT in Plate 1 of that work). Using simultaneous satellite observations of electrons at two altitudes in the cusp, Lin et al. (1986) showed that some parallel potential drop is needed to explain the enhanced energy flux observed in the lower-altitude electrons, although we do not know whether their enhanced energy flux events are related to cusp moving auroral structures.

In this study, we demonstrated a one-to-one correspondence between the moving cusp auroral structure measured at a wavelength of 630.0 nm and the electron precipitation region responsible for its creation. This result will help develop the quantitative relation between the red line aurora intensity and soft precipitating electrons. In Regions *b* and *c*, the integral electron energy flux increased to 2×10^{11} to 4×10^{11} eV cm^{-2} s^{-1} sr^{-1}, and the average energies at those times were approximately 2×10^2 eV (see the black dots in the top two panels of Figure 3 for the corresponding interval). From the intensities of Auroras *B* and *C*, we know that the electron precipitation with these characteristics created a 4.5 to 5 kR red line emission.

Conclusions

Using data from the high-sensitivity all-sky imager at Longyearbyen, Svalbard, and particle data from the DMSP F18 spacecraft, we examined particle precipitation features in the early and final stages of a moving cusp auroral structure. We focused on two structures that occurred near noon, separated by approximately 3 min, during a southwestward IMF condition. The second structure occurred when the IMF turned further southward. Immediately after the appearance of the latter structure, the two auroral structures were adjacently situated, and the DMSP F18 spacecraft passed through these regions.

To delineate the boundaries of different precipitating regions in the DMSP particle data, we introduced the ratio of the integral ion number flux to the integral electron number flux. Our investigation shows that the ion precipitation in the young cusp aurora had a high energy flux, while that in the old aurora had a very low energy flux. Regarding the electron precipitation, the features in both regions were found to be very similar: the energy fluxes at approximately 100 eV often exceeded 1×10^9 eV cm^{-2} s^{-1} sr^{-1} eV^{-1} in both regions. From this study, we suggest that the precipitation of the electrons contained in the open flux tube is maintained at a high flux over a certain period of time from its beginning and the high flux of the electron precipitation is controlled by a mechanism independent of the ion precipitation.

Abbreviations

ACE: Advanced Composition Explorer; DMSP: Defense Meteorological Satellite Program; EMCCD: electron multiplier charge-coupled device; FTE: flux transfer event; GSE: geocentric solar ecliptic; GSM: geocentric solar magnetospheric; IMF: interplanetary magnetic field; MLAT: magnetic latitude; MLT: magnetic local time; SuperDARN: Super Dual Auroral Radar Network.

Competing interests

The authors declare that they have no competing interests.

Authors' contributions

ST proposed the topic and analyzed the data. HK and YO facilitated the data analysis process and helped in data interpretation. HK and YO also collaborated with the corresponding author in the construction of manuscript. All authors read and approved the final manuscript.

Acknowledgements

We would like to thank Takeshi Aoki of the University of Electro-Communications for his maintenance of the all-sky imager. We also thank Fred Sigernes and Margit Dyrland for their help at the University Centre in Svalbard, Longyearbyen. Discussion with Dai Hasegawa of the University of Electro-Communications is acknowledged. The DMSP particle detectors were designed by Dave Hardy of AFRL, and the data were obtained from JHU/APL. GEOTAIL magnetic field data were provided by T. Nagai through DARTS at the Institute of Space and Astronautical Science, JAXA, in Japan. ACE magnetic field and plasma data were provided by N. Ness and D. McComas through the NASA CDAWeb. This work was supported by JSPS KAKENHI Grant Number 22340143.

Author details

[1]Department of Communication Engineering and Informatics, University of Electro-Communications, Chofu, Tokyo 182-8585, Japan. [2]Center for Space Science and Radio Engineering, University of Electro-Communications, Chofu, Tokyo 182-8585, Japan. [3]Now at Department of Geophysics, Graduate School of Science, Kyoto University, Kyoto 606-8502, Japan. [4]National Institute of Polar Research, Tachikawa, Tokyo 190-8518, Japan.

References

Burch JL (1985) Quasi-neutrality in the polar cusp. Geophys Res Lett 12:469–472

Fairfield DH (1971) Average and unusual locations of the Earth's magnetopause and bow shock. J Geophys Res 76(28):6700–6716, doi:10.1029/JA076i028p06700

Fasel GJ (1995) Dayside poleward moving auroral forms: a statistical study. J Geophys Res 100(A7):11891–11905, doi:10.1029/95JA00854

Hardy DA, Schmidt LK, Gussenhoven MS, Marshall FJ, Yeh HC, Shumaker TL, Huber A, Pantazis J (1984) Precipitating electron and ion detectors (SSJ/4) for block 5D/flights 4–10 DMSP satellites: Calibration and data presentation. Tech Rep AFGL-TR-84-0317, Air Force Geophys Lab, Hanscom Air Force Base, Mass

Hardy DA, Gussenhoven MS, Holeman E (1985) A statistical model of auroral electron precipitation. J Geophys Res 90(A5):4229–4248, doi:10.1029/JA090iA05p04229

Haerendel G, Paschmann G, Sckopke N, Rosenbauer H, Hedgecock PC (1978) The frontside boundary layer of the magnetosphere and the problem of reconnection. J Geophys Res 83(A7):3195–3216

Hosokawa K, Taguchi S, Ogawa Y, Aoki T (2013a) Periodicities of polar cap patches. J Geophys Res 118:447–453, doi:10.1029/2012JA018165

Hosokawa K, Taguchi S, Ogawa Y, Sakai J (2013b) Two-dimensional direct imaging of structuring of polar cap patches. J Geophys Res 118:6536–6543, doi:10.1002/jgra.50577

Hosokawa K, Taguchi S, Shiokawa K, Otsuka Y, Ogawa Y, Nicolls M (2014) Global imaging of polar cap patches with dual airglow imagers. Geophys Res Lett 41:1–6, doi:10.1002/2013GL058748

Khan H, Cowley SWH (1999) Observations of the response time of high-latitude ionospheric convection to variations in the interplanetary magnetic field using EISCAT and IMP-8 data. Ann Geophys 17:1306–1335, doi:10.1007/s00585-999-1306-8

Lin CS, Burch JL, Winningham JD (1986) Near-conjugate observations of polar cusp electron precipitation using DE 1 and DE 2. J Geophys Res 91(A10):11186–11202, doi:10.1029/JA091iA10p11186

Lockwood MP, Sandholt E, Cowley SWH, Oguti T (1989) Interplanetary magnetic field control of dayside auroral activity and the transfer of momentum across the dayside magnetopause. Planet Space Sci 37(11):1347–1365, doi:10.1016/0032-0633(89)90106-2

Lockwood M, Smith MF (1994) Low and middle altitude cusp particle signatures for general magnetopause reconnection rate variations: 1. Theory J Geophys Res 99(A5):8531–8553, doi:10.1029/93JA03399

Milan SE, Lester M, Cowley SWH, Moen J, Sandholt PE, Owen CJ (1999) Meridian-scanning photometer, coherent HF radar, and magnetometer observations of the cusp: a case study. Ann Geophys 17:159–172

Oksavik K, Moen J, Carlson HC (2004) High-resolution observations of the small-scale flow pattern associated with a poleward moving auroral form in the cusp. Geophys Res Lett 31:L11807, doi:10.1029/2004GL019838

Oksavik K, Moen J, Carlson HC, Greenwald RA, Milan SE, Lester M, Denig WF, Barnes RJ (2005) Multi-instrument mapping of the small-scale flow dynamics related to a cusp auroral transient. Ann Geophys 23:2657–2670

Onsager TG, Kletzing CA, Austin JB, MacKiernan H (1993) Model of magnetosheath plasma in the magnetosphere: Cusp and mantle particles at low-altitudes. Geophys Res Lett 20:479–482, doi:10.1029/93GL00596

Owen CJ, Fazakerley AN, Carter PJ, Coates AJ, Krauklis IC, Szita S, Taylor MGGT, Travnicek P, Watson G, Wilson RJ, Balogh A, Dunlop MW (2001) Cluster PEACE observations of electrons during magnetospheric flux transfer events. Ann Geophys 19:1509–1522

Roble RE, Rees MH (1977) Time-dependent studies of the aurora: effects of particle precipitation on the dynamic morphology of ionospheric and atmospheric properties. Planet Space Sci 25(11):991–1010

Russell CT, Elphic RC (1978) Initial ISEE magnetometer results: magnetopause observations. Space Sci Rev 22:681–715, doi:10.1007/ BF00212619

Sakai J, Hosokawa K, Taguchi S, Ogawa Y (2014) Storm-time enhancements of 630.0 nm airglow associated with polar cap patches. J Geophys Res 119:2214–2228, doi:10.1002/2013JA019197

Sandholt PE, Deehr CS, Egeland A, Lybekk B, Viereck R, Romick GJ (1986) Signatures in the dayside aurora of plasma transfer from the magnetosheath. J Geophys Res 91(A9):10,063–10,079, doi:10.1029/JA091iA09p10063

Sandholt PE, Lockwood M, Oguti T, Cowley SWH, Freeman KSC, Lybekk B, Egeland A, Willis DM (1990) Midday auroral breakup events and related energy and momentum transfer from the magnetosheath. J Geophys Res 95(A2):1039–1060, doi:10.1029/JA095iA02p01039

Sandholt PE, Farrugia CJ (2003) Does the aurora provide evidence for the occurrence of antiparallel magnetopause reconnection? J Geophys Res 108(A12):1466, doi:10.1029/2003JA010066

Southwood DJ (1987) The ionospheric signature of flux transfer events. J Geophys Res 92(A4):3207–3213, doi:10.1029/JA092iA04p03207

Spreiter JR, Stahara SS (1980) A new predictive model for determining solar wind-terrestrial planet interactions. J Geophys Res 85(A12):6769–6777, doi:10.1029/JA085iA12p06769

Taguchi S, Sugiura M, Winningham JD, Slavin JA (1993) Characterization of the IMF B_Y-dependent field-aligned currents in the cleft region based on DE 2 observations. J Geophys Res 98(A2):1393–1407, doi:10.1029/92JA01014

Taguchi S, Hosokawa K, Suzuki S, Tawara A, Frey HU, Matzka J, Yukimatu AS, Sato N (2010) Plasma flow during the brightening of proton aurora in the cusp. J Geophys Res 115:A10308, doi:10.1029/2010JA015535

Taguchi S, Hosokawa K, Ogawa Y, Aoki T, Taguchi M (2012) Double bursts inside a poleward-moving auroral form in the cusp. J Geophys Res 117:A12301, doi:10.1029/2012JA018150

Yang YF, Lu JY, Wang JS, Peng Z, Zhou L (2013) Influence of interplanetary magnetic field and solar wind on auroral brightness in different regions. J Geophys Res 118:209–217, doi:10.1029/2012JA017727

4

Expected geoneutrino signal at JUNO

Virginia Strati[1,2*], Marica Baldoncini[1,3], Ivan Callegari[2], Fabio Mantovani[1,3], William F McDonough[4], Barbara Ricci[1,3] and Gerti Xhixha[2]

Abstract

Constraints on the Earth's composition and on its radiogenic energy budget come from the detection of geoneutrinos. The Kamioka Liquid scintillator Antineutrino Detector (KamLAND) and Borexino experiments recently reported the geoneutrino flux, which reflects the amount and distribution of U and Th inside the Earth. The Jiangmen Underground Neutrino Observatory (JUNO) neutrino experiment, designed as a 20 kton liquid scintillator detector, will be built in an underground laboratory in South China about 53 km from the Yangjiang and Taishan nuclear power plants, each one having a planned thermal power of approximately 18 GW. Given the large detector mass and the intense reactor antineutrino flux, JUNO aims not only to collect high statistics antineutrino signals from reactors but also to address the challenge of discriminating the geoneutrino signal from the reactor background. The predicted geoneutrino signal at JUNO is $39.7^{+6.5}_{-5.2}$ terrestrial neutrino unit (TNU), based on the existing reference Earth model, with the dominant source of uncertainty coming from the modeling of the compositional variability in the local upper crust that surrounds (out to approximately 500 km) the detector. A special focus is dedicated to the $6° \times 4°$ local crust surrounding the detector which is estimated to contribute for the 44% of the signal. On the basis of a worldwide reference model for reactor antineutrinos, the ratio between reactor antineutrino and geoneutrino signals in the geoneutrino energy window is estimated to be 0.7 considering reactors operating in year 2013 and reaches a value of 8.9 by adding the contribution of the future nuclear power plants. In order to extract useful information about the mantle's composition, a refinement of the abundance and distribution of U and Th in the local crust is required, with particular attention to the geochemical characterization of the accessible upper crust where 47% of the expected geoneutrino signal originates and this region contributes the major source of uncertainty.

Keywords: Geoneutrino flux; JUNO experiment; Earth reference model; Earth composition; Heat-producing elements; Reactor antineutrinos

Background

The first experimental evidence of geoneutrinos, i.e., electron antineutrinos produced in beta decays along the ^{238}U and ^{232}Th decay chains, was claimed by the Kamioka Liquid scintillator Antineutrino Detector (KamLAND) Collaboration in 2005 (KamLAND Collaboration 2005), which ushered in a new method for exploring the Earth's interior and provided constraints on the planet's composition and specifically its radiogenic element budget (Fiorentini et al. 2007). The geoneutrino energy spectrum contains in it distinctive contributions from U and Th, each one resulting from different rates and shapes of their decays (see Figures three and five of Fiorentini et al. 2007) and from concentrations and spatial distributions of these elements inside the Earth.

Geoneutrinos are measured in liquid scintillation detectors via the inverse beta decay (IBD) reaction on free protons:

$$\bar{\nu}_e + p \rightarrow e^+ + n$$

whose energy threshold of 1.806 MeV means that only a small fraction of the antineutrinos produced from the U and Th decay chains are detectable. The IBD detection event in a liquid scintillator produces two flashes of light: the annihilation flash, from electron-positron interaction, followed by the deuterium formation flash, which is 2.2 MeV of light that follows some 200 μs later. The delayed coincidence of these two flashes of light provides the critical identification of the antineutrino

* Correspondence: strati@fe.infn.it
[1]Department of Physics and Earth Sciences, University of Ferrara, Via Saragat 1, 44121 Ferrara, Italy
[2]Legnaro National Laboratories, INFN, Viale dell'Università, 2, 35020 Legnaro, Italy
Full list of author information is available at the end of the article

interaction and eliminates most background events. The KamLAND and Borexino experiments recently reported 116^{+28}_{-27} geoneutrino events over 2,991 days (KamLAND Collaboration 2013) and 14.3 ± 4.4 geoneutrino events in 1,353 days (Borexino Collaboration 2013), respectively. Differences in the detection rates reflect the detector sizes, with the KamLAND detector being approximately 1 kton and the Borexino detector 0.3 kton.

The most significant source of background for geoneutrino measurements is due to reactor antineutrinos, i.e., electron antineutrinos emitted during the beta decays of fission products from ^{235}U, ^{238}U, ^{239}Pu, and ^{241}Pu burning. Approximately 30% of the reactor antineutrino events are recorded in the geoneutrino energy window extending from the IBD threshold up to the endpoint of the ^{214}Bi beta decay spectrum (3.272 MeV) (Fiorentini et al. 2010). The Terrestrial Neutrino Unit (TNU), which is the signal that corresponds to one IBD event per 10^{32} free protons per year at 100% efficiency, is used to compare the different integrated spectral components (i.e., antineutrinos from U, Th, and reactors) measured by the detectors or just beneath the Earth's surface.

In the past decade, reactor antineutrino experiments played a decisive role in unraveling the neutrino puzzle, which currently recognizes three flavor eigenstates (ν_e, ν_μ, and ν_τ), each of which mixes with three mass eigenstates (ν_1, ν_2, and ν_3) via three mixing angles (θ_{12}, θ_{13}, and θ_{23}). The quantities that govern the oscillation frequencies are two differences between squared masses (i.e., $\delta m^2 = m_2^2 - m_1^2 > 0$ and $\Delta m^2 = m_3^2 - (m_1^2 + m_2^2)/2$). Central to neutrino studies is understanding the neutrino mass hierarchy (i.e., $\Delta m^2 > 0$ or $\Delta m^2 < 0$) (Capozzi et al. 2014; Ge et al. 2013).

Massive (>10 kton) detectors such as the Jiangmen Underground Neutrino Observatory (JUNO) (Li 2014) and Reno-50 (Kim 2013) experiments are being constructed at medium baseline distances (a few tens of kilometers) away from bright reactor antineutrino fluxes in order to assess significant physics goals regarding the neutrino properties, in the first place, the mass hierarchy. These experiments intend also to obtain subpercent precision measurements of neutrino oscillation parameters and along the way make observations of events of astrophysical and terrestrial origin.

JUNO is located (N 22.12°, E 112.52°) in Kaiping, Jiangmen, Guangdong Province (South China), about 53 km away from the Yangjiang and Taishan nuclear power plants, which are presently under construction. The combined thermal power of these two units is planned to be on the order of 36 GW (Li and Zhou 2014) (Figure 1). The JUNO experiment is designed as a liquid scintillator detector of 20 kton mass that will be built in a laboratory some 700 m underground (approximately 2,000 m water equivalent). This amount of overburden will attenuate the cosmic muon flux, which contributes to the overall detector background signal, but this overburden is significantly less than that at the KamLAND and Borexino experiments. The detector energy response and the spatial distribution of the reactor cores are the most critical features affecting the experimental sensitivity (Li et al. 2013) required to achieve the intended physics goals.

The goal of this present study is to predict the geoneutrino signal at JUNO on the basis of an existing reference Earth model (Huang et al. 2013), together with an estimate of the expected reactor antineutrino signal. Since a significant contribution to the expected geoneutrino signal comes from U and Th in the continental crust surrounding the site, we follow past approaches to study the local contribution (Coltorti et al. 2011; Fiorentini et al. 2012; Huang et al. 2013; Huang et al. 2014), with a particular interest in focusing on the closest 6° × 4° grid surrounding the detector. We define this latter region as the local crust (LOC) (Figure 1).

Methods

The geoneutrino signal expected at JUNO is calculated adopting the same methodology and the same inputs of the reference Earth model developed by Huang et al. (2013). It provides a description of the abundances and distribution of the heat-producing elements (HPEs; i.e., U, Th, and K) in the Earth's crust, along with their uncertainties. According to this model, the silicate portion of the Earth is composed of five dominant reservoirs: the depleted mantle (DM), the enriched mantle (EM), the lithospheric mantle (LM), the continental crust (CC), and the oceanic crust (OC). The continental crust is dominantly composed of the lower crust (LC), middle crust (MC), and upper crust (UC), and it is overlain by shallow layers of sediments (Sed) which also cover the OC.

The surface geoneutrino flux is calculated by dividing the Earth's surface in 1° × 1° tiles that are projected vertically into discrete volume cells, and each cell is assigned with physical and chemical states. Just for the sake of computing flux, the 1° × 1° tiles are further subdivided into many subcells with the same properties of the parent tile. The number of subcells is progressively bigger approaching the detector location with the aim of not introducing any bias due to discretization.

The total crustal thickness of each cell and its associated uncertainty correspond, respectively, to the mean and the half range of three crustal models obtained from different approaches: the global crustal model based on reflection and refraction data 'CRUST 2.0' (Bassin et al. 2000; Laske et al. 2001), the global shear velocity model of the crust and upper mantle 'CUB 2.0' (Shapiro and Ritzwoller 2002), and the high-resolution map of Moho (crust-mantle boundary) depth based on the gravity field data 'GEMMA' (Reguzzoni and Tselfes 2009; Reguzzoni

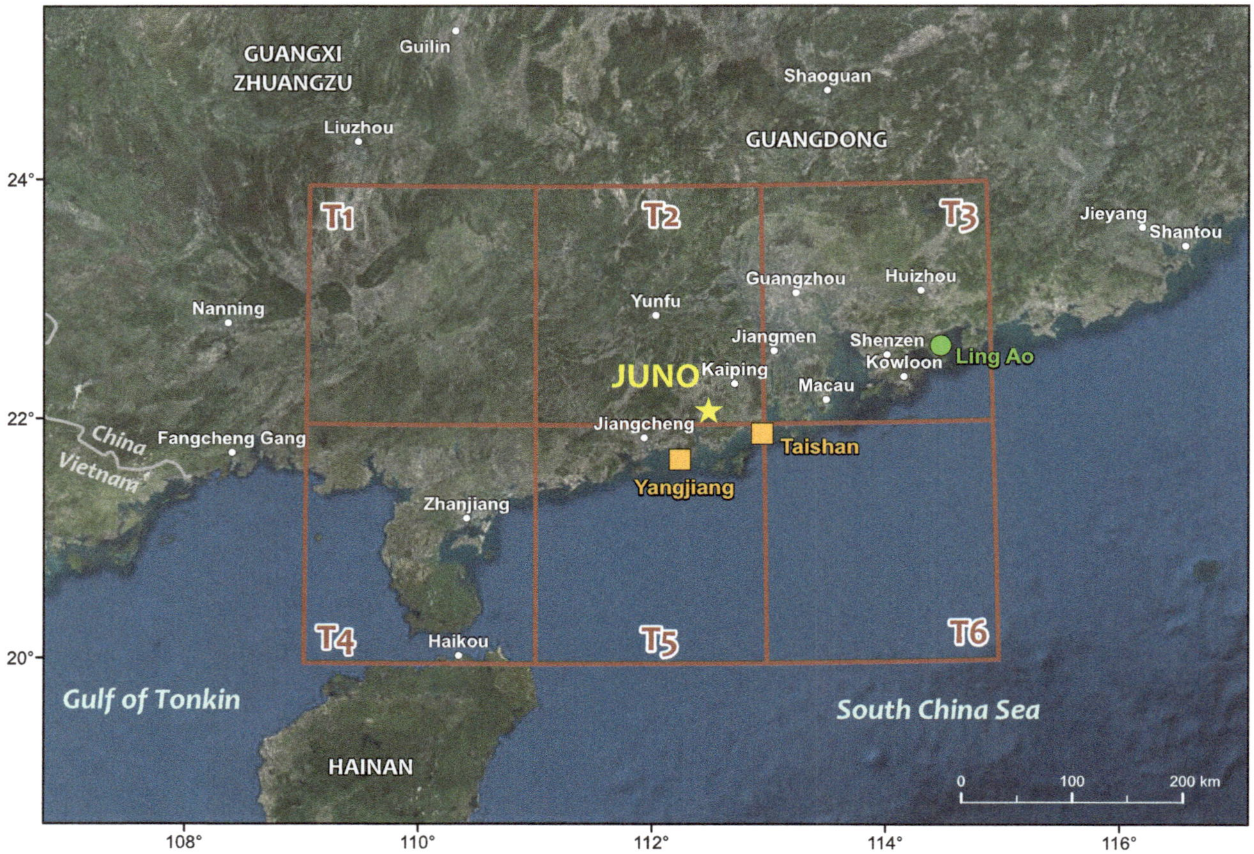

Figure 1 Map of LOC surrounding JUNO. JUNO (yellow star) located in Kaiping, Jiangmen, Guangdong Province (South China) and the planned (orange square) and operational (green circle) nuclear power plants. The six 2° × 2° tiles (dark red lines) define the LOC.

and Sampietro 2015). The reference model incorporates the relative proportional thickness of the crustal layers along with density and elastic properties (compressional and shear wave velocity) reported in CRUST 2.0. The same information is adopted for the Sed layer using the global sediment map of Laske and Masters (1997). In Figure 2, the thicknesses of the continental crust layers in the 24 cells constituting the LOC for JUNO are reported. Their total crustal thickness ranges between 26.3 and 32.3 km with an uncertainty for each cell of approximately 7%.

The HPE abundances in the Sed, OC, and UC layers are assumed to be relatively homogenous and correspond to the values reported in Table three of Huang et al. (2013). The ratio between the felsic and mafic components in the deep CC (MC and LC) is inferred from seismic velocity data, and these data are in turn used to estimate the U and Th content of each cell of the reference crustal model. Focusing on the LOC, the central values of U abundance in MC and LC vary in the range 0.8 to 1.2 μg/g and 0.3 to 0.1 μg/g, respectively. The Th/U ratio in the deep CC of the LOC is typically approximately 5 as compared to a bulk silicate Earth ratio of 3.9 or a bulk CC ratio just greater than 4.0; the higher Th/U ratio in the deep CC is likely due to the greater upward mobility of U during dehydration reactions that accompany granulite facies metamorphism of the deep CC.

In the reference model of Huang et al. (2013), the LM corresponds to the portion of the Earth between the Moho discontinuity and an assumed standard depth of 175 km beneath the surface. The thickness of this unit in the LOC ranges between 143 and 149 km, and its composition is modeled from the database reported in McDonough (1990) and the update in Huang et al. (2013). In our calculation, we adopt for the LM the U and Th abundances of $0.03^{+0.05}_{-0.02}$ and $0.15^{+0.28}_{-0.10}$ μg/g, respectively (Huang et al. 2013).

The sublithospheric mantle extends down from the base of the lithosphere to the core-mantle boundary and is divided in two spherically symmetric domains, the DM and the EM, whose density profiles are derived from the Preliminary Reference Earth Model (PREM) (Dziewonski and Anderson 1981). Adopting a mass ratio $M_{DM}/M_{EM} = 4.56$ (Huang et al. 2013), we calculate the

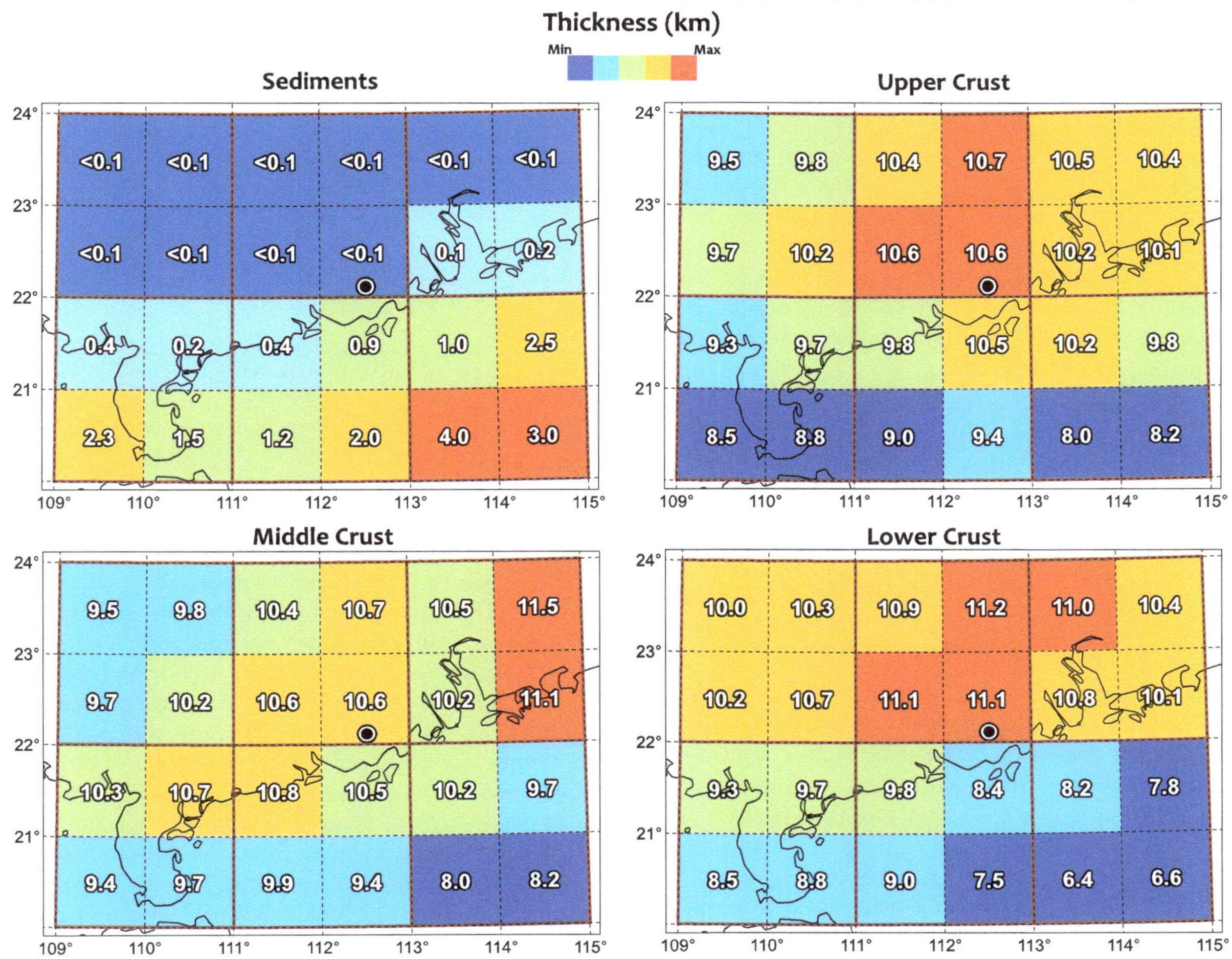

Figure 2 Thicknesses of the four crustal layers in the LOC. The thicknesses in kilometers of the Sed, UC, MC, and LC layers are reported for each of the 24 cells constituting the LOC surrounding JUNO (black circle), with color coding to illustrate gradients in thickness.

masses of these two reservoirs $M_{DM} = 3.207 \times 10^{24}$ kg and $M_{EM} = 0.704 \times 10^{24}$ kg. In a survey of the midocean ridge basalts (MORB), Arevalo and McDonough (2010) reported the lognormal-based average abundances of uranium ($U_{MORB} = 80$ ng/g) and thorium ($Th_{MORB} = 220$ ng/g) and, from this, calculated the $U_{DM} = 8$ ng/g and $Th_{DM} = 22$ ng/g based on a simple melting model. Based on these assumptions, the U_{EM} can be calculated:

$$U_{EM} = \frac{m_{BSE} - m_C}{M_{EM}} - U_{DM}\frac{M_{DM}}{M_{EM}},$$

where $m_{BSE} = 8.1 \times 10^{16}$ kg is the U mass in the bulk silicate earth (BSE) (McDonough and Sun 1995) and $m_C = 3.1 \times 10^{16}$ kg is the total U mass in the crust (Huang et al. 2013). The mantle geoneutrino signals reported in Table 1 are calculated with $U_{DM} = 8$ ng/g and $U_{EM} = 34$ ng/g together with $(Th/U)_{DM} = 2.8$ and $(Th/U)_{EM} = 4.8$.

Table 1 Geoneutrino signals from U and Th expected in JUNO

	S (U)	*S* (Th)	*S* (U + Th)
Sed CC	$0.48^{+0.06}_{-0.06}$	$0.16^{+0.02}_{-0.02}$	$0.64^{+0.06}_{-0.06}$
UC	$14.6^{+3.5}_{-3.4}$	$3.9^{+0.5}_{-0.5}$	$18.5^{+3.6}_{-3.4}$
MC	$4.7^{+3.0}_{-1.8}$	$1.7^{+1.6}_{-0.8}$	$6.8^{+3.6}_{-2.3}$
LC	$0.9^{+0.7}_{-0.4}$	$0.4^{+0.7}_{-0.2}$	$1.5^{+1.0}_{-0.6}$
Sed OC	$0.08^{+0.02}_{-0.02}$	$0.03^{+0.01}_{-0.01}$	$0.11^{+0.02}_{-0.02}$
OC	$0.05^{+0.02}_{-0.02}$	$0.01^{+0.01}_{-0.01}$	$0.06^{+0.02}_{-0.02}$
Bulk crust	$21.3^{+4.8}_{-4.2}$	$6.6^{+1.9}_{-1.2}$	$28.2^{+5.2}_{-4.5}$
CLM	$1.3^{+2.4}_{-0.9}$	$0.4^{+1.0}_{-0.3}$	$2.1^{+2.9}_{-1.3}$
Total lithosphere	$23.2^{+5.9}_{-4.8}$	$7.3^{+2.4}_{-1.5}$	$30.9^{+6.5}_{-5.2}$
DM	4.2	0.8	5.0
EM	2.9	0.9	3.8
Grand total	$30.3^{+5.9}_{-4.8}$	$9.0^{+2.4}_{-1.5}$	$39.7^{+6.5}_{-5.2}$

The inputs for the calculations are taken from Huang et al. (2013), and the signals from the different reservoirs indicated in the first column are in TNU.

Results and discussion

In Table 1, we summarize the contributions to the expected geoneutrino signal at JUNO produced by U and Th in each of the reservoirs identified in the model. The central value and the asymmetric uncertainties are, respectively, the median and 1σ errors of a positively skewed distribution obtained from Monte Carlo simulation. This approach was developed for the first time by Huang et al. (2013) in order to combine the Gaussian probability density function of geophysical and (some) geochemical inputs, together with the lognormal distributions of U and Th abundances observed in the felsic and mafic rocks of MC and LC.

The total geoneutrino signal at JUNO is $G = 39.7^{+6.5}_{-5.2}$ TNU where the 1σ error only recognizes the uncertainties of the inputs of the lithosphere, which are mainly due to the uncertainties in the composition of the rocks and subsequently to the geophysical inputs. The predicted mantle contribution at JUNO is assumed to be $S_M \approx 9$ TNU (Huang et al. 2013). The expected geoneutrino signal from the mantle is essentially model dependent, and it is estimated according to a mass balance argument. Uncertainty in the assumed mantle model is much less than the predicted for the lithosphere (e.g., $\delta G \approx \pm 6$ TNU). An extensive discussion of different mantle structures is described in Šrámek et al. (2013), which considers a range of geoneutrino signals for different mantle models.

Thus, a future refinement of the abundances and distribution of HPEs in the UC surrounding the JUNO detector is strongly recommended, as this region provides approximately 47% of the total geoneutrino signal (G) and is a significant contributor to the total uncertainty.

Plotting the cumulative geoneutrino signal as a function of the distance from JUNO for the different Earth reservoirs (Figure 3), we observe that half of the total signal comes from U and Th in the regional crust that lies within 550 km of the detector. Since the modeling of the geoneutrino flux is based on $1° \times 1°$ cells, we study the signal produced in LOC subdivided in six $2° \times 2°$ tiles (Figure 1).

The geoneutrino signals from U and Th in the lithosphere of each tile are reported in Table 2 with their uncertainties. The main contribution (27% of G) comes from tile T2 in which the JUNO experiment is located (Figure 1). The thick UC in this tile, which is covered by a very shallow layer of Sed (Figure 2), is predicted to give a signal of $7.6^{+1.5}_{-1.4}$ TNU. Therefore, a refined study of the U and Th content of the UC in tile T2 is a high-value target for improving the accuracy and precision of the predicted geoneutrino signal at JUNO. Evaluating the antineutrino signal requires knowledge of several ingredients necessary for modeling the three antineutrino life stages: production, propagation to the detector site, and detection in liquid scintillation detectors via the IBD reaction. The propagation and detection processes are independent from the source of the particles, and we modeled these two stages using the oscillation parameters from Ge et al. (2013) and the IBD cross section from Strumia and Vissani (2003). The spectral parameters for U and Th geoneutrinos are from Fiorentini et al. (2007), and the modulation of these fluxes are based on Huang et al. (2013). Reactor antineutrino production is calculated adopting the data from a worldwide reference model from Baldoncini et al. (2015). Reported in Figure 4 are the energy distributions of geoneutrinos and reactor antineutrino signals in two different scenarios: in the full energy region, $R_{OFF} = 95.3^{+2.6}_{-2.4}$ TNU is obtained with data from the worldwide commercial reactors operating in 2013 and $R_{ON} = 1,566^{+111}_{-100}$ TNU, including the Yangjiang (17.4 GW) and Taishan (18.4 GW) nuclear power plants operating at an 80% annual average load factor (Baldoncini et al. 2015). In the geoneutrino energy window (i.e., 1.806 to 3.272 MeV), the reactor signals are $S_{OFF} = 26.0^{+2.2}_{-2.3}$ and $S_{ON} = 354^{+45}_{-41}$ TNU (Table 3).

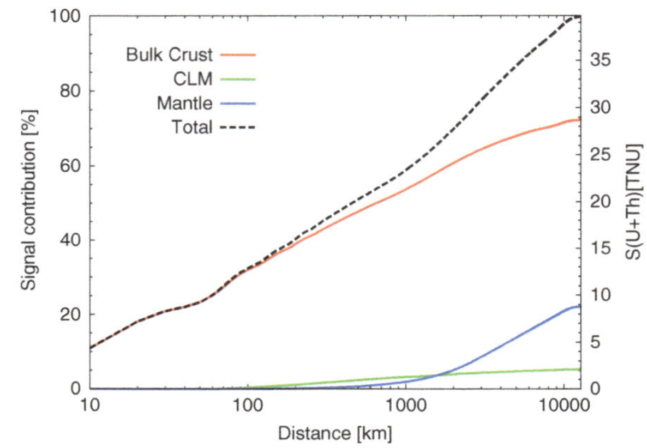

Figure 3 Geoneutrino signal contribution. The cumulative geoneutrino signal and the percentage contributions of the bulk crust, continental lithospheric mantle (CLM), and mantle are represented as functions of the distance from JUNO.

Table 2 Geoneutrino signals from six tiles of the LOC

Tile	S (U)	S (Th)	S (U + Th)	Percentage
T1	$0.4^{+0.1}_{-0.1}$	$0.1^{+0.1}_{-0.1}$	$0.5^{+0.1}_{-0.1}$	3.0
T2	$8.1^{+1.9}_{-1.7}$	$2.6^{+0.8}_{-0.5}$	$10.8^{+2.1}_{-1.8}$	62.1
T3	$1.1^{+0.3}_{-0.2}$	$0.4^{+0.2}_{-0.1}$	$1.5^{+0.3}_{-0.3}$	8.6
T4	$0.3^{+0.1}_{-0.1}$	$0.1^{+0.1}_{-0.1}$	$0.4^{+0.1}_{-0.1}$	2.2
T5	$2.5^{+0.5}_{-0.5}$	$0.7^{+0.2}_{-0.1}$	$3.2^{+0.6}_{-0.5}$	18.2
T6	$0.8^{+0.2}_{-0.2}$	$0.2^{+0.1}_{-0.1}$	$1.0^{+0.2}_{-0.2}$	5.9

The expected geoneutrino signals from U and Th contained in the lithosphere (CC+CLM) of the six tiles reported in Figure 1 are expressed in TNU. In the last column contributions in percentage are reported.

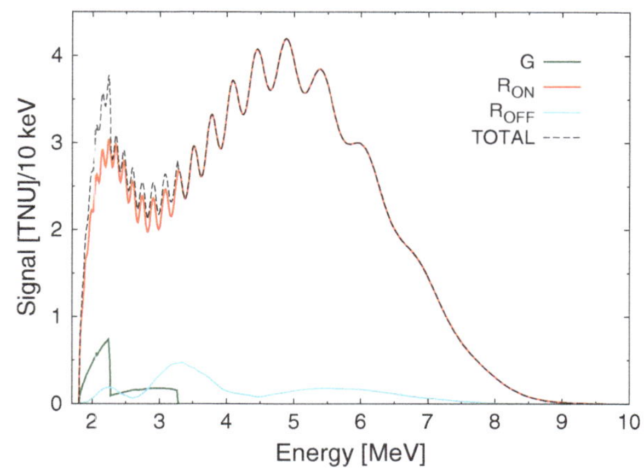

Figure 4 Antineutrino energy spectra expected at JUNO. Geoneutrino energy spectrum (green) is reported together with the energy reactor antineutrino spectra computed considering the commercial reactors operating all over the world in 2013 (cyan) and adding the contribution of the Yangjiang and Taishan nuclear power plants (red). The reactor antineutrino spectra are computed assuming normal hierarchy and neutrino oscillation. The total spectrum (black dashed lines) is obtained assuming the R_{ON} scenario.

Assuming a scenario whereby JUNO's signal does not have a background signal from Yangjiang and Taishan nuclear power plants, the ratio of $S_{OFF}/G = 0.7$, which compares to a value of 0.6 for the Borexino detector (Baldoncini et al. 2015). Considering only the statistical uncertainties, in the R_{OFF} scenario, JUNO is an excellent experiment for geoneutrino measurements reaching a 10% accuracy on the geoneutrino signal in approximately 105 days (assuming a $C_{17}H_{28}$ liquid scintillator composition, a 100% detection efficiency, and reactor antineutrinos as the sole source of background), given 576 geoneutrino events per year for a target mass of 14.5×10^{32} free protons. This optimistic expectation does not take into account the uncertainties of S_{OFF} and the background due to production of cosmic muon spallation, accidental coincidences, and radioactive contaminants in the detector.

Table 3 Geoneutrino and reactor antineutrinos signals at JUNO

	S (TNU)
Local contribution	$17.4^{+3.3}_{-2.8}$
Far-field crust	$13.4^{+3.3}_{-2.4}$
Mantle	8.8
Grand total of geoneutrinos	$39.7^{+6.5}_{-5.2}$
Reactors OFF	$26.0^{+2.2}_{-2.3}$
Reactors ON	354^{+45}_{-41}

The total geoneutrino signal (*G*) is the sum of the contributions from the local lithosphere (S_{LOC}), from the rest of the lithosphere (i.e., far-field crust, S_{FFC}), and from the mantle (S_M). The reactor antineutrino signal in the geoneutrino window is calculated from the data for commercial reactors operating all over the world in 2013 (S_{OFF}) and adding the contribution of the Yangjiang (17.4 GW) and Taishan (18.4 GW) nuclear power plants (S_{ON}) (Baldoncini et al. 2015). All the signals are expressed in TNU.

Conclusions

Designed as a 20 kton liquid scintillator detector, the JUNO experiment will collect high statistics for antineutrino signals from reactors and from the Earth. In this study, we focused on predicting the geoneutrino signal using the Earth reference model of Huang et al. (2013). The contribution originating from naturally occurring U and Th in the 6° × 4° LOC surrounding the JUNO detector (Figure 1) was determined. The main results of this study are summarized as follows:

- The thicknesses of the Sed, UC, MC, and LC layers of the 24 1° × 1° cells of the LOC are reported (Figure 2). The Moho depth of the continental LOC ranges between 26.3 and 32.3 km, and the uncertainty for each 1° × 1° cell is of the order of 7%.
- The total and local geoneutrino signals at JUNO are $G = 39.7^{+6.5}_{-5.2}$ and $S_{LOC} = 17.4^{+3.3}_{-2.8}$ TNU, respectively. The asymmetric 1σ errors are obtained from the Monte Carlo simulations and account only for uncertainties from the lithosphere. The major source of uncertainty comes from predicting the abundances and distribution of U and Th in local crustal rocks.
- High-resolution seismic data acquired in the LOC can improve the present geophysical model of the crust and CLM, of which the latter is assumed to have a homogenous depth of 175 ± 75 km. The CLM composition is derived from data for U and Th abundances inferred from the peridotite xenoliths, and its geoneutrino signal is of $2.1^{+2.9}_{-1.3}$ TNU.
- The HPEs in the regional crust extending out to 550 km from the detector produce half of the total expected geoneutrino signal (Figure 3). The U and Th in the 2° × 2° tile that hosts JUNO produces $10.8^{+2.1}_{-1.8}$ TNU corresponding to 27% of *G*. Since this region is characterized by a thick UC, which gives $7.6^{+1.5}_{-1.4}$ TNU, a refined geophysical and geochemical model of the UC of this tile is highly desired.
- The reactor signal in the geoneutrino window assuming two scenarios is $S_{OFF} = 26.0^{+2.2}_{-2.3}$ TNU with the 2013 reactor operational data only and $S_{ON} = 355^{+44}_{-41}$ TNU when the contributions of the Yangjiang and Taishan nuclear power plants are added. There is a potential to achieve up to 10% accuracy on geoneutrinos after 105 days of data accumulation, under conditions of the Yangjiang and Taishan nuclear power plants being off.

The JUNO experiment has the potential to reach a milestone in geoneutrino science, although some technical challenges must be addressed to minimize background (e.g., production of cosmic muon spallation, accidental coincidences, and radioactive contaminants in the detector). Assuming $S_{OFF}/G = 0.7$, JUNO can collect hundreds of low-background geoneutrino events in less than a year under optimal conditions. A future refinement of the U and Th distribution and abundance in the LOC is strongly recommended. Such data will lead to insights on the radiogenic heat production in the Earth, the composition of the mantle, and constraints on the chondritic building blocks that made the planet.

Abbreviations

BSE: bulk silicate Earth; CC: continental crust; CLM: continental lithospheric mantle; DM: depleted mantle; EM: enriched mantle; HPEs: heat-producing elements; IBD: inverse beta decay; JUNO: Jiangmen Underground Neutrino Observatory; LC: lower crust; LM: lithospheric mantle; MC: middle crust; OC: oceanic crust; Sed: sediments; TNU: terrestrial neutrino unit; UC: upper crust.

Competing interests

The authors declare that they have no competing interests.

Authors' contributions

VS and FM proposed and conceived the study and produced the geoneutrino data with WFM. MB and BR produced the reactor antineutrino data. IC and GX participated in the data analysis and interpretation of the results. VS took the lead in designing and composing the manuscript, and all the authors contributed to it. All authors read and approved the final manuscript.

Authors' information

The authors have a geological and physics background. VS and MB are graduate students attending PhD course in Physics at Physics and Earth Sciences Department of the University of Ferrara. FM and BR are researchers of Physics and Earth Sciences Department of the University of Ferrara and INFN. IC and GX are researchers at Legnaro National Laboratories. WFM is a professor of the Department of Geology at the University of Maryland.

Acknowledgements

We are grateful to R. L. Rudnick and Y. Huang for their fruitful discussions on crustal modeling of geoneutrino fluxes. We appreciate the observations on the geoneutrino signal predictions from S. Dye, G. Fiorentini, L. Ludhova, and H. Watanabe. We thank J. Mandula for the valuable help in compiling the nuclear reactor database. We wish to thank two anonymous reviewers for their detailed and thoughtful reviews. This work was partially supported by the Istituto Nazionale di Fisica Nucleare (INFN) through the ITALRAD Project, by the University of Ferrara through the research initiative 'Fondo di Ateneo per la Ricerca scientifica FAR 2014' and partially by the U.S. National Science Foundation Grants EAR 1067983/1068097.

Author details

[1]Department of Physics and Earth Sciences, University of Ferrara, Via Saragat 1, 44121 Ferrara, Italy. [2]Legnaro National Laboratories, INFN, Viale dell'Università, 2, 35020 Legnaro, Italy. [3]Ferrara Section, INFN, Via Saragat 1, 44121 Ferrara, Italy. [4]Department of Geology, University of Maryland, 237 Regents Drive, College Park, MD 20742, USA.

References

Arevalo R, McDonough WF (2010) Chemical variations and regional diversity observed in MORB. Chem Geol 271(1–2):70–85. doi:10.1016/j.chemgeo.2009.12.013

Baldoncini M, Callegari I, Fiorentini G, Mantovani F, Ricci B, Strati V, Xhixha G (2015) Reference worldwide model for antineutrinos from reactors. Physical Review D 91(6):065002

Bassin C, Laske G, Masters TG (2000) The current limits of resolution for surface wave tomography in North America. EOS Trans AGU 81:F897

Borexino Collaboration (2013) Measurement of geo-neutrinos from 1353 days of Borexino. Phys Lett B 722(4–5):295–300. doi:10.1016/j.physletb.2013.04.030

Capozzi F, Lisi E, Marrone A (2014) Neutrino mass hierarchy and electron neutrino oscillation parameters with one hundred thousand reactor events. Phys Rev D 89(1):013001

Coltorti M, Boraso R, Mantovani F, Morsilli M, Fiorentini G, Riva A, Rusciadelli G, Tassinari R, Tomei C, Di Carlo G, Chubakov V (2011) U and Th content in the Central Apennines continental crust: a contribution to the determination of the geo-neutrinos flux at LNGS. Geochim Cosmochim Acta 75(9):2271–2294. doi:10.1016/j.gca.2011.01.024

Dziewonski AM, Anderson DL (1981) Preliminary reference Earth model. Phys Earth Planet In 25:297–356. doi:10.1016/0031-9201(81)90046-7

Fiorentini G, Fogli G, Lisi E, Mantovani F, Rotunno A (2012) Mantle geoneutrinos in KamLAND and Borexino. Physical Review D 86 (3). doi:10.1103/PhysRevD.86.033004

Fiorentini G, Ianni A, Korga G, Lissia M, Mantovani F, Miramonti L, Oberauer L, Obolensky M, Smirnov O, Suvorov Y (2010) Nuclear physics for geo-neutrino studies. Physical Review C 81(3). doi:10.1103/PhysRevC.81.034602

Fiorentini G, Lissia M, Mantovani F (2007) Geo-neutrinos and Earth's interior. Phys Rep 453(5–6):117–172. doi:10.1016/j.physrep.2007.09.001

Ge S-F, Hagiwara K, Okamura N, Takaesu Y (2013) Determination of mass hierarchy with medium baseline reactor neutrino experiments. J High Energ Phys 2013(5):1–23. doi:10.1007/JHEP05(2013)131

Huang Y, Chubakov V, Mantovani F, Rudnick RL, McDonough WF (2013) A reference Earth model for the heat-producing elements and associated geoneutrino flux. Geochem Geophys Geosyst 14(6):2023–2029. doi:10.1002/ggge.20129

Huang Y, Strati V, Mantovani F, Shirey SB, McDonough WF (2014) Regional study of the Archean to Proterozoic crust at the Sudbury Neutrino Observatory (SNO+), Ontario: predicting the geoneutrino flux. Geochem Geophys Geosyst 15(10):3925–3944. doi:10.1002/2014gc005397

KamLAND Collaboration (2005) Measurement of neutrino oscillation with KamLAND: evidence of spectral distortion. Phys Rev Lett 94(8):081801

KamLAND Collaboration (2013) Reactor on-off antineutrino measurement with KamLAND. Phys Rev D 88(3):033001

Kim S-B (2013) Proposal for RENO-50. In: International workshop on RENO-50, Seoul National University, Korea, 13–14 June 2013

Laske G, Masters TG (1997) A global digital map of sediment thickness. Eos Trans AGU 78:F483

Laske G, Masters TG, Reif C (2001) CRUST 2.0: a new global crustal model at 2 × 2 degrees. http://igppweb.ucsd.edu/~gabi/crust2.html

Li Y-F, Cao J, Wang Y, Zhan L (2013) Unambiguous determination of the neutrino mass hierarchy using reactor neutrinos. Phys Rev D 88(1):013008

Li Y-F, Zhou Y-L (2014) Shifts of neutrino oscillation parameters in reactor antineutrino experiments with non-standard interactions. Nucl Phys B 888(0):137–153. doi:10.1016/j.nuclphysb.2014.09.013

Li YF (2014) Overview of the Jiangmen Underground Neutrino Observatory (JUNO). arXiv:1402.6143 [physics.ins-det] http://arxiv.org/pdf/1402.6143

McDonough WF (1990) Constraints on the composition of the continental lithospheric mantle. Earth Planet Sci Lett 101(1):1–18. doi:10.1016/0012-821x(90)90119-i

McDonough WF, Sun S-S (1995) The composition of the Earth. Chem Geol 120:223–253. doi:10.1016/0009-2541(94)00140-4

Reguzzoni M, Sampietro D (2015) GEMMA: an earth crustal model based on GOCE satellite data. Int J Appl Earth Observation Geoinform 35(Part A (0)):31–43. doi:10.1016/j.jag.2014.04.002

Reguzzoni M, Tselfes N (2009) Optimal multi-step collocation: application to the space-wise approach for GOCE data analysis. J Geodes 83(1):13–29. doi:10.1007/s00190-008-0225-x

Shapiro NM, Ritzwoller MH (2002) Monte-Carlo inversion for a global shear-velocity model of the crust and upper mantle. Geophys J Int 151:88–105. doi:10.1046/j.1365-246X.2002.01742.x

Šrámek O, McDonough WF, Kite ES, Lekić V, Dye ST, Zhong S (2013) Geophysical and geochemical constraints on geoneutrino fluxes from Earth's mantle. Earth Planet Sci Lett 361:356-366. doi:10.1016/j.epsl.2012.11.001

Strumia A, Vissani F (2003) Precise quasielastic neutrino/nucleon cross-section. Phys Lett B 564(1–2):42–54. doi:10.1016/s0370-2693(03)00616-6

5

Auroral substorms as an electrical discharge phenomenon

Syun-Ichi Akasofu

Abstract

During the last 50 years, we have made much progress in studying auroral substorms (consisting of the growth phase, the expansion phase, and the recovery phase). In particular, we have *quantitatively* learned about auroral substorms in terms of the global energy input–output relationship. (i) What powers auroral substorms? (ii) Why is there a long delay (1 h) of auroral activities after the magnetosphere is powered (growth phase)? (iii) How much energy is accumulated and unloaded during substorms? (iv) Why is the lifetime of the expansion phase so short (1h)? (v) How is the total energy input–output relationship? (vi) Where is the magnetic energy accumulated during the growth phase? On the basis of the results obtained in (i)–(vi), we have reached the following crucial question: (vii) how can the unloaded energy produce a secondary dynamo, which powers the expansion phase? Or more specifically, how can the accumulated magnetic energy get unloaded such that it generates the earthward electric fields needed to produce the expansion phase of auroral substorms? It is this dynamo and the resulting current circuit that drive a variety of explosive auroral displays as electrical discharge phenomena during the expansion phase, including the poleward advance of auroral arcs and the electrojet. This chain of processes is summarized in Section 4.2. This is the full version of work published by Akasofu (2015).

A tentative answer to this crucial question is attempted. Phase occurs impulsively seems to be that the magnetosphere within a distance of 10 Re becomes inflated and unstable ($\beta \sim 1.0$), when the accumulated energy W during the growth phase (at the rate of about $\varepsilon = 5 \times 10^{18}$ erg/s in about 1.5 h) reaches 2×10^{22}—or at most 10^{23}—ergs. Thus, the magnetosphere unloads and dissipates the energy in order to stabilize itself by deflating at the rate of about 5×10^{18} erg/s (mainly as the Joule heat in the ionosphere), resulting in an impulsive (1 h, 2×10^{22} ergs $\div$ 3.5×10^{18} erg/s) expansion phase. The deflating process results in a dynamo in a thin magnetic shell near the earthward end of the current sheet by separating electrons from protons and produces an earthward electric field of more than ~10 mV/m. The separated electrons are discharged along the circuit of the expansion phase, constituting an electrical discharge currents of 5×10^6 A and causing brightening an arc, the first indication of the onset of the expansion phase.

Keywords: Auroral substorms; Expansion phase; Magnetospheric substorms; Cosmical electrodynamics

Introduction

Auroral substorms have been one of the main subjects in magnetospheric physics for several decades after the first publication on this subject (Akasofu 1964) and the first observational confirmation by a satellite (Frank et al. 1982). We have been learning how the solar wind–magnetosphere–ionosphere system causes auroral substorms, taking a variety of approaches by a large number of researchers in the past and present.

Correspondence: nbauer@iarc.uaf.edu
International Arctic Research Center, University of Alaska Fairbanks, Fairbanks, AK, USA

In studying auroral substorms, it is important to recognize that auroral substorms are basically various manifestations of an *electrical discharge* process. Thus, it is natural to discuss processes of auroral substorms in terms of an *input–output relationship*, and more specifically, of power supply (dynamo), transmission (currents and their circuits), and dissipation (auroral phenomena). In fact, in his paper titled "The second approach to cosmical electrodynamics," Alfven (1967) emphasized a new approach beyond the conventional MHD and stated: "It is important to note that in many cases the physical basis of the phenomena is better understood if the discussion is centered on the picture of the current lines." Further: "We may say that the first new principle is associated with a "thaw" of the frozen-in field lines." Alfven (1977) also noted: "Hence in order to understand the properties of a current-carrying plasma we must take account of the properties of the whole circuit in which the current flows."

Indeed, as Alfven (1986) warned: "In low density plasma the concept of frozen-in lines of force is questionable.—may be grossly misleading if applied to the magnetosphere of the earth."

Thus, in this paper, the concept of "frozen-in field lines" and of "moving field lines" with plasmas is avoided and not considered. The discussion is mainly based on electric currents and their circuits. However, it will be shown that physics of auroral substorms can be understood reasonably well by this approach because it is simply due to the fact that auroral substorms are manifestations of an electrical discharge phenomenon.

Auroral substorms consist of three phases: growth, expansion, and recovery. Based on the above premise, it is worthwhile to consider a conceptual circuit for each of the three phases of auroral substorms; they are shown in Fig. 1 and will be discussed in the following sections. The conceptual circuit of the magnetosphere was originally proposed by Bostrom (1974).

Power supply, transmission, and dissipation

Power supply

The primary dynamo process occurs as the solar wind blows along the boundary of the magnetosphere (the magnetopause) across the merged field lines between the interplanetary magnetic field (IMF) and the earth's magnetic field. Thus, it is called the S–M dynamo here and is shown at the top of each circuit in Fig. 1; see also Fig. 2a.

The power ε of this dynamo is given by $\varepsilon(t) = VB^2\sin^4(\theta/2)\iota^2$, where V, B and θ are, respectively, the solar wind speed, the IMF magnitude, and the polar angle of the IMF vector; ι is a constant (cf. Akasofu, 1981); note that ε is not just a parameter, but has the dimension of power (erg/s or watts). A typical power ε during substorms is about 5×10^{18} erg/s (5×10^{11} w).

Convection and the DD current

This primary S–M dynamo generates a voltage V of about 100 kV and more across the morning–evening side of the magnetosphere [Fig. 2a and b], which is transmitted to the auroral oval (and causes the region 1/2 current systems and the oval) and drives also a large-scale convection of plasmas in the ionosphere and the magnetosphere [Fig. 2b]. It may be noted here that the convection does not have to be interpreted as a magnetic reconnection effect. The potential drop across the magnetotail to the ionosphere or the potential drop transmitted from the magnetotail can drive it.

This ionospheric/magnetospheric convection drives a *two-cell* convection of the ionospheric plasma, as indicated by the flow lines (the equipotential lines), as illustrated in Fig. 2c, and as a result also, a two-cell electric current flows along the convection flow [but in the opposite direction (the Hall current)]. This process was suggested by Dungey (1961) and Axford and Hines (1961); it may be noted that Dungey (1961) did not consider this process as an energy production process and further did not mention the term 'reconnection.' The

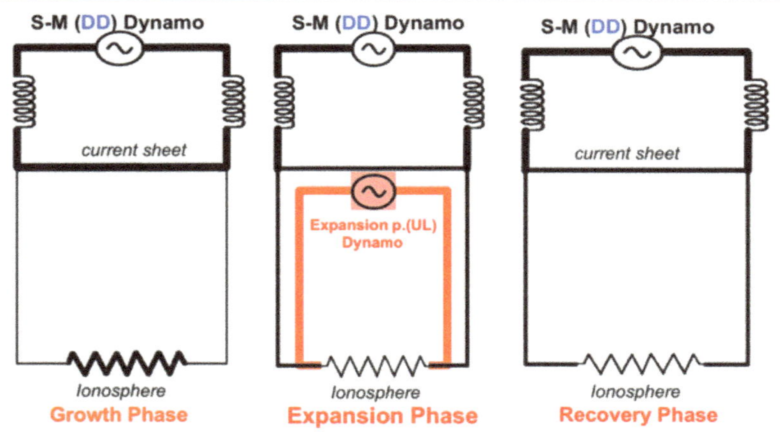

Fig. 1 Conceptual circuit for the three phases of auroral substorms: the primary (S–M) dynamo is shown at the *top*. During the growth phase, the ionosphere cannot dissipate much increasing power ε, and thus, the magnetosphere accumulates magnetic energy in its inductive circuit. During the expansion phase, a new internal dynamo, the expansion phase dynamo (shown in *red*), is established as a result of unloading the accumulated magnetic energy and drives the unloading (UL) current; its 3-D circuit is shown in Fig. 6. The streaming current-carrying electrons along the magnetic field lines ionize the ionosphere, producing active auroras and increasing its conductivity during the expansion phase. The recovery phase is driven only by the S–M dynamo

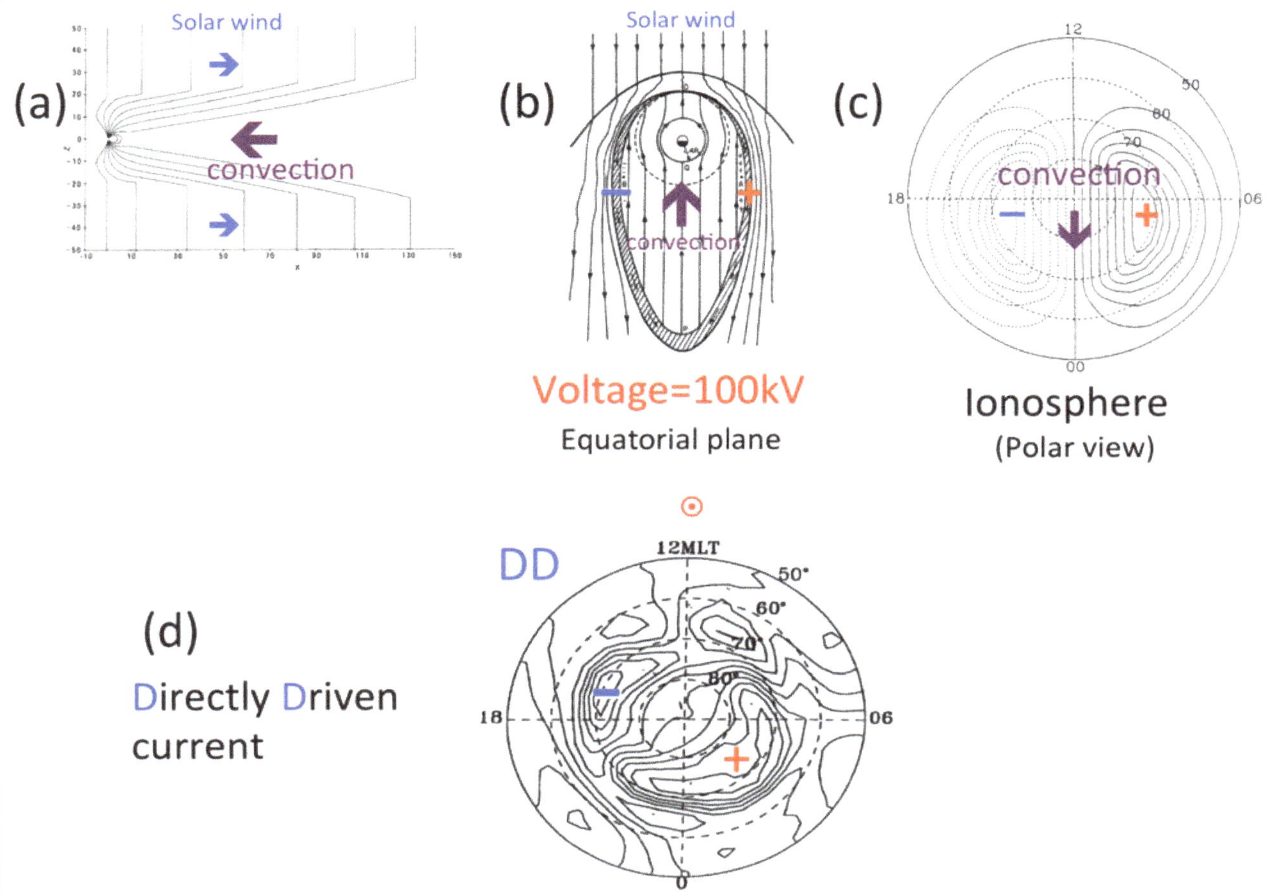

Fig. 2 The figure illustrates how the S–M dynamo works. **a** As the solar wind blows across the merged field lines between the IMF and the earth's magnetic field; the resulting ±eV × B force sets up the dynamo. **b** The dynamo produces a voltage of about 100 kV or more, which drives a convective motion of plasma in the ionosphere and the magnetosphere. **c** This convective motion flows along the equipotential contour. The resulting two-cell DD current flows along the equipotential contours, but in the opposite direction of the plasma flow. **d** The actual DD current pattern is greatly distorted because of the nonuniform and anisotropic conductivity of the ionosphere

current thus produced in the ionosphere is called the *directly driven* (DD) current because it is directly driven by the S–M dynamo power, as proved later. The two cells is greatly distorted from an ideal condition by the nonuniform and anisotropic conductivity (Fig. 2d).

The 3-D current circuit suggested by Bostrom (1974) cannot easily be proven in its entirety, so that it is necessary to infer the current configuration by combining ionospheric (ground-based observations) and the field-aligned currents and the equatorial currents (based on satellite observations) [Akasofu, 1992].

In this paper, magnetic records from the international six meridian chains of magnetometers (Fig. 3a) are analyzed to obtain the ionospheric part of the substorm current system (Kamide et al. 1982); an example of the current distribution is shown in Fig. 3b.

The substorm index auroral electrojet (AE) responds to the S–M dynamo power ε when it exceeds 10^{18} erg/s, so that a substorm begins when the power ε becomes above 10^{18} erg/s and end when ε becomes less than 10^{18} erg/s. This can be shown by comparing the AE index with the polar cap index ($\sqrt{B_x^2 + B_y^2}$, determined on the basis of magnetic records in the polar cap stations) which is more sensitive to the dynamo power ε than the AE index when the power is less than 10^{18} erg/s [Fig. 3c].

Separation of the DD and UL component

In order to understand characteristics of the three phases of auroral substorms and the associated energy flow, in particular of the expansion phase, it is instructive to consider conceptually that the magnetosphere is a water (energy) bucket with an inlet spout of the solar wind energy, together with a faucet corresponding to the IMF polar angle (θ) and two outlet spouts (Fig. 4). The amount of the output flow from the first outlet tends to follow the input flow from the faucet after the water

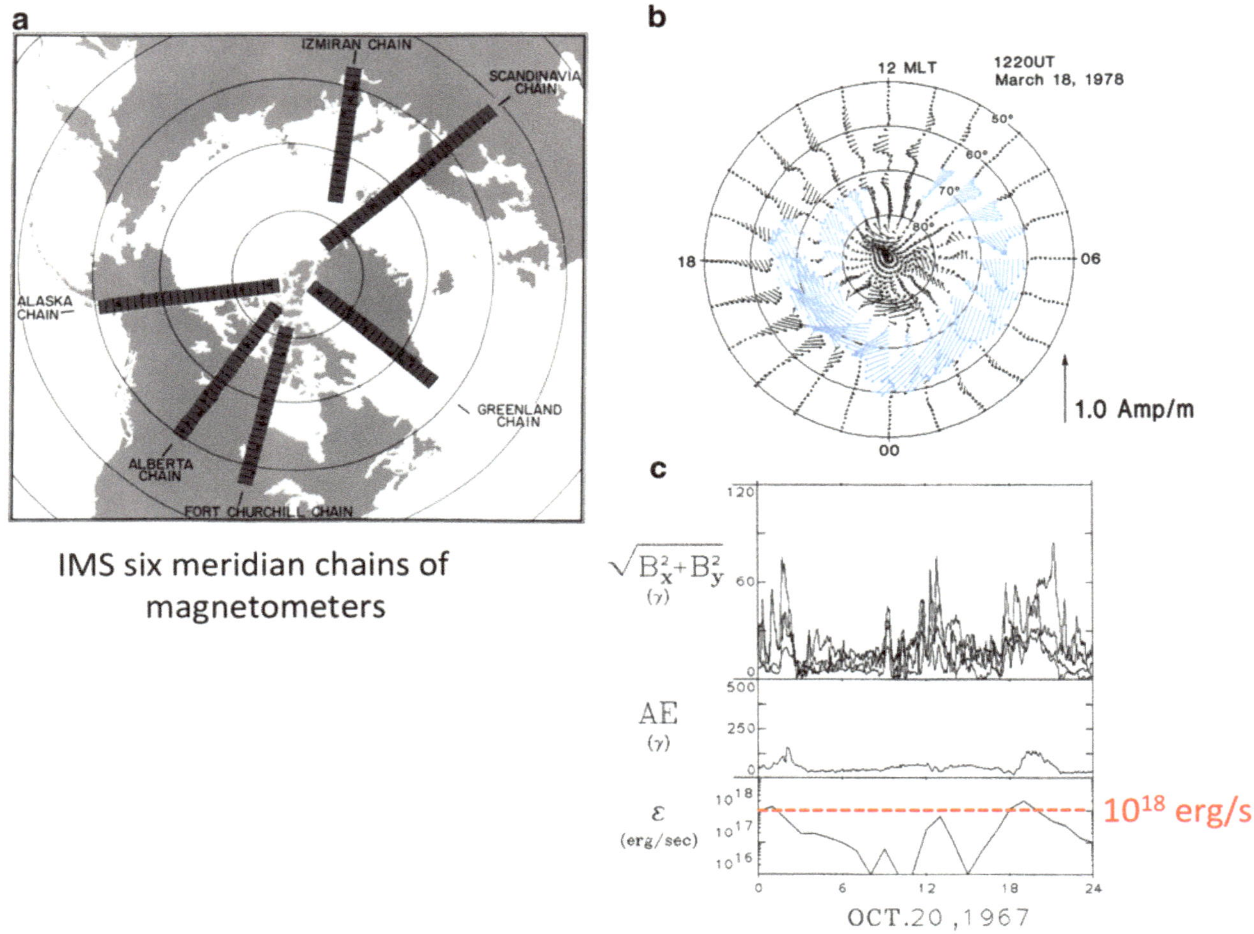

Fig. 3 a The six meridian chains of magnetometers which were set up by international efforts. **b** An example of changes of the ionospheric current during a typical substorm (Kamide et al. 1982). **c** The comparison between the polar cap index (which reveals magnetic disturbances inside the polar cap), the AE index and the power ε. Note that the AE index responds to the power ε when it exceeds 10^{18} erg/s

level reaches the spout; this output flow from it drives the ionospheric/magnetospheric *ExB* convection and is dissipated by the DD current in the ionosphere as the Joule heat.

The other spout has a receiving tippy bucket (corresponding to the accumulation of magnetic energy) which tips when it is full or the spring attached to it gives away. It is this impulsive flipped flow (energy) which is responsible for the expansion phase and causes the expansion phase of auroral substorms for about 1 h; the current driven by the tippy bucket produces the *unloading* (UL) current in the ionosphere, which is a single-cell current system (Fig. 5) [Akasofu et al. 1965]; it will be shown later the limit of amount of water (energy) is about 2×10^{22} ergs. At the peak of the expansion phase, the UL current can exceed the DD current (see Fig. 9), indicating that the energy for the expansion phase is not just an enhancement of the S–M dynamo power and thus once stored before expansion onset. Thus, the growth phase is not caused by a system delay (Akasofu, 2013; p. 7222 [18]) and thus is the loading period.

However, since magnetometers record magnetic fields of both the DD and UL currents together (see Fig. 4), it is necessary to separate the DD and UL currents in order to study the expansion phase. The method of natural orthogonal component [MNOC] is used to separate the DD (two-cell) and UL (single-cell) currents (Fig. 5); Sun et al. (1998). Unlike the two-cell DD current associated with the ionosphere/magnetospheric convection flow, the UL current is a *single-cell* current. It will be shown later (Fig. 9) that the UL current is independent of the two-cell DD current and is not just an enhancement of the DD current and thus requires a new internal dynamo (the expansion phase dynamo, Fig. 1) and a new circuit (shown in red in Figs. 1 and 6 for the expansion phase). It is crucial to recognize the difference between the DD and UL currents in an effort to understand substorms, in particular the expansion phase.

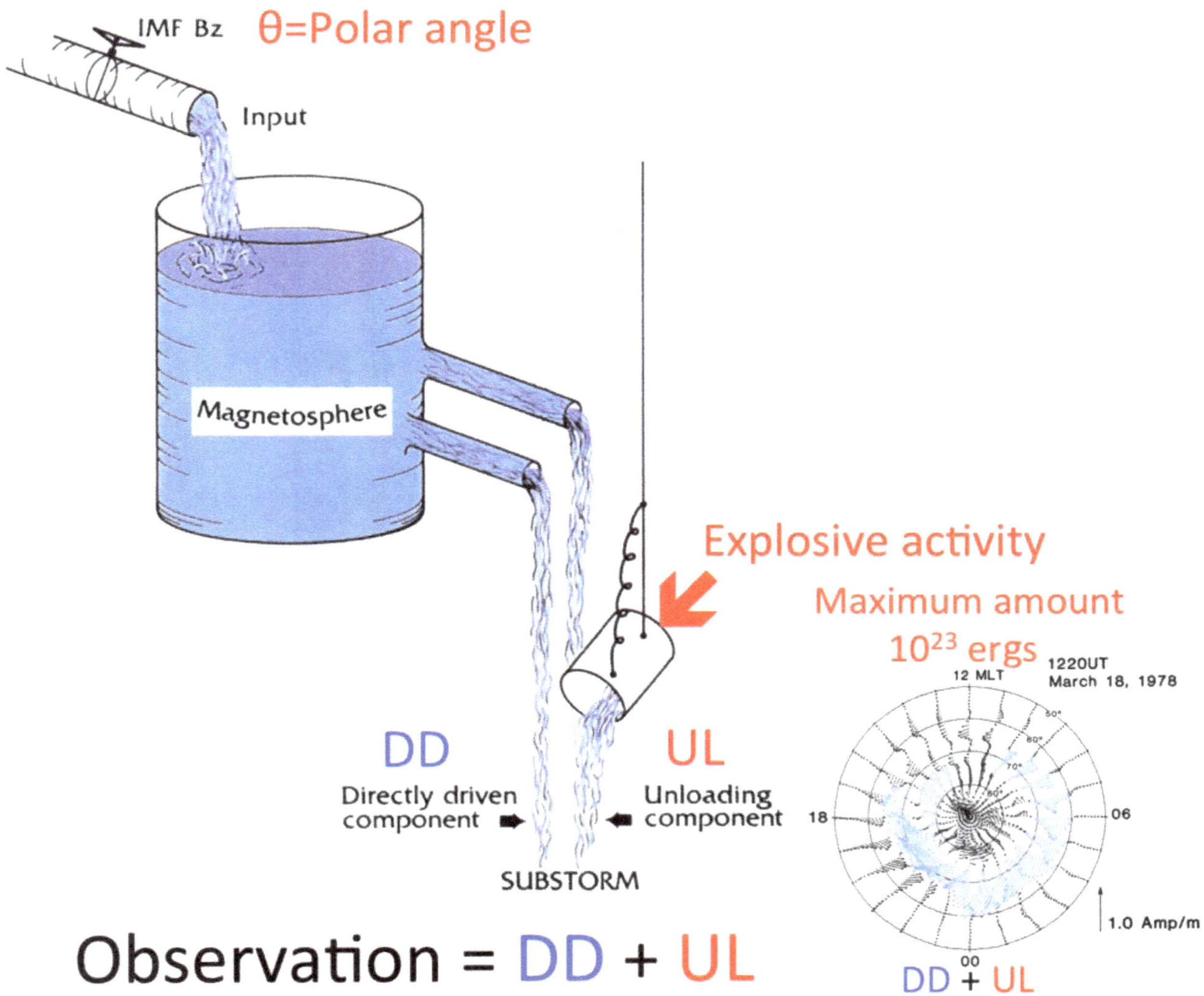

Fig. 4 Conceptual illustration of the energy flow in the magnetosphere–ionosphere coupling system. Since magnetometers record magnetic fields of both the DD and UL currents together, it is necessary to find a way to separate them in studying the expansion phase (Sun et al. 1998)

The UL current system for the expansion phase

Figure 6 shows the UL current system and its whole circuit. The UL current is the ionospheric part of the 3-D current during the expansion phase. The 3-D current system, suggested by Bostrom (1964), consists of two components, the azimuthal and meridional components; note that meridional component is a sheet current which is needed for a curtain-like structure of auroral arcs.

The 3-D current system can be driven by an earthward-directed electric field on the equatorial plane where $\boldsymbol{E} \circ \boldsymbol{J} < 0$ (cf. Akasofu, 2003a); this fact is crucial in understanding the cause of the expansion phase (Synthesis). When such an earthward electric field is transmitted to the ionosphere, it is directed southward in the northern hemisphere. The resulting current is mainly the Hall current, $\boldsymbol{E} \cdot \boldsymbol{J}_h = 0$, (the auroral electrojet); in the ionosphere $\boldsymbol{E} \circ \boldsymbol{J}_p > 0$, where the J_p denotes the Pedersen current. The conceptual circuit for the UL current system is also shown in Fig. 6.

Dissipation

The ionosphere is the main sink of the energy which causes auroral substorms, so that it is crucial to determine the global energy dissipation rate δ, the total dissipation, and their time variations over the whole polar ionosphere in order to learn *how much energy an auroral substorm consumes* (the particle precipitation is about one tenth of the Joule heating; Ahn et al. 1983, 1989). Fortunately, we can determine both quantities on the basis of the current distribution obtained by ground-based magnetometer (Kamide et al., 1982), incoherent scatter radars and the superDARN radar network (Bristow and Jensen, 2007). Figure 7 shows an example of the

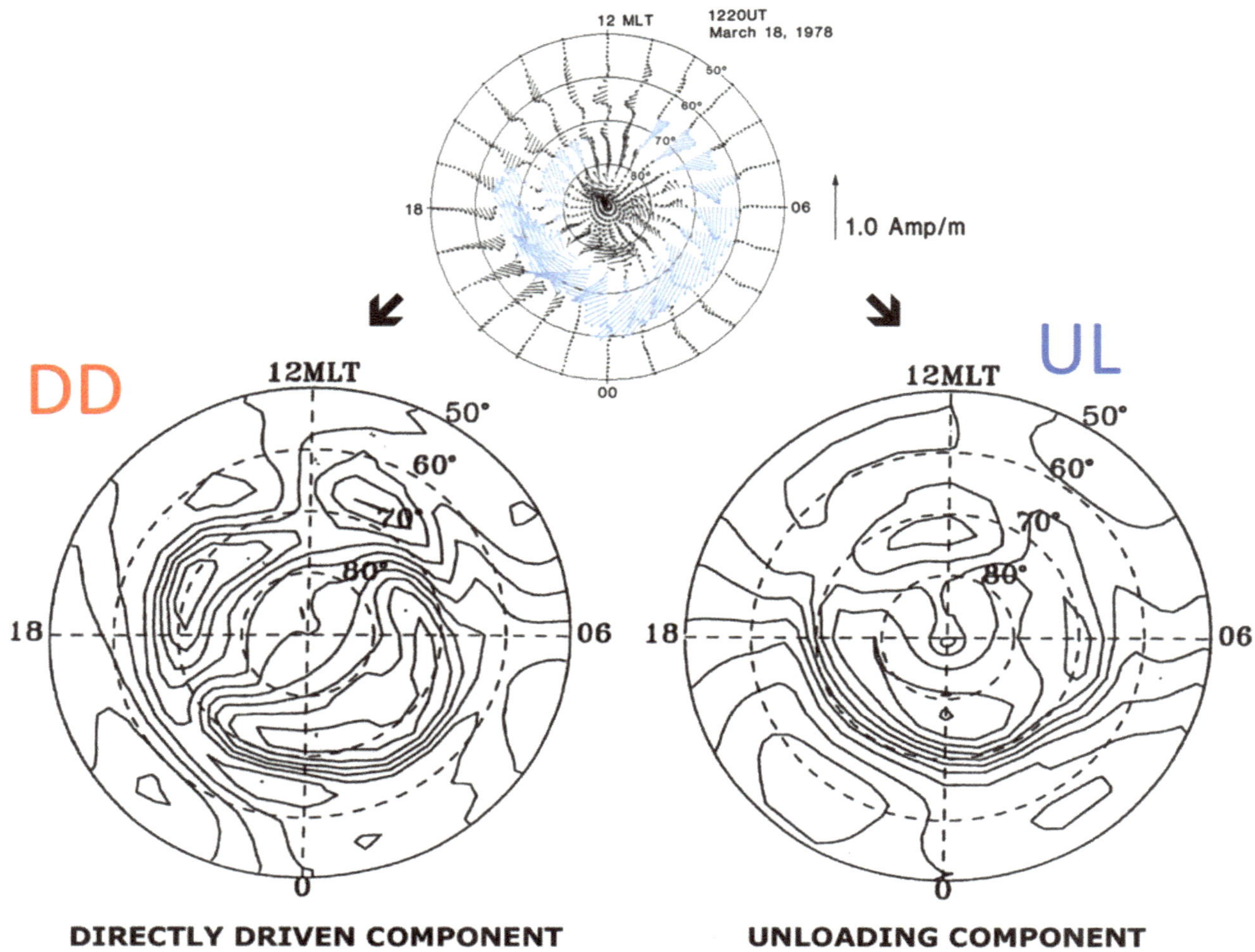

Fig. 5 Separation of the DD and UL components from the observed current system. Note that the DD current system is a two-cell system, while the UL current system is a single-cell system. The MNOC method is a sort of 2-D Fourier analysis, separating the first and second modes

distribution of the dissipation rate (the Joule heat production rate) over the polar region and its time variations of the globally integrated δ for a moderate substorm (Ahn et al. 1983, 1989). A typical value of the dissipation rate is $\delta = 5 \times 10^{18}$ erg/s and the total dissipation $\int \delta(t)dt$ is 2×10^{22} ergs ($=5 \times 10^{18}$ erg/s $\times$ 1 h), respectively. Note also that the transition from the growth phase to the expansion phase is sudden, indicating that the growth phase is not a system delay.

Figure 8 shows how the DD and UL currents vary during the course of a typical substorm. One can see that the DD current is very weak during the growth phase, indicating that the dissipation is small. The UL current appears for less than 1 h and decays quickly during the expansion phase. The DD current intensifies during the expansion phase because the conductivity is increased during the expansion phase. The DD current remains during the recovery phase, after the end of the expansion phase (and also after the end of the UL current) until the power becomes less than 10^{18} ergs/s.

Relationship among the power ε(t), DD(t), and UT(t)

Now, let us examine quantitatively what is shown in Fig. 8. We examine here the relationship among the power $\varepsilon(t)$, DD(t), and UL(t) for two moderate and typical substorms. The results are shown in Fig. 9. For this purpose, note that the dissipation rate δ (the Joule heat production rate) in the ionosphere is roughly proportional to the current intensity J because the Joule heating rate is given by $J^2 / \sigma = J(J / \sigma) \propto J$ because the conductivity σ is produced by the field-aligned currents (streaming electrons), which is connected to the ionospheric currents J. Thus, DD(t) and UL(t) are approximately proportional to their dissipation rate $\delta(t)$; this is an interesting and unique characteristic of the magnetosphere–ionosphere system.

First of all, the time variations of the power $\varepsilon(t)$ and DD(t) current [proportional to the dissipation $\delta(t)$] have similar time variations within the accuracy of the data set (ε based only on the solar wind data set, J deduced

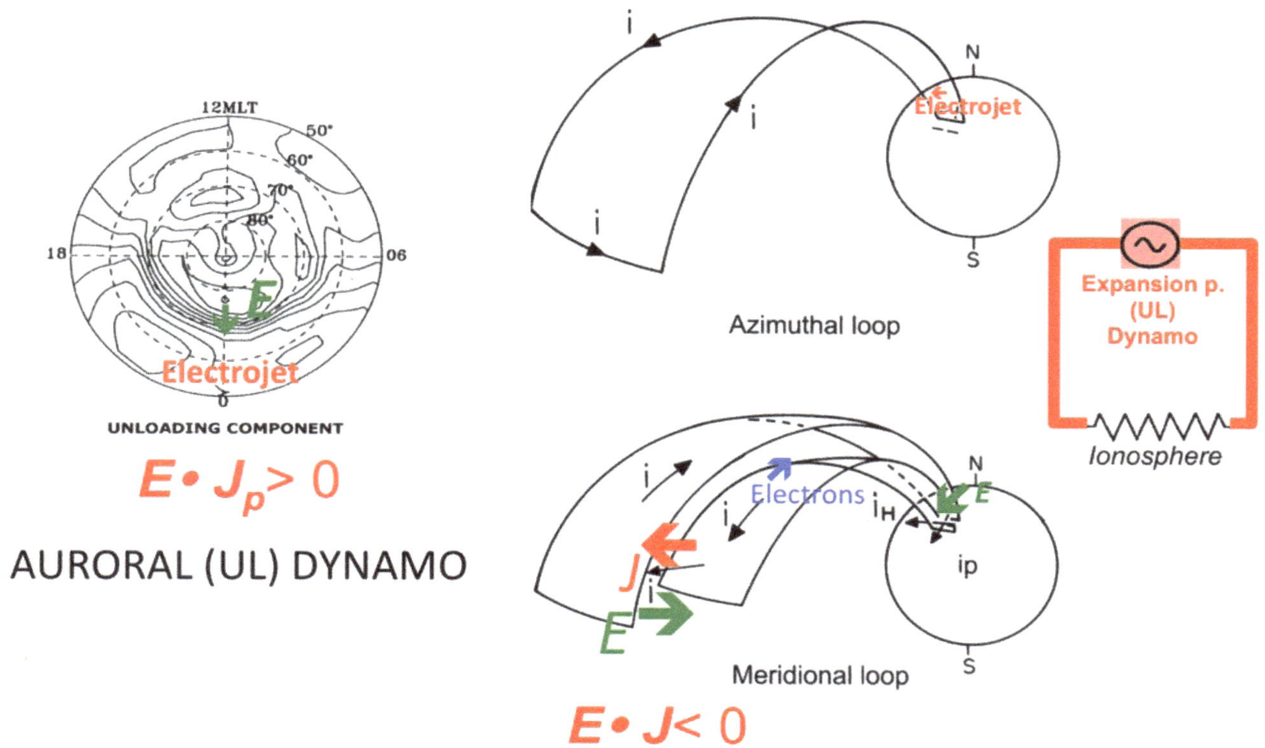

Fig. 6 The UL current is the ionospheric part of Bostrom's 3-D current circuit during the expansion phase [mainly the Hall current] (Bostrom. 1964). It can be criven by an earthward-directed electric field on the equatorial plane where $\boldsymbol{E}\,\boldsymbol{J} < 0$. Note that in the ionosphere, $\boldsymbol{E}\,\boldsymbol{J}_p > 0$

only from the ground data set), confirming that the DD current is primarily driven by the S–M dynamo. On the other hand, unlike the DD current, the UL current develops impulsively and lasts for about 1 h during the expansion phase. Its changes are independent of changes of ε and DD, indicating that the UL current is driven by an independent dynamo (the expansion phase dynamo, Fig. 1) which is different from the S–M dynamo. A typical dissipation rate δ is about 5×10^{18} erg/s and is thus *similar to the average value of power* ε. Substorms continue until the power ε becomes less than 10^{18} erg/s.

The above study suggests that $\int \varepsilon(t)dt = \int \delta(t)dt$ for each isolated substorm within the accuracy of the data. If there are any additional energy process other than ε (if magnetic reconnection is such a case) then the amount is too small to be treated in the analysis process.

Variety of the development of substorms

Individual substorms develop differently. In fact, no two substorms are alike. Examining a number of substorms, it is suggested that a variety of the development of substorms depend on many factors, such as ε(t), δ(t), and σ(t), as well as pre-substorm conditions of the magnetosphere, including the degree of the conductivity of the ionosphere and of stability of the current-carrying plasma in the plasma sheet at substorm onset (Synthesis), corresponding to the strength of the spring (stability of the current sheet) attached to the tippy bucket in Figs. 4 and 9. Indeed, not all substorms have a typical expansion phase (Fig. 8).

Why does the aurora flare up?

It is now possible to discuss the input–output relationship on the basis of what we have learned in the above. The scheme for this purpose is illustrated in Fig. 10. This study of the global input–output relationship is only the first and preliminary attempt in substorm studies, and all the quantities used for this purpose should be improved in future studies.

What powers auroral substorms?

It was shown in Power supply that the solar wind–magnetosphere interaction in terms of the input–output relationship is given by ε as a first approximation. It was shown there that the substorm index AE responds to ε when ε exceeds 10^{18} erg/s. A substorm lasts as long as the power ε is above 10^{18} erg/s.

Why is there a long delay of auroral activities after the magnetosphere is powered (the growth phase)?

As shown in Figs. 7 and 8, there is a significant delay of as much as about 1 h before expansion phase onset after the power ε is increased. This delay is not like an

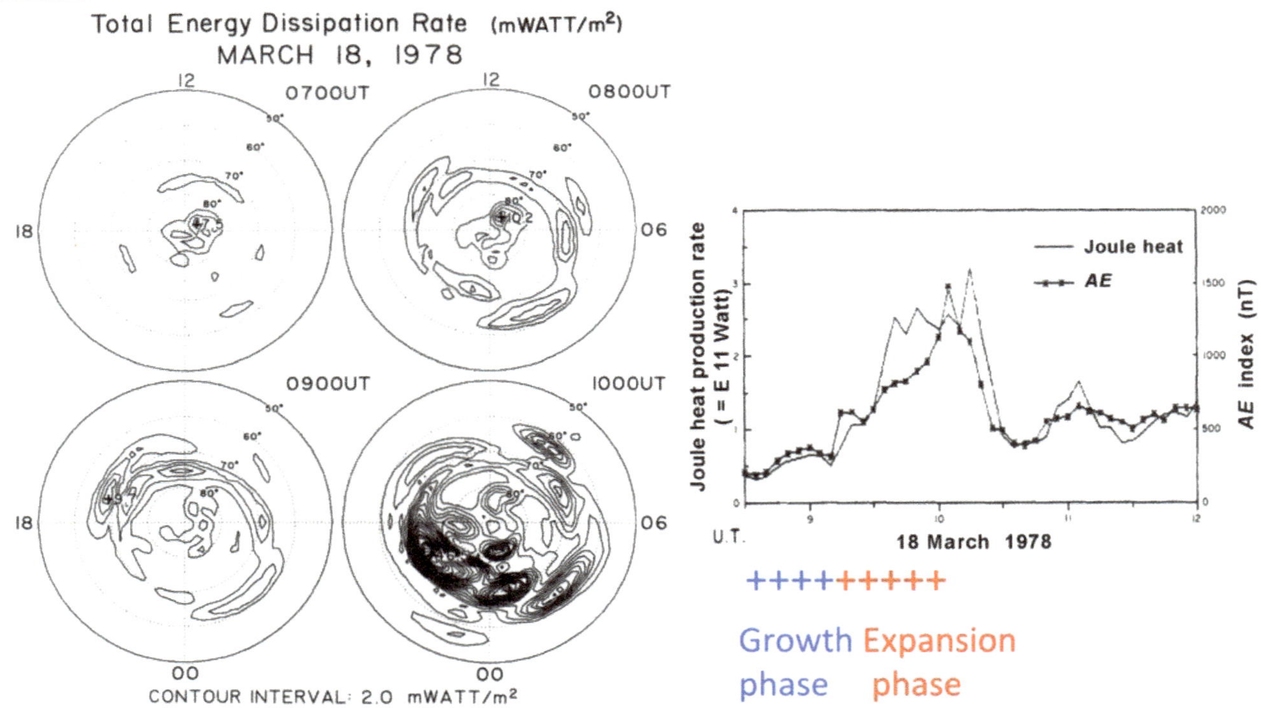

Fig. 7 An example of the energy dissipation rate (the Joule heat production rate) during a moderate substorm (Ahn et al. 1983, 1989). (*Left*) The global distribution of dissipation, including a quiet time and the maximum epoch of a substorm. (*Right*) The time variations of the globally integrated heat production (dissipation) rate δ(t), together with the AE index; both the growth and expansion phases are indicated

ordinary electrical system delay. Figure 8 shows that the DD current is very weak and does not increase much until the onset time of the expansion phase, indicating that the dissipation remains small during the growth phase. Further, the transition from the growth phase to expansion onset is rather sudden, rather than a gradual increase from onset of the growth phase (Fig. 7). Furthermore, the peak of the UL current can be greater than that of the DD current. These facts confirm that the energy for the expansion phase is once stored.

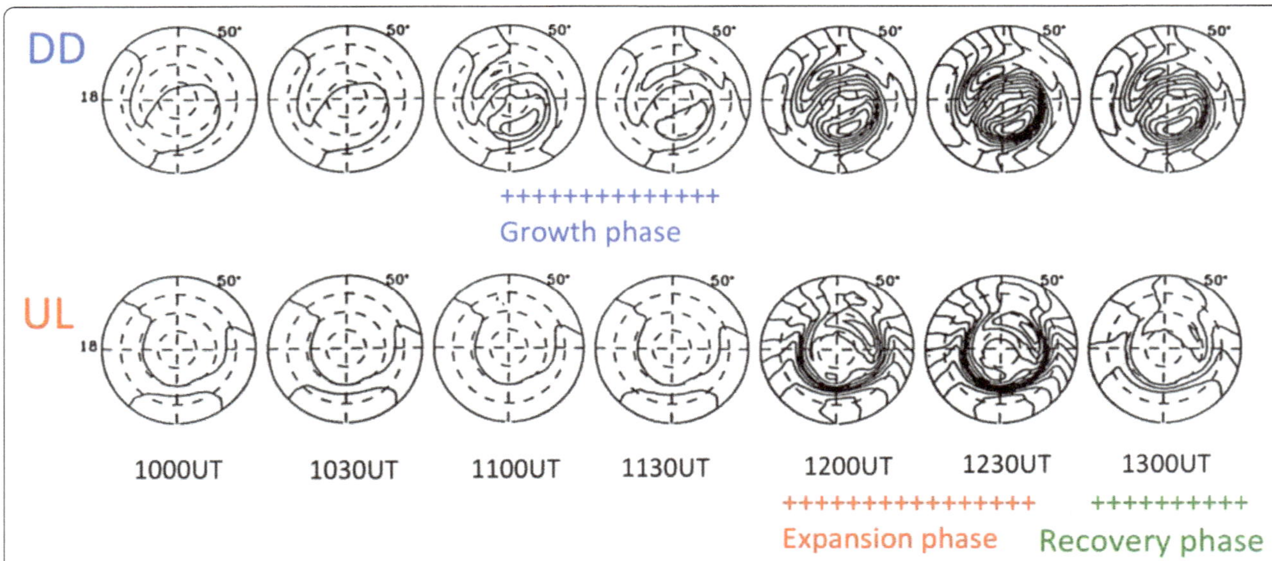

Fig. 8 The time variations of the DD and UL currents during the course of a substorm (Sun et al. 1998)

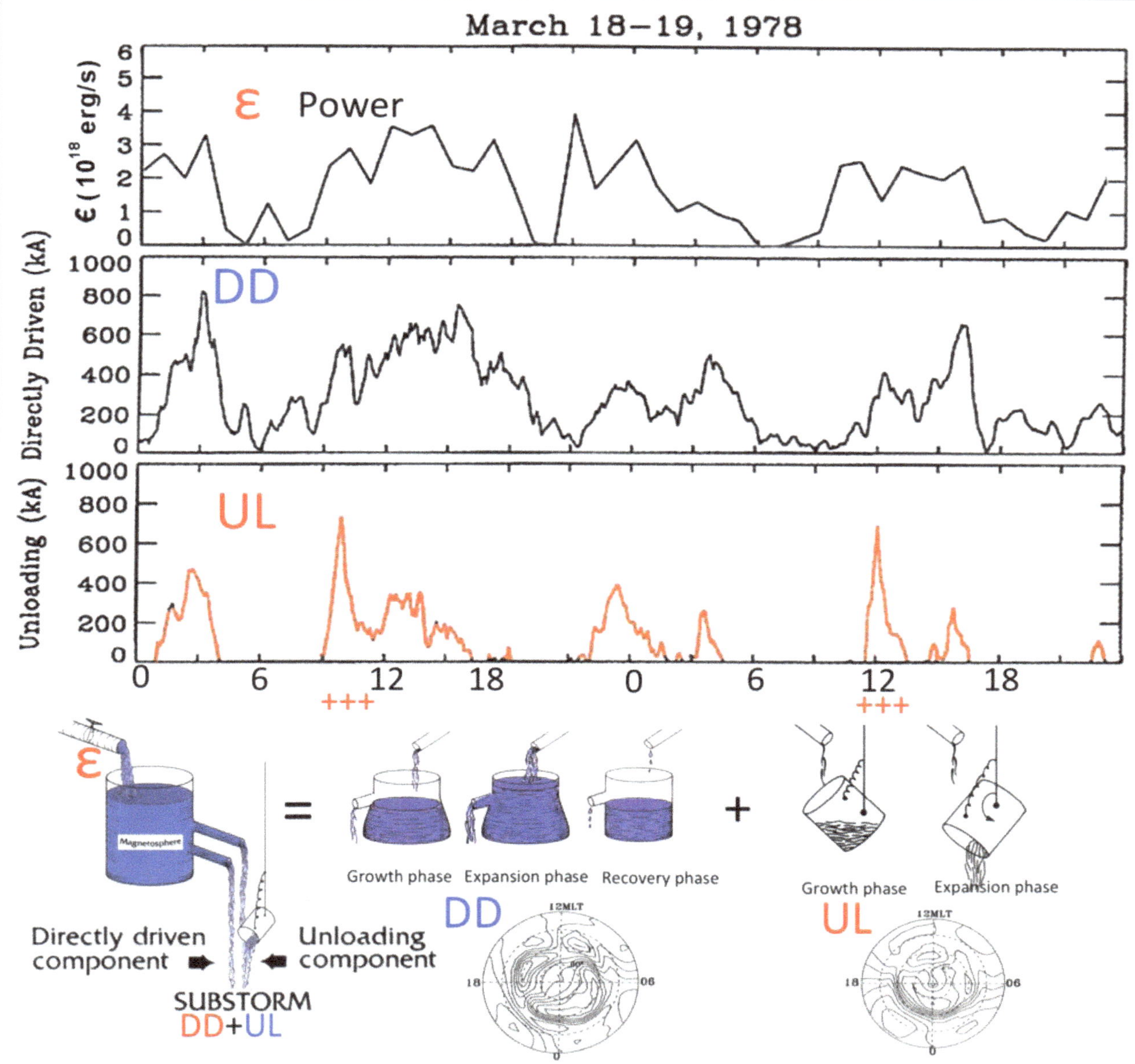

Fig. 9 The relationship among the power ε(t), the intensity of the DD(t), and UL(t) currents, together with the schematic representation of the energy flow and the DD and UL current patterns (Sun et al. 1998)

Input		Output
Power supply	Transmission	Dissipation
Dynamo	Current	Auroral substorm
ε (erg/s)	(Circuit)	δ (erg/s)

Fig. 10 Scheme in discussing the input–output relationship

It can be inferred that this delay is caused by the fact that both the ionization (caused by the field-aligned currents [streaming electrons]) in the ionosphere and the resulting conductivity are low during the growth phase (before expansion onset), resulting in a poor connection between the magnetosphere and the ionosphere and also in little dissipation (the Joule heat production) in the ionosphere. Therefore, the power ε generated by the S–M dynamo has no choice but is accumulated in the inductive circuit of the magnetosphere during the growth phase as magnetic energy W [$=(1/2)\ J^2\ L$],where

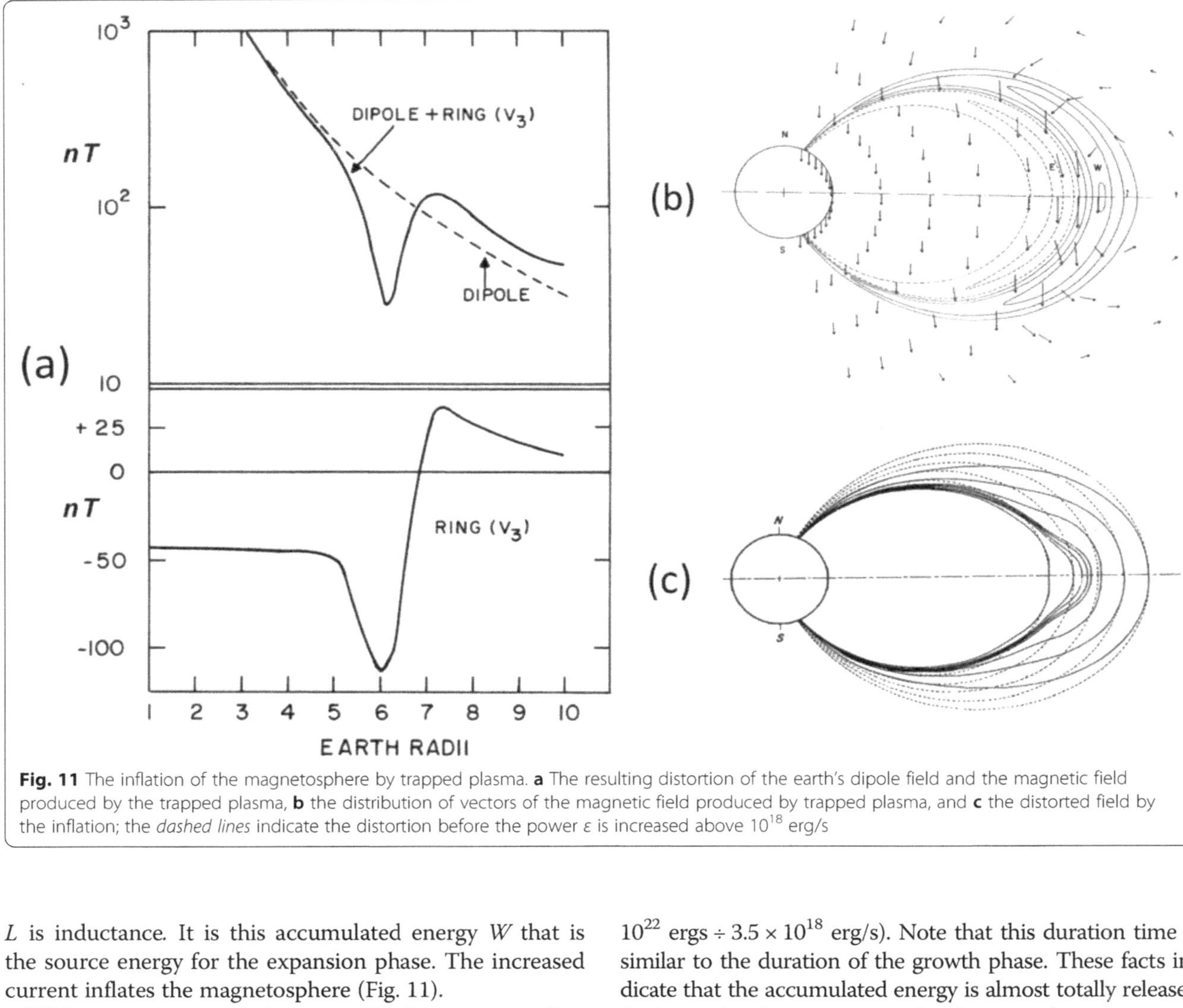

Fig. 11 The inflation of the magnetosphere by trapped plasma. **a** The resulting distortion of the earth's dipole field and the magnetic field produced by the trapped plasma, **b** the distribution of vectors of the magnetic field produced by trapped plasma, and **c** the distorted field by the inflation; the *dashed lines* indicate the distortion before the power ε is increased above 10^{18} erg/s

L is inductance. It is this accumulated energy W that is the source energy for the expansion phase. The increased current inflates the magnetosphere (Fig. 11).

With respect to magnetic energy W, the corresponding tentative value of the relevant quantities is $J = 5 \times 10^6$ A, $L = 1.6 \times 10^2$ H.

Note also that $\varepsilon = VJ = 5 \times 10^{18}$ erg/s = 100 kV × 5 × 10^6 A.

How much energy is accumulated and unloaded (dissipated) during substorms?

Since the growth phase lasts for about 1 h, the amount of the accumulated magnetic energy W is about 2.0 × 10^{22} (3.5 × 10^{18} erg/s × 1.5 h)—or at most 10^{23}—ergs; for the value of 3.5 × 10^{18} erg/s, see The expansion phase (UL) dynamo power.

Why is the lifetime of the expansion phase so short?

As mentioned earlier, the expansion phase is very brief, lasting for about 1 h, because the accumulated energy during the growth phase can be spent in about 1.5 h (2 × 10^{22} ergs ÷ 3.5 × 10^{18} erg/s). Note that this duration time is similar to the duration of the growth phase. These facts indicate that the accumulated energy is almost totally released during each substorm, except for the case when ε is greater than δ. Thus, we can now understand why the expansion phase is brief (for about 1 h), although the S–M dynamo may still operating after the expansion phase (Figs. 7 and 8).

Where is the magnetic energy accumulated during the growth phase?

An important question in relating to the expansion phase is then *where the magnetic energy W during the growth phase is accumulated and stored in the inductive circuit of the magnetosphere.*

Most magnetospheric physicists have long considered that all the magnetic energy is stored in the tail region of the magnetosphere, where the field lines are nearly antiparallel, so that it could be unloaded by the process of magnetic reconnection.

However, first of all, *even an overestimated* magnetic energy 6.5 × 10^{21} ergs in the tail region between 10 and

20 Re ($B^2/8\pi \times$ volume; $B = 30$ nT) is less than what a single substorm consumes, namely 2×10^{22} ergs; see Akasofu (2013); p. 7223 [30]), although magnetic reconnection and auroral substorms are often reported to occur within 20 Re (cf. Angelopoulos et al. 2008). A serious problem in considering magnetic reconnection is (surprisingly) that even such a simple "back-of-envelope" estimate has not been made in the past. Therefore, there is no basis to consider a priori that all the magnetic energy must be accumulated in the magnetotail and then unloaded for auroral substorms by magnetic reconnection.

Further, on the basis of the magnetic configuration of the magnetosphere, it is expected that much of the Poynting flux (directed perpendicular to $\boldsymbol{B}$) produced by the power ε on the magnetopause is directed toward a dipole-like *main body* of the magnetosphere (within 10 Re), rather than toward the tail region. This suggests that the current in the current sheet is stronger near the earthward end than the tailward end. This process of accumulation of the magnetic energy by increased current in the current sheet causes *inflation* of the main body of the magnetosphere.

Figure 11 shows a second-order computation of the deviation (inflation) of the earth's dipole field resulting from currents by a group of trapped particles centered around 6 Re (Akasofu et al. 1961); the resulting trend of inflation is expected to be similar to that caused by the plasma sheet current, regardless of the current carrier. In Fig. 11, although W is only 2.9×10^{21} ergs (about one tenth of the needed expansion phase energy 2×10^{22} ergs), the local magnetic pressure ($B^2/8\pi$) becomes close to the plasma pressure (nkT), namely $\beta \sim 1.0$, so that B decreases greatly inside the current sheet. This value of β suggests that the magnetic field configuration which is storing the accumulated energy reaches the maximum amount of energy it can hold and begins to lose control of its inflated configuration (Cheng, 2004). It can also be seen in Fig. 11 that the magnetic field produced by the stored energy of 2.9×10^{21} ergs produces almost 50 nT a ring current which is the minimum value of Dst value for a storm.

Indeed, the fact that the expansion phase occurs with the accumulated energy of 2×10^{22} ergs indicate that even the main body of the magnetosphere (within less than 10 Re) is expected to become unstable when the magnetic energy W reaches about 2×10^{22} (or at most 10^{23}) ergs.

Therefore, the reason why the expansion phase occurs is that even the main body of the magnetosphere cannot store more than 2×10^{22}—or at most 10^{23}—ergs, and tries to unload the accumulated energy, destabilizing itself. This is another reason that the magnetotail (where the field intensity is much weaker than in the main body) cannot store the magnetic energy of more than 2×10^{22} ergs; if the lobe field could accumulate 2×10^{22} ergs between 10 and 20 Re, the magnetic field intensity would have to increase 30 ~ 50 % from a quiet value; there has been no report of such a change of the lobe field.

This point is a crucial one in understanding why the expansion phase even occurs. The reason why the explosive expansion phase occurs is that when the stored (excess) energy W reaches 2×10^{22} ergs in the main body of the magnetosphere, the magnetosphere becomes unstable, and as a result, unloads it in the process of stabilizing itself. It may indeed be this process of unloading the total energy accumulated during the growth phase, which is manifested as the expansion phase. As a result, the magnetosphere is deflated (going back to pre-growth phase condition) as shown in Fig. 12.

This limitation of the amount of the accumulated energy can also be demonstrated by examining magnetic storms during which a relatively high ε is maintained almost constantly for more than 10 h, while substorms occur repeatedly and successively during such periods. Thus, in such cases, it is likely that ε greatly exceeds δ even when the conductivity is relatively high, so that the energy can be accumulated even though the ionosphere becomes conductive. Thus, when the accumulated energy reaches the limit (at most, 10^{23} ergs), the magnetosphere repeatedly unloads the energy; this situation is schematically shown in Fig. 12.

This limit suggests the upper limit of capability of the intensity of the plasma sheet current (which produces the magnetic energy expressed by $W = [(1/2)J^2L]$. Because of this limit, the peak value of the AE index is about 1000 nT, and Dst is less than 50 nT for most substorms. On the other hand, during major storms, the AE index can exceed 2000 nT, suggesting a major structural change in the inner magnetosphere (including the earthward advance of both the ring current and the plasma current sheet); in fact, the auroral oval expands greatly during major storms. This is an indication that the energy for major storms is stored at a closer distance to the earth than that for medium substorms, say less than 8 Re. Figure 13 shows the Dst index (removing the earth's induction effect of 200 nT) and a high-latitude magnetometer record during the most intense geomagnetic storms during the last century (February 11, 1958). The equatorward boundary of the auroral oval was observed to be as low as at gm. lat. 45 ° ($L = 2.8$) before the expansion phase, which was one of the most spectacular ones in recent years. It is very unlikely that field lines of $L = 2.8$ can be stretched beyond a distance of even 8 Re before substorm onset. Undoubtedly, substorms will occur well before the field lines can be stretched that far. It is also important to note that the onset can occur at $L = 2.8$.

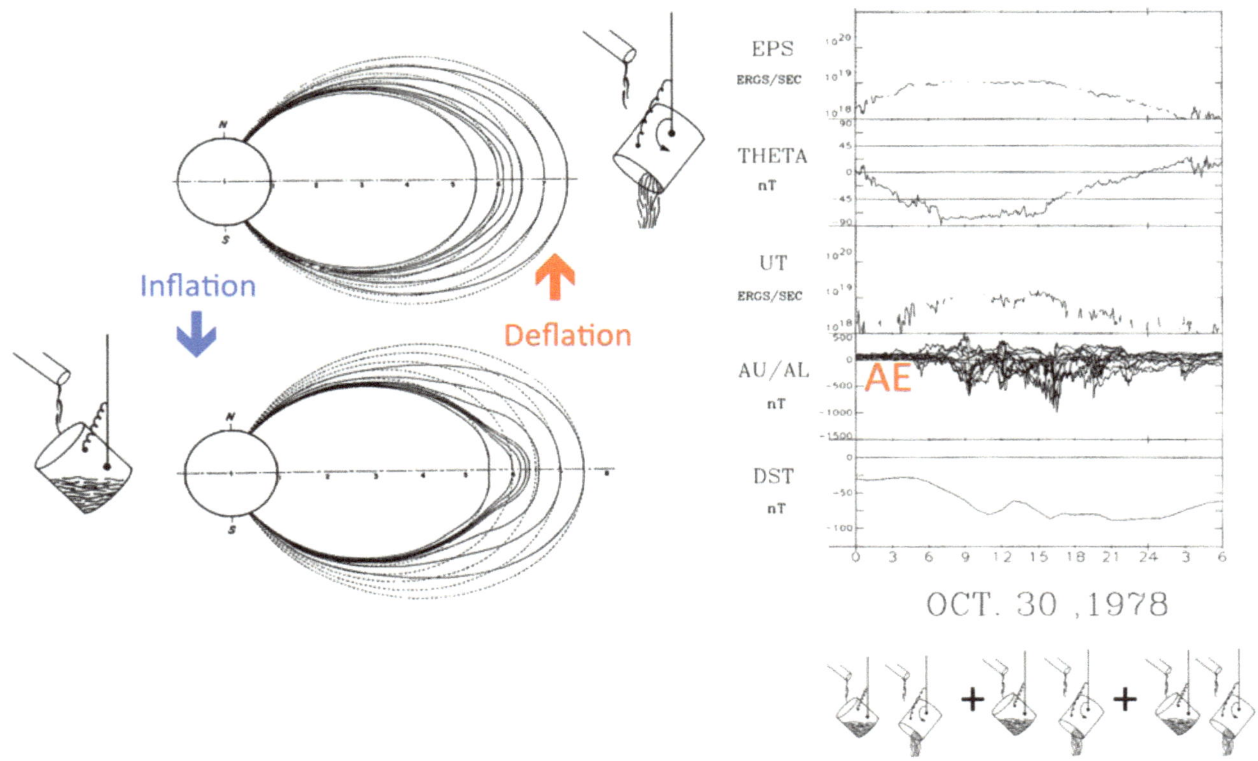

Fig. 12 (*Left*) The figure shows how the magnetosphere is inflated and deflated; the magnetic energy in this case is 2.9×10^{21} ergs (almost one tenth of the substorm energy) and $\beta \sim 1.0$ at a distance of 6 Re. The expansion phase occurs when the magnetosphere unloads the accumulated energy and becomes deflated. (*Right*) When a moderately high S–M power ε is maintained almost constantly for more than 10 h, substorms occur repeatedly and successively, suggesting that the magnetosphere can accumulate only a limited amount of magnetic energy (at most, 10^{23} ergs) and become unstable and thus unloading repeatedly the accumulated magnetic energy

The expansion phase (UL) dynamo power

Unlike the current system, it is not possible to determine the distribution of the dissipation rate for the DD and UL current. However, from Fig. 9, it can be inferred that the dissipation rate of the UL current is a little greater than that of the DD current. Thus, let us assume that the dissipation rate of the UL current is 3.5×10^{18} erg/s. It is the power of the UL dynamo power. Thus, the voltage is estimated to be 140 kV.

How can the unloaded energy produce a secondary dynamo which powers the expansion phase?

Basic requirement

Now, since the magnetic field configuration in the main body (within 10 Re) of the magnetosphere is far from antiparallel, it is unlikely that magnetic reconnection can occur there; indeed, satellite observations show that magnetic reconnection is rare within a distance of 10 Re (Ge and Russell, 2006). Thus, unless proven otherwise, it is necessary to find processes other than magnetic reconnection in converting the magnetic energy for the energy of the expansion phase. In fact, as Fig. 11c shows, the magnetic field is absent in the middle of the intensified current sheet.

It was concluded in The expansion phase (UL) dynamo power that the expansion phase requires a dynamo of the power 3.5×10^{18} erg/s in generating Bostrom's current. The conversion process must be able to generate the currents in Bostrom's 3-D current. Thus, the search for the cause of the expansion phase may begin by identifying the nature of the dynamo for the expansion phase, which causes the unloading process of the accumulated magnetic energy.

The UL current (the auroral electrojet, directed westward) is the ionospheric part of the 3-D current system, which can be driven by a southward electric field of more than 100 mV/m (100 V/km) and perhaps about 250 mV/m in the ionosphere during the peak of the expansion phase. As mentioned earlier and shown in Fig. 6, in Bostrom's circuit, the only place where $\boldsymbol{E} \cdot \boldsymbol{J} < 0$ is located in the equatorial part of the circuit (actually, a thin magnetic shell near the earthward end of the current sheet, The UL current system for the expansion phase) in the meridional component [Akasofu, 2003a], not in the ionosphere where $\boldsymbol{E} \cdot \boldsymbol{J} > 0$, indicating that the expansion phase dynamo must be located in the magnetosphere.

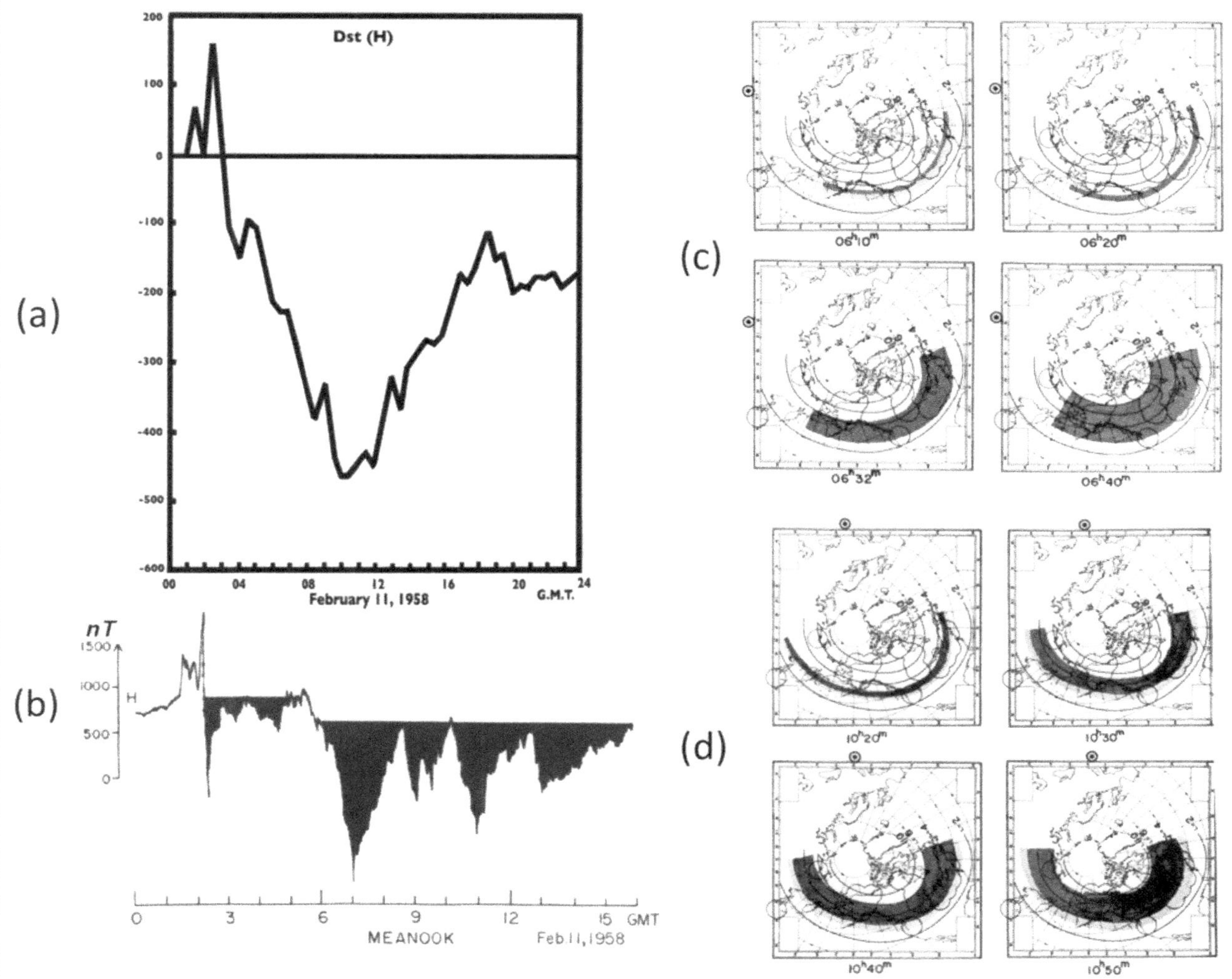

Fig. 13 **a** The Dst index of February 11, 1958 storm; **b** the magnetic record from Meanook, Canada; and **c**, **d** changes of the auroral distribution during two substorms during the storm. Note that the auroral oval expanded as far as gm. lat. 45 ° before the substorm onsets

Synthesis

Synthesizing what we have learned in the above, we attempt to consider a possible process for the expansion phase. It is emphasized here this section describes only an exercise. First of all, in order to produce electric currents for a discharge, it is necessary to separate electrons from protons (either chemically [a battery] or by force [a dynamo]). Further, in generating the UL current, the electric field thus produced must be directed earthward, as mentioned in The UL current system for the expansion phase. One possibility is as follows (Figs. 14 and 15).

When the power ε is increased, the current in the plasma sheet is increased during the growth phase, inflating the magnetosphere (accumulating magnetic energy. When the accumulated energy $W = [(1/2)J^2L]$ reaches 2×10^{22}—or at most 10^{23}—ergs, the magnetosphere becomes unstable and unload the energy in order to stabilize itself). *It is this very process which causes the spectacular expansion phase.*

The magnetic energy, which inflates the inner magnetosphere, is caused by the *sheet current in the plasma sheet.* Thus, the fact that the magnetosphere becomes unstable (because of the accumulated energy) means that the current in the current sheet becomes unstable. The limit of stability (breakpoint) of the current in the plasma sheet may be equivalent to that of the strength of the spring attached to the tippy bucket (Fig. 4). In fact, such an instability was observed at the time of expansion onset (Fig. 16).

Thus, in the process of stabilizing itself, the magnetosphere must *reduce the accumulated magnetic energy* $W = [(1/2)J^2L]$ *by reducing the current* J in the plasma sheet. As a result, the magnetosphere deflates itself. The reduced magnetic energy W must be consumed for

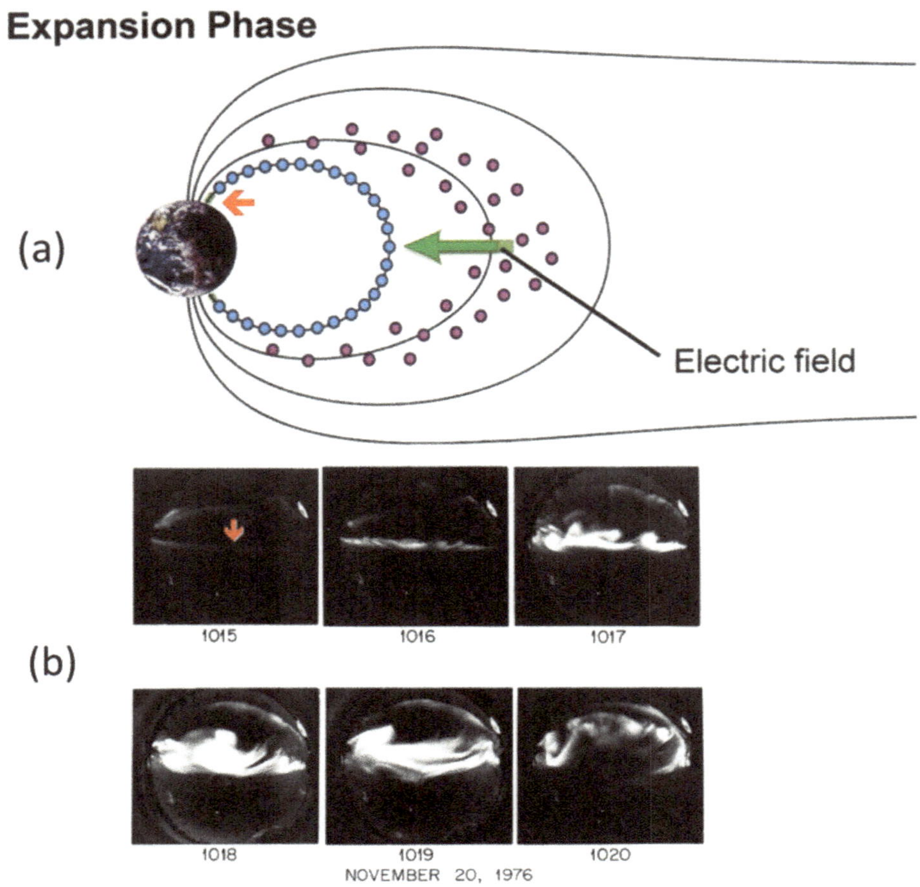

Fig. 14 **a** Possible charge separation which is caused by the deflating magnetosphere. The resulting electric field is directed toward the earth (the electron layer may correspond to the inverted V in the ionosphere); **b** an example of substorm onset

the expansion phase. In other words, the accumulated magnetic energy must be consumed to generate the expansion phase dynamo. Thus, the deflation process must generate the expansion phase dynamo, namely the separation of electrons from protons (Fig. 14).

The deflation of the magnetosphere increases B inside the current sheet; there is little magnetic energy inside the current sheet (see Fig. 11c). Thus, it is expected that $\partial Bz/\partial t > 0$ there during the expansion phase, when W is reduced. The resulting electric field can be very roughly estimated to be about $([-\partial B/\partial t)\int \partial y] = 150$ mV/m (=50 nT/20 min × 20 Re = 1.26×10^{10} emu). This value of the electric field is larger than what is observed by a satellite (Fig. 16). The deflation will end in about 1 h by spending the accumulated magnetic energy, explaining the short life of the expansion phase. Bostrom (1974) estimated the time constant of the UL circuit to be 8 min (=50H/0.1Ω), so that the dissipation rate δ can reach the peak value in a rather short time. This may also explain a sharp increase of the UL current at the beginning and the explosive nature of the expansion phase.

As mentioned earlier, the reduction of the current may be caused by developing plasma instabilities. If plasma instabilities can occur in reducing the current in the current sheet, the frozen-in magnetic field lines condition is expected to break down at the critical moment of expansion onset. Microscopically, the deflating magnetosphere brings electrons (gyrating tightly around the magnetic field lines) toward the earth, but not protons and thus separating both electrons and protons in the plasma sheet, and thus also generating an earthward electric field in a thin shell near the earthward end of the current (plasma) sheet. This process does not require the whole plasma motion ($\boldsymbol{v} \times \boldsymbol{B}$). As shown earlier, a simple estimate of the electric field is about 150 mV/m.

This process of charge separation was suggested by Lui and Kamide (2003), as shown schematically in Fig. 14. The separated electrons in a thin magnetic shell

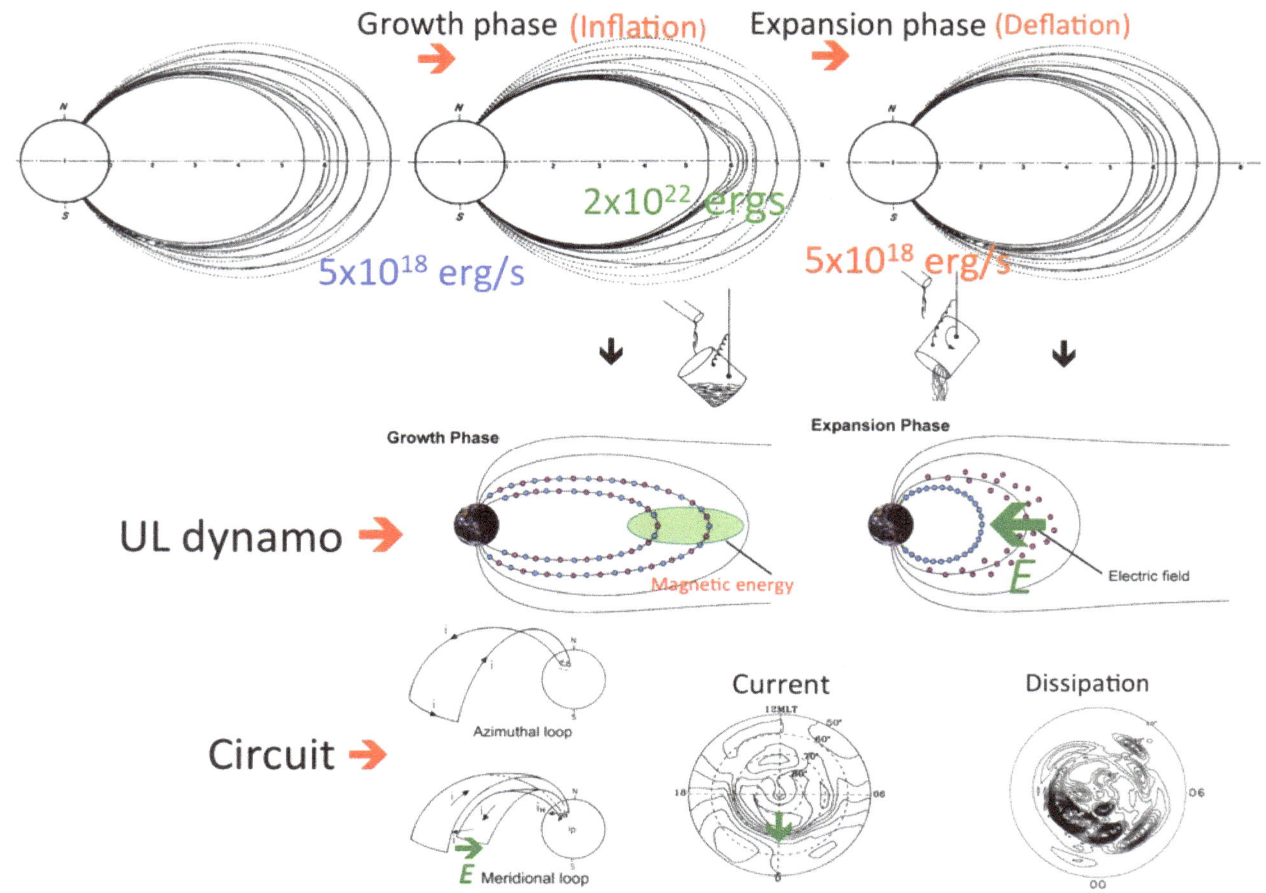

Fig. 15 A schematic illustration to show how the deflating magnetosphere might cause a charge separation which can result in an earthward electric field (directed southward in the ionosphere). Because of this separation, the frozen-in field line condition may break down. This electric field can generate Bostrom's 3-D current, including the UL current in the ionosphere

thus produced will be discharged along the magnetic field lines toward the ionosphere, producing the field-aligned current sheet and brightening an auroral arc as an initial indication of expansion onset. The reason why the separation occurs near the earthward end of the plasma sheet is related to the fact that a typical onset arc lies near the equatorward boundary of the auroral oval.

Further, when the earthward electric field is communicated to the ionosphere by the streaming electrons, it becomes a southward electric field that can drive the UL current (the auroral electrojet). This process continues until the deflation is complete. Figure 15 schematically shows an overall possible chain of processes which could lead to substorm onset. Thus, this suggested chain of processes requires breakdown of the condition of the frozen-in field line during the expansion phase. The generation of the earthward electric field requires new theoretical and observational studies without using the concept of frozen-in field lines condition.

In summary, the proposed chain of processes described here is

(1) The magnetosphere accumulates magnetic energy in the main body (its inductive circuit) during the growth phase (1 h) because the ionoshere cannot dissipate the power. Thus, the magnetosphere is inflated.
(2) When the accumulated energy reaches 2×10^{22} ergs or at most 10^{23} ergs, the current in the current sheet develops plasma instabilities and becomes unstable.
(3) Thus, the current is reduced, and the magnetosphere is deflated and the accumulated magnetic energy is unloaded.
(4) During the deflation (unloading) process (1 h), a charge separation occurs, separating electrons from protons.
(5) The separation of the charges causes an earthward electric field near the earthward end of the current

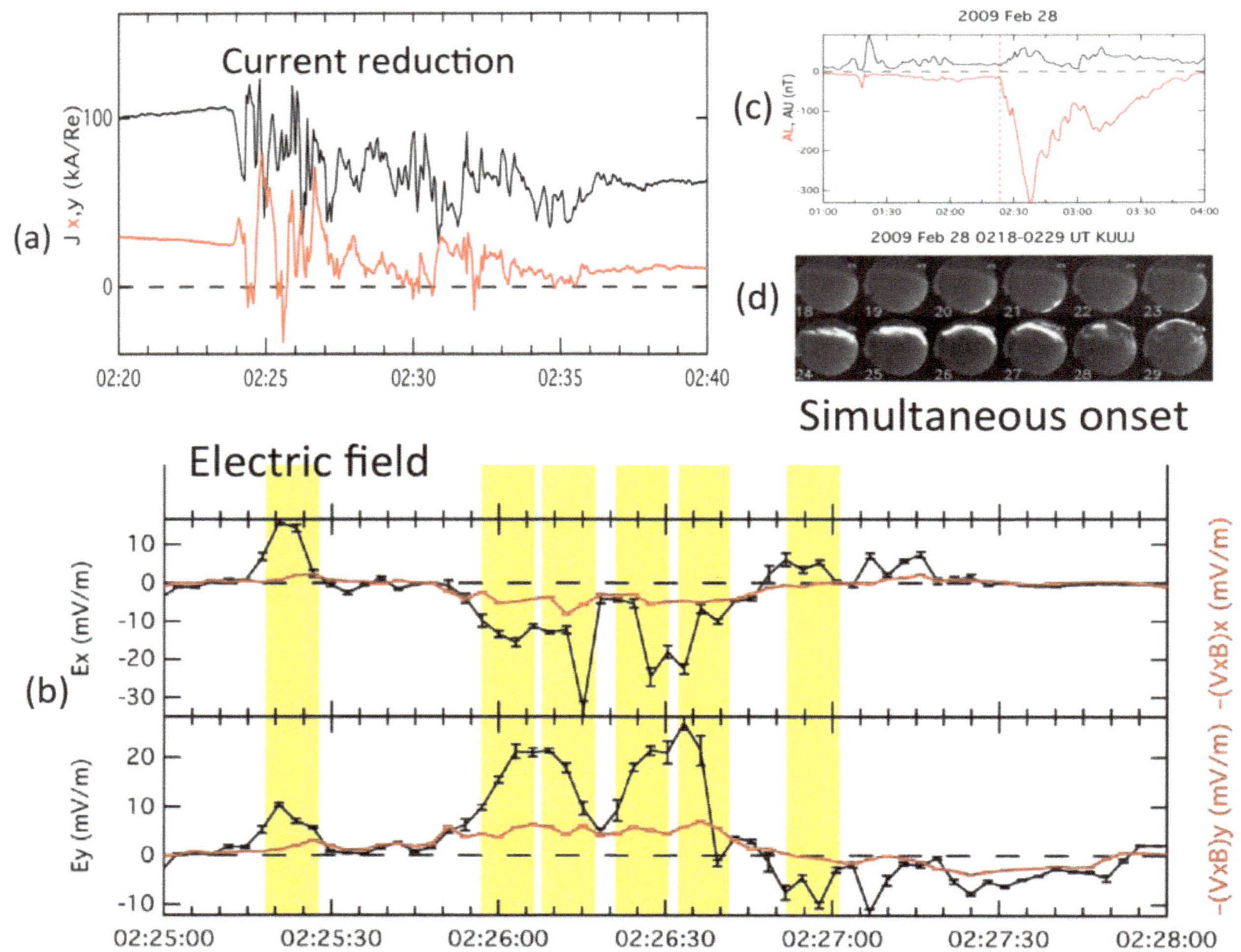

Fig. 16 A satellite (THEMIS) observation of changes of various relevant quantities at the time of the occurrence of the expansion phase onset (Lui, 2011)

sheet. Thus, the unloaded magnetic energy is consumed in producing the earthward electric field (and thus the expansion phase dynamo).

(6) The electric current thus produced causes a great variety of auroral activities during the expansion phase as an electrical discharge process.

Supporting satellite observation

There is an important set of satellite observations which are consistent with the above chain of processes *at 8.1 Re* (Lui, 2011); Fig. 16. One of the THEMIS satellites observed: (i) plasma instabilities, (ii) a electric current reduction, (iii) break down of the frozen-in field line condition, (iv) an earthward electric field of 20 ~ 30 mV/m, (v) simultaneously with onset of an auroral substorms. The intensity of this observed electric field is less than that drives the UL current in the ionosphere. These observational facts are consistent with the synthesis described in the above. However, other cause–effect relationships even for the same set (or other set) of the facts should be explored.

Triggering?

When the magnetosphere accumulates about or more than it can hold, the magnetosphere is likely to be ready to unload the accumulated energy. The reduction of the current in the current sheet could be induced by a stimulation, namely by triggering. Busty bulk flows (BBFs) could trigger some substorms, although they do not have enough energy causing substorms. This subject will be discussed further in Auroral displays.

On the other hand, it is known that a significant number of substorms occur at about the time of 'northward turning' of the IMF some time *after* a southward turning [*after* ε becomes above 10^{18} erg/s] (Lyons et al. 2001). In these cases, it is expected that a northward turning (*external cause*) can trigger the expansion onset by the reduction of

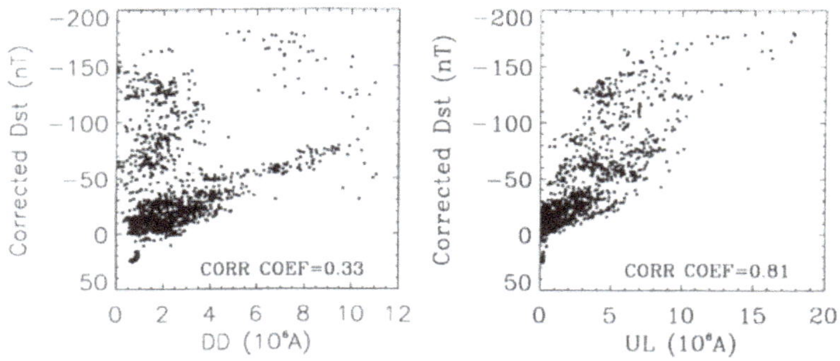

Fig. 17 The relationship between the Dst index with DD and UL

ε because ε and the current in the current sheet are reduced at the time of 'northward turning.' It is known that interplanetary shock waves can also trigger substorms, if they are accompanied by a southward IMF.

Other studies

Since the cause of the expansion phase is such an important subject, a large number of observational and theoretical efforts have been made. A few recent examples are satellite observations, such as GEOTAIL, THEMIS, IMAGE, and others (cf. Machida et al. 1994; Ohtani et al. 2002; Perraut et al. 2003; Shiokawa et al. 2005; Coumans et al. 2007; Angelopoulos et al., 2008), theoretical studies (cf. Cheng, 2004; Haerendel 2007, Harendel 2008, Haerendel 2009; Israelevich et al. 2008), and numerical simulation studies (cf. Birn et al., 2012); there have been many simultaneous observations between satellite and ground-based observations (cf. Nishimura et al. 2010) in addition to

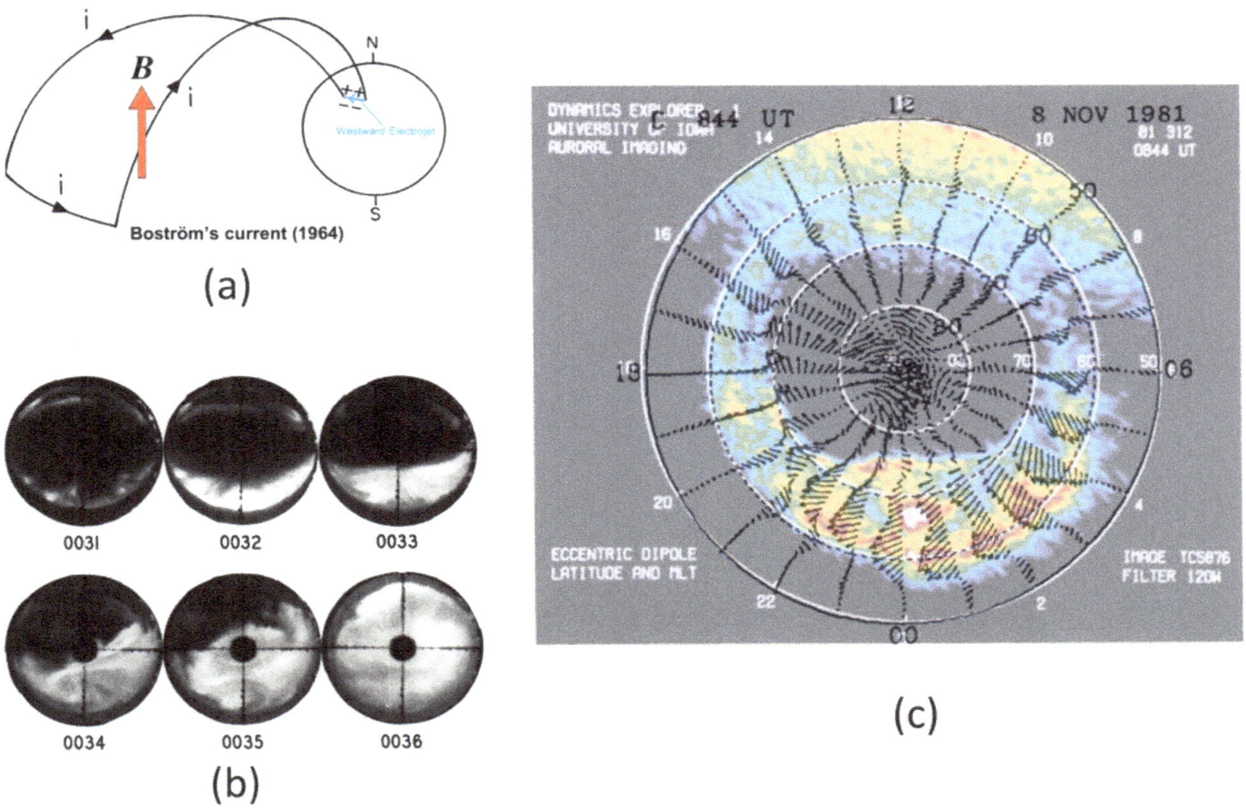

Fig. 18 **a** A part of the UL circuit which produces a positive Bz field (*red arrow*) which is mainly responsible for the poleward expansion of auroras and the electrojet, **b** a series of all-sky photographs (an example of the poleward expansion), **c** an expanded aurora system, and also the extended auroral electrojet (Courtesy of Y. Kamide)

ground-based optical observations (cf. Henderson, 2009) and superDARN radar observations (cf. Lyons et al. 2010). However, it is regrettably not the intent of this paper to review these works and many others since it is beyond the capability of the present author.

On magnetic reconnection

Magnetic reconnection has been mentioned earlier a few times. It may be worthwhile to summarize them here. This theory is based on the premise that the magnetotail (where the field is nearly antiparallel) has sufficient magnetic energy to be converted into the energy of substorms by magnetic reconnection. However, it was shown in (Where is the magnetic energy accumulated during the growth phase?) that the magnetotail does not have sufficient energy.

Further, magnetic reconnection is supposed to generate a tail-wide plasma flow so that it is supposed to produce the DD current (not the UL current), but the DD current is not mainly responsible for the expansion phase. The expansion phase requires an electrical discharge along Bostrom's circuit. Further, even if the expected flows by magnetic reconnection are narrow and short-lived, such as those observed BBFs, they do not have enough energy (10^{21} ergs or less). At most, they might *only trigger* some substorms (but not all) by inducing the abovementioned instability (Triggering).

Above all, however, the theory, in particular simulation studies, has to rely on fictitious resistivity called "effective resistivity" because the theory is based on the "frozen-in field lines" concept, in spite of the fact that Alfven warned in as early as 1967 that it should not be applied to magnetospheric physics.

So far, the "magnetic field lines" approach has not provided quantities which can be compared and discussed with those given in this paper. In fact, for example, there is no way to discuss quantitatively our observed result of the impulsiveness of the expansion phase with simulation results so long as they rely on effective resistivity which is supposed to determine how fast magnetic reconnection is supposed to occur and has so far not demonstrated as a physical quantity.

Some future problems

Storm–substorm relationship

The storm–substorm relation issue is one of the controversial ones in the past. A substorm is a mini-storm (Akasofu et al. 2007) so that it is clear that a storm period is the period when intense substorms occur frequently. Thus, the issue should be the difference of the degree of the contribution between a DD type plasma convection and the expansion phase. This can be examined by comparing the contribution of DD and UL on the Dst, indicating that the expansion phase contribution is greater than the convection; Fig. 17. However, this is not a settled issue.

Auroral displays

(a) The causes of poleward advance of auroras during the expansion phase

As the name of the expansion phase indicates, the active auroral arc system in the auroral oval advances rapidly poleward with a speed of 200 m/s (Fig. 18b). The shift of the poleward boundary of the oval during the expansion phase is about 500~700 km, from the onset location gm. lat. 63 ° to as far as 70 °. In the past, this feature was interpreted simply as a result of the so-called dipolarization. However, the

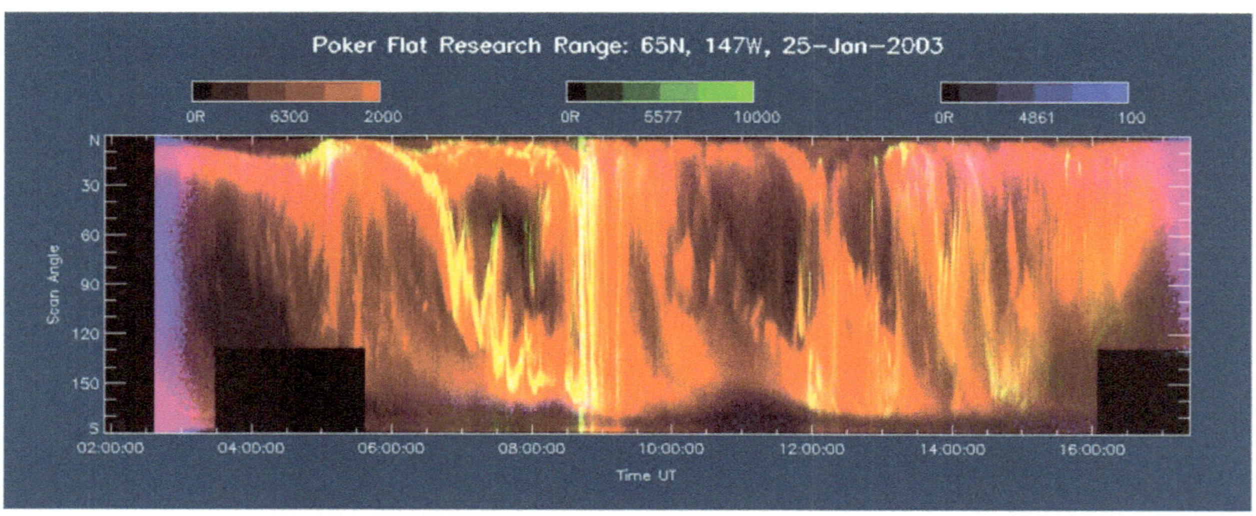

Fig. 19 An example of meridian scanning photometer (MSP) record, showing equatorward shift of the auroral oval and substorm onsets (the poleward expansion)

situation is obviously not just returning to the condition before the substorm onset (about 63 °), so that the amount of reconnected flux may not be enough to explain the poleward expansion.

Such a long-range poleward shift requires a magnetic flux of more than 10^{11} nT km^2. The observed shift will be possible when the magnetic flux of the azimuthal component of the 3-D current (~50nT × 2.1 × 10^9 km^2) is added (Fig. 18a); the magnitude of the vertical component of the azimuthal field (so-called positive bays, observe in middle and low latitudes during the expansion phase) is about 50 nT. In fact, it is known that the auroral electrojet shifts rapidly poleward with the front of the expanding auroral system during the expansion phase (Fig. 18c) so that the earthward end of Bostrom's circuit must advance poleward during the expansion phase. However, this phenomenon has not been studied.

(b) Auroral features during the growth phase

Auroral features during the growth phase have not been discussed until recently, except that the auroral oval tends to expand before substorm onset. Akasofu et al. (2011) showed that this expansion does not seem to have a clear relationship with the so-called southward turning of the IMF. In fact, it is known that just before the arrival of westward traveling surges, auroral arcs tend to shift equatorward (Akasofu, 1968, Figure 28, p. 43). Further, during the growth phase, activities of arcs in the equator half of the oval and arcs in the poleward half of the oval seem to be different. The auroral arc which brightens first during a typical substorm onset is located near the equatorward

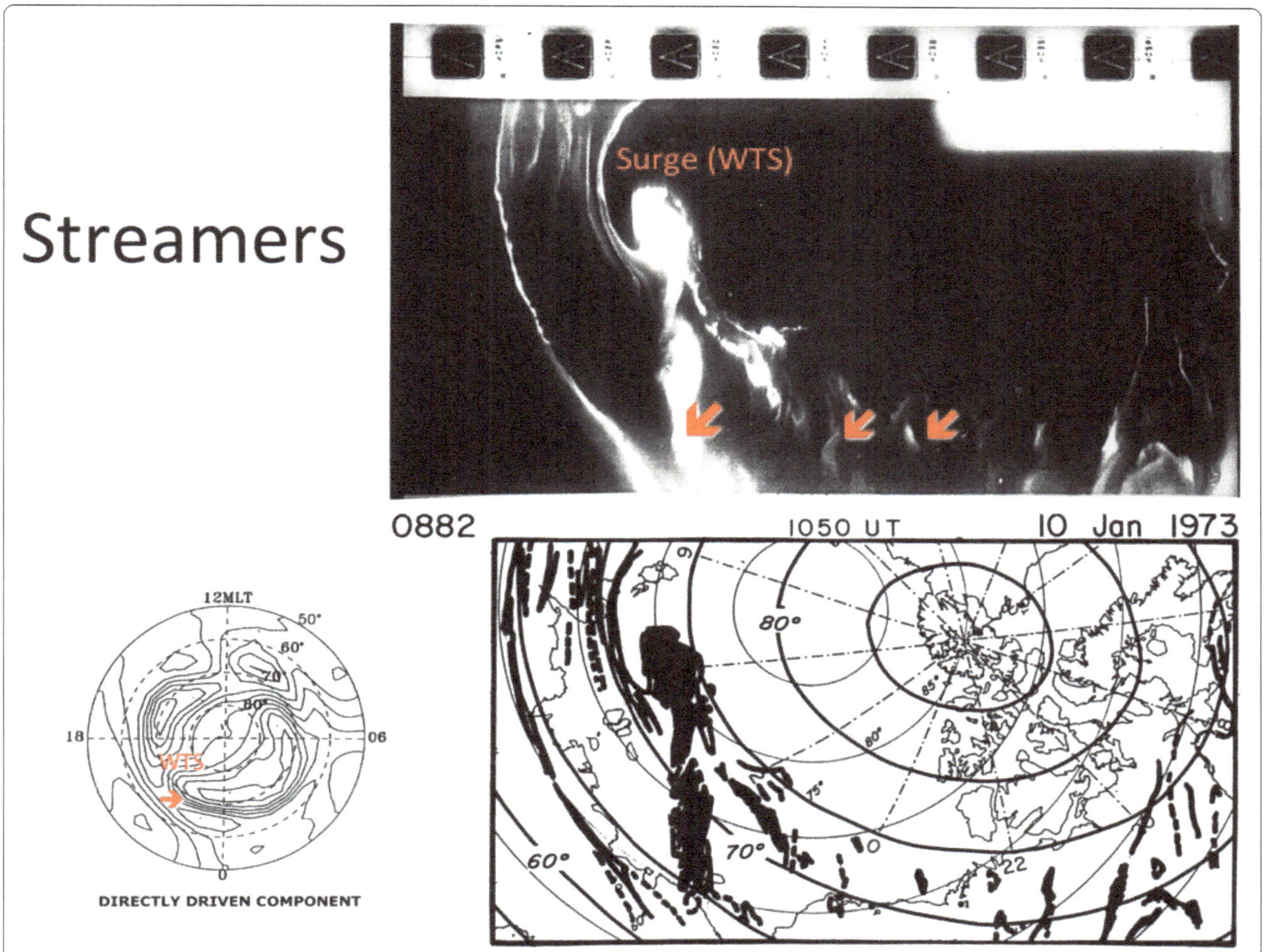

Fig. 20 An example of DMSP images of a westward traveling surge and associated streamers. Note that there is no streamer in the polar cap, indicating that streamers do not occur in the main part of the polar cap. The average DD current pattern is also shown to indicate the western end of the electrojet which is the head of westward traveling surges

boundary of the auroral oval. However, arcs in the poleward half do not seem to show any systematic change during the growth phase (Fig. 19). In fact, this is one of the reasons why auroral substorms are not likely to be caused by magnetic reconnection which is supposed to occur at a distance of about 20 Re.

(c) Streamers

Recently, Nishimura et al. (2010) and Lyons et al. (2010) showed that when 'streamers' reach arcs near the equatorward half of the oval, the expansion phase begins. Streamers originate often from westward traveling surges (Fig. 20), indicating that a substorm is already in progress. If they trigger a substorm, it may be a secondary substorm. On the other hand, if 'streamers' could be confirmed to be caused by BBFs, there is an interesting possibility that some BBFs may trigger substorms (Triggering). If this would be indeed the case, the main energy for a substorm must be located with 10 Re for the unloading, namely a destabilization effect.

Concluding remarks

In this paper, it is shown that the "current lines" approach suggested by Alfven et al. (1967) has given us some quantitative understanding of the input–output relation and the associated flow of energy in causing auroral substorms.

It is emphasized here that the attempted synthesis is only a first attempt based on what we have learned so far. It is hoped that the requirements needed to explain the expansion phase are at least useful for future studies, regardless of theories to explain the expansion phase.

Obviously, however, all the quantities presented here are very tentative ones, and they may not be totally consistent among them. Thus, the purpose of this paper is in part to emphasize the need for a quantitative input–output approach and the "current lines" approach, instead of "magnetic field lines" approach, in studying auroral substorms.

Competing interests

The author declares that he has no competing interests.

Acknowledgements

I would like to thank all the magnetospheric physicists in the past and the present for their efforts in making a great progress in substorm research during the last 50 years. I owe especially Dr. Y. Kamide, Dr. Sun Wei, and Dr. B.-H. Ahn, A.T.Y. Lui, and C.-I. Meng for their long collaboration in pursuing the electric current/circuit approach.

I would like to dedicate this paper to Sydney Chapman and Hannes Alfven for their guidance during my early research life and also to Lou Frank for his great success in excellent imaging of auroral substorms.

Personal note

It may be worthwhile to mention why the four statements made by Alfven were mentioned in the introduction. It is because those statements have become the basis of this paper.

I met Hannes Alfven for the first time when Sydney Chapman asked me to join him to attend the Birkeland Symposium held in Norway in 1967. I heard about the great controversy between Chapman and Alfven (cf. Akasofu 2003b) by others before, so that I was a little apprehensive to meet Hannes. However, unexpectedly, he mentioned nothing about the debate. He was anxious to tell me that the concept of "frozen-in magnetic field lines" and "moving field lines" concepts should not be considered in studying auroral substorms and solar flares (except for solar interior); I told him that I have just begun to learn MHD from his book "Cosmical Electrodynamics". He emphasized that I should take "current lines" approach, instead of "magnetic field lines" approach. His paper on this subject was published in the conference proceedings (which I quoted in this paper). I invited him to visit Alaska in 1974; at that time, he told me that his statements were realized in a study of the acceleration of auroral electrons; MHD does not allow for an electric field along magnetic field lines. Similarly, the breakdown of the "frozen-in magnetic field lines" condition (as observed during the expansion phase) may be essential in understanding the onset of the expansion phase. When I met him a few times in Stockholm, he was complaining about my slow progress. However, in order to take the current lines approach, we had to set up six meridian chains of magnetometers to begin with. The analysis of the records required developing a few computer codes with my colleagues. It was as late as the end of 1990s, when we succeeded in separating DD and UL. Then, it took many more years to digest the results. Unfortunately, the "current lines" approach has not been popular in our particular field, so much that many readers of this paper must encounter many unfamiliar terms, such as dynamo, power, circuits, inductance, and even the term electric currents for some. However, the "current lines" approach is one of the ways of studying substorms. Actually, the basis of this paper is not more than what is in physics 101, but the "current lines" approach has provided, I believe, some of the basic aspects of physics of auroral substorms, which has so far not been demonstrated by "the magnetic field lines" approach.

References

Ahn B-H, Akasofu S-I, Kamide Y (1983) The Joule heat production rate and the particle energy injection rate as a function of the geomagnetic indices AE and AL. J Geophys Res 88:62756. doi:10.1029/JA088iA08p0675

Ahn B-H et al (1989) The auroral energy deposition over the polar ionosphere during substorms. Planet Space Sci 37:239. doi:1016/0032-0633(89)90022-6

Akasofu, S.-I., J. C. Cain and S. Champman (1961) The magnetic field of a model radiation belt, numercally computed. J Geophys Res 66:4013–4026.

Akasofu S-I (1964) The development of the auroral substorm. Planet Space Sci 12:273

Akasofu S-I, Chapman S, Mend C-I (1965) The polar electrojet. J Atmos Terr Phys 27:1301

Akasofu, S.-I. (1968) *Polar and magnetospheric substorms*, D. Reidel Pub. Co. Dordrecht-Holland, Holland.

Akasofu S-I (1981) Energy coupling between the solar wind and the magnetosphere. Space Sci Rev 28:121

Akasofu S-I (1992) A confirmation of the validity of the electric current distribution determined by a ground-bases magnetometer network. Geophys J Inst 109:191–196

Akasofu, S.-I., *Exploring the Secrets of the Aurora*, Springer, 2007.

Akasofu S-I (2003a) A source of auroral electrons and the magnetospheric current systems. J Geophys Res 108:2006. doi:10.1029/2002 JA009547 A4

Akasofu S-I (2003b) Chapman and Alfven: A rigorous mathematical physicist versus an inspirational experimental physicist, EOS, 84, 269–274.

Akasofu S-I, Meng C-I, Lui TY (2011) Importance of auroral features in the search for substorm onset processes. J Geophys Res 116:A02209. doi:10.1029/2010JA016079

Akasofu S-I (2013) Where is the magnetic energy for the expansion phase of auroral substorms accumulated? J Geophys Res 118:1–7. doi:10.1002/2013JA019042

Alfven, H (1967) *The Birkeland Symposium on Aurora and Magnetic Storms*, ed. by A. Egeland and J. Holtet, Centre National de la Recherche Scientific, 439–444
Alfven H (1977) Electric currents in cosmic plasmas. Rev Gephys And Space Phys 15:271–284
Alfven, H (1986) Double layers and circuits in astrophysics, IEEE Transaction on Plasma Science, Vol.PS-14, No. 6, 779–796
Angelopoulos V et al (2008) Tail reconnection triggering substorm onset. Science 321:931–933
Axford WI, Hines CO (1961) A unified theory of high-latitude geophysical phenomena and geomagnetic storms. Can J Phys 39:1433
Birn, J., K. Schinder and M. Hesse (2012) Magnetotail aurora connection: the role of thin current sheet, 337–346, in Auroral Phenomenology and magnetospheric processes: Earth and other planets, ed. by A. Keiling, E. Donovon, F. Bagenal and T. Karlsson, AGU Monograph 197. Washington, D.C.
Bostrom R (1964) A model of the auroral electrojet. J Geophys Res 69:4983–4999
Bostrom, R. (1974), Ionosphere-magnetosphere coupling, 45, in *Magnetospheric physics*, ed. by B. M. McCormac, Dordrecht, D. Reidel-Holland Publishing Co., Holland
Bristow WA, Jensen P (2007) A superposed epoch study of SuperDARN convection observations during substorms. J Geophys Res 112:A06232. doi:10.1029/2006JA012049
Cheng CZ (2004) Physics of substorm growth phase, onset and depolarization. Space Sci Rev 113:207
Coumans V, Blockx C, Gerard J-C, Hubert B, Connors M (2007) Global morphology of substorm growth phase observed by the IMAGER-SI12 imager. J Geophys Res 112:A11211. doi:10.1029/2007JA012329
Dungey JW (1961) Interplanetary magnetic field and the auroral zones. Phys Res Lett 6:7. doi:doi.1103/PhysRevLett. 6, 47
Frank LA, Craven JD, Burch JL, Williams DJ (1982) Polar view of the earth's aurora with Dynamic Explorer. Geophys Res Lett 9:1001
Ge YS, Russell CT (2006) Polar survey of magnetic field in near tail: reconnection rare inside R_E. Geophys Res Lett 33:L02101. doi:1029/2005GL024574
Haerendel G (2007) Auroral arcs as sites of magnetic stress release. J Geophys 112:A09214. doi:10.1029/2007JA012378
Haerendel G (2008) Auroral arcs as current transformers. J Geophys Res 113:A07205. doi:10.1029/2007JA012947
Haerendel G (2009) Poleward arcs of the auroral oval; during substorms and the inner edge of the plasma sheet. J Gephys Res 114:A06214. doi:10.1029/2009JA014138
Henderson, M. G. (2009) Observational evidence for an inside-out substorm onset scenario, Ann. Geophys 27:2129–2140
Israelevich PL, Ershkovich AI, Oran R (2008) Current carriers in the bifurcated tail current sheet: ions or electrons. J Geopys Res 113:A04215. doi:10.1029/2007JA012541
Kamide Y et al (1982) Global distribution of ionospheric and field-aligned currets during substorms as determined from six IMS meridian chains of magnetometers: initial results. J Geophys Res 87(A10):8228–8240. doi:10.1029/JA087iA10po8228
Lui ATY, Kamide Y (2003) A fresh perspective of the substorm current system and its dynamo. Geophys Res Lett 30:1958. doi:10.1029/2003GL017835
Lui ATY (2011) Reduction of the cross-tail current during near-Earth depolarization with multisatellite observations. J Geophys Res 116:A12239. doi:10.1029/2010JA016078, doi:10. 1029/2011JA017107 (2011b),116, A03211
Lyons LR et al (2001) Timing of substorm signatures during the November 24, 1996, Geospace Environment Modeling event. J Geophys Res 106:349–359
Lyons LA, Nishimura Y, Shi Y, Zou S, Kim H-J, Angelopolous V, Heinselman C, Nicolls MJ, Fornacon K-H (2010) Substorm triggering by a new plasma intrusion: incoherent-scatter radar observations. J Geophys Res 115:A07223. doi:10.1029/2009JA015168
Machida S, Mukai T, Saito Y, Hirahara M, Obara T, Nishida A, Terasawa T, Maezawa K (1994) Plasma distribution functions in the earth's magnetotail ($X_{GSM} \sim -42$ Re) at the time of a magnetospheric substorm: GEOTAIL/LEP observation. Geophys Res Lett 21(11):1027–1030
Nishimura Y, Lyons L, Zou S, Angelopoulos V, Mende S (2010) Substorm triggering by new plasma intrusion: THEMIS all-sky imager observations. J Geophys Res 115:A07222. doi:10.1029/2009JA015166
Ohtani S-I, Takahashi K, Lui ATY (2002) Electron dynamics in the current disruption region. J Geophys Res 107(A10):1322. doi:10.1029/2001JA009236
Perraut S, OI C, Roux A, Park G, Chua D, Hoshino M, Mukai T, Nagai T (2003) Substorm expansion phase: observations from Geotail, polar and IMAGE network. J Geophys Res 108(A4):1159. doi:10.1029/2002JA009376
Shiokawa K, Yago K, Yumoto K, Baishev DG, Solovyev SI, Rich EJ, Mende SB (2005) Ground and satellite observations of substorm onset arc. J Geophys Res 110:A12225. doi:10.1029/2005JA011281
Sun W, Xu S-Y, Akasofu S-I (1998) Mathematical separation of directly-driven and unloading components in ionospheric equivalent currents during substorms. J Gephys Res 103(A6):11695–11700. doi:10.1029/97JA03458

6

$\delta^{18}O_{sw}$ estimate for *Globigerinoides ruber* from core-top sediments in the East China Sea

Keiji Horikawa[1*], Tomohiro Kodaira[2], Jing Zhang[1] and Masafumi Murayama[3]

Abstract

The paired analyses of the Mg/Ca ratio and oxygen isotopic composition ($\delta^{18}O_c$) of surface-dwelling planktonic foraminifera have become a widely used method for reconstructing the oxygen isotopic composition of ambient seawater ($\delta^{18}O_{sw}$) as a robust proxy for surface salinity. *Globigerinoides ruber* (*G. ruber*) is a mixed-layer dweller, and its fossil shell is an ideal archive for recording past sea surface water conditions, such as those caused by variability in the East Asian summer monsoon (EASM). Here, we investigate the validity of shell-derived $\delta^{18}O_{sw}$ estimates for *G. ruber* using core-top sediments from the East China Sea (ECS). First, we determined a local $\delta^{18}O_{sw}$–salinity equation for the eastern part of the ECS in July [$\delta^{18}O_{sw} = -7.74 + 0.23 \times salinity$]. Then, we calculated $\delta^{18}O_{sw}$ from core-top $\delta^{18}O_c$ and Mg/Ca values in *G. ruber* using the $\delta^{18}O_c$–temperature equation of Bemis et al. (Paleoceanography 13(2):150–160, 1998) and the Mg/Ca–temperature equation of Hastings et al. (EOS 82:PP12B-10, 2001). The core-top $\delta^{18}O_{sw}$ and salinity were estimated to be in the ranges of −0.2 to +0.39 ‰ and 33.7 to 34.5, respectively, which fall close to the local $\delta^{18}O_{sw}$–salinity regression line. The core-top data showed that the Mg/Ca–temperature calibration by Hastings et al. (EOS 82:PP12B-10, 2001) and the $\delta^{18}O_c$–temperature equation by Bemis et al. (Paleoceanography 13(2):150–160, 1998) are appropriate for calculating $\delta^{18}O_{sw}$ in the ECS. Furthermore, we measured core-top Ba/Ca ratios of *G. ruber* ($Ba/Ca_{G.\ ruber}$), which ranged from 0.66 to 2.82 μmol mol^{-1}. There was not a significant relationship between the salinity and $Ba/Ca_{G.\ ruber}$ ratios due to the highly variable $Ba/Ca_{G.\ ruber}$ data. Given the seawater Ba/Ca data and the published partition coefficient for Ba ($D_{Ba} = 0.15$–0.22), pristine $Ba/Ca_{G.\ ruber}$ ratios at northern Okinawa Trough sites should be less than 0.84 μmol mol^{-1}. Anomalously high $Ba/Ca_{G.\ ruber}$ ratios (>0.84 μmol mol^{-1}) might be attributable to contamination by sedimentary barite adherent on fossil shells. Therefore, further evaluation of the $Ba/Ca_{G.\ ruber}$ ratio as a paleo-salinity proxy requires diethylene triamine pentaacetic acid (DTPA)-cleaned Ba/Ca data that can minimize the influence of barite contamination.

Keywords: Oxygen isotope composition of seawater; Reconstruction of sea surface salinity; Mg/Ca–temperature; *Globigerinoides ruber*; East China Sea

Background

An East Asian summer monsoon (EASM) precipitation zone is formed by the convergence of a westerly flow to the north of the Tibetan Plateau and a southerly monsoon flow over eastern China, which gradually moves northward from South China and reaches northern China in late July (Qian and Lee 2000). This EASM precipitation has caused anomalous climatic events (both wet and dry) on an inter-decadal scale, influencing the agricultural production and economy in Asia (Wang 2006). Paleoclimate records show that anomalous dry periods, when summer monsoons weakened, caused crop failure and led to the collapse of some Chinese dynasties (Yancheva et al. 2007). Therefore, EASM variability, its effect on remote areas, and the mechanisms behind this variability have been investigated intensively as important climate change issues (Wang et al. 2005; Chen et al. 2008; Wang et al. 2008; Chang et al. 2009; Liu et al. 2014).

The EASM brings a large amount of precipitation to South China and the drainage area of the Changjiang River (Yangtze River) and induces a large influx of Changjiang River runoff to the East China Sea (ECS), forming Changjiang diluted water (CDW) in the

* Correspondence: horikawa@sci.u-toyama.ac.jp
[1]Graduate School of Science and Engineering for Research, University of Toyama, Gofuku 3190, Toyama 930-8555, Japan
Full list of author information is available at the end of the article

Changjiang River Estuary (Zhang et al. 1990; Chen et al. 1994) (Fig. 1). The CDW extends farther offshore, driven by the southerly and southwesterly monsoons from June to August, and advection of the CDW causes a significant decrease in the sea surface salinity (SSS) of the northern Okinawa Trough (Chang and Isobe 2003; Lee and Chao 2003). Consequently, the northern Okinawa Trough is considered an oceanic area where EASM-related variability in SSS can be seen (Chang and Isobe 2003).

In such areas, the oxygen isotopic compositions of seawater ($\delta^{18}O_{sw}$) have been estimated from paired analyses

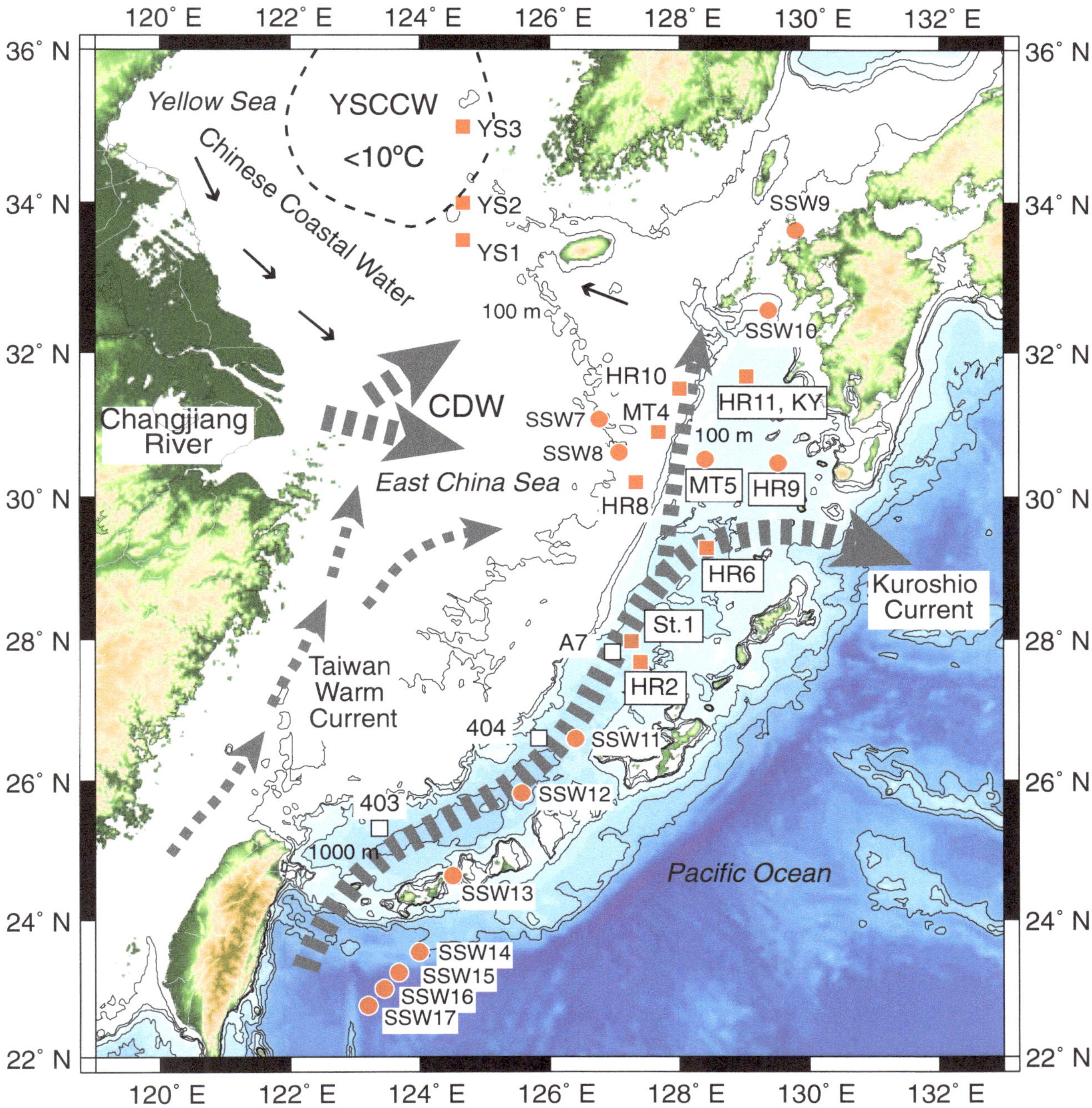

Fig. 1 Map of sample locations. Locations for surface seawater samples (*red circles*) and surface–subsurface seawater samples (*red squares*) are shown together with the present current system (Ichikawa and Beardsley 2002). Surface sediments were taken from HR11, MT5, HR9, HR6, HR2, and St.1. Locations for sediment core at KY, A7, 403, and 404 (Sun et al. 2005; Lin et al. 2006; Chen et al. 2010; Kubota et al. 2010) are also shown. CDW and YSCCW represent the Changjiang diluted water and Yellow Sea Central Cold Water, respectively. The bathymetric contours represent 100, 200, 300, 1000, 2000, and 3000 m

of Mg/Ca temperatures and oxygen isotopes ($\delta^{18}O_c$) in the surface-dwelling foraminifera *Globigerinoides ruber* (*G. ruber*) to reconstruct EASM-related variability in SSS (Sun et al. 2005; Kubota et al. 2010). However, $\delta^{18}O_{sw}$ estimates from paired analyses of Mg/Ca temperatures and $\delta^{18}O_c$ in planktonic foraminifera entail high levels of uncertainty (>0.2 ‰) due to the propagation error of $\delta^{18}O_c$ and Mg/Ca measurements and Mg/Ca–temperature calibration (Schmidt 1999; Rohling 2007). Furthermore, the lack of a local $\delta^{18}O_{sw}$–salinity equation causes additional uncertainties in salinity estimates calculated from $\delta^{18}O_{sw}$ values. Additionally, if the habitat of *G. ruber* shifts seasonally, independent of EASM variability, shell-derived $\delta^{18}O_{sw}$ estimates also change, and interpreting EASM variability will be complicated. Therefore, a reliable reconstruction of SSS with respect to EASM variability requires adequate Mg/Ca–temperature, $\delta^{18}O_c$–temperature, and local $\delta^{18}O_{sw}$–salinity calibration equations and an understanding of the habitat season and depth of *G. ruber* in the ECS.

In this study, we first derive a local $\delta^{18}O_{sw}$–salinity equation for the northern Okinawa Trough to provide a basis for salinity estimates calculated from $\delta^{18}O_{sw}$ for *G. ruber*. Then, using ECS core-top sediments, we calculate the core-top $\delta^{18}O_{sw}$ from Mg/Ca ratios and $\delta^{18}O_c$ for *G. ruber* and compare them with the local $\delta^{18}O_{sw}$–salinity line. We find that the Mg/Ca–temperature calibration by Hastings et al. (2001) and the $\delta^{18}O_c$–temperature equation by Bemis et al. (1998) are appropriate for calculating $\delta^{18}O_{sw}$ in the ECS. Finally, we present the Ba/Ca ratios of *G. ruber* from ECS core-top sediments for the first time and discuss the possibility of using shell Ba/Ca ratios as an independent paleo-salinity proxy in the ECS.

Oceanographic setting

The ECS is a marginal sea in the northwestern Pacific bounded by China, Taiwan, Ryukyu Island, Kyushu, and the Korean Peninsula (Fig. 1). The Yellow Sea is located in the northern part of the ECS, with most (>70 %) of both the Yellow Sea and ECS located above the continental shelf (<200 m water depth). The deep Okinawa Trough (~2000 m water depth) is a back-arc basin that occupies the remaining southeastern part of the ECS (Ichikawa and Beardsley 2002).

The Yellow Sea and the ECS receive the saline and warm Kuroshio water that enters the ECS along the east of Taiwan and flows northeastward along the shelf slope in the ECS (Fig. 1). The annual cycle of Kuroshio volume transport has an estimated maximum of 24 Sv (1 Sv = 1×10^6 m^3 s^{-1}) in the summer and a minimum of 20 Sv in the autumn (Lee et al. 2001). In addition, the Taiwan warm current (TWWC), which originates from the Kuroshio, flows into the ECS from the South China Sea (SCS) through the Taiwan Strait off western Taiwan, with an annual mean northward transport of 0.78 Sv (Jan et al. 2006; Zhang et al. 2014) (Fig. 1). In the autumn and winter, Changjiang River runoff decreases (1.0×10^4 m^3 s^{-1} in January) (Yanagi 1994), and the CDW flows southwestward along the Chinese coast as a narrow band; the saline water (>34) of Kuroshio origin enters the Yellow Sea (Chang and Isobe 2003). However, during the summer, when the Changjiang River supplies a large amount of freshwater to the ECS (4.8×10^4 m^3 s^{-1} in July) (Yanagi 1994), the CDW starts to extend offshore and spread east- and northeastward. Then, the eastward-flowing CDW mixes with the TWWC and the Kuroshio water in the southern and eastern parts of the ECS, respectively (Chang and Isobe 2003) (Fig. 2a).

Since the annually averaged precipitation and evaporation are near equilibrium in the ECS and Yellow Sea and the Changjiang River runoff accounts for 90 % of the total river discharge (with an annual mean of ~3.0 × 10^4 m^3 s^{-1}), the Changjiang River discharge is considered the dominant freshwater source in the ECS (Beardsley et al. 1985; Chen et al. 1994; Yanagi 1994). This Changjiang River water extends offshore, and therefore, the water mass characteristics of the offshore ECS can be roughly described by the binary mixing of the cooler, less saline CDW and the warmer, more saline Kuroshio water (Zhang et al. 1990; Ichikawa and Beardsley 2002; Zhang et al. 2007). In contrast, there are other local water masses in the coastal ECS and the Yellow Sea, such as the TWWC and the Yellow Sea Central Cold Water (YSCCW) (Fig. 1). Therefore, the water mass characteristics and $\delta^{18}O_{sw}$–salinity relationships of the coastal ECS are more complicated than those of the offshore ECS, as discussed in a later section.

The northern Okinawa Trough, which is the main focus of this study, shows distinct seasonal changes in SSS (e.g., site HR11, SSS = 33.2–34.7) due to the advection of the CDW, in comparison with the central Okinawa Trough (e.g., site HR2, SSS = 34.4–34.8) (Figs. 1 and 2a). Sea surface temperature (SST) distribution also varies seasonally owing to monsoonal winds (Lee and Chao 2003). The climatological mean annual SST is 22.2 °C near site HR11 in the northern Okinawa Trough, with a maximum in August (28.1 °C) and a minimum in February (17.5 °C). In the central Okinawa Trough, the mean annual SST is 25.0 °C near site HR2, with a maximum in August (28.9 °C) and a minimum in February (21.5 °C) (Japan Oceanographic Data Center (JODC), available at http://www.jodc.go.jp/).

Methods

Seawater and sediment samples

Seawater samples were collected using a CTD (conductivity, temperature, and depth) Carousel multisampling

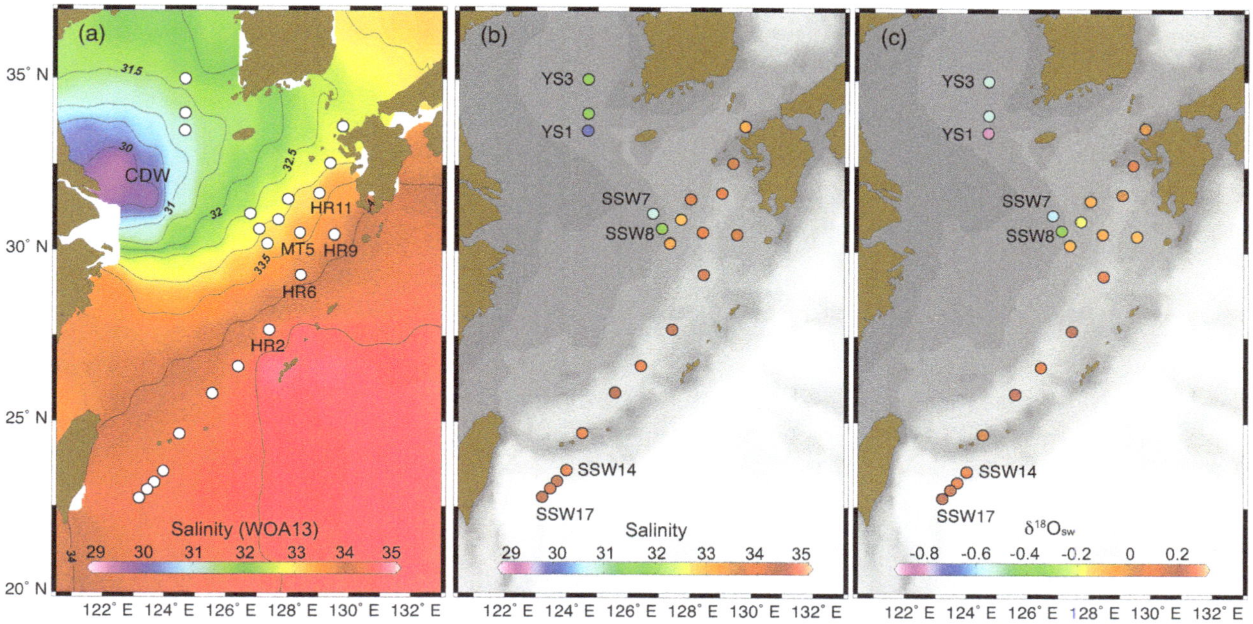

Fig. 2 Surface distribution of salinity and $\delta^{18}O_{sw}$. **a** SSS in July–September averaged statistically for 1955–2012. The data is from the World Ocean Atlas 2013 (Zweng et al. 2013). *Circles* represent locations for surface seawater samples (see Fig. 1 and Table 1). **b** SSS and **c** $\delta^{18}O_{sw}$ of seawater in early July 2013. The lowest SSS and $\delta^{18}O_{sw}$ values were found at YS1 within the CDW

system (Sea-Bird, model SBE 9plus) during the KH-13-4 cruise in early July 2013 (R/V Hakuho Maru) (Table 1). The CTD system has 24 attached Niskin bottles, each with a volume of 12 L. Subsurface seawater samples were taken with the Niskin bottles, and surface water samples were taken using a bucket (~0 m depth) or pumped from below the ship (~5 m depth). All seawater samples were transferred into glass vials for salinity and $\delta^{18}O_{sw}$ measurements, and the samples for the $\delta^{18}O_{sw}$ measurements were stored at 4 °C until laboratory analysis. Salinity was measured with an onboard Autosal laboratory salinometer (Model 8400B, Guildline Instruments Ltd., Canada) and is reported using the practical salinity scale. Sampling vials for salinity were prepared according to Joint Global Ocean Flux Study (JGOFS) protocols. The Autosal was standardized using International Association for the Physical Sciences of the Oceans (IAPSO) standard seawater.

Three sampling stations (YS1–3) are located in the southwest area off the Korean Peninsula, and 19 other stations (MT, HR, and SSW sites) are located in the continental shelf slope area of the ECS and in the Pacific Ocean (Fig. 1). We report salinity and $\delta^{18}O_{sw}$ data from these stations, and the vertical profiles of both datasets were obtained for nine stations (Fig. 1 and Table 1). Surface sediment samples were taken by a multiple corer at six sites along the central to northern parts of the Okinawa Trough from water depths of 709 to 1675 m during the KH-13-4 and KT-12-25 cruises (Fig. 1 and Table 2). Sediment samples were subsampled onboard with a 1 cm resolution and stored in a refrigerator until analysis in the laboratory.

The lithology of the sediment cores (~30 cm core length) was characterized as foraminifer-bearing silty clay or clayey silt with no visible turbidities or erosional surfaces. In general, the central to northern parts of the Okinawa Trough preserve thick late Holocene sediments due to their high sedimentation rates (10–50 cm kyr^{-1}), except in shelf slope regions (Ujiié et al. 2003; Sun et al. 2005; Kubota et al. 2010). All core-top $\delta^{18}O_c$ values of *G. ruber* were isotopically lighter than −2.0 ‰, which is consistent with the late Holocene values in the central to northern Okinawa Trough (Table 2) (Sun et al. 2005; Kubota et al. 2010). Therefore, we conclude that the core-top sediments used in this study represent at least late Holocene sediments. Site HR2 was located at a depth below the present calcite saturation horizon (CSH) in the ECS (~1600 m) (Fig. 6a). The Mg/Ca ratios from this site should be treated with caution regarding the partial dissolution of *G. ruber* shells, which will be discussed in a later section.

Analysis of the oxygen isotopic composition of seawater ($\delta^{18}O_{sw}$)

We measured $\delta^{18}O_{sw}$ with a stable isotope mass spectrometer (PRISM, Micromass UK, Ltd.) at the University of Toyama. Oxygen isotopic analysis was carried out using the automated H_2O–CO_2 equilibrium method

Table 1 Salinity and $\delta^{18}O_{sw}$ data at each station in the ECS

Depth (m)	Temperature (°C)	Salinity	$\delta^{18}O_{sw}$ (‰, VSMOW)	Std.dev (‰)	Site area
YS1 (30 June 2013, 124° 39.66′ E, 33° 29.80′ N)					Yellow Sea–ECS shelf site
0	23.40	29.89	−0.81	0.03	
5	22.95	30.60	−0.67	0.03	
10	22.83	31.51	−0.47	0.02	
20	17.10	32.82	−0.18	0.02	
40	10.17	33.05	−0.12	0.02	
YS2 (30 June 2013, 124° 40.07′ E, 34° 00.40′ N)					Yellow Sea–ECS shelf site
0	22.90	31.37	−0.54	0.03	
5	22.80	31.50	−0.52	0.02	
10	22.76	31.86	−0.43	0.03	
20	17.14	32.35	−0.35	0.04	
30	13.54	32.57	−0.29	0.03	
YS3 (1 July 2013, 124° 40.04′ E, 34° 59.64′ N)					Yellow Sea–ECS shelf site
0	23.90	31.88	−0.53	0.03	
5	23.12	31.86	−0.52	0.03	
10	17.29	31.87	−0.48	0.02	
20	13.73	32.09	−0.44	0.02	
30	7.71	32.21	−0.42	0.02	
HR8 (2 July 2013, 127° 19.84′ E, 30° 12.68′ N)					Yellow Sea–ECS shelf site
0	26.70	33.40	−0.06	0.02	
20	25.50	33.90	0.07	0.02	
40	24.09	34.34	0.15	0.02	
60	21.33	34.58	0.27	0.02	
70	19.44	34.61	0.26	0.02	
MT4 (2 July 2013, 127° 41.25′ E, 30° 54.42′ N)					Yellow Sea–ECS shelf site
0	26.10	33.10	−0.18	0.02	
6	25.81	33.16	−0.19	0.02	
10	25.79	33.37	−0.14	0.04	
30	24.44	33.81	0.00	0.03	
60	21.56	34.19	0.10	0.02	
100	19.08	34.64	0.20	0.02	
HR10 (2 July 2013, 128° 00.21′ E, 31° 30.22′ N)					Yellow Sea–ECS shelf site
0	26.10	33.75	−0.04	0.02	
5	25.87	33.68	−0.08	0.02	
10	25.80	33.82	−0.02	0.02	
20	25.49	34.07	0.05	0.02	
40	24.96	34.20	0.09	0.02	
59	22.91	34.39	0.17	0.04	
80	21.01	34.59	0.22	0.02	
99	19.86	34.65	0.24	0.03	
HR11 (2 July 2013, 129° 01.85′ E, 31° 40.60′ N)					Okinawa Trough site
0	26.90	33.88	0.01	0.02	
49	23.76	34.30	0.15	0.02	

Table 1 Salinity and $\delta^{18}O_{sw}$ data at each station in the ECS *(Continued)*

75	21.96	34.56	0.23	0.02	
99	20.27	34.69	0.24	0.04	
HR6 (8 July 2013, 128° 25.43′ E, 29° 17.42′ N)					Okinawa Trough site
0	29.70	34.22	0.12	0.05	
20	29.57	34.23	0.11	0.02	
50	28.63	34.35	0.23	0.02	
75	27.21	34.52	0.27	0.04	
99	23.69	34.66	0.33	0.04	
HR2 (9 July 2013, 127° 24.03′ E, 27° 40.97′ N)					Okinawa Trough site
0	29.40	34.25	0.18	0.02	
5	29.63	34.24	0.13	0.02	
10	29.36	34.19	0.14	0.03	
20	29.01	34.17	0.14	0.02	
50	26.32	34.40	0.24	0.02	
75	23.86	34.61	0.32	0.02	
100	22.47	34.74	0.35	0.02	
HR9 (7 July 2013, 129° 31.29′ E, 30° 28.26′ N)					Okinawa Trough site
0	28.50	33.73	−0.07	0.03	
MT5 (8 July 2013, 128° 23.41′ E, 30° 31.22′ N)					Okinawa Trough site
0	28.50	33.91	−0.03	0.02	
SSW07 (2 July 2013, 126° 45.99′ E, 31° 04.86′ N)					Yellow Sea–ECS shelf site
5	23.20	30.80	−0.59	0.04	
SSW08 (2 July 2013, 127° 04.33′ E, 30° 37.06′ N)					Yellow Sea–ECS shelf site
5	25.30	31.80	−0.36	0.02	
SSW09 (7 July 2013, 129° 47.01′ E, 33° 38.63′ N)					Off-Kyushu site
5	24.10	33.33	0.00	0.02	
SSW10 (7 July 2013, 129° 22.50′ E, 32° 34.00′ N)					Off-Kyushu site
5	23.60	33.71	0.12	0.05	
SSW11 (9 July 2013, 126° 24.20′ E, 26° 36.00′ N)					Okinawa Trough site
5	28.80	34.00	0.12	0.02	
SSW12 (9 July 2013, 125° 33.84′ E, 25° 49.26′ N)					Okinawa Trough site
5	29.20	34.38	0.21	0.03	
SSW13 (9 July 2013, 124° 30.06′ E, 24° 38.41′ N)					Okinawa Trough site
5	28.80	33.81	0.03	0.02	
SSW14 (10 July 2013, 123° 59.02′ E, 23° 32.27′ N)					Kuroshio site
5	29.30	34.06	0.06	0.02	
SSW15 (10 July 2013, 123° 40.09′ E, 23° 14.07′ N)					Kuroshio site
5	29.30	34.33	0.12	0.09	
SSW16 (10 July 2013, 123° 26.97′ E, 22° 59.89′ N)					Kuroshio site
5	29.40	34.23	0.15	0.02	
SSW17 (10 July 2013, 123° 11.60′ E, 22° 45.00′ N)					Kuroshio site
5	29.40	34.34	0.20	0.02	

Table 2 Details of the core-top sediments and metal/Ca and $\delta^{18}O$ values from *Globigerinoides ruber* (sensu stricto)

Core site	Latitude	Longitude	Water depth (m)	$\delta^{18}O$ (VPDB ‰)		Mg/Ca (mmol mol^{-1})	±SD	Ba/Ca (μmol mol^{-1})	±SD	Mn/Ca (μmol mol^{-1})	±SD
KH13-4-HR2MC	27° 40.79′ N	127° 23.95′ E	1675	−2.93	Ave	4.09	0.18	0.72	0.74	7.88	5.5
					Run #1	4.05		0.66		0.74	
					Run #2	3.94		0.78		6.50	
					Run #3	4.03		1.53		11.58	
					Run #4	4.35		2.25		12.71	
KT12-25-St.1	27° 58′ N	127° 16′ E	1153	−2.52	Ave.	4.20	0.03	0.77	0.01	23.25	21.3
					Run #1	4.22		0.78		8.18	
					Run #2	4.18		0.77		38.31	
KH13-4-HR6MC	29° 17.35′ N	128° 25.18′ E	1065	−2.47	Ave.	3.97	0.25	0.79	0.20	4.37	4.4
					Run #1	3.68		1.04		1.92	
					Run #2	4.13		0.66		9.49	
					Run #3	4.09		0.92		1.72	
KH13-4-HR9MC	30° 28.19′ N	129° 31.06′ E	709	−2.78	Ave.	4.19	0.25	0.81	1.09	53.01	13.7
					Run #1	3.91		0.81		48.40	
					Run #2	4.36		1.11		68.44	
					Run #3	4.30		2.82		42.18	
KH13-4-MT5MC	30° 31.23′ N	128° 23.38′ E	823	−2.27	Ave.	3.61	0.22	0.88	0.11	44.50	7.7
					Run #1	3.41		0.72		51.73	
					Run #2	3.44		0.95		46.37	
					Run #3	3.68		0.97		46.21	
					Run #4	3.89		0.87		33.67	
KH13-4-HR11MC	31° 40.57′ N	129° 01.99′ E	725	−2.42	Ave.	3.30	0.02	0.85	0.06	44.86	37.0
					Run #1	3.31		0.89		18.67	
					Run #2	3.29		0.80		71.05	
A7	27° 49.2′ N	126° 58.7′ E	1264	−2.35		4.06					
MD012403	25° 17′ N	123° 10′ E	1420	−2.51		4.40					
MD012404	26° 38.84′ N	125° 48.75′ E	1397	−2.60		4.24					
KY07-04-01	31° 38.35′ N	128° 56.64′ E	725	−2.50		3.50					

Table 2 Details of the core-top sediments and metal/Ca and $\delta^{18}O$ values from *Globigerinoides ruber* (sensu stricto)

No dissolution-corrected Mg/Ca-derived SST			Dissolution-corrected Mg/Ca-derived SST							
SST[a] (°C)	SST[b] (°C)	SST[c] (°C)	SST[a] (°C)	$\Delta[CO_3^{2-}]$ (μmol kg^{-1})	Corrected Mg/Ca (mmol mol^{-1})	Calcification SST	Shell-derived $\delta^{18}O_{sw}$ (VSMOW ‰)	Calcification salinity	Calcification season	Calcification depth (m)
26.7	24.4	29.4	28.7	−2.3	4.90	28.5	0.22	34.43	July–Sept	0–10
27.0	24.6	29.7	28.0	5.9	4.60	26.6	0.27	34.50	May–Oct	0–30
26.4	24.1	29.0	27.3	7.3	4.30	26.3	0.19	34.41	May–Oct	0–30
27.0	24.6	29.6	27.1	12.9	4.24	27.3	0.01	34.01	June–Sept	0–30
25.3	23.2	27.9	25.7	11.1	3.75	25.1	0.17	34.17	May–Oct	0–30
24.3	22.3	26.9	24.5	12.7	3.37	25.3	−0.20	33.76	June–Sept	0–30
26.6	24.3	29.3	27.9	4.2	4.55	25.9	0.37	34.54	May–Oct	0–30
27.5	25.1	30.2	29.0	1.7	5.01	26.5	0.39	34.43	May and Oct	0–30
27.1	24.7	29.8	28.6	2.1	4.84	26.9	0.21	34.38	May–Oct	0–30
24.9	22.9	27.6	25.2	12.7	3.57	25.6	−0.14	33.68	June–Sept	0–20

Decreases in Mg/Ca ratios due to calcite dissolution were calculated using critical thresholds for the dissolution (14 μmol kg^{-1} $\Delta[CO_3^{2-}]$) and sensitivity of Mg/Ca ratio to $\Delta[CO_3^{2-}]$ (0.05 mmol mol^{-1} per μmol^{-1} kg^{-1}) (Johnstone et al. 2011). Core-top Mg/Ca and $\delta^{18}O$ data from A7, MD012403, MD012404, KY07-04-01 were from Sun et al. (2005), Lin et al. (2006), Chen et al. (2010), and Kubota et al. (2010), respectively

[a]SST = ln(Mg/Ca/0.38)/0.089, Hastings et al. (2001)

[b]SST = ln(Mg/Ca/0.34)/0.102, Anand et al. (2003)

[c]SST = ln(Mg/Ca/0.3)/0.089, Lea et al. (2000)

(Epstein and Mayeda 1953). The stable isotope ratios are given as conventional δ values (‰), and the analytical precision was ±0.02 ‰ (1σ) for $\delta^{18}O$. Oxygen isotopic composition is expressed relative to Vienna Standard Mean Ocean Water (VSMOW).

$\delta^{18}O$ and Mg/Ca analyses of *G. ruber*

Core-top sediments (0–1 cm) were wet washed through a 63 μm sieve and then dried in an oven at 50 °C. To minimize ontogenic and growth rate effects on shell geochemistry, specimens of *G. ruber* (sensu stricto form; 20–30 shells) were handpicked from the 250–355 μm size fraction. Planktonic foraminifera were gently crushed and rinsed three times with ultrapure water and methanol (super special grade, Wako Pure Chemical Industries, Ltd.) to remove adherent clay particles. The shells were then split into two fractions for $\delta^{18}O_c$ and metal/Ca measurements. $\delta^{18}O_c$ values were obtained using a Finnigan MAT 253 mass spectrometer at the Center for Advanced Marine Core Research (CMCR), Kochi University and calibrated in accordance with standard NBS 19. The precision of these measurements was better than ±0.08 ‰ (1σ) for $\delta^{18}O_c$.

The samples for metal/Ca analysis were cleaned according to the procedure developed for trace element analysis (Boyle and Keigwin 1985; Rosenthal et al. 1997). In brief, samples underwent a multistep process consisting of initial rinses in ultrapure water and methanol, followed by treatments with hot reducing and oxidizing solutions, transferred into new acid-leached microcentrifuge tubes (1.5 mL), and finally leached with a dilute ultrapure nitric acid solution (0.001 M HNO_3, TAMAPURE-AA-100 from Tama Chemicals, Ltd.). The sample solution was then dissolved with a Sc-spiked

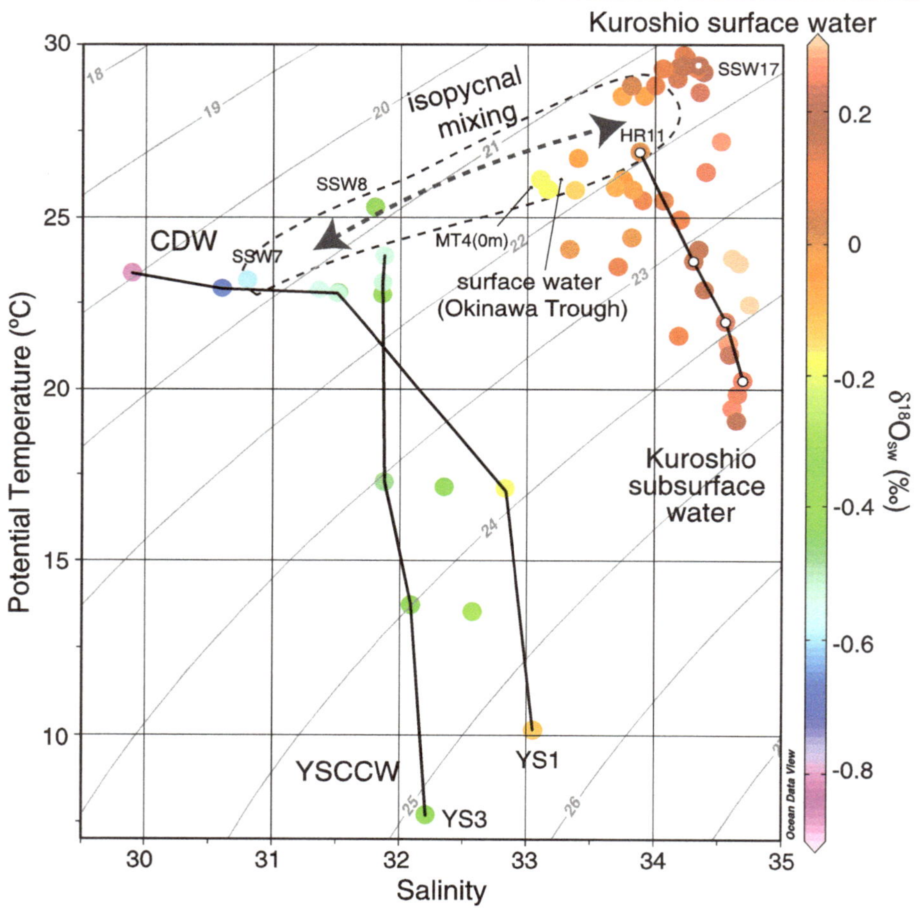

Fig. 3 Potential temperature and salinity (*T–S*) plot in the ECS. The *T–S* diagram shows four distinct water masses in early July: the Kuroshio surface water, Kuroshio subsurface water, YSCCW, and CDW. The *color in circles* indicates the $\delta^{18}O_{sw}$ value of the seawater. The Kuroshio surface and subsurface waters are characterized by higher $\delta^{18}O_{sw}$ values, whereas the CDW is characterized by relatively low $\delta^{18}O_{sw}$ values. Surface water properties in the northern part of the Okinawa Trough can be explained by isopycnal mixing with the CDW and Kuroshio surface water. The potential density curves (σ_θ = 19–27) are also shown. The *T–S* diagram was made using Ocean Data View (Schlitzer R, Ocean Data View, http://odv.awi.de, 2014)

dilute ultrapure nitric acid solution (2 % HNO_3) to obtain a Ca concentration of 10 ± 1 μg g^{-1}. All clean work was conducted in laminar flow benches or a clean room under trace metal clean conditions.

The metal/Ca ratio was determined with a Thermo Scientific ELEMENT 2 sector field inductively coupled plasma mass spectrometer (ICP–MS) at the University of Toyama, operated in a low-resolution mode ($m/\Delta m$ = 300) (Marchitto 2006). We analyzed ^{24}Mg, ^{26}Mg, ^{43}Ca, ^{44}Ca, ^{88}Sr, ^{137}Ba, and ^{138}Ba during the same run to determine Mg/Ca for temperature reconstructions and measured ^{56}Fe (in a mid-resolution mode) and ^{55}Mn to monitor contamination by clay minerals and diagenetic coatings (Table 2). Element counts were converted into molar ratios by the intensity ratio method based on a series of matrix-matched standard solutions. The accuracy and precision of Mg/Ca ratios were confirmed by analyses of the $CaCO_3$ reference materials BAM RS3 and ECRM 752-1. The repeated analyses gave Mg/Ca ratios of 0.786 ± 0.008 (n = 100, 1σ, RSD = 1.0 %) and 3.92 ± 0.06 (n = 24, 1σ, RSD = 1.5 %), respectively, which were within the reported value for BAM RS3 (0.791 ± 0.03) and slightly higher than the reported value for ECRM 752-1 (3.824 ± 0.095) (Greaves et al. 2008). Samples often show high Mn/Ca ratios (>100 μmol mol^{-1}) due to the presence of diagenetic coatings that were not removed during the cleaning process. Such samples were rejected. To further minimize the impact of sample heterogeneity and the analytical error of metal/Ca ratios, we report replicate measurements of metal/Ca ratios on re-picked *G. ruber* shells (Table 2).

Results and discussion

$\delta^{18}O_{sw}$-salinity relationship in the eastern ECS

In early July, surface water samples showed the highest SSS (34.3) at site SSW17 in the Kuroshio water area and the lowest SSS (29.9) at YS1 in the Yellow Sea (Fig. 2b and Table 1). The $\delta^{18}O_{sw}$ ranged from −0.81 to 0.21 ‰, and relatively low $\delta^{18}O_{sw}$ values were observed in the Yellow Sea, the ECS shelf, and the northern part of the Okinawa Trough (Fig. 2c). Spatial distributions of salinity and $\delta^{18}O_{sw}$ values were almost similar to the summer (July–September) salinity distribution averaged statistically over 1955–2012 (Zweng et al. 2013) (Fig. 2a), suggesting that the CDW spread east- and northeastward in the southern Yellow Sea and the northern Okinawa Trough during the sampling period.

A temperature and salinity plot (T–S diagram) of the surface and subsurface waters (0–100 m) in the Yellow Sea and the ECS shows four typical water masses: (1) the Kuroshio surface water in the subtropical Pacific Ocean (e.g., SSW14–17), (2) the Kuroshio subsurface

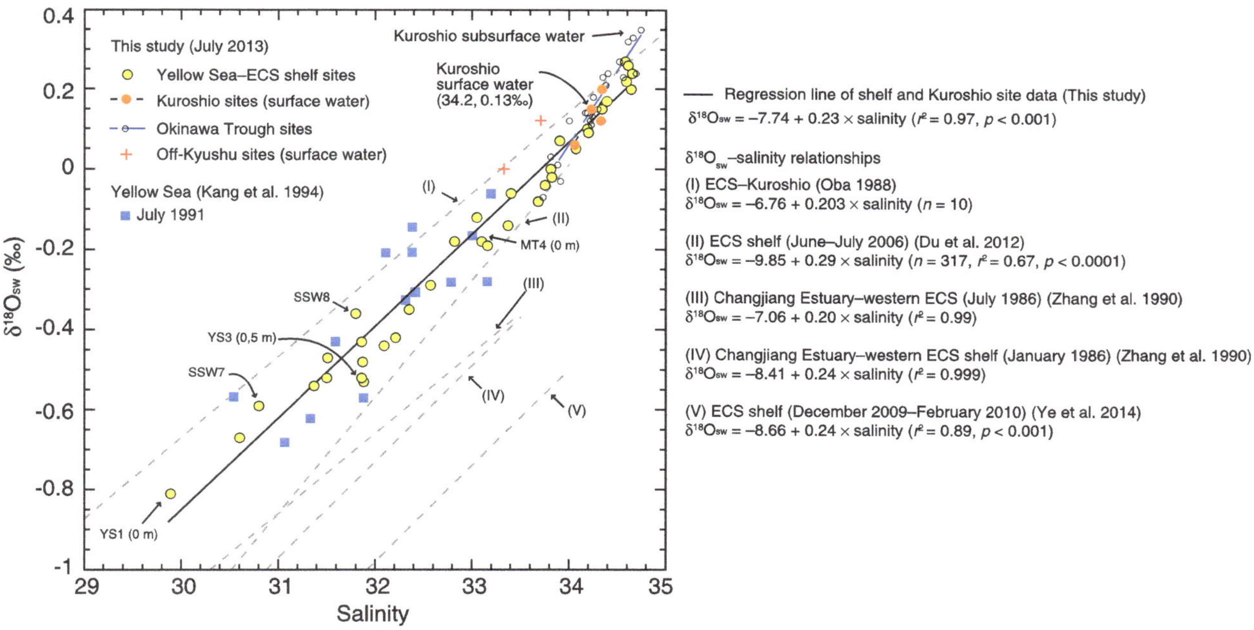

Fig. 4 $\delta^{18}O_{sw}$–salinity plot of shallow seawaters in the ECS. $\delta^{18}O_{sw}$–salinity plot of all shallow seawater (0–100 m) data in this study (early July 2013). The data are classified into four site groups: Yellow Sea–ECS shelf, Kuroshio, Okinawa Trough, and off-Kyushu sites (Table 1). The *black* regression line is derived from the data in the Yellow Sea–ECS shelf and Kuroshio sites. Since the surface waters in the northern Okinawa Trough are a mixture of the CDW and Kuroshio surface water, $\delta^{18}O_{sw}$ and salinity data from northern Okinawa Trough sites fall on this regression line. Regression lines (*dotted lines*) and equations from the literature are also shown. Regression lines III, IV, and V were developed in mainly coastal and shelf waters near the Changjiang Estuary. Although these regression lines represent freshwater end-members that are similar to our regression line, the end-member of the seawater shows lower $\delta^{18}O_{sw}$ values than that of the Kuroshio surface water (34.2 and 0.13 ‰; SSW14–17)

water in the Okinawa Trough (identified by a salinity of ~34.4 and a $\delta^{18}O_{sw}$ of 0.2–0.5 ‰) (Kang et al. 1994; Kim et al. 2005), (3) the YSCCW (<10 °C) in the bottom water (30 m) at YS3, and (4) the less saline CDW in the surface water at YS1 (Fig. 3). The *T–S* diagram also shows that the majority of the shallow water data in the Okinawa Trough fall on the isopycnal mixing line connecting the CDW and Kuroshio surface water (Fig. 3). Indeed, the $\delta^{18}O_{sw}$ values and salinity of the shallow seawater (0–100 m) are strongly correlated, indicating that the data can be explained by a binary mixing of freshwater and saline seawater (Fig. 4).

To derive a local $\delta^{18}O_{sw}$–salinity relationship around the northern Okinawa Trough, sampling sites were classified by geographic and oceanographic characteristics (Table 1). Sites SSW14–17, which are outside the ECS and under the influence of the Kuroshio surface water, were defined as Kuroshio sites. Data from the Yellow Sea and the ECS were divided into three groups: Yellow Sea–ECS shelf sites at depths shallower than 150 m, Okinawa Trough sites deeper than 150 m, and off-Kyushu sites influenced by local freshwater inputs from Kyushu (Fig. 4 and Table 1). Given that the eastward-flowing CDW mixes with the warmer, more saline Kuroshio water near the shelf edge in the summer (Chang and Isobe 2003), salinity and $\delta^{18}O_{sw}$ values in the Kuroshio surface water should reflect the end-member for saline water (Kang et al. 1994). Therefore,

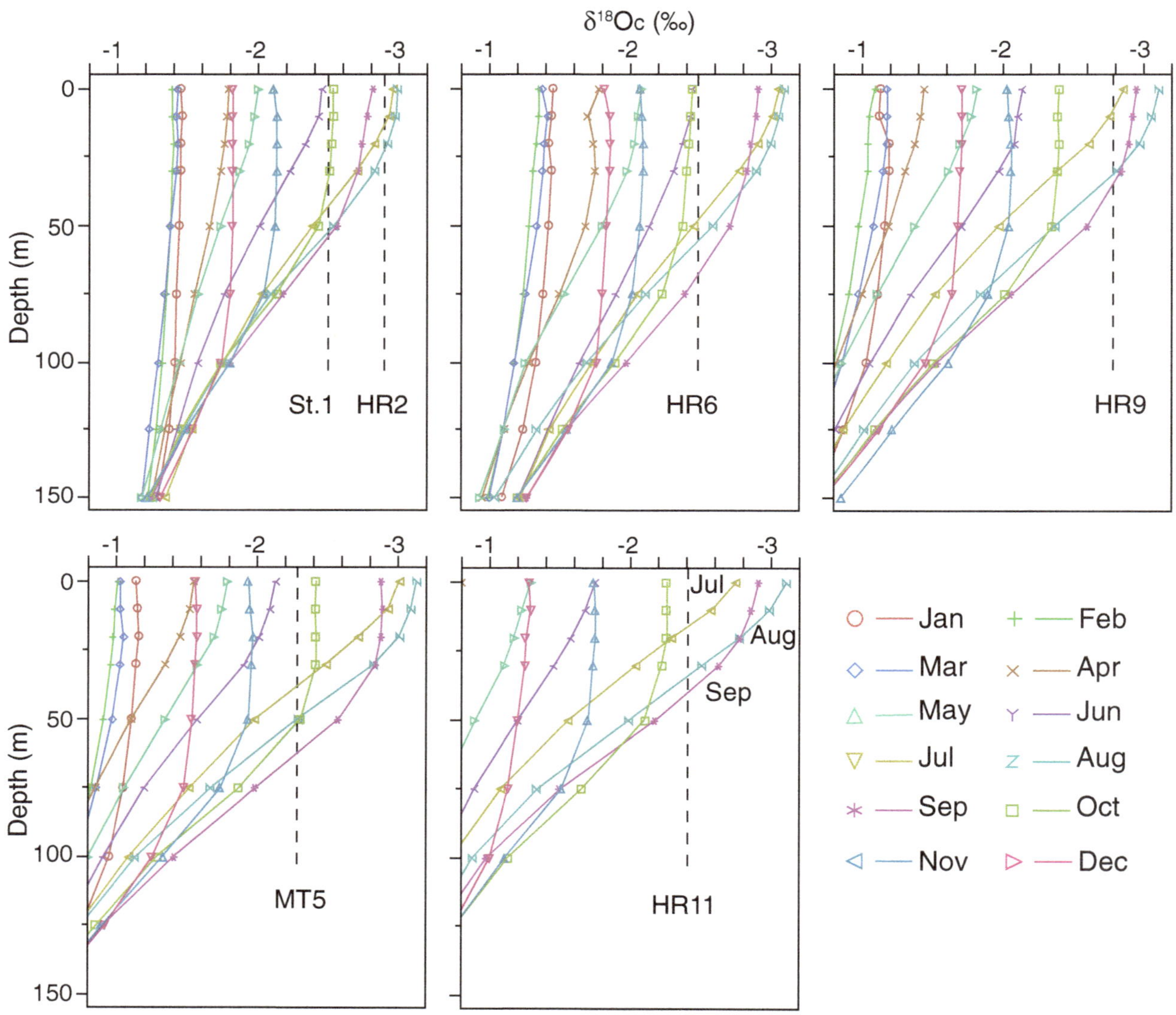

Fig. 5 Comparison of monthly variation in the predicted $\delta^{18}O_c$ values with the core-top $\delta^{18}O_c$ values. The *solid lines* show the monthly variation of the predicted $\delta^{18}O_c$ values within the water column. The *dotted lines* show the core-top $\delta^{18}O_c$ values of *G. ruber*. The predicted $\delta^{18}O_c$ values are calculated from temperature and salinity data (http://www.jodc.go.jp/) using a calcite equilibrium equation (Bemis et al. 1998). See details in the text

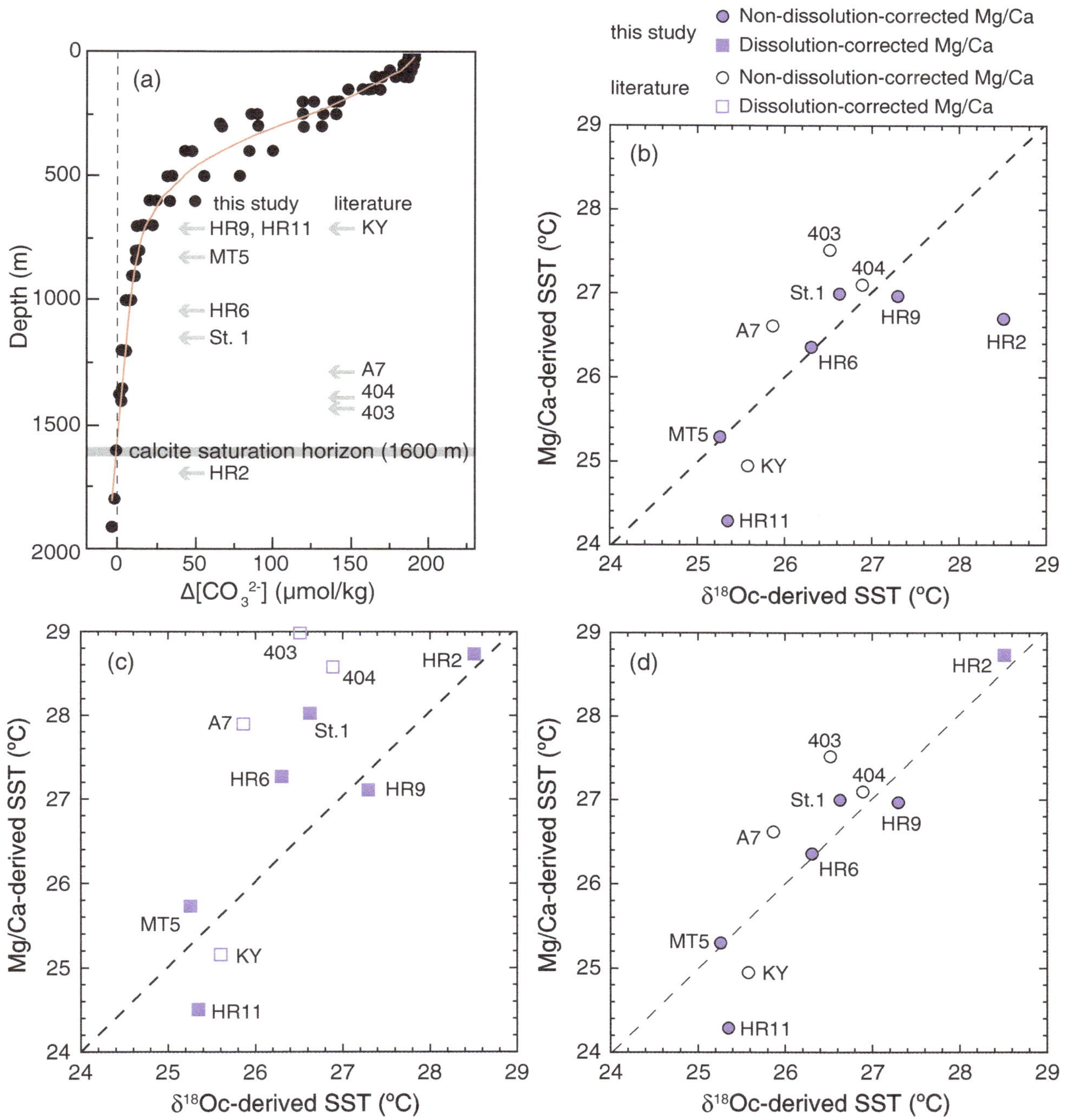

Fig. 6 Comparison of $\delta^{18}O_c$-derived calcification temperature with Mg/Ca-derived temperature. **a** Water column profile of $\Delta[CO_3^{2-}]$ in the ECS. Calculation of the calcite saturation state $\Delta[CO_3^{2-}]$ was accomplished via subtracting the carbonate ion concentration $[CO_3^{2-}]$ at saturation from the in situ $[CO_3^{2-}]$. To compute the in situ $[CO_3^{2-}]$ with the program CO2sys.xls (Lewis and Wallace 1998), we used total alkalinity, TCO_2, pH, and hydrographic water column data in the ECS (NODC Accession 0109919, http://data.nodc.noaa.gov/nodc/archive/metadata/approved/). *Arrows* indicate the water depths of cores analyzed in this study and by Sun et al. (2005), Lin et al. (2006), Chen et al. (2010), and Kubota et al. (2010). **b** $\delta^{18}O_c$-derived calcification temperature and non-dissolution-corrected Mg/Ca-derived temperature plot. **c** $\delta^{18}O_c$-derived calcification temperature and dissolution-corrected Mg/Ca-derived temperature plot. **d** $\delta^{18}O_c$-derived calcification temperature and Mg/Ca-derived temperature plot. The dissolution-corrected Mg/Ca ratio was used for the deepest site, HR2, where $\Delta[CO_3^{2-}] < 0$ μmol kg^{-1}. The other data were non-dissolution-corrected Mg/Ca ratios

taking into consideration this mixing of two water masses in the eastern part of the ECS, we derived the following $\delta^{18}O_{sw}$–salinity equation from the summer Yellow Sea–ECS shelf and Kuroshio site data:

$$\delta^{18}O_{sw} = -7.74\,(\pm 0.4) + 0.23\,(\pm 0.01) \times salinity\,(r^2 = 0.97, p < 0.001, n = 40) \quad (1)$$

The quoted errors of the slope and intercept are 95 % confidence intervals. This equation includes the surface (0–30 m water depth) and subsurface (40–100 m water depth) data. The surface data (0–30 m) alone result in the same equation within the 95 % confidence intervals of Eq. (1).

Equation (1) indicates a projected freshwater end-member of −7.74 ‰ (±0.4 ‰). This value is consistent with the $\delta^{18}O$ values (−8.4 to −7.1 ‰) for Changjiang River water in January and July (Zhang et al. 1990). However, the regression lines of the $\delta^{18}O_{sw}$–salinity relationships developed for the Changjiang River mouth and the western part of the ECS indicate lower $\delta^{18}O_{sw}$ values in the salinity range of 30 to 34 [$\delta^{18}O_{sw} = -8.41 + 0.24 \times$ salinity in January (line IV of Fig. 4) and $\delta^{18}O_{sw} = -7.06 + 0.20 \times$ salinity in July (line III of Fig. 4)]. Similar regression lines were developed for the Yellow Sea and the ECS [$\delta^{18}O_{sw} = -8.66 + 0.24 \times$ salinity in winter (line V of Fig. 4) and $\delta^{18}O_{sw} = -10.7 + 0.27 \times$ salinity in summer] (Ye et al. 2014). These regression lines do not pass through the end-member values of the Kuroshio surface water (salinity = 34.2–34.4 and $\delta^{18}O_{sw}$ = 0.1–0.2 ‰), indicating that the Kuroshio surface water was not the source of saline water for these areas (Fig. 4). Since these data were mainly from coastal and shelf waters near the Changjiang Estuary, the less saline, $\delta^{18}O$-depleted TWWC may be the dominant source of saline water rather than the Kuroshio water. Therefore, $\delta^{18}O_{sw}$–salinity relationships in the coastal ECS are highly variable and differ from our equation.

Like our regression line, regression lines that pass through the saline Kuroshio surface water were derived for the ECS shelf covering a broad sampling area for June–July (line II of Fig. 4) (Du et al. 2012), for the southern Yellow Sea in July (Kang et al. 1994), and for the area from the ECS to off the southern Japan Islands (Oba 1990) (line I of Fig. 4). Our regression line occupied an intermediate area between these regression lines. Furthermore, our equation is derived from the data obtained in the eastern part of the ECS where the eastward-flowing CDW mixes with the Kuroshio water in the summer. Therefore, our $\delta^{18}O_{sw}$–salinity equation should be representative of the northern Okinawa Trough during the EASM season.

Even if the freshwater end-member changed from −7.74 to −9.0 ‰ (given the averaged *y*-intercept of lines II, IV, and V as the potential change in the riverine $\delta^{18}O$), this change in amplitude does not yield a significant difference between the two calibration lines within the salinity range of 33 to 34.2. In addition, since end-member values of the CDW and the Kuroshio surface water may not have changed significantly during the Holocene in comparison with the glacial–interglacial cycles (Wang et al. 2008), our local $\delta^{18}O_{sw}$–salinity equation may provide an approximate relationship between the salinity and $\delta^{18}O_{sw}$ in the eastern part of the ECS, at least during the Holocene.

Core-top $\delta^{18}O_c$ and Mg/Ca ratios for *G. ruber*

The core-top $\delta^{18}O_c$ values for *G. ruber* ranged from −2.9 to −2.3 ‰ in the central to northern parts of the Okinawa Trough (Fig. 1 and Table 2). The lightest value was observed at HR2, the southernmost site, whereas the heaviest value was found at MT5 near the CDW advection area. Core-top Mg/Ca ratios of *G. ruber* ranged from 3.29 to 4.35 mmol mol^{-1} (Table 2). The Mg/Ca value at HR2 (4.09 mmol mol^{-1}, 1675 m water depth), which is the deepest site, was lower than that at the nearby shallower site St.1 (4.20 mmol mol^{-1}, 1153 m), even though HR2 had the lowest $\delta^{18}O_c$ value of *G. ruber* (−2.9 ‰) (Table 2). Since site HR2 is below the CSH (~1600 m) (Fig. 6a), the Mg/Ca ratio at HR2 seems to be lowered due to calcite dissolution (discussed below). Core-top Mg/Ca ratios, including the data from the literature (Sun et al. 2005; Lin et al. 2006; Chen et al. 2010; Kubota et al. 2010), tend to be higher in the southern sites (e.g., HR2 and St.1) and lower in the northern sites (HR11 and MT5) (Table 2), which is consistent with the distribution of the summer SST. In contrast, core-top $\delta^{18}O_c$ values correlate poorly with the Mg/Ca ratios ($r^2 = 0.19$), indicating that $\delta^{18}O_c$ values are strongly influenced by $\delta^{18}O_{sw}$.

Calcification season and calcification temperature of *G. ruber*

The foraminifera shell Mg/Ca response to temperature is biologically modulated and dependent on pH and/or [CO_3^{2-}] (Lea et al. 1999; Russell et al. 2004). Therefore, many species- and basin-specific Mg/Ca–temperature equations have been developed, some of which take into consideration the potential effect of dissolution (Lea et al. 2000; Dekens et al. 2002; Anand et al. 2003; McConnell and Thunell 2005). There is, however, no equation for *G. ruber* in the ECS, and previous studies conducted in the ECS have converted Mg/Ca values of *G. ruber* to SST using a Mg/Ca–temperature calibration equation [T (°C) = ln(Mg/Ca/0.38)/0.089] developed for *G. ruber* based on the SCS core-top sediments by Hastings et al. (2001) (Sun et al. 2005; Lin et al. 2006; Chen et al. 2010; Kubota et al. 2010). Although this calibration equation yields temperatures corresponding to the warm summer months, it has not been fully investigated whether there is consistency between $\delta^{18}O_c$- and Mg/Ca-derived temperatures, even though reconciled estimates of temperatures are crucial for calculating $\delta^{18}O_{sw}$.

First, to calculate calcification temperatures ($\delta^{18}O_c$-derived SSTs), we identified the calcification season and depth based on a comparison of core-top $\delta^{18}O_c$ values with predicted $\delta^{18}O_c$ values. Predicted $\delta^{18}O_c$ values were calculated using the following $\delta^{18}O_c$–temperature equation assuming oxygen isotopic equilibrium in foraminiferal calcite:

$$T(^{\circ}C) = 14.9\text{–}4.8\left(\delta^{18}O_c\text{–}\left(\delta^{18}O_{sw}\text{–}0.27‰\right)\right) \quad (2)$$

This $\delta^{18}O_c$–temperature equation was determined from culturing experiments of a symbiont-bearing species, *Orbulina universa*, grown under high-light conditions (Bemis et al. 1998). The high-light conditions (>380 μEinstein m^{-2} s^{-1}) in the culturing experiments take into consideration symbiont effects on shell $\delta^{18}O_c$ (i.e., high photosynthetic activity forms ^{18}O-depleted shells). The applicability of this equation to the surface-dwelling, symbiont-bearing species *G. ruber* has been confirmed (Bemis et al. 1998; Benway et al. 2006). Temperature and salinity data averaged statistically for 1930–2003 in a 1° × 1° (latitude × longitude) grid provided by JODC were used for the calculation. Salinity was converted to $\delta^{18}O_{sw}$ values using Eq. (1). The term –0.27 ‰ corrects for the $\delta^{18}O$ difference between VSMOW and Pee Dee Belemnite (PDB).

Figure 5 shows a comparison of the predicted vertical profiles of $\delta^{18}O_c$ values for each month with core-top $\delta^{18}O_c$ values of *G. ruber* at each site. Among the six core-top datasets, HR2 and HR9 showed relatively lighter $\delta^{18}O_c$ values that are approximately equal to the $\delta^{18}O_c$ values in the surface water for July–September. In contrast, the other sites (St.1, HR6, MT5, and HR11) were marked by isotopically heavier values than the peak summer $\delta^{18}O_c$ in the surface water, suggesting a large contribution of *G. ruber* shells formed in deeper waters (>50 m) or in surface water in May, June, and/or October (Fig. 5). Since *G. ruber* (sensu stricto) lives predominantly in the top 20 m of the subtropical North Pacific (Kuroyanagi and Kawahata 2004), calcification of shells in deeper water (>50 m) should be negligible. Furthermore, sediment trap data from the ECS, which unfortunately do not include data for September, show that the fluxes of *G. ruber* from May to October account for 80 % of the total fluxes each year and the fluxes in May alone account for 20 % of the yearly fluxes (Xu et al. 2005). Based on these data, we regard the dominant habitat depth of *G. ruber* in the ECS as the top 0 to 30 m (mixed layer) and contend that the *G. ruber* shell geochemistry in the ECS records the weighted mean surface ocean conditions from May to October.

To calculate the calcification temperatures and salinity of the seawater in which *G. ruber* shells were formed, we computed seasonal average (May–October or July–September) $\delta^{18}O_c$ values and compared these predicted values to core-top $\delta^{18}O_c$ values (Table 2). The calcification seasons and depths at HR2 and HR11 were identified as July–September and 0–10 m and June–September and 0–30 m, respectively. The $\delta^{18}O_c$-derived SSTs at HR2 and HR11 were calculated as 28.5 and 25.3 °C, respectively (Fig. 6 and Table 2). The calculated calcification SSTs at all sites ranged from 25.1 to 28.5 °C in the central to northern Okinawa Trough (Table 2). It is noteworthy that the calcification season of *G. ruber* was not necessarily uniform across the ECS; data from sites HR2 and HR9 clearly show a warmer seasonal bias (close to the peak summer SST), in contrast to other site data (Figs. 1 and 5). The core-top $\delta^{18}O_c$ data shows that Eq. (2) provides better estimates for calcification depth, season, and temperature that agree with sediment trap and plankton tow data (Kuroyanagi and Kawahata 2004; Xu et al. 2005).

Although we applied the $\delta^{18}O_c$–temperature equation by Bemis et al. (1998) to the core-top data, there are other $\delta^{18}O_c$–temperature relationships that yield various estimates for calcification temperature. For instance, $\delta^{18}O_c$–temperature equations by Kim and O'Neil (1997) and Mulitza et al. (2003) yield an offset of approximately 1 °C to the calcification temperatures calculated by Eq. (2). If these equations are applied to our core-top data with the assumption of $\delta^{18}O$ equilibrium, calcification temperatures at HR2 and HR9 will exceed the peak summer SST. Therefore, we should consider $\delta^{18}O$ disequilibrium in foraminiferal calcite (i.e., vital effect) for at least these sites. However, since the vital effect for *G. ruber* has not been determined precisely (0 to –1.0 ‰) (Niebler et al. 1999), applying these equations will introduce significant uncertainties in calcification temperature estimates.

Assessment of *G. ruber* Mg/Ca-derived temperatures

G. ruber is one of the species of planktonic foraminifera most susceptible to dissolution (Thunell and Honjo 1981). Partial dissolution of *G. ruber* shells can decrease Mg/Ca ratios owing to the higher solubility of Mg-rich calcite (Dekens et al. 2002). The ECS is the most undersaturated marginal sea with respect to calcite, and the CSH is shallow (~1600 m) (Fig. 6a). In our dataset, site HR2 is below the CSH, and the published core-top data from site A7, MD012404 (hereafter 404), and MD012403 (hereafter 403) were also derived from bottom water conditions with low calcite saturation states of <5 μmol kg^{-1} ($\Delta[CO_3^{2-}] = [CO_3^{2-}]_{in\ situ} - [CO_3^{2-}]_{saturation}$) (Fig. 6a; see caption of Fig. 6a for more details). According to data from core-top *G. ruber* tests of a depth transect in the Ontong Java Plateau, Mg/Ca ratios of *G. ruber* started to decrease with decreasing $\Delta[CO_3^{2-}]$ (−0.05 mmol mol^{-1} per µmol^{-1} kg^{-1}) from an initial $\Delta[CO_3^{2-}]$ of 14 µmol kg^{-1} (Johnstone et al. 2011). This finding is a cause for concern that ECS Mg/Ca data obtained under bottom water conditions with low $\Delta[CO_3^{2-}]$ may be influenced by dissolution (Fig. 6a).

Therefore, we assessed the impact of calcite dissolution on Mg/Ca ratios in the ECS by applying critical thresholds for the dissolution (14 µmol kg^{-1} $\Delta[CO_3^{2-}]$) and the sensitivity of the Mg/Ca ratio to $\Delta[CO_3^{2-}]$ (0.05 mmol mol^{-1} per µmol^{-1} kg^{-1}) (Johnstone et al. 2011). We calculated $\Delta[CO_3^{2-}]$ for the bottom waters at each core site and estimated the dissolution-corrected Mg/Ca ratios for all core sites (Table 2). Figure 6 shows a comparison of the $\delta^{18}O_c$-derived SSTs with non-dissolution-corrected and dissolution-corrected Mg/Ca-derived SSTs (Fig. 6b, c). Both Mg/Ca ratios were converted to SSTs using the SCS calibration equation of Hastings et al. (2001).

The shallowest site, HR9, where the $\Delta[CO_3^{2-}]$ value is 12.9 µmol kg^{-1}, showed good agreement with the non-dissolution-corrected Mg/Ca- and $\delta^{18}O_c$-derived SSTs (Fig. 6b). It is noteworthy that core-top data at depths ranging from 1000 to ~1400 m (HR6, St.1, A7, 404, and 403) also showed agreement with the non-dissolution-corrected Mg/Ca- and $\delta^{18}O_c$-derived SSTs, even though the $\Delta[CO_3^{2-}]$ values were lower than the critical threshold for dissolution (10–20 µmol kg^{-1}) (Regenberg et al. 2009; Johnstone et al. 2011) (Fig. 6b). The deepest site, HR2 (1675 m water depth), showed that the non-dissolution-corrected Mg/Ca-derived SST underestimates the $\delta^{18}O_c$-derived SST by 1.8 °C (Fig. 6b), and the dissolution-corrected Mg/Ca-derived SST agrees well with the $\delta^{18}O_c$-derived SST (Fig. 6c). However, this dissolution correction induced a larger offset to the $\delta^{18}O_c$-derived SSTs for other sites (1000 to ~1400 m water depths) (Fig. 6c).

The Mg/Ca–temperature calibration equation of Hastings et al. (2001) was developed for *G. ruber* from SCS core-top samples recovered from depths above 2000 m. As in the ECS, the deep waters in the SCS are characterized by low $\Delta[CO_3^{2-}]$ with the CSH at a water depth of 2500 m (Regenberg et al. 2009). Presumably, samples recovered from depths of 1500 to 2000 m (<15 µmol kg^{-1} $\Delta[CO_3^{2-}]$) for the calibration equation might have been influenced by partial dissolution under low $\Delta[CO_3^{2-}]$. Therefore, we consider that the SCS calibration equation potentially involves a correction for calcite dissolution under low $\Delta[CO_3^{2-}]$. Consequently, the correction for calcite dissolution is required only for samples obtained below the CSH in the ECS.

Thus, in this study, we integrated data of dissolution-corrected (for HR2 from below the CSH) and non-dissolution-corrected (for other cores above the CSH) Mg/Ca values to obtain Mg/Ca-derived SSTs (Fig. 6d). Although HR11 and 403 are still offset from the $\delta^{18}O_c$-derived SST by 1 °C, the other core-top data showed good agreement with both SST estimates ($r^2 = 0.83$, $p < 0.001$, except HR11 and 403) (Fig. 6d). This SCS calibration equation reconciles Mg/Ca-derived and $\delta^{18}O_c$-derived SSTs in the ECS, both of which reflect SSTs in the warmer season months. However, other calibration equations (Lea et al. 2000; Anand et al. 2003; McConnell and Thunell 2005) do not yield temperatures corresponding to the warmer season months and do not agree well with $\delta^{18}O_c$-derived SSTs (Table 2).

$\delta^{18}O_{sw}$ estimate for *G. ruber* in core-top sediments

Core-top $\delta^{18}O_{sw}$ values were calculated from Mg/Ca-derived SSTs and core-top $\delta^{18}O_c$ values using Eq. (2). The core-top data of the shell-derived $\delta^{18}O_{sw}$ and the salinity, including the published data from KY, A7, 404, and 403 (Sun et al. 2005; Lin et al. 2006; Chen et al. 2010; Kubota et al. 2010), are plotted in Fig. 7a. Since the estimation of the shell-derived $\delta^{18}O_{sw}$ is affected by the uncertainty in the $\delta^{18}O_c$ (±0.11 ‰) and the Mg/Ca-derived temperature (±1 °C, corresponding to approximately ±0.26 ‰) (Schmidt 1999; Rohling 2007), the reconstructed $\delta^{18}O_{sw}$ introduces a large propagation error of ±0.28 ‰ (1σ). Although the regression line of the core-top $\delta^{18}O_{sw}$ data has a steeper gradient and a lower *y*-intercept, given 1σ error bars of $\delta^{18}O_{sw}$ estimates the data points fall primarily on the local $\delta^{18}O_{sw}$–salinity regression line of the eastern part of the ECS (Fig. 7a). Indeed, sites KY and HR11 in the northern Okinawa Trough were marked by the lowest shell-derived $\delta^{18}O_{sw}$ values, whereas St.1 and A7 in the central Okinawa Trough showed the highest shell-derived $\delta^{18}O_{sw}$ values. Sites HR9 and MT5 had intermediate $\delta^{18}O_{sw}$ values between the above two datasets (Fig. 7a). The core-top $\delta^{18}O_{sw}$ values were roughly distributed as expected from the peak summer SSS at each site.

Although site MT5 is in an area with a lower summer SSS than that of site HR9, the shell-derived $\delta^{18}O_{sw}$ value at MT5 (0.17 ‰) was higher than that at HR9 (0.01 ‰) (Fig. 7a). One potential explanation is that *G. ruber* shells at HR9 were formed in the surface waters during the peak summer season when SSS decreases distinctly, whereas *G. ruber* at MT5 was not abundant in the peak summer season, as inferred from the $\delta^{18}O_c$ data (Fig. 5). These data tell us that the difference of 0.16 ‰ in the shell-derived $\delta^{18}O_{sw}$ may be caused by changes in the seasonal habitat of *G. ruber*.

This core-top investigation shows that realistic $\delta^{18}O_{sw}$ values can be reconstructed by the Mg/Ca–temperature calibration of Hastings et al. (2001) and the $\delta^{18}O_c$–

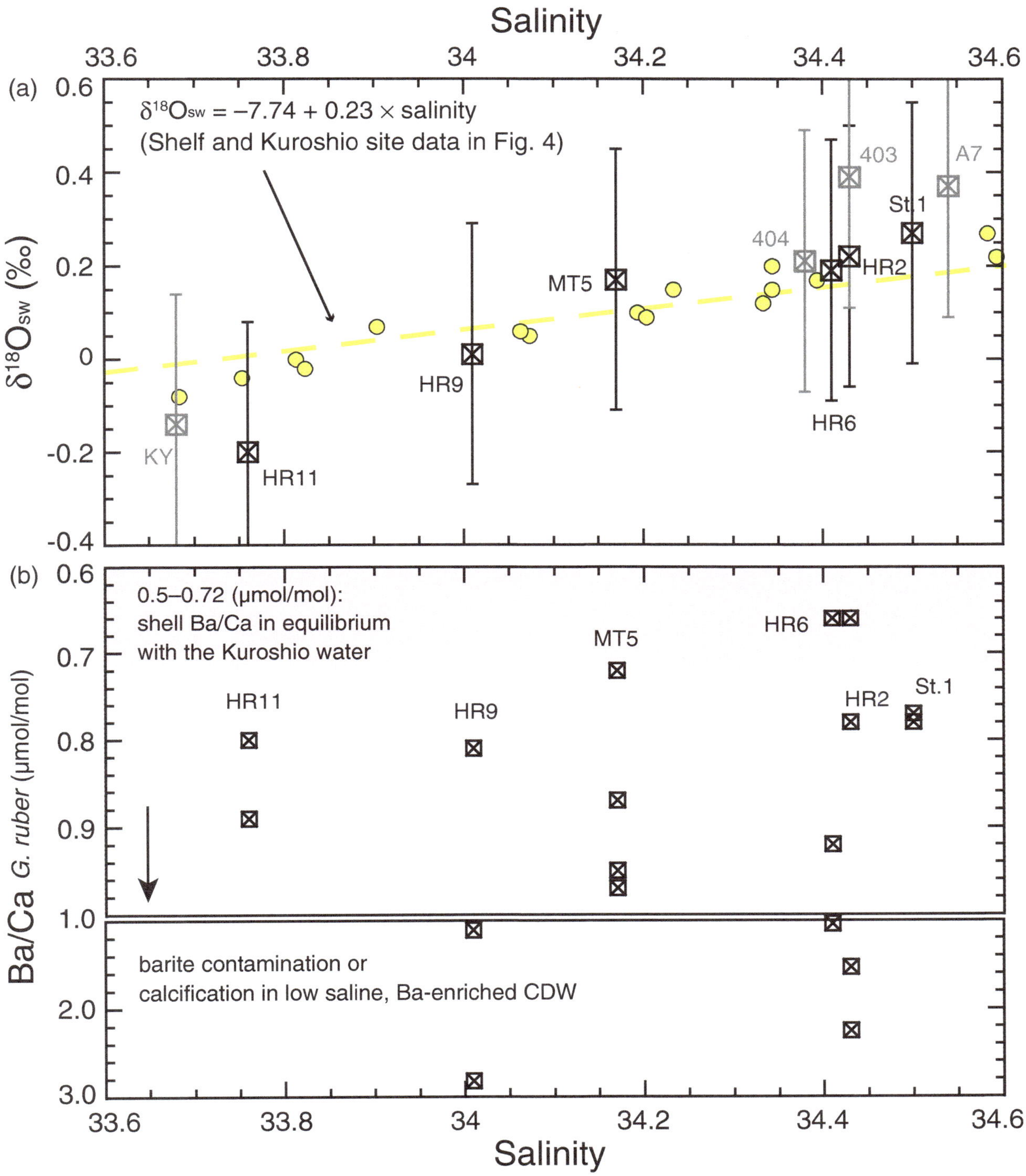

Fig. 7 Salinity, shell-derived $\delta^{18}O_{sw}$, and Ba/Ca ratios of core-top *G. ruber* tests. **a** Shell-derived $\delta^{18}O_{sw}$ and calcification salinity plot of core-top samples in this study and the literature. Shell-derived $\delta^{18}O_{sw}$ values are calculated using the Mg/Ca-derived temperature and core-top $\delta^{18}O_c$ of *G. ruber*. The data from KY, A7, 403, and 404 (*gray squares*) are from Kubota et al. (2010), Sun et al. (2005), Lin et al. (2006), and Chen et al. (2010), respectively. *Error bars* show the propagation error of ±0.28 ‰ (1σ). Seawater data (*green circles*) and their regression line (green broken line, Fig. 4) are also shown with the core-top data. Core-top data fall close to the local $\delta^{18}O_{sw}$–salinity regression line. **b** $Ba/Ca_{G.\ ruber}$ ratios and calcification salinity plot of all core-top samples in this study. The replicated analyses of $Ba/Ca_{G.\ ruber}$ ratios at each core site show high variability, resulting in the range from 0.66 to 2.82 μmol mol^{-1}. The calculated shell Ba/Ca ratio in equilibrium with the Kuroshio water is 0.5–0.72 μmol mol^{-1}. Lower values at HR2 and HR6 are within this range. Higher shell Ba/Ca ratios than this Kuroshio water range might be attributable to barite contamination or calcification in low saline, Ba-enriched CDW

temperature equation of Bemis et al. (1998), although $\delta^{18}O_{sw}$ estimates introduce a large error. Furthermore, this finding indicates that robust reconstruction of EASM-related variability in SSS might be possible by conducting multiple replicates of paired $\delta^{18}O_c$ and Mg/Ca analyses to decrease the large uncertainty. In addition, the utility of potential paleo-salinity proxies, such as the foraminiferal Ba/Ca ratio and the organic compound-specific δD, should be investigated further (Rohling 2007).

Possibility of using Ba/Ca ratios of *G. ruber* as a proxy for surface salinity

In general, the desorption of Ba^{2+} from suspended sediments in rivers results in a very high riverine [Ba^{2+}] relative to seawater. If there is a single river source, conservative mixing of river water and ambient seawater produces a linear inverse correlation between salinity and [Ba^{2+}] (Bahr et al. 2013). Ba^{2+} incorporation into living planktonic foraminifera shells is linearly dependent on the [Ba^{2+}] of the water, with a constant partition coefficient for Ba ($D_{Ba} = 0.15$) (Lea and Spero 1994; Hönisch et al. 2011). *G. ruber* populations can thrive in low-salinity surface water influenced by river discharge (Schmuker and Schiebel 2002; Arnold and Parker 2003). This ecological preference of *G. ruber* and the incorporation of Ba into foraminifera shells makes Ba/Ca ratios in fossil *G. ruber* a proxy for the influence of river discharge on salinity (Weldeab et al. 2007; Schmidt and Lynch-Stieglitz 2011).

In the case of the ECS, the Changjiang River accounts for approximately 90 % of the total river discharge into the ECS, with large quantities of Ba^{2+} (~360 nmol L^{-1}) (Beardsley et al. 1985; Qu et al. 1993; Yanagi 1994). Based on our preliminary [Ba^{2+}] data in the ECS measured by an isotope dilution method (Klinkhammer and Chan 1990), the surface water at SSW8 (30.80 salinity) showed a high [Ba^{2+}] of 57.0 nmol L^{-1} (Ba/Ca ratio of 6.31 μmol mol^{-1}), and the Kuroshio surface water (SSW14–17, average 34.2) showed a low [Ba^{2+}] of 32.9 nmol L^{-1} (Ba/Ca of 3.27 μmol mol^{-1}). The [Ba^{2+}] data for the Kuroshio water were consistent with previously reported data from Kuroshio waters (29.9–33.5 nmol L^{-1}) (Sugiyama et al. 1984). Other surface samples from the northern Okinawa Trough fall within these [Ba^{2+}] and Ba/Ca ranges, showing a linear relationship between $Ba/Ca_{seawater}$ and salinity (Horikawa et al., personal communication).

Our Ba/Ca ratios in *G. ruber* (hereafter $Ba/Ca_{G.\ ruber}$) from core-top sediments ranged from 0.66 to 2.82 μmol mol^{-1} (Fig. 7b). The lowest Ba/Ca value of 0.66 μmol mol^{-1} was observed at sites HR2 and HR6 in the central Okinawa Trough. Given the $Ba/Ca_{seawater}$ ratio of 3.27 μmol mol^{-1} in the Kuroshio surface water and the likely D_{Ba} of 0.15–0.22 (Hall and Chan 2004; Hönisch et al. 2011), the calculated shell Ba/Ca ratio in equilibrium with the Kuroshio water should be 0.5 to 0.72 μmol mol^{-1}. Indeed, HR2 and HR6 values were within the expected range for Kuroshio surface water. In contrast, HR11, where the lowest SSS among the studied sites was observed, had relatively higher $Ba/Ca_{G.\ ruber}$ ratios (0.80–0.89 μmol mol^{-1}) (Fig. 7b). Given the $Ba/Ca_{seawater}$ ratio of 3.8 μmol mol^{-1} (at HR10, 0 m) and the D_{Ba} of 0.15 to 0.22, the shell Ba/Ca ratio in equilibrium should be 0.57 to 0.84 μmol mol^{-1}. The core-top $Ba/Ca_{G.\ ruber}$ ratios observed at HR11 were consistent with the upper range of expected values, probably reflecting the influence of the Ba-enriched CDW (i.e., low-salinity waters).

However, all core-top $Ba/Ca_{G.\ ruber}$ data at six sites do not show a significant relationship with the salinity, but replicate data at each site showed highly variable $Ba/Ca_{G.\ ruber}$ ratios, some of which exceeded the values at site HR11 (Fig. 7b). Although we are unable to adequately explain such highly variable $Ba/Ca_{G.\ ruber}$ ratios at this time, one likely reason is sample heterogeneity. In this study, to obtain averaged $Ba/Ca_{G.\ ruber}$ ratios for each site, we re-picked 20–30 shells of *G. ruber* more than twice after picking the samples labeled run #1 (Table 2). Although the samples from run #1 gave reasonable $Ba/Ca_{G.\ ruber}$ ratios (0.66–0.89 μmol mol^{-1}) at each site except HR6, the samples from runs #2 to #4 yielded anomalously high $Ba/Ca_{G.\ ruber}$ ratios compared with the samples from run #1. Given that there were few well-preserved, clean *G. ruber* tests for these sites and that replicate measurements of Mn/Ca and Mg/Ca ratios were within an acceptable range, we argue that samples from runs #2 to #4 may have involved impure *G. ruber* tests in which barite crystallized on fossil shell surfaces.

Finally, based on the available dataset of $Ba/Ca_{seawater}$ in the ECS, we propose that pristine $Ba/Ca_{G.\ ruber}$ ratios at northern Okinawa Trough sites would be less than 0.84 μmol mol^{-1}. A possible reason for higher core-top $Ba/Ca_{G.\ ruber}$ ratios (>0.84 μmol mol^{-1}) is contamination by sedimentary barite adherent on fossil shells. Our preliminary data suggest that $Ba/Ca_{G.\ ruber}$ data as a paleo-salinity proxy should be derived from well-preserved, clean *G. ruber* tests or diethylene triamine pentaacetic acid (DTPA)-cleaned *G. ruber* tests that can minimize the influence of barite contamination (Lea and Boyle 1991; Hall and Chan 2004). Since the $Ba/Ca_{G.\ ruber}$ records from the northern Okinawa Trough sediments may potentially identify high river discharge events related to stronger EASMs, assessment of foraminiferal Ba/Ca should be continued in the ECS.

Conclusions

The main findings of this study are as follows:

(1) Shallow seawater samples (0–100 m) taken from the ECS in early July showed a strong correlation between $\delta^{18}O_{sw}$ and salinity. The *T–S* diagram in the surface waters in the ECS shows that the eastward-flowing CDW mixes with the saline Kuroshio waters. Based on this finding, we derived the following $\delta^{18}O_{sw}$–salinity equation from the Yellow Sea–ECS shelf and Kuroshio site data:

$$\delta^{18}O_{sw} = -7.74(\pm 0.4) + 0.23(\pm 0.01) \times salinity\left(r^2 = 0.97, p < 0.001, n = 40\right).$$

This local $\delta^{18}O_{sw}$–salinity equation might be representative of the northern Okinawa Trough during the EASM season.

(2) We found that the dominant habitat depth of *G. ruber* is within the top 0 to 30 m in the ECS. The calcification season of *G. ruber* is mainly during the warm summer months (May–October) but may not necessarily be uniform across the ECS. We confirmed that the Mg/Ca–temperature calibration by Hastings et al. (2001) yields temperatures corresponding to the warmer season months, as expected from sediment trap data. The Mg/Ca-derived SSTs agreed with calcification temperatures calculated by Bemis et al. (1998). Site HR2, where the bottom water is undersaturated with calcite, required a correction for calcite dissolution.

(3) We found that core-top data for shell-derived $\delta^{18}O_{sw}$ and salinity fall primarily on our local $\delta^{18}O_{sw}$–salinity regression line giving 1σ error of $\delta^{18}O_{sw}$ estimates. The Mg/Ca–temperature calibration by Hastings et al. (2001) and the $\delta^{18}O_c$–temperature equation by Bemis et al. (1998) should be appropriate for calculating $\delta^{18}O_{sw}$ in the ECS.

(4) $Ba/Ca_{G.\ ruber}$ from core-top sediments ranged from 0.66 to 2.82 μmol mol^{-1}. There was not a significant relationship between salinity and $Ba/Ca_{G.\ ruber}$ due to the highly variable $Ba/Ca_{G.\ ruber}$ data. Given the seawater Ba/Ca data and the published partition coefficient for Ba (D_{Ba} = 0.15–0.22), pristine $Ba/Ca_{G.\ ruber}$ ratios at northern Okinawa Trough sites should be less than 0.84 μmol mol^{-1}. One possible reason for higher $Ba/Ca_{G.\ ruber}$ ratios (>0.84 μmol mol^{-1}) is contamination by sedimentary barite adherent on fossil shells. Further evaluation of the $Ba/Ca_{G.\ ruber}$ ratio as a paleo-salinity proxy requires DTPA-cleaned Ba/Ca data that can minimize the influence of barite contamination.

Abbreviations

EASM: East Asian summer monsoon; ECS: East China Sea; CDW: Changjiang diluted water; SSS: sea surface salinity; TWWC: Taiwan warm current; SCS: South China Sea; YSCCW: Yellow Sea Central Cold Water; SST: sea surface temperature; JODC: Japan Oceanographic Data Center; CTD: conductivity, temperature, and depth; JGOFS: Joint Global Ocean Flux Study; IAPSO: International Association for the Physical Sciences of the Oceans; CSH: calcite saturation horizon; VSMOW: Vienna Standard Mean Ocean Water; CMCR: Center for Advanced Marine Core Research; PDB: Pee Dee Belemnite; DTPA: diethylene triamine pentaacetic acid.

Competing interests

The authors declare that they have no competing interests.

Authors' contributions

KH conceived and designed the study. TK carried out the measurements of the seawater $\delta^{18}O$, foraminiferal $\delta^{18}O_c$, and metal/Ca ratios under the supervision of Professors MM and JZ. All authors analyzed and discussed the data. All authors contributed toward writing the manuscript and read and approved the final manuscript.

Authors' information

KH is an associate professor at the University of Toyama. He is a paleoceanographer and a geochemist who has worked on SST reconstruction around Japan using the Mg/Ca–temperature proxy and the analysis of deep-sea water mass circulation using neodymium isotopes. JZ is also a geochemist and a professor at the University of Toyama with considerable experience in geochemical analysis. TK is a Ph.D. student supervised by KH and JZ. MM is a professor at the Center for Advanced Marine Core Research (CMCR), Kochi University.

Acknowledgements

We acknowledge the tremendous support of the KH-13-4 and KT-12-25 shipboard scientists and staff. This study was performed under the cooperative research program of the Center for Advanced Marine Core Research (CMCR), Kochi University (#13B050). We thank the chief editors Ryuji Tada, Hodaka Kawahata, Yasufumi Iryu, and two anonymous reviewers for their significant help in improving this manuscript.

Author details

[1]Graduate School of Science and Engineering for Research, University of Toyama, Gofuku 3190, Toyama 930-8555, Japan. [2]Graduate School of Science and Engineering for Education, University of Toyama, Gofuku 3190, Toyama 930-8555, Japan. [3]Center for Advanced Marine Core Research, Kochi University, B200 Monobe, Nankoku 783-8502, Japan.

References

Anand P, Elderfield H, Conte MH (2003) Calibration of Mg/Ca thermometry in planktonic foraminifera from a sediment trap time series. Paleoceanography 18(2):1050. doi:10.1029/2002PA000846

Arnold A, Parker W (2003) Biogeography of planktonic Foraminifera. In: Gupta BKS (ed) Modern Foraminifera. Springer, Netherlands, pp 103–122. doi:10.1007/0-306-48104-9_7

Bahr A, Schönfeld J, Hoffmann J, Voigt S, Aurahs R, Kucera M, Flögel S, Jentzen A, Gerdes A (2013) Comparison of Ba/Ca and $\delta^{18}O_{WATER}$ as freshwater proxies: A multi-species core-top study on planktonic foraminifera from the vicinity of the Orinoco River mouth. Earth Planet Sci Lett 383:45–57. doi:10.1016/j.epsl.2013.09.036

Beardsley RC, Limeburner R, Yu H, Cannon GA (1985) Discharge of the Changjiang (Yangtze River) into the East China Sea. Cont Shelf Res 4(1–2):57–76. doi:10.1016/0278-4343(85)90022-6

Bemis BE, Spero HJ, Bijima J, Lea DW (1998) Reevaluation of the oxygen isotopic composition of planktonic foraminifera: Experimental results and revised paleotemperature equations. Paleoceanography 13(2):150–160

Benway HM, Mix AC, Haley BA, Klinkhammer GP (2006) Eastern Pacific Warm Pool paleosalinity and climate variability: 0–30 kyr. Paleoceanography 21(3):PA3008. doi:10.1029/2005PA001208

Boyle EA, Keigwin LD (1985) Comparison of Atlantic and Pacific paleochemical records for the last 215,000 years: changes in deep ocean circulation and chemical inventories. Earth Planet Sci Lett 76(1–2):135–150

Chang P-H, Isobe A (2003) A numerical study on the Changjiang diluted water in the Yellow and East China Seas. J Geophys Res 108(C9):3299. doi:10.1029/2002JC001749

Chang Y-P, Chen M-T, Yokoyama Y, Matsuzaki H, Thompson WG, Kao S-J, Kawahata H (2009) Monsoon hydrography and productivity changes in the East China Sea during the past 100,000 years: Okinawa Trough evidence (MD012404). Paleoceanography 24(3):PA3208. doi:10.1029/2007PA001577

Chen C, Beardsley RC, Limeburner R, Kim K (1994) Comparison of winter and summer hydrographic observations in the Yellow and East China Seas and adjacent Kuroshio during 1986. Cont Shelf Res 14(7–8):909–929. doi:10.1016/0278-4343(94)90079-5

Chen F, Yu Z, Yang M, Ito E, Wang S, Madsen DB, Huang X, Zhao Y, Sato T, Birks HJB, Boomer I, Chen J, An C, Wünnemann B (2008) Holocene moisture evolution in arid central Asia and its out-of-phase relationship with Asian monsoon history. Quat Sci Rev 27(3–4):351–364. doi:10.1016/j.quascirev.2007.10.017

Chen MT, Lin XP, Chang YP, Chen YC, Lo L, Shen CC, Yokoyama Y, Oppo DW, Thompson WG, Zhang R (2010) Dynamic millennial-scale climate changes in the northwestern Pacific over the past 40,000 years. Geophys Res Lett 37(23):L23603. doi:10.1029/2010GL045202

Dekens PS, Lea DW, Pak DK, Spero HJ (2002) Core top calibration of Mg/Ca in tropical foraminifera: Refining paleotemperature estimation. Geochem Geophys Geosyst 3(4):1–29. doi:10.1029/2001GC000200

Du J, Chen M, Cao J, Qiu Y, Tong J, Ma Q, Yang J (2012) Oxygen isotope in seawater and its hydrological implication in the southern Yellow Sea and the East China Sea. Oceanol Limnol Sin 43(6):1057–1066 (in Chinese with English abstract)

Epstein S, Mayeda T (1953) Variation of O18 content of waters from natural sources. Geochim Cosmochim Acta 4(5):213–224. doi:10.1016/0016-7037(53)90051-9

Greaves M, Caillon N, Rebaubier H, Bartoli G, Bohaty S, Cacho I, Clarke L, Cooper M, Daunt C, Delaney M, deMenocal P, Dutton A, Eggins S, Elderfield H, Garbe-Schoenberg D, Goddard E, Green D, Groeneveld J, Hastings D, Hathorne E, Kimoto K, Klinkhammer G, Labeyrie L, Lea DW, Marchitto T, Martinez-Boti MA, Mortyn PG, Ni Y, Nuernberg D, Paradis G et al (2008) Interlaboratory comparison study of calibration standards for foraminiferal Mg/Ca thermometry. Geochem Geophys Geosyst 9(8):Q08010. doi:10.1029/2008gc001974

Hall JM, Chan LH (2004) Ba/Ca in Neogloboquadrina pachyderma as an indicator of deglacial meltwater discharge into the western Arctic Ocean. Paleoceanography 19(1):PA1017. doi:10.1029/2003PA000910

Hastings DW, Whitko A, Kienast M, Steinke S (2001) A comparison of three independent paleotemperature estimates from a high resolution record of deglacial SST records in the tropical South China Sea. Eos 82:PP12B–10

Hönisch B, Allen KA, Russell AD, Eggins SM, Bijma J, Spero HJ, Lea DW, Yu J (2011) Planktic foraminifers as recorders of seawater Ba/Ca. Mar Micropaleontol 79(1–2):52–57. doi:10.1016/j.marmicro.2011.01.003

Ichikawa H, Beardsley R (2002) The Current System in the Yellow and East China Seas. J Oceanogr 58(1):77–92. doi:10.1023/A:1015876701363

Jan S, Sheu DD, Kuo H-M (2006) Water mass and throughflow transport variability in the Taiwan Strait. J Geophys Res 111(C12):C12012. doi:10.1029/2006JC003656

Johnstone HJH, Yu J, Elderfield H, Schulz M (2011) Improving temperature estimates derived from Mg/Ca of planktonic foraminifera using X-ray computed tomography–based dissolution index, XDX. Paleoceanography 26(1):PA1215. doi:10.1029/2009PA001902

Kang D-J, Chung CS, Kim SH, Hong GH, Kim K-R (1994) Oxygen isotope characteristics of seawaters in the Yellow Sea. La mer 32:279–284

Kim S-T, O'Neil JR (1997) Equilibrium and nonequilibrium oxygen isotope effects in synthetic carbonates. Geochim Cosmochim Acta 61(16):3461–3475. doi:10.1016/S0016-7037(97)00169-5

Kim K-R, Cho Y-K, Kang D-J, Ki J-H (2005) The origin of the Tsushima Current based on oxygen isotope measurement. Geophys Res Lett 32(3):L03602. doi:10.1029/2004GL021211

Klinkhammer GP, Chan LH (1990) Determination of barium in marine waters by isotope dilution inductively coupled plasma mass spectrometry. Anal Chim Acta 232(0):323–329. doi:10.1016/S0003-2670(00)81249-0

Kubota Y, Kimoto K, Tada R, Oda H, Yokoyama Y, Matsuzaki H (2010) Variations of East Asian summer monsoon since the last deglaciation based on Mg/Ca and oxygen isotope of planktic foraminifera in the northern East China Sea. Paleoceanography 25(4):PA4205. doi:10.1029/2009pa001891

Kuroyanagi A, Kawahata H (2004) Vertical distribution of living planktonic foraminifera in the seas around Japan. Mar Micropaleontol 53(1–2):173–196. doi:10.1016/j.marmicro.2004.06.001

Lea DW, Boyle EA (1991) Barium in planktonic foraminifera. Geochim Cosmochim Acta 55(11):3321–3331

Lea DW, Spero HJ (1994) Assessing the reliability of paleochemical tracers: Barium uptake in the shells of planktonic foraminifera. Paleoceanography 9(3):445–452. doi:10.1029/94PA00151

Lea DW, Mashiotta TA, Spero HJ (1999) Controls on magnesium and strontium uptake in planktonic foraminifera determined by live culturing. Geochim Cosmochim Acta 63(16):2369–2379

Lea DW, Pak DK, Spero HJ (2000) Climate impact of Late Quaternary equatorial Pacific sea surface temperature variations. Science 289(5485):1719–1724. doi:10.1126/science.289.5485.1719

Lee H-J, Chao S-Y (2003) A climatological description of circulation in and around the East China Sea. Deep-Sea Res PT II 50(6–7):1065–1084. doi:10.1016/S0967-0645(03)00010-9

Lee TN, Johns WE, Liu C-T, Zhang D, Zantopp R, Yang Y (2001) Mean transport and seasonal cycle of the Kuroshio east of Taiwan with comparison to the Florida Current. J Geophys Res 106(C10):22143–22158. doi:10.1029/2000JC000535

Lewis E, Wallace DWR (1998) CO2SYS–Program Developed for CO_2 System Calculations ORNL/CDIAC-105 (Carbon Dioxide Information Analysis Center, Oak Ridge National Laboratory, U.S. Department of Energy, Oak Ridge, TN)

Lin Y-S, Wei K-Y, Lin I-T, Yu P-S, Chiang H-W, Chen C-Y, Shen C-C, Mii H-S, Chen Y-G (2006) The Holocene Pulleniatina Minimum Event revisited: Geochemical and faunal evidence from the Okinawa Trough and upper reaches of the Kuroshio current. Mar Micropaleontol 59(3–4):153–170. doi:10.1016/j.marmicro.2006.02.003

Liu Z, Wen X, Brady EC, Otto-Bliesner B, Yu G, Lu H, Cheng H, Wang Y, Zheng W, Ding Y, Edwards RL, Cheng J, Liu W, Yang H (2014) Chinese cave records and the East Asia Summer Monsoon. Quat Sci Rev 83:115–128. doi:10.1016/j.quascirev.2013.10.021

Marchitto TM (2006) Precise multielemental ratios in small foraminiferal samples determined by sector field ICP-MS. Geochem Geophys Geosyst 7(5):Q05P13. doi:10.1029/2005GC001018

McConnell MC, Thunell RC (2005) Calibration of the planktonic foraminiferal Mg/Ca paleothermometer: Sediment trap results from the Guaymas Basin, Gulf of California. Paleoceanography 20(2):PA2016. doi:10.1029/2004PA001077

Mulitza S, Boltovskoy D, Donner B, Meggers H, Paul A, Wefer G (2003) Temperature:$\delta^{18}O$ relationships of planktonic foraminifera collected from surface waters. Palaeogeogr Palaeocl 202(1–2):143–152. doi:10.1016/S0031-0182(03)00633-3

Niebler HS, Hubberten HW, Gersonde R (1999) Oxygen isotope values of planktic foraminifera: A tool for the reconstruction of surface water stratification. In: Wefer G (ed) Fischer G. Use of Proxies in Paleoceanography, Springer Berlin Heidelberg, pp 165–189. doi:10.1007/978-3-642-58646-0_6

Oba T (1990) Paleoceanographic information obtained by isotopic measurement of individual foraminiferal specimens. In: Wang P, Lao Q, He Q (eds) Proceedings of the First International Conference on Asian Marine Geology. China Ocean Press, Beijing, pp 169–180

Qian W, Lee D-K (2000) Seasonal march of Asian summer monsoon. Int J Climatol 20:1371–1386

Qu CH, Chen CZ, Yang JR, Wang LZ, Lu YL (1993) Geochemistry of dissolved and particulate elements in the major rivers of China (The Huanghe, Changjiang, and Zhunjiang rivers). Estuaries 16(3):475–487. doi:10.2307/1352595

Regenberg M, Steph S, Nürnberg D, Tiedemann R, Garbe-Schönberg D (2009) Calibrating Mg/Ca ratios of multiple planktonic foraminiferal species with $\delta^{18}O$-calcification temperatures: Paleothermometry for the upper water column. Earth Planet Sci Lett 278(3–4):324–336. doi:10.1016/j.epsl.2008.12.019

Rohling EJ (2007) Progress in paleosalinity: Overview and presentation of a new approach. Paleoceanography 22(3):PA3215. doi:10.1029/2007PA001437

Rosenthal Y, Boyle EA, Labeyrie L (1997) Last glacial maximum paleochemistry and deepwater circulation in the Southern Ocean: Evidence from foraminiferal cadmium. Paleoceanography 12(6):787–796

Russell AD, Hönisch B, Spero HJ, Lea DW (2004) Effects of seawater carbonate ion concentration and temperature on shell U, Mg, and Sr in cultured planktonic foraminifera. Geochim Cosmochim Acta 68(21):4347–4361. doi:10.1016/j.gca.2004.03.013

Schmidt GA (1999) Error analysis of paleosalinity calculations. Paleoceanography 14(3):422–429. doi:10.1029/1999PA900008

Schmidt MW, Lynch-Stieglitz J (2011) Florida Straits deglacial temperature and salinity change: Implications for tropical hydrologic cycle variability during

the Younger Dryas. Paleoceanography 26(4):PA4205. doi:10.1029/2011PA002157
Schmuker B, Schiebel R (2002) Planktic foraminifers and hydrography of the eastern and northern Caribbean Sea. Mar Micropaleontol 46(3–4):387–403. doi:10.1016/S0377-8398(02)00082-8
Sugiyama M, Matsui M, Nakayama E (1984) Direct determination of barium in sea water by inductively coupled plasma emission spectrometry. J Oceanogr 40:295–302
Sun Y, Oppo DW, Xiang R, Liu W, Gao S (2005) Last deglaciation in the Okinawa Trough: Subtropical northwest Pacific link to Northern Hemisphere and tropical climate. Paleoceanography 20(4):PA4005. doi:10.1029/2004PA001061
Thunell RC, Honjo S (1981) Calcite dissolution and the modification of planktonic foraminiferal assemblages. Mar Micropaleontol 6(2):169–182. doi:10.1016/0377-8398(81)90004-9
Ujiié Y, Ujiié H, Taira A, Nakamura T, Oguri K (2003) Spatial and temporal variability of surface water in the Kuroshio source region, Pacific Ocean, over the past 21,000 years: evidence from planktonic foraminifera. Mar Micropaleontol 49(4):335–364
Wang B (2006) The Asian Monsoon. Springer Praxis, Chichester
Wang Y, Cheng H, Edwards RL, He Y, Kong X, An Z, Wu J, Kelly MJ, Dykoski CA, Li X (2005) The Holocene Asian Monsoon: Links to solar changes and North Atlantic climate. Science 308(5723):854–857. doi:10.1126/science.1106296
Wang Y, Cheng H, Edwards RL, Kong X, Shao X, Chen S, Wu J, Jiang X, Wang X, An Z (2008) Millennial- and orbital-scale changes in the East Asian monsoon over the past 224,000 years. Nature 451(7182):1090–1093
Weldeab S, Lea DW, Schneider RR, Andersen N (2007) 155,000 years of west African monsoon and ocean thermal evolution. Science 316(5829):1303–1307. doi:10.1126/science.1140461
Xu X, Yamasaki M, Oda M, Honda MC (2005) Comparison of seasonal flux variations of planktonic foraminifera in sediment traps on both sides of the Ryukyu Islands, Japan. Mar Micropaleontol 58(1):45–55. doi:10.1016/j.marmicro.2005.09.002
Yanagi T (1994) Material transport in the Yellow/East China Seas. Bull Coast Oceanogr 31:239–256
Yancheva G, Nowaczyk NR, Mingram J, Dulski P, Schettler G, Negendank JFW, Liu J, Sigman DM, Peterson LC, Haug GH (2007) Influence of the intertropical convergence zone on the East Asian monsoon. Nature 445(7123):74–77
Ye F, Deng W, Xie L, Wei G, Jia G (2014) Surface water $\delta^{18}O$ in the marginal China seas and its hydrological implications. Est Coast Shelf Sci 147:25–31. doi:10.1016/j.ecss.2014.05.033
Zhang J, Letolle R, Martin JM, Jusserand C, Mouchel JM (1990) Stable oxygen isotope distribution in the Huanghe (Yellow River) and the Changjiang (Yangtze River) estuarine systems. Cont Shelf Res 10(4):369–384. doi:10.1016/0278-4343(90)90057-S
Zhang L, Liu Z, Zhang J, Hong GH, Park Y, Zhang HF (2007) Reevaluation of mixing among multiple water masses in the shelf: An example from the East China Sea. Cont Shelf Res 27(15):1969–1979. doi:10.1016/j.csr.2007.04.002
Zhang W-Z, Chai F, Hong H-S, Xue H (2014) Volume transport through the Taiwan Strait and the effect of synoptic events. Cont Shelf Res. doi:10.1016/j.csr.2014.07.010
Zweng MM, Reagan JR, Antonov JI, Locarnini RA, Mishonov AV, Boyer TP, Garcia HE, Baranova OK, Johnson DR, Seidov D, Biddle MM (2013) World Ocean Atlas 2013, Volume 2: Salinity. NOAA Atlas NESDIS 74. Silver Spring, MD

7

Historical and paleo-tsunami deposits during the last 4000 years and their correlations with historical tsunami events in Koyadori on the Sanriku Coast, northeastern Japan

Daisuke Ishimura[1*] and Takahiro Miyauchi[2]

Abstract

Large tsunamis occurring throughout the past several hundred years along the Sanriku Coast on the Pacific coast of northeastern Japan have been documented and observed. However, the risk of large tsunamis like the tsunami generated by the 2011 off the Pacific coast of Tohoku earthquake could not be evaluated from previous studies, because these studies lacked evidence of historical and paleo-tsunami deposits on the coastline. Thus, we first identified event deposits, which are candidates for tsunami deposits, from excavating surveys conducted on the coastal marsh in Koyadori on the Sanriku Coast, northeastern Japan. Second, we determined the physicochemical sediment properties of the deposits (roundness of grains, color, wet and dry densities, and loss on ignition) and established their geochronology by radiocarbon dating and tephra analysis. Third, we identified event deposits as tsunami deposits, based on their sedimentary features and origin, sedimentary environment, paleo-shoreline, and landowner interviews. In this study, we report 11 tsunami deposits (E1–E11) during the past 4000 years, of which E1, E2, E3, and E4 were correlated with the 2011 Tohoku-oki tsunami, the 1896 Meiji Sanriku tsunami, the 1611 Keicho Sanriku tsunami, and the 869 Jogan tsunami, respectively. From age data and the number of tsunami deposits in the trench, we estimated that tsunamis larger than the 1896 Meiji Sanriku tsunami occur and hit the study area on average every 290–390 years. However, historical tsunami correlations revealed variable tsunami occurrence, indicating diverse tsunami generation and/or the combination of several types of large earthquakes from different sources around the Japan Trench.

Keywords: Historical and paleo-tsunami deposits; Sanriku Coast; 2011 Tohoku-oki tsunami; Historical tsunami correlation; Geochronology

Background

Historical and paleo-tsunami research and its application to geophysical study

The 2011 off the Pacific coast of Tohoku earthquake (2011 Tohoku-oki earthquake) (M_w 9.0) that occurred on March 11, 2011, triggered a large tsunami (2011 Tohoku-oki tsunami) along the east coast of Japan, causing severe damage and loss of life. The Pacific coastal region, in particular, the Sanriku Coast (Fig. 1), ranks among the highest tsunami risk areas in Japan, in terms of both severity and frequency (Watanabe 1998). However, we could not assess the risk of large tsunamis like the 2011 Tohoku-oki tsunami because we lack sufficient knowledge about large earthquakes and the tsunami history in the Japan Trench. Even after the 2011 Tohoku-oki tsunami, the long-term tsunami history of the Sanriku Coast remains obscured by inadequate field data (Sugawara et al. 2012). Moreover, the geological and geophysical problems revealed by the 2011 Tohoku-oki earthquake,

* Correspondence: ishimura@irides.tohoku.ac.jp
[1]Disaster Science Division, International Research Institute of Disaster Science, Tohoku University, 468-1 Aza-aoba, Aramaki, Aoba-ku, Sendai, Miyagi 980-0845, Japan
Full list of author information is available at the end of the article

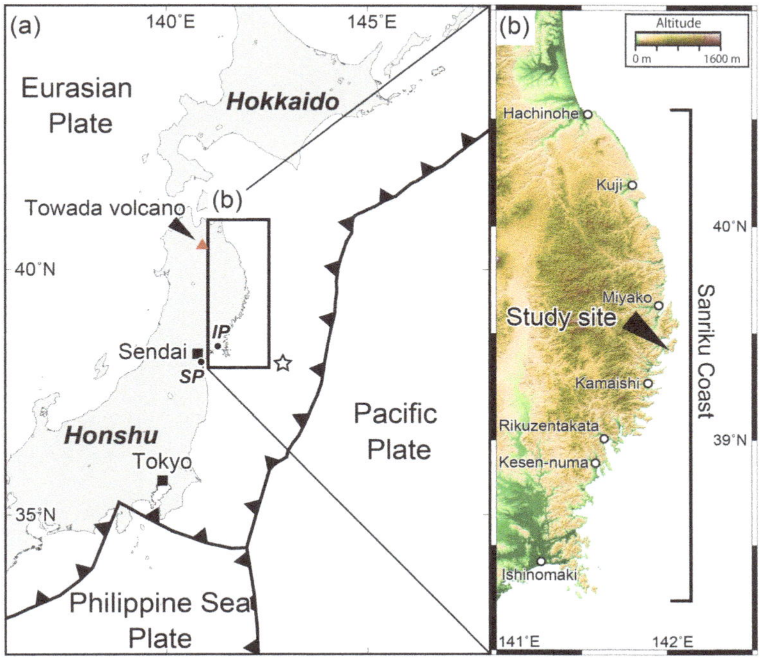

Fig. 1 Study site location. **a** Plate tectonic map in and around northeastern Japan. The star indicates the epicenter location of the 2011 Tohoku-oki earthquake determined by the Japan Meteorological Agency. *SP* Sendai Plain, *IP* Ishinomaki Plain. **b** Study area and geomorphology along the Sanriku Coast

including mechanisms of tsunami generation (Kawamura et al. 2012; Tappin et al. 2014) and large earthquakes (Ikeda et al. 2012; Goldfinger et al. 2013; Rajendran 2013), remain unsolved. Recent studies have suggested that submarine mass failure (e.g., submarine landslide) may have contributed to large tsunamis along the Japan Trench (Kawamura et al. 2012; 2014; Tappin et al. 2014). To resolve these problems, we require long-term solid evidence (e.g., historical and paleo-tsunami deposits).

The best sources of precise long-term tsunami data are coastal lowlands, in particular, marshes (Minoura and Nakaya 1991; Witter et al. 2003; Sawai et al. 2009; Shennan et al. 2014). In some regions (Hokkaido of Japan, Alaska of USA, and the North Island of New Zealand), coseismic and post-seismic crustal movements are recorded in sediments as lithological and biological changes (Witter et al. 2003; Sawai and Nasu 2005; Hamilton and Shennan 2005; Hayward et al. 2005). Thus, in this study, we excavated trenches at a coastal marsh in Koyadori, in the middle part of the Sanriku Coast (Fig. 1). The aim was to provide new geological evidence of historical and paleo-tsunami deposits.

Study site

Koyadori is located in the central part of the Sanriku Coast, the easternmost part of Honshu Island (Figs. 1 and 2). Approximately 200 km east of Koyadori, the Pacific plate subducts underneath the Eurasian Plate, where rupture areas of historical and observed earthquakes have been identified on the plate boundary (Earthquake Research Committee Headquarters for Earthquake Research Promotion Prime Minister's Office, Government of Japan 1999). In Koyadori, the mouth of the valley is closed by beach ridges. Prior to the 2011 Tohoku-oki tsunami, which hit the valley (Fig. 3), the study site was used as paddy fields.

The geology differs on both sides of the valley (Fig. 2; Yoshida et al. 1984). The east side comprises Early Cretaceous hornblende-biotite granodiorite, granite, granite porphyry, and tonalite. The west side comprises dacite to rhyolite lava and pyroclastic rock deposited during the Early Cretaceous. Ishimura et al. (2014) drilled cores at the study site (Fig. 4), revealing visible tephra layers such as the Towada–Chuseri tephra (To–Cu: 6 ka; Machida and Arai 2003) and Oguni Pumice (7.3–7.4 ka; Ishimura et al. 2014) erupted from the Towada Volcano (Fig. 1).

Tsunami history and previous study of tsunami deposits

Large historical earthquakes and tsunamis have been instrumentally recorded during the last several decades and have been recorded in historical documents and legends during the last 1300 years (Utsu 2004). Many earthquake-generated tsunamis have hit the areas around Koyadori (Table 1), nine of which were large events with runup heights exceeding a few meters (Table 2). The 2011 Tohoku-oki tsunami exhibits the largest runup height measured between AD 2011 and 1896. In contrast, based on the estimated runup heights, the largest tsunami occurring between AD 1611 and 1856 was probably the 1611 Keicho Sanriku tsunami (Table 2). Candidates of

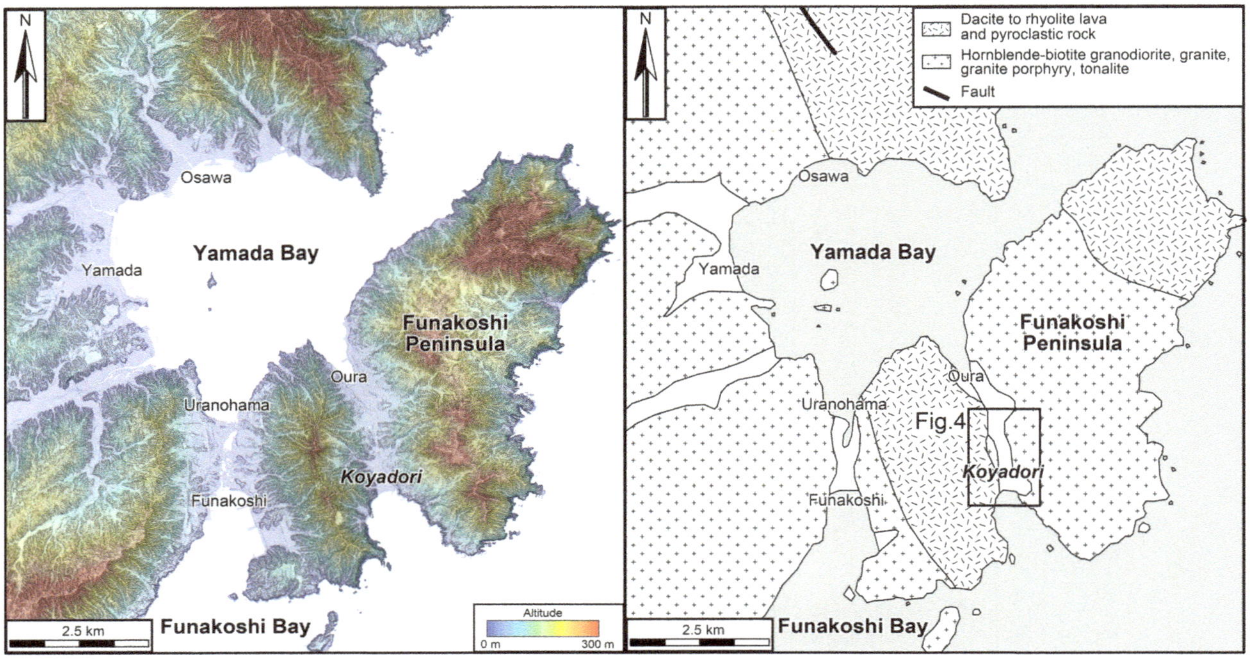

Fig. 2 Topography and geology around the Funakoshi Peninsula. Base map is based on the 5 m mesh DEM supplied by the Geospatial Information Authority of Japan. Geology is modified after Yoshida et al. (1984)

historical tsunamis older than AD 1611 are the 869 Jogan tsunami and the 1454 Kyotoku tsunami. The 869 Jogan tsunami, identified in the Sendai and Ishinomaki Plains (Fig. 1) (Minoura and Nakaya 1991; Minoura et al. 2001; Sawai et al. 2007; Shishikura et al. 2007), has been noted as the penultimate large tsunami event around these plains (after the 2011 Tohoku-oki earthquake and tsunami), because its inundation in the Sendai Plain was similar to that of the 2011 Tohoku-oki tsunami (Sugawara et al. 2012; Namegaya and Satake 2014). However, the northern and southern distribution limits of the 869 Jogan tsunami deposits have not yet been determined (Sugawara et al. 2012). Thus, elucidating whether the 869 Jogan tsunami reached the Sanriku Coast is essential for tsunami risk assessment and seismological study. Consequently, this information is urgently required (Sugawara et al. 2012). If the 869 Jogan tsunami and 2011 Tohoku-oki tsunami were of similar severity, a large tsunami is also likely to have struck the areas along the Sanriku Coast in AD 869. In recent times, the 1454 Kyotoku tsunami has been identified from tsunami deposits in the Sendai and Ishinomaki Plains (Sawai et al. 2012) and historical documents (Namegaya and Yata 2014). However, at our study site, information on the Kyotoku tsunami deposits and the size of the tsunami has not yet been obtained.

Historical and paleo-tsunami deposits along the Sanriku Coast have been studied at several sites by The Headquarters for Earthquake Research Promotion (2006; 2007; 2008; 2009; and 2010), Haraguchi et al. (2006a, b; 2007), Haraguchi and Goya (2007), Imaizumi et al. (2007), Torii et al. (2007), and Haraguchi and Ishibe (2009) before the 2011 event. However, historical tsunami deposits were identified at only one of the onshore sites at Rikuzentakata (Fig. 1). This site has revealed four historical tsunami deposits during the past 700 years, the latest of which was correlated with the 1960 Chile tsunami (The Headquarters for Earthquake Research Promotion 2007). The other events are not well correlated with historical tsunamis. Recent sediments, in particular, those deposited during the past 1000 to 2000 years, are badly preserved at other onshore sites (Torii et al. 2007; The Headquarters for Earthquake Research Promotion 2010). Therefore, historical tsunami deposits are poorly understood along the Sanriku Coast.

Methods

Geomorphological classification

Initial mapping of geomorphic surfaces around Koyadori was based on interpretation of 1:8000- and 1:10,000-scale aerial photographs taken by the Geospatial Information Authority of Japan before and after the 2011 Tohoku-oki tsunami, and anaglyph images prepared from 1 m and 5 m mesh DEM (Digital Elevation Model) provided by the Geospatial Information Authority of Japan and the Iwate Prefecture.

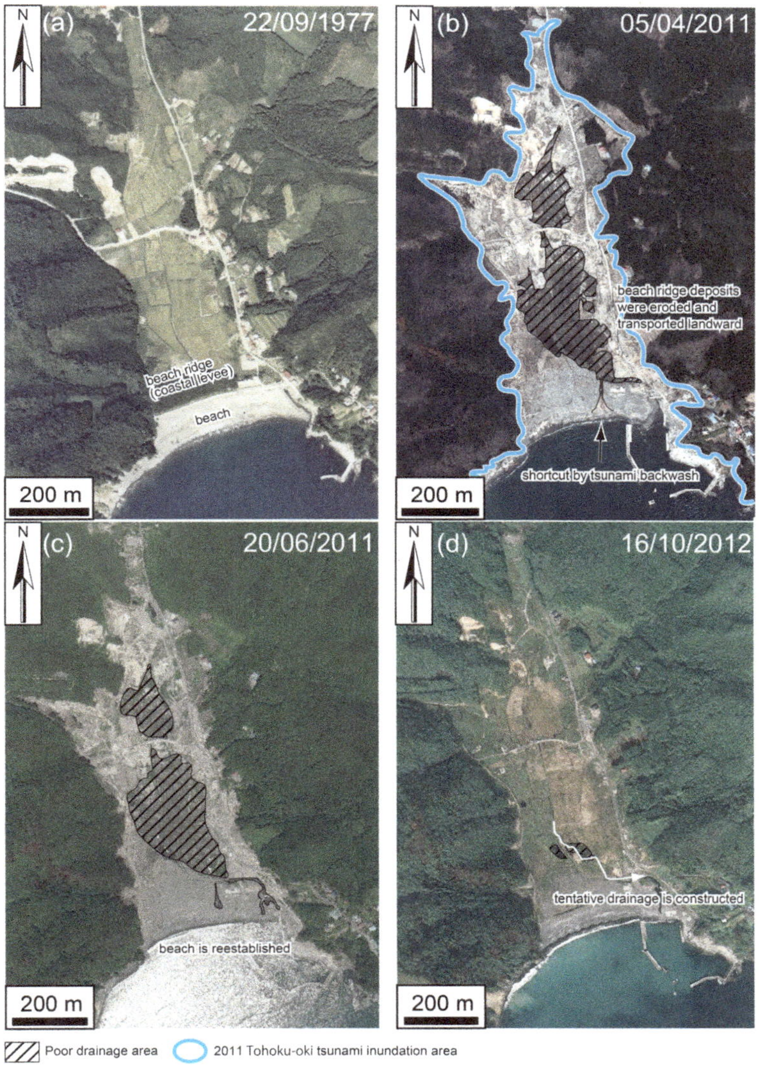

Fig. 3 Aerial photographs around Koyadori before and after the 2011 Tohoku-oki earthquake. Aerial photos (**a**)–(**d**) were taken on September 1977, March 2011, June 2011, and October 2012, respectively, by the Geospatial Information Authority of Japan

Trench survey

In December 2012, we excavated a 12 m long, 3 m wide, and 2 m deep trench (KYD-trench) approximately 3 m above sea level (a.s.l.) and 300 m distant from the shoreline (Figs. 4 and 5a, b). We logged and took photographs of each trench wall. Block samples (50 cm long, 10 cm wide, and 5 cm deep) were taken from the west, east, and south walls. The block samples of the west wall overlapped, while those of the other walls were separate (Fig. 6). We also sampled deposits of each event from the east wall for particle roundness analysis (Fig. 6). In 2013, a construction company excavated an additional canal (the *canal-trench*; approximate length, width, and depth 300, 1, and 0.5 m, respectively) as a tentative drainage for field restoration (Fig. 4). To confirm the continuity of each event, we logged a 150 m-long section of the canal-trench wall and sampled sediments for particle roundness analysis.

Dry and wet density, color, and loss on ignition measurement

We sampled the entire 7 cm^3 cube (each side = 2.2 cm) from the block samples and measured their wet and dry bulk densities. The color of wet sediments in cubic samples is quantified by the L*, a*, and b* parameters measured by the Soil Color Reader SPAD-503 instrument (Konica Minolta Sensing, Inc.). The a* and b* parameters specify the red (+) to green (−) and yellow (+) to blue (−) content, respectively, while L* represents lightness (0 = black, 100 = white). The loss on ignition (LOI) was conducted in each block sample following Bos et al. (2012) at 3–6 cm

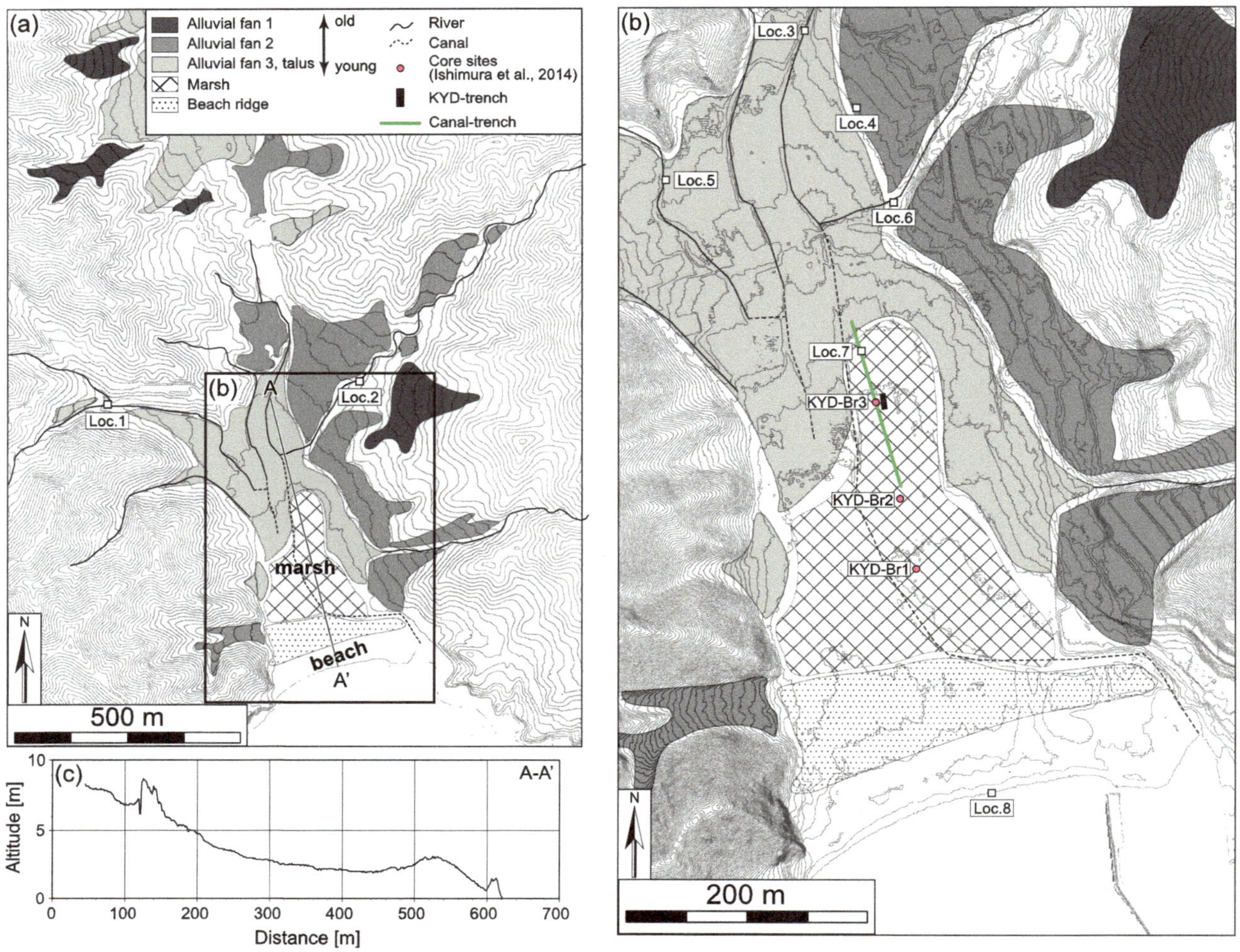

Fig. 4 Geomorphological classification around Koyadori. Contour maps are based on the 1 m mesh DEM supplied by the Iwate Prefecture. The contour interval in (**a**) and (**b**) is 5 m and 1 m, respectively. **c** Topographic profile along A–A′ line based on the 1 m mesh DEM

intervals, although this sampling was restricted to peat and peaty silt.

Particle roundness

Furthermore, to reveal the origins of the event deposits and to confirm tsunami deposits, we sampled fluvial and beach sediments (Locations 1–8) (Fig. 4) for particle roundness analysis in 2012 and 2013. Samples were washed and dry-sieved through 2 mm mesh and the gravels were divided into six roundness categories (very angular, angular, sub-angular, sub-rounded, rounded, and well-rounded).

Radiocarbon dating

Radiocarbon dating (30 samples) was conducted by accelerator mass spectroscopy (AMS) at the Institute of Accelerator Analysis Ltd. and Geo Science Laboratory. The obtained age data were calibrated using the OxCal 4.2 program (Ramsey 2009) and the calibration curve IntCal13 (Reimer et al. 2013).

Tephra analysis

The Towada-a tephra (To-a) (AD915; Machida and Arai 2003) is a useful indicator of the 869 Jogan tsunami deposits in the Sendai and Ishinomaki Plains (Minoura and Nakaya 1991; Sawai et al. 2007; Shishikura et al. 2007). From the presence and distribution of To-a along the southern Sanriku Coast, Ishimura et al. (2014) suggested that To-a had also been deposited at the central Sanriku Coast. Therefore, we conducted a cryptotephra analysis to identify the invisible To-a horizon.

Each block sample was sampled at 3–6 cm intervals. These samples were washed using 60 μm nylon mesh and dry-sieved using 124 μm nylon mesh. Thin sections made with the 60–124 μm fractions revealed volcanic glass contents. The refractive index of volcanic glass shards, which is useful for identifying widespread tephras in Japan

Table 1 Historical tsunamis along the Sanriku Coast during AD 1611–2011

Date	Name	Latitude [°N]	Longitude [°E]	Magnitude
Earthquakes in observed records (after 1896)				
11 March 2011	Tohoku-oki	38.1	142.86	9.0
26 September 2003	Tokachi-oki	41.78	144.08	8.0
16 May 1968	Tokachi-oki	40.73	143.58	7.9
23 May 1960	Chile	−38.17	−72.57	9.5
4 March 1952	Tokachi-oki	41.8	144.13	8.2
3 November 1936	Miyagiken-oki	38.26	142.07	7.5
3 March 1933	Showa Sanriku	39.23	144.52	8.1
1 November 1915	Miyagiken-oki	38.3	142.9	7.5
5 August 1897	Miyagkeni-oki	38.3	143.3	7.7
15 June 1896	Meiji Sanriku	39.5	144	8.2
Earthquakes in documented records (before 1896)				
23 August 1856	Ansei	41	142.3	7.7
17 February 1793	Kansei	38.5	144.5	8.2
29 January 1763	Horeki	41	142.5	7.7
13 April 1677	Empo	41	143	7.9
2 December 1611	Keicho	39	144	8.1

Based on Utsu (2004). We selected major earthquakes generating tsunamis hitting the Sanriku Coast

(Machida and Arai 2003), was measured with a refractive index measuring system (RIMS 2000: Kyoto Fission Track Co., Ltd.). The RIMS system measures volcanic glass shards to an accuracy of ±0.0002 (Danhara et al. 1992). The major element compositions were analyzed by energy-dispersive spectrometry using an electron probe microanalysis (EPMA) system (Horiba Emax Energy EX-250) at the FURUSAWA Geological Survey. The major elements were measured by scanning a 4 μm grid of the targeted grain under a counting time of 150 s and accelerating voltage of 15 kV. The beam current and diameter were 0.3 nA and 150 nm, respectively. The atomic number effect was corrected by the ZAF procedure.

Results

2011 Tohoku-oki tsunami and its deposits

The inundation and runup heights of the 2011 Tohoku-oki tsunami at Koyadori ranged from 13 to 18 m a.s.l. and

Table 2 Historical tsunamis' runup height around Koyadori

Tsunami event	Age [AD]	Koyadori [m]	Oura [m]	Funakoshi [m]	Uranohama [m]	Yamada [m]	Osawa [m]	References
Measured runup height								
Tohoku-oki	2011	26.0–29.4	7.7–11.3	13.5–16.1	13.6–14.9	8.7–10.4	5.8–8.9	Haraguchi and Iwamatsu (2011)
Tokachi-oki	1968			3.3		1.1	0.7	Watanabe (1998)
Chile	1960	No inundation[a]		3.5–4.4		3.5	4.0	Iwate Prefecture (1969)
Showa Sanriku	1933	6.6	1.2–4.2	3.7–5.0	3.0	3.0–4.0	1.1–2.5	Earthquake Research Institute, the University of Tokyo (1934)
Meiji Sanriku	1896	15.0	6.0	10.0–12.0		6.0	5.5	Unohana and Ota (1988)
Estimated runup height based on historical documents								
Ansei	1856	3	2	4–6		2	3	Tsuji and Ueda (1995), Hatori (2000)
Kansei	1793		2	3–4		2–3	2	Hatori (2009)
Empo	1677					2–4		Hatori (1975), Tsuji and Ueda (1995),
Keicho	1611	20–25		10–12		5–8		Hatori (2009)

We selected earthquakes generating tsunamis (runup height is larger than a few meters around Koyadori). Blank space means no data
[a]Based on interviews of landowners

Fig. 5 Photographs around the trench sites. **a** The KYD- and canal-trench sites, **b** the KYD-trench, **c** the canal-trench, **d** erosion of beach ridges, **e** tsunami boulder, and **f** 2011 tsunami deposits

from 26 to 29 m a.s.l., respectively (Table 2; Haraguchi and Iwamatsu 2011). Figure 3 shows the landform changes before and after the 2011 event. Immediately following the 2011 Tohoku-oki tsunami (April 2011), the beach was not yet re-established and the beach ridges may have been shortened by the tsunami backwash (Fig. 3a, b). After the beach was restored in June 2011, the shortcut channel was filled with beach deposits (Fig. 3c). The poor drainage area remained until October 2012 (Fig. 3d). The Tohoku-oki tsunami hit the coastal levee originating from beach ridges, destroying it and the trees on it, and eroding it to a depth of 1–1.5 m (Fig. 5d). The eroded materials were transported landward and deposited as tsunami deposits. Approximately 9 m a.s.l. and 600 m landward, a boulder was recognizable as a tsunami deposit because of the attached oyster shells (Fig. 5e). Tsunami deposits composed of sand and gravel sourced from the beach and beach ridges were found up to 600 m landward in December 2012 (Fig. 5f).

Description of the KYD-trench

Deposits in the trench wall were divided into five facies (event deposits, marsh deposits, channel fill deposits, artificial fill deposits, and cultivation soil), based on their sediment structure, continuity, and composition (Fig. 6). All the walls contained marsh deposits and interbedded event deposits.

The event deposits are composed of coarse sand and granule, and are traceable in the trench (Figs. 6 and 7). Deposits showing good continuity, horizontal sedimentation, erosional features, and loading structure at the base were considered as potential candidates for tsunami deposits, and were labeled E1 (youngest) to E11 (oldest). The characteristics of each event deposit are presented in Table 3. The E1 deposits (the 2011 Tohoku-oki tsunami deposits) are divisible into two units. The lower unit is composed of coarse sediments (pebble to coarse sand) with normal grading. The upper unit comprises finer sediments (granule to medium sand) and is partially laminated. The E2 deposits are thin and composed of granule to coarse sand. The E2 layer is interbedded with cultivation soil, indicating partial disturbance by cultivation. The E3 deposits are well traceable and characterized by a bluish color (Fig. 6). As these deposits thicken from grids E–3 to S–3, they also become coarser (cobble to pebble). However, while the basal contact is very sharp, the upper contact is partially disturbed by cultivation. The E4 deposits are partially disturbed and eroded by channel deposits. In general, their compositions are fine (granule to fine sand), although some parts contain pebble to cobble gravels. The E5 deposits are interbedded with low-LOI organic sediments (Fig. 7). Their basal contact is sharp, but their upper contact is disturbed. They are intermittently distributed because of plant bioturbation. The E6 deposits are well traceable in the KYD-trench (Fig. 6), with very sharp basal contacts and a loading structure at the bottom of the layer. The E7 deposits are also well traceable and some of them have eroded the E8 deposits (Fig. 6). The E7 layer also shows a loading structure at the bottom. The E8 deposits are intermittently distributed because of erosion by the E7 deposits. Below the E8 deposits,

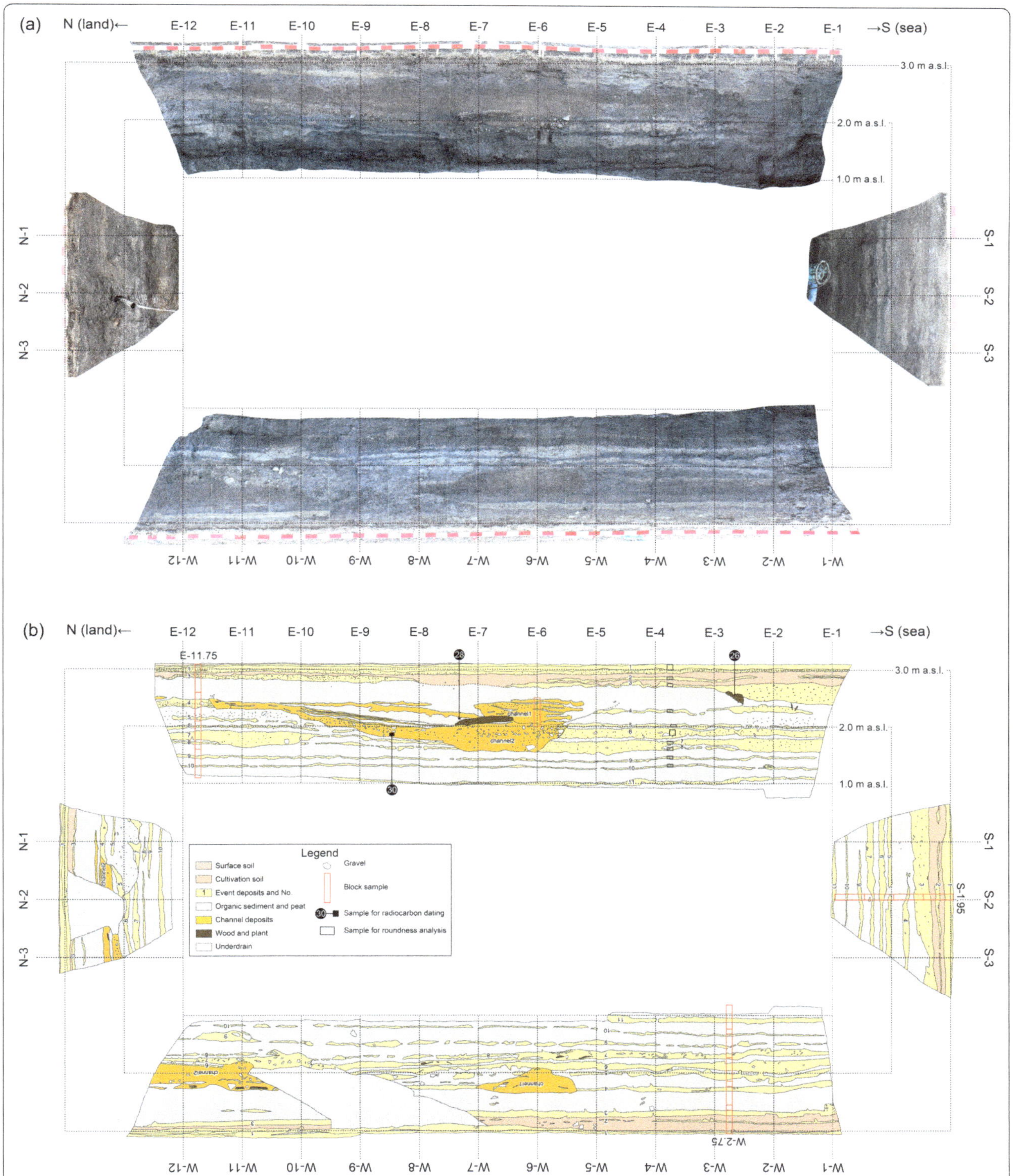

Fig. 6 Picture and sketch of the KYD-trench walls. **a** Photo mosaics of the KYD-trench wall. **b** Sketch of the KYD-trench wall

the upper and basal contacts of the event deposits have been disturbed by plant bioturbation. The E9 and E10 deposits are very thin and similarly disturbed by plant bioturbation, but are nonetheless traceable in the KYD-trench (Fig. 6). The E11 deposits are traceable in the southern half of the KYD-trench and distributed under

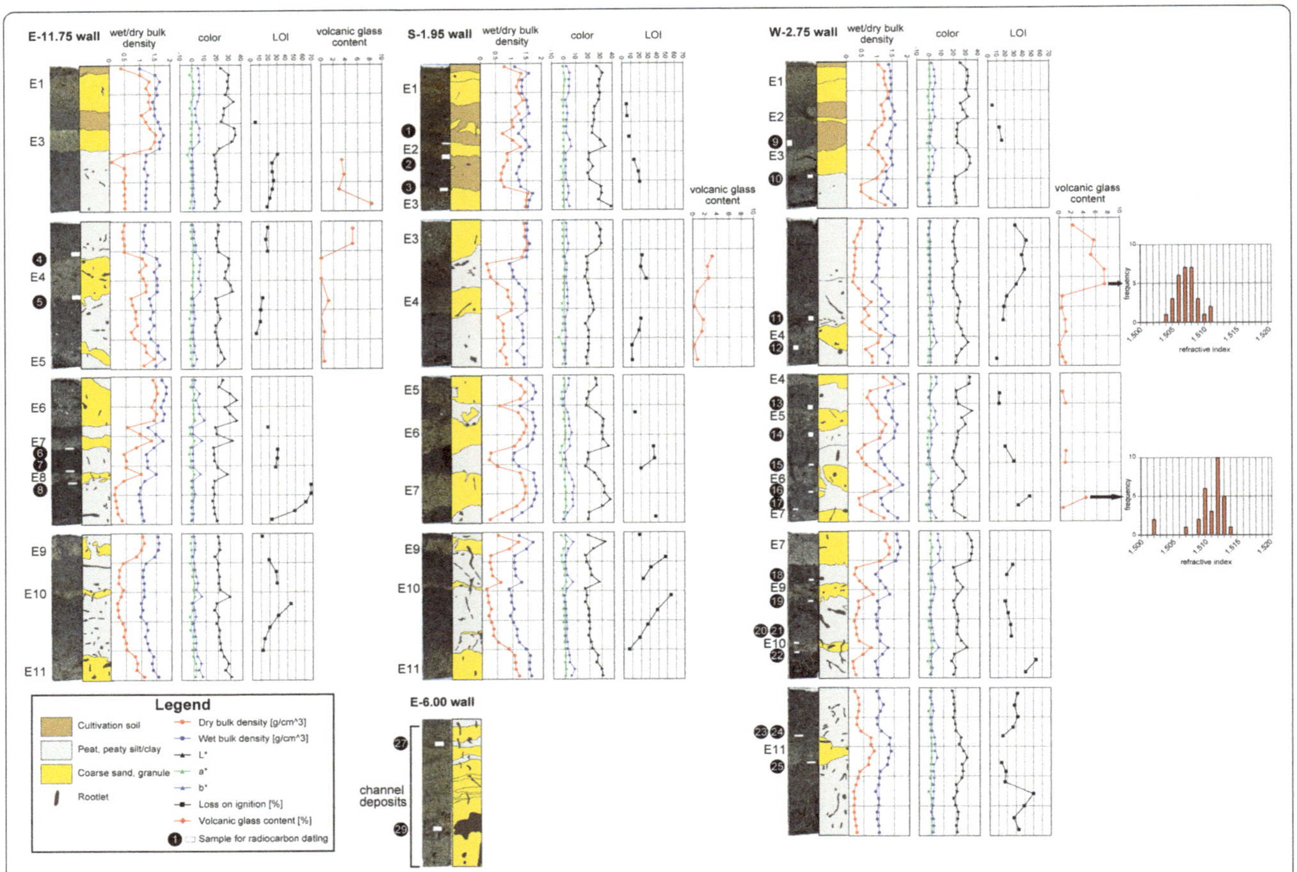

Fig. 7 Tephra analysis, color, dry/wet bulk density, and loss on ignition of the KYD-trench wall samples

the trench bottom in the northern half (Fig. 6). The E11 deposits are of medium thickness and contain fine grains (coarse to medium sand).

The marsh deposits are composed of plant remains and organic sediments. Their densities are inversely correlated with their LOIs and indirectly indicate their organic carbon content and degree of decomposition (Fig. 7). Color, in particular, the L* and b* parameters, is correlated with density, whereas the LOI fluctuates between event deposits. Macroscopically, the LOI decreases from the

Table 3 Characteristics of event deposits in the KYD-trench

Event deposits	General thickness	Comparison of grain size among event deposits	General grain size	Measured thickness [cm]	Upper contact	Basal contact	Gravel content [wt%]	Roundness (well-rounded + rounded) [%]
E1	Medium	Coarse	Upper: granule to medium sand, Lower: pebble to coarse sand	3–20	–	Sharp	10–28	45–66
E2	Thin	Medium	Granule to coarse sand	1–8	Disturbed	Sharp	8–19	13–53
E3	Thick	Coarse	Granule to coarse sand	2–38	Sharp	Sharp	11–55	48–86
E4	Medium	Fine	Granule to fine sand	3–16	Disturbed	Sharp	2–18	32–41
E5	Medium	Medium	Granule to coarse sand	2–13	Disturbed	Sharp	7–24	30–56
E6	Thick	Medium	Granule to coarse sand	1–24	Sharp	Sharp	18–35	16–62
E7	Thick	Medium	Granule to coarse sand	2–28	Disturbed/sharp	Sharp	2–31	26–67
E8	Thin	Fine	Coarse to medium sand	1–7	Disturbed	Disturbed	8–10	16–45
E9	Thin	Fine	Granule to medium sand	2–18	Disturbed	Disturbed/sharp	8–17	12–42
E10	Thin	Fine	Coarse to medium sand	1–9	Disturbed	Disturbed	3–11	14–43
E11	Medium	Fine	Coarse to medium sand	2–13	Disturbed	Disturbed	3–4	50–61

trench bottom to the E4 deposits and increases from the E4 deposits to the E3 deposits.

The channel fill deposits exhibit two cross-sectional geometries and compositions, categorized as Channel 1 and Channel 2 (Fig. 6). Channel 1 is distributed from grid N–5 to N–10 and from grid W–5.5 to W–7 (Fig. 6). From the altitude of the channel bottom in both walls, the flow direction of Channel 1 was determined as east to west. Sediments are finer in Channel 1 than in Channel 2, comprising coarse sand to fine pebbles, and interbedded with peaty silt. Channel 2 is distributed from grid N–1 to W–10 and from grid E–5.5 to E–11.5 and flows from northwest to southeast (Fig. 6). The composition is poorly sorted pebble to cobble.

The artificial fill deposits with a buried PVC pipe, distributed from grid W–7 to W–11 and from grid N–1.5 to N–2.5 (Fig. 6), were identified from interviews with landowners as underdrains constructed 40–50 years ago.

The cultivation soil is distinguished from marsh deposits by its different particle composition, color, and texture. This soil type is interbedded between the E3 and E1 deposits (Fig. 6). Event deposits, marsh deposits, and cultivation soil are also easily distinguishable by their density and color (Fig. 7). The dry bulk density of cultivation soil is intermediate between low-density marsh deposits and high-density event deposits.

According to radiocarbon dating of these marsh and channel deposits (Figs. 6, 7, and 8; Table 4), the sediments in the KYD-trench wall provide a continuous record since approximately 4000 cal. BP.

Description of the canal-trench

Event deposits (the E1, E2, and E3 deposits), other event deposits, cultivation soil/peat, and debris flow deposits were identified in the canal-trench (Figs. 4 and 9). Debris flow deposits are distinguished by poorly sorted gravel beds interfingered with cultivation soil/peat in the geometric cross-section (Fig. 5c). Although the E1 deposits are traceable, the original thickness of the E1 layer has been obscured by artificial modification following the 2011

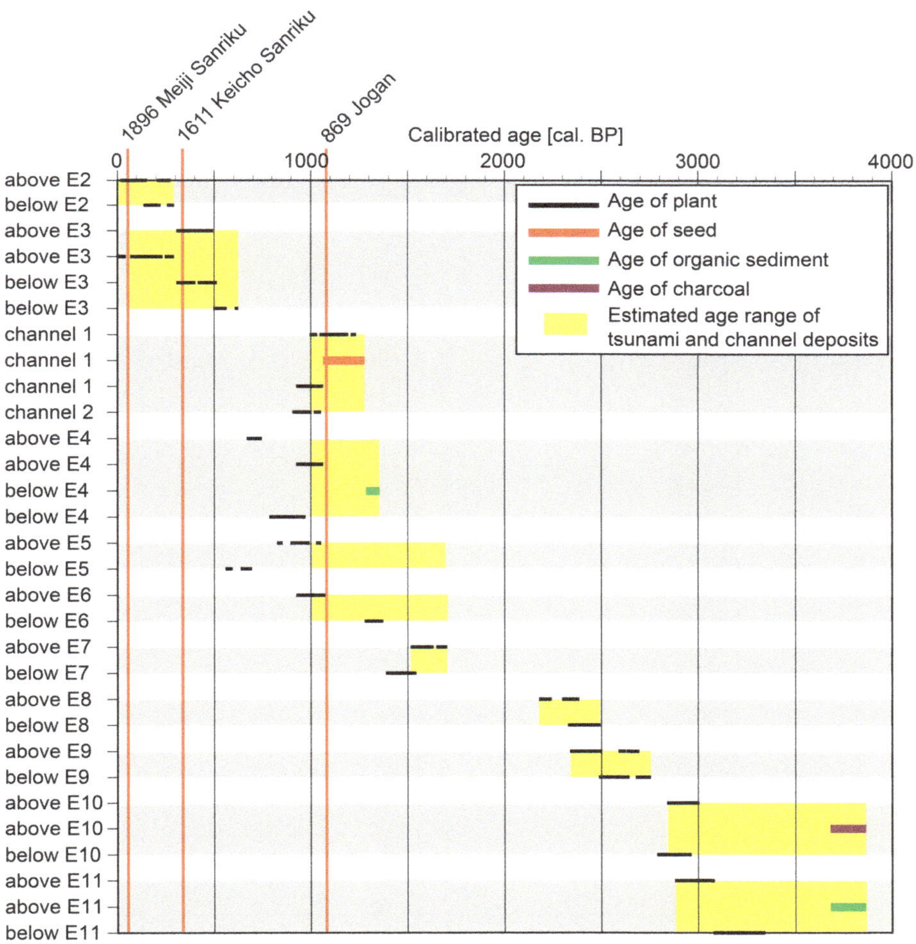

Fig. 8 Event age diagram in the KYD-trench

Table 4 Radiocarbon ages and calibrated ages

Sample no.	Sample name	Labo no.	Sample position	Grid no.	Depth [m]	Material	$\delta^{13}C$ [‰]	Conventional ^{14}C age [yrBP]	Calibrated age (2σ) [calBP]
1	KYD-TrS-1	IAAA-131322	Above E2	S-1.95	0.27	Plant	−25.3	110 ± 20	270–210 (27.4 %), 150–20 (67.9 %)
2	KYD-TrS-2	IAAA-131323	Below E2	S-1.95	0.31–0.33	Plant	−26.7	190 ± 20	290–260 (19.6 %), 220–140 (52.9 %), 20– (22.9 %)
3	KYD-TrS-3	Beta-357404	Above E3	S-1.95	0.43	Plant	−28.8	340 ± 30	490–310 (95.4 %)
4	KYD-TrE-1	IAAA-131320	Above E4	E-11.75	0.60–0.62	Plant	−28.0	790 ± 20	740–670 (95.4 %)
5	KYD-TrE-2	IAAA-131321	Below E4	E-11.75	0.74–0.76	Organic sediment	−25.8	1420 ± 20	1350–1290 (95.4 %)
6	KYD-TrE-3	Beta-341796	Below E7	E-11.75	1.23–1.24	Plant	−24.8	1570 ± 30	1540–1390 (95.4 %)
7	KYD-TrE-4	Beta-341572	Above E8	E-11.75	1.31–1.32	Plant	−25.1	2320 ± 30	2380–2300 (90.1 %), 2240–2180 (5.3 %)
8	KYD-TrE-5	Beta-341797	Below E8	E-11.75	1.35–1.36	Plant	−26.8	2360 ± 30	2490–2330 (95.4 %)
9	KYD-TrW-1	IAAA-131324	Above E3	W-2.75	0.27–0.29	Plant	−25.5	150 ± 20	290–250 (15.7 %), 230–130 (48.1 %), 120–70 (13.1 %), 40–0 (18.5 %)
10	KYD-TrW-2	Beta-339851	Below E3	W-2.75	0.39–0.41	Plant	−27.7	370 ± 30	510–420 (55.0 %), 400–310 (40.4 %)
11	KYD-TrW-3	Beta-339855	Above E4	W-2.75	0.84–0.86	Plant	−29.1	1090 ± 30	1060–930 (95.4 %)
12	KYD-TrW-4	Beta-339856	Below E4	W-2.75	0.94–0.96	Plant	−25.7	990 ± 30	970–890 (57.4 %), 880–790 (38.0 %)
13	KYD-TrW-5	Beta-339857	Above E5	W-2.75	1.10–1.12	Plant	−26.8	1030 ± 30	1050–1030 (2.8 %), 990–900 (91.9 %), 850–830 (0.7 %)
14	KYD-TrW-6	Beta-341794	Below E5	W-2.75	1.19–1.21	Plant	−26.3	700 ± 30	690–640 (77.5 %), 590–560 (17.9 %)
15	KYD-TrW-7	Beta-339858	Above E6	W-2.75	1.30–1.31	Plant	−26.1	1100 ± 30	1070–930 (95.4 %)
16	KYD-TrW-8	Beta-341795	Below E6	W-2.75	1.39–1.40	Plant	−26.4	1420 ± 30	1370–1280 (95.4 %)
17	KYD-TrW-9	Beta-339859	Above E7	W-2.75	1.45–1.46	Plant	−27.6	1680 ± 30	1700–1650 (10.2 %), 1630–1520 (85.2 %)
18	KYD-TrW-10	Beta-339861	Above E9	W-2.75	1.66–1.67	Plant	−26.9	2410 ± 30	2690–2630 (11.2 %), 2620–2590 (2.9 %), 2500–2340 (81.3 %)
19	KYD-TrW-11	Beta-341798	Below E9	W-2.75	1.73–1.74	Plant	−24.5	2530 ± 30	2750–2680 (35.8 %), 2640–2490 (59.6 %)
20	KYD-TrW-12	Beta-339862	Above E10	W-2.75	1.87–1.88	Plant	−27.2	2810 ± 30	3000–2840 (95.4 %)
21	KYD-TrW-13	Beta-340343	Above E10	W-2.75	1.87–1.88	Charcoal	−25.4	3500 ± 30	3860–3690 (95.4 %)
22	KYD-TrW-14	Beta-341799	Below E10	W-2.75	1.92–1.93	Plant	−28.0	2780 ± 30	2960–2790 (95.4 %)
23	KYD-TrW-15	Beta-339863	Above E11	W-2.75	2.16–2.17	Plant	−26.1	2870 ± 30	3080–2880 (95.4 %)
24	KYD-TrW-16	IAAA-131669	Above E11	W-2.75	2.16–2.17	Organic sediment	−23.5	3500 ± 30	3860–3690 (95.4 %)
25	KYD-TrW-17	Beta-341800	Below E11	W-2.75	2.25–2.26	Plant	−25.8	3020 ± 30	3340–3140 (92.0 %), 3130–3110 (1.4 %), 3100–3080 (2.0 %)
26	KYD-Tr-a	Beta-357405	Below E3	E-2.40-2.60	0.50–0.70	Plant	−28.2	500 ± 30	620–610 (0.7 %), 560–500 (94.7 %)
27	KYD-Tr-b	Beta-339852	Channel 1	E-6.00	0.68–0.70	Plant	−27.0	1190 ± 30	1230–1210 (2.9 %), 1190–1050 (89.0 %), 1030–1000 (3.5 %)
28	KYD-Tr-c	Beta-339853	Channel 1	E-7.30	0.95	Seed (*Juglans sp.*)	−25.3	1240 ± 30	1270–1070 (95.4 %)
29	KYD-Tr-d	Beta-357406	Channel 1	E-6.00	0.94–0.99	Plant	−28.9	1090 ± 30	1060–930 (95.4 %)
30	KYD-Tr-e	Beta-339854	Channel 2	E-8.45	1.30	Plant	−25.1	1040 ± 30	1050–1020 (5.2 %), 1000–910 (90.2 %)

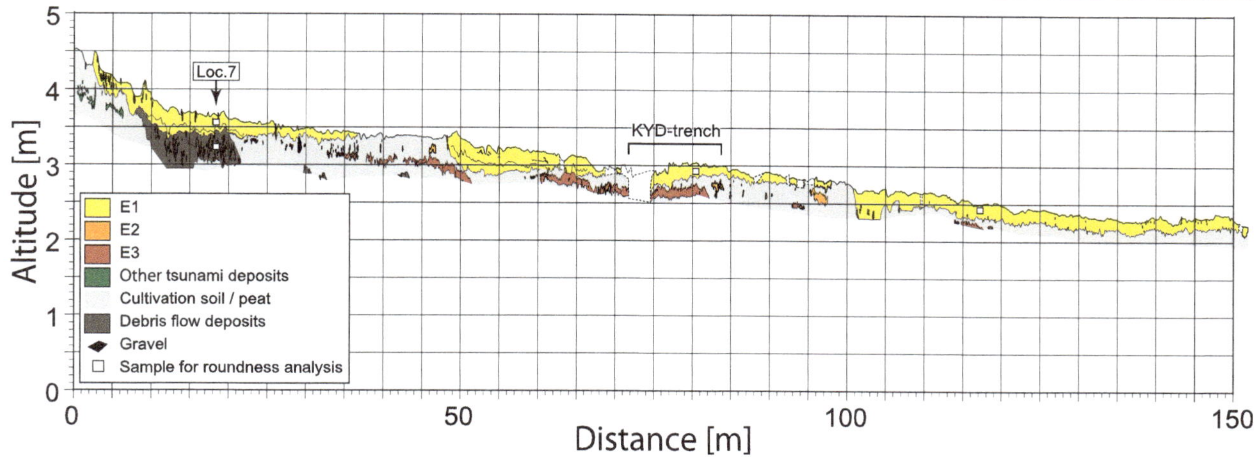

Fig. 9 Sketch of the canal-trench

event. The E2 deposits are intermittently distributed and some of them have been modified by cultivation. The E3 deposits are traceable and partially disturbed. The E3 layer is less than 20 cm thick, decreasing in the landward direction.

Particle roundness

Event deposits were sampled from the KYD- and canal-trenches (Figs. 6 and 9). Modern beach and river deposits, and debris flow deposits, were sampled from the canal-trench and outcrops (Figs. 4 and 9). Next, the origins of the event deposits were inferred from their roundness measures. Modern river and debris flow deposits consist almost entirely of angular/very angular and sub-rounded/sub-angular gravels, with no well-rounded/rounded gravels (Fig. 10a–c). Conversely, modern beach deposits contain well-rounded/rounded and sub-rounded/sub-angular gravels; angular/very angular gravels are absent (Fig. 10d). The E1 and E3 deposits (Fig. 10e, f) are similar to modern beach deposits, with high contents of well-rounded/rounded and sub-rounded/sub-angular gravels. The clear roundness differences between modern river and beach deposits are shown in the triangular diagram of Fig. 10g. The roundness composition of all event deposits in the KYD-trench is shown in Fig. 10h. In all samples, the proportion of well-rounded/rounded gravel contents exceeds 10 %, while the angular/very angular gravel content is below 40 %. Unlike the terrestrial deposits, all event deposits contain beach gravels.

Tephra analysis

From the radiocarbon dating, we determined that the invisible To-a (AD915 (1035 cal. BP)) lies between the E3 and E5 deposits. The volcanic glass contents in each trench wall sample increase after the E4 deposition (Fig. 7). In particular, in the east and west walls, the volcanic glass content suddenly increases and gradually decreases from the lower to upper parts, indicating an invisible tephra horizon. However, this trend is absent in the south wall, probably because it has been eroded by the E3 deposits.

The origins of the volcanic glass shards were determined from their refractive indices. Above the E4 deposits, the region of highest volcanic glass content, the glass refractive index ranges from 1.504 to 1.511 (mode: 1.507–1.508). Below the E4 deposits, the refractive index ranges from 1.509 to 1.514 (mode: 1.512). Furthermore, we analyzed the major element compositions of the volcanic glass shards above the E4 deposits (Table 5).

Discussion

Identification of tsunami deposits

The roundness similarities between the event and beach deposits (Fig. 10) indicate that event deposits were transported from beach and beach ridges to the inland trench sites. Landward transport from the sea is expected in tsunami and storm events. The general characteristics of tsunami and storm deposits have been reported by many researchers (Morton et al. 2007; Kortekaas and Dawson 2007; Switzer and Jones 2008; Goff et al. 2012; Phantuwongraj and Choowong 2012). On average, tsunami deposits are generally thinner than storm deposits (Morton et al. 2007; Phantuwongraj and Choowong 2012), and sedimentary structure is less common in tsunami deposits than in storm deposits (Morton et al. 2007; Kortekaas and Dawson 2007; Switzer and Jones 2008; Goff et al. 2012). The basal contact of both sediments is unconformable or erosional (Morton et al. 2007; Kortekaas and Dawson 2007; Switzer and Jones 2008; Goff et al. 2012; Phantuwongraj and Choowong 2012), although tsunami deposits sometimes show a loading structure (Goff et al.

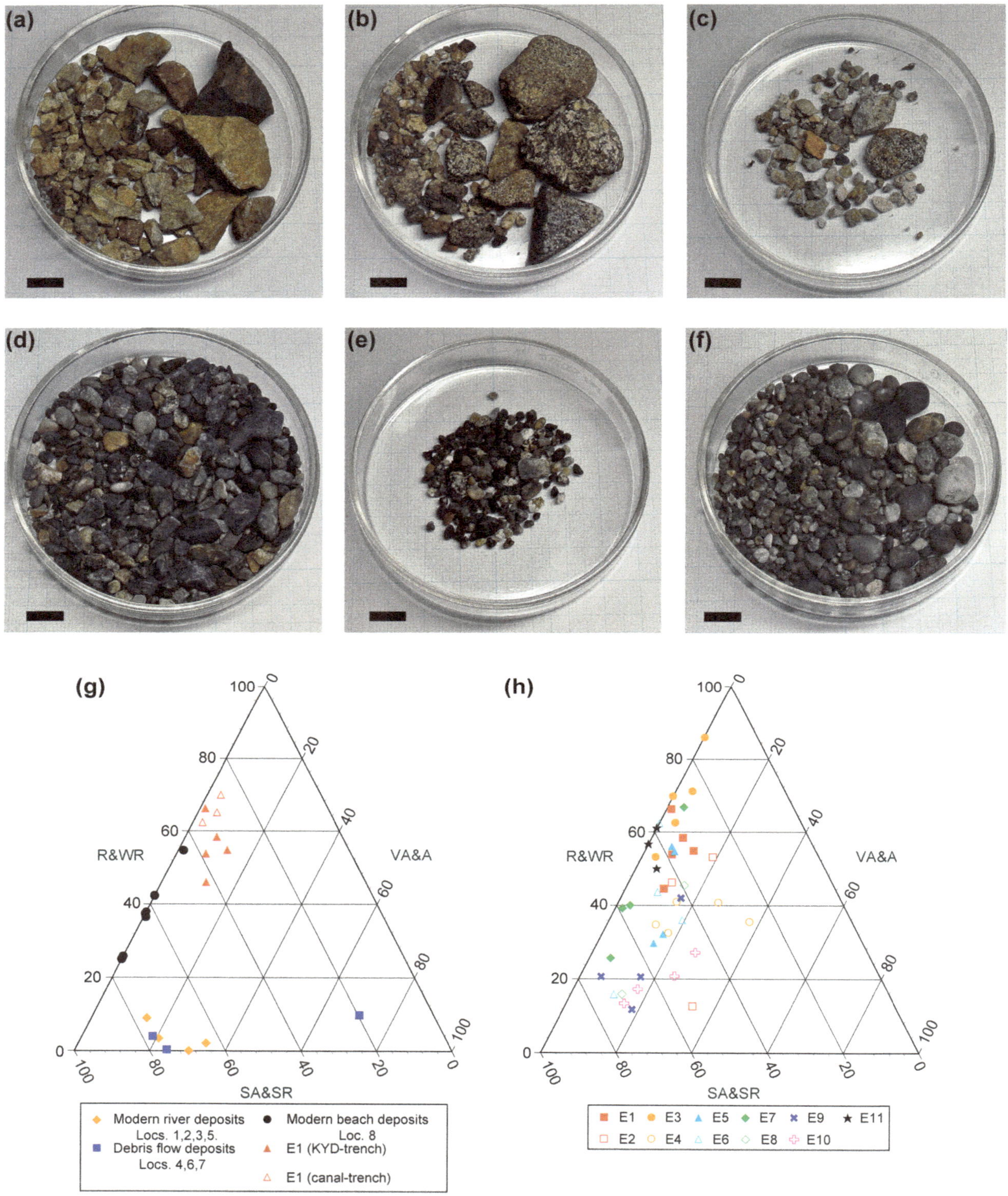

Fig. 10 Photographs and roundness of tsunami deposits, channel deposits, debris flow deposits, modern beach deposits, and modern river deposits. **a** Modern river deposits (Loc. 1). **b** Modern river deposits (Loc. 2). **c** Debris flow deposits in the canal-trench (Loc. 7). **d** Modern beach deposits (Loc. 8). **e** The E1 deposits in the south wall of the KYD-trench. **f** The E3 deposits in the south wall of the KYD-trench. **g** Comparison of the 2011 Tohoku-oki tsunami deposits, beach deposits, modern river deposits, and debris flow deposits; **h** tsunami deposits in the KYD-trench wall. Scale in the photographs is 1 cm. In the triangular diagrams, x-axis indicates well-rounded/rounded, y-axis indicates sub-rounded/sub-angular, and z-axis indicates very angular/angular

Table 5 Major element compositions of volcanic glass shards

	SiO_2	TiO_2	Al_2O_3	FeO	MnO	MgO	CaO	Na_2O	K_2O	*n*	Total[a]
ONM-GS1-C.42	77.0	0.4	12.7	1.8	0.1	0.4	2.1	4.1	1.4	15	97.5
Ishimura et al. (2014)	0.1	0.1	0.1	0.1	0.1	0.1	0.1	0.1	0.0		0.6
NTR-Br-34.93	77.0	0.4	12.8	1.8	0.1	0.4	2.1	4.1	1.4	15	96.1
Ishimura et al. (2014)	0.3	0.1	0.2	0.1	0.1	0.0	0.1	0.3	0.1		0.7
To-a (35)	77.50	0.36	12.73	1.62	0.09	0.38	1.81	3.90	1.37	19	98.41
Aoki and Machida (2006)	0.7	0.0	0.4	0.2	0.0	0.1	0.2	0.1	0.1		1.4
To-a (36)	77.69	0.36	12.74	1.66	0.09	0.35	1.80	3.99	1.31	8	98.53
Aoki and Machida (2006)	0.6	0.1	0.3	0.1	0.0	0.0	0.1	0.2	0.0		1.3
To-a (37)	76.17	0.42	13.41	1.89	0.09	0.38	1.99	4.08	1.56	18	92.89
Aoki and Machida (2006)	0.3	0.0	0.2	0.1	0.0	0.0	0.1	0.1	0.1		1.1
KYD-TrW sec. 2 20–25 cm	77.1	0.4	12.7	1.8	0.1	0.4	2.1	4.1	1.4	15	94.4
(This study)	0.3	0.1	0.2	0.2	0.1	0.0	0.0	0.1	0.1		1.3

Number on upper line shows a mean value and that on lower line shows a standard deviation. Measured values are recalculated to 100 % on a water-free basis. Results of Aoki and Machida (2006) are shown for comparison and numbers in brackets show sample number in the reference
[a]Raw data before recalculations

2012). On transect scales (several hundred meters), the cross-shore geometries of tsunami and storm deposits are characterized by "broad thin drapes with tabular or landward thinning" and "narrow thick deposits with abrupt landward thinning," respectively (Morton et al. 2007). These characteristics of tsunami deposits are recognized in the event deposits in the trenches. In the KYD-trench (length = 12 m), all event deposits are generally thinner than 20 cm and appear as draped or eroded paleo-surfaces. Some of them exhibit a loading structure. In the canal-trench (length = 150 m), the E1 and E3 deposits appear as draped deposits, with landward thinning in the E3 deposits. Furthermore, the KYD-trench is located 300 m inland from the beach, and landowners reported no storm deposits in the trench sites during the past 40–50 years. In contrast, the paleo-shoreline after the To–Cu deposition (about 6 ka) is estimated to be at least on the seaside of the KYD-Br1 site (Fig. 4). The elevations of primary To–Cu tephra within the KYD-Br1 to KYD-Br3 cores are −1.60, −1.82, and −3.08 m a.s.l., respectively, showing landward deepening, and sediments deposited after the To–Cu deposition are consistent with non-marine environments such as marsh (Ishimura et al. 2014). From these data, we consider that the present depositional setting (beach ridge and behind marsh) around the trench sites was already established by 6 ka. Therefore, prior to the To–Cu fall, the paleo-surface topography places the paleo-shoreline on the seaside of the KYD-Br1 core site. The features of the event deposits and the geomorphological settings from 6 ka to the present, together with the responses of interviewed landowners regarding recent events, indicate that all event deposits in the KYD-trench are sourced from tsunamis rather than storms.

Ages of tsunami deposits and correlation to historical tsunami event

Radiocarbon dating (Fig. 8, Table 4) suggests that the deposits in the KYD-trench are relatively close in age with no large age gap. Radiocarbon dating of event deposits is performed on plant fragments (such as reeds), because these constitute the youngest material in a sampled horizon. Plant materials in the trench are likely to be fragments of in situ plants, but downward invasion of roots and underground stems should not be ruled out. Thus, the ages of the plant material were assumed to represent the youngest ages of the sampled horizons. In contrast, a charcoal and a hard-shell plant seed (*Juglans sp.*) (Samples No. 21 and 28; Table 4) are assumed to be transported materials, whose ages mark the older age limit of the sampled horizons. Organic sediments (Samples No. 5 and 24; Table 4) are older than plant fragments, consistent with our radiocarbon dating interpretations. The true age of the sediment is expected to lie between the ages of the plant and other materials. Ishimura et al. (2014) identified To–Cu (6 ka) tephra and Oguni Pumice (7.3 to 7.4 ka) in the KYD-Br3 core drilled next to the KYD-trench at depths of 4.41–5.98 m (total thickness of primary and secondary tephra) and 8.55–8.60 m, respectively. The horizons and ages of these sediments are consistent with the radiocarbon-dated geochronology of the KYD-trench determined in this study.

We estimated the event ages based on radiocarbon dating and the above criteria (Fig. 8, Table 6). The tsunami deposit events are labeled E1–E11 in order of increasing age, considering the age constraint of the next older event. Since the ages of the E5 and E6 deposits were not determined by radiocarbon dating, we interpolated their ages using the radiocarbon ages of the E4 and E7 deposits.

Table 6 Estimated ages of tsunami deposits and their correlation with historical tsunami events

	Age [BP]	Most probable historical event	Tephra
E1		2011 Tohoku-oki Earthquake Tsunami	
E2	Modern–290	1896 Meiji Sanriku Tsunami	
E3	0–620	1611 Keicho Sanriku Tsunami	
E3/E4			To-a (AD 915)
E4	1000–1350	869 Jogan Tsunami?(no historical document around Koyadori)	
E5	1000–1700		
E6	1000–1700		
E7	1520–1700		
E8	2180–2490		
E9	2340–2750		
E10	2840–3860		
E11	2880-3860		

From these chronological estimations, we can correlate the E2, E3, and E4 deposits to historical tsunamis because these deposits are younger than AD 600.

The ages of the E2 deposits range from modern times to 290 cal. BP (i.e., they are younger than AD 1660). Certainly, the E2 deposits can be correlated to one event among the 1611, 1677, 1793, 1856, 1896, and 1933 events (Table 2). The runup heights (Table 2) and stratigraphic position of the E2 deposits suggest a correlation with the 1933 Showa Sanriku tsunami and the 1896 Meiji Sanriku tsunami, because the height of the beach ridge at Koyadori was approximately 5 m a.s.l. in both events. Although both tsunamis inundated up to the trench sites, only single-event deposits were identified from AD 2011 to 1660. Tsunami deposits can be absent for several reasons stated as follows: 1) disturbance and/or removal by cultivation, 2) erosion by succeeding tsunami events, 3) sediment availability, and 4) tsunami size. The first cause is easily explained. If sediments were deposited by the 1896 Meiji Sanriku and the 1933 Showa Sanriku tsunamis, the latter deposits would first be disturbed and removed by cultivation processes. In this case, we would correlate the preserved tsunami deposits to the 1896 Meiji Sanriku tsunami. Regarding the second cause, the 1896 deposits might have been eroded by the 1933 deposits. However, the E2 deposits show no clear base erosion in either the KYD- or canal-trenches, and no remnants of eroded tsunami deposits are evident between the E2 and E3 deposits. Thus, the second cause is inconsistent with the observations. Meanwhile, the third cause is inconsistent with the study site setting. If a large tsunami, with height exceeding that of the beach ridge, hits Koyadori, sediments of beach and beach ridge must be transported landward because there is much sediment in the coast and the beach was re-established a few months after the 2011 event. The forth cause, tsunami size, relates to the transportation and preservation of tsunami deposits. The inundation heights were larger in the 1896 event than in the 1933 event (Table 2). Thus, we can easily expect that the volume of the 1896 tsunami deposits exceeded that of the 1933 deposits. This also suggests that the 1896 tsunami deposits were better preserved than the 1933 tsunami deposits. From these considerations, we inferred that the E2 deposits are correlated to the 1896 Meiji Sanriku tsunami.

Considering the above correlation of the E2 deposits, the age of the E3 deposits was estimated as 54–620 cal. BP (AD 1896–1330). Thus, the E3 deposits can be correlated to one event among the 1454, 1611, 1677, 1793, and 1856 events (Table 2). Based on the tsunami runup height of these events (Table 2), the E3 deposits are most probably associated with the 1611 Keicho Sanriku tsunami. The E3 deposits are thick and composed of coarse materials (Table 3), and are traceable in the canal-trench (Fig. 9). Assuming a similar depositional setting from about 6 ka onward, we considered that the feature differences (thickness and grain size) among event deposits roughly indicate the tsunami size. The features of the E3 deposits (Table 3) suggest a large, very energetic tsunami. According to a local legend (Imamura 1934), the 1611 Keicho Sanriku tsunami inundated and surged through the Koyadori–Oura pass (Fig. 2). This indicates that the 1611 Keicho Sanriku tsunami was at least as high as the 2011 Tohoku-oki tsunami, since the latter failed to reach the geomorphic pass.

According to the radiocarbon dates of the channel deposits, the E4 deposits are aged 1000–1350 cal. BP (AD 950–600), and possibly correlate with the 869 Jogan tsunami. By targeting our tephra analysis at the To-a (AD 915) horizon, we determined an absolute timing for the E4 deposits. The increased content of volcanic glass above the E4 deposits (Fig. 7) suggests a tephra fall after the E4 sedimentation. The refractive index of volcanic glass shards above the E4 deposits ranged from 1.504 to 1.511 (mode: 1.507–1.508), which includes the To-a tephra range (Machida and Arai 2003; Ishimura et al. 2014). Similarly, the chemical compositions of volcanic glass shards were consistent with previously reported To-a compositions (Aoki and Machida 2006; Ishimura et al. 2014). From these data, we inferred that the To-a tephra fell between the E4 and E3 deposits, and we assigned the E4 deposits to the 869 Jogan tsunami. This identification based on radiocarbon dating and tephra provides significant information on the size and source of the 869 Jogan tsunami and earthquake, indicating that this tsunami reached the middle part of the Sanriku Coast and its inundation area was possibly as large as the 2011 event. Since the Jogan tsunami is not reported in historical records around

Koyadori and insufficient information is available for regionally and chronologically identifying the tsunami deposits along the Sanriku Coast, this finding requires confirmation in paleographical and geological researches.

Tsunami ages and their intervals

Conclusive age estimates and correlations of historical tsunami events are summarized in Table 6. Although some ambiguity of the ages remains, we calculated the average interval of tsunami occurrence as 290–390 years. Before considering the approximate age intervals of tsunami events, we need to discuss the preservation potential of tsunami deposits at this site. Szczucinski (2012) and Spiske et al. (2013) mentioned the preservation potential of tsunami deposits in tropical and temperate climate regions, respectively, and showed that the characteristics of tsunami deposits (thickness and sedimentary structure) degrade over time. Spiske et al. (2013) emphasized the significance of the preservation potential in assessing the intervals and frequencies of tsunamis, because tsunami deposits are not necessarily preserved in whole inundated areas. They identified five determining factors of preservation potential as follows: 1) composition and genetic type of the tsunami deposits, 2) coastal topography and depositional environment, 3) co- and post-seismic uplift or subsidence, 4) climate, and 5) anthropogenic modification. In Koyadori, tsunami deposits originated from beach and beach ridge deposits and are coarser than those reported in Szczucinski (2012) and Spiske et al. (2013), indicating larger resistance to post-tsunami surface processes. As mentioned above, the sedimentary environment has remained largely unchanged since 6 ka, and the beach ridge and behind-marsh environment have maintained accommodation space for tsunami and marsh deposits. The 2011 event was followed by co- and post-seismic subsidence (Ozawa et al. 2011), enhancing the preservation environment of tsunami deposits. The Sanriku Coast has a temperate climate and experiences fewer and weaker storms and high tide events (such as typhoons) than the western part of Japan. According to interviews with landowners, no storm deposits have settled in the trench sites during the past 40–50 years. Artificial modification is limited to deposits younger than E3 at this site. Moreover, 2011 tsunami deposits were found in pits and coring surveys conducted around the KYD-trench in 2013 and 2014. These deposits were clearly identifiable, despite being partially bioturbated by grass and reed. Such vegetation covered the tsunami deposits, preventing erosion and removal by post-tsunami surface processes. Even in the event of dense bioturbation, tsunami deposits are easily identified by their grain composition, size, and roundness, which widely differ from those of background deposits (e.g., peat and debris flow deposits). Therefore, we conclude that the preservation potential of tsunami deposits is very high in Koyadori. Consequently, the calculated average interval probably truly reflects the interval and frequency of large tsunamis.

The calculated average interval (290–390 years) is shorter than that obtained for the Sendai and Ishinomaki Plains (Sawai et al. 2007; 2012; Shishikura et al. 2007), reflecting the high frequency of large tsunamis causing destructive damage along the Sanriku Coast. However, if we have correctly correlated the historical deposits to the historical tsunami events, we can state the age intervals from the E1 to E4 deposits as 115, 285, and 742 years, respectively. This variability probably indicates the diversity of the tsunami generation mechanism (e.g., large earthquake, tsunami earthquake, submarine mass failure, and tsunami of distant origin) and/or the combination of several types of large earthquakes from different sources around the Japan Trench.

On the other hand, the size of historical and paleo-tsunamis can be estimated from our results because the 1896 Meiji Sanriku tsunami inundated the KYD-trench site and transported tsunami deposits there. In contrast, neither the 1968 Tokachi-oki tsunami (nearby source, runup height approximately 3 m around Koyadori; Table 2) nor the 1960 Chile tsunami (distant source, runup height approximately 4 m around Koyadori; Table 2) inundated, perhaps because they were blocked by beach ridges (height approximately 5 m a.s.l.). Furthermore, the environmental setting at the study site has been established since approximately 6 ka. These observations preliminarily suggest that tsunamis larger than the 1896 Meiji Sanriku tsunami occur at the calculated average interval, providing a first step for assessing the risk and size of tsunamis along the Sanriku Coast. To understand the tsunami generation mechanism and earthquakes along the Japan Trench, we require detailed information of ages, intervals, and sizes of historical and paleo-tsunamis at multiple sites.

Conclusions

We identified eleven tsunami deposits, including the 2011 tsunami deposits, based on sedimentary structure and continuity in two trenches and comparisons of the roundness of the gravel composing the event deposits. Radiocarbon dating and tephra analysis allowed us to establish the geochronology in the KYD-trench wall sediments and to correlate tsunami deposits with historical tsunami events. The four younger tsunami deposits (the E1–E4 deposits) are correlated with the 2011 Tohoku-oki tsunami, the 1896 Meiji Sanriku tsunami, the 1611 Keicho Sanriku tsunami, and the 869 Jogan tsunami events, respectively. The average interval of tsunami occurrence at Koyadori is estimated at 290–390 years based on continuous records in the KYD-trench. However, the age intervals between the E1 to E4 deposits are variable (E1/E2: 115 years, E2/E3: 285 years, E3/E4: 742 years), likely reflecting the diversity of the tsunami generation mechanism and/or different earthquake sources around the Japan Trench. By correlating the historical tsunami runup height data with extant tsunami deposits, we could preliminarily estimate the sizes of paleo-tsunamis at the study site. In the future study, we need to confirm our tsunami correlations by correcting many geological data along the Sanriku Coast. Ultimately, we aim to assess tsunami risk and understand the earthquake phenomena around the Japan Trench.

Competing interests

The authors declare that they have no competing interests.

Authors' contributions

DI conducted preliminary surveys, chose the study site, collected samples, performed experiments, and wrote the paper. TM designed and directed the project and gave helpful comments for the paper. Both authors read and approved the final manuscript.

Acknowledgements

We are grateful to Kazuomi Hirakawa, Toshifumi Imaizumi, Shuji Yoshida, Heitaro Kaneda, Tomoo Echigo, and Shinsuke Okada for their comments and helps in the fieldwork. We thank Hiroyuki Tustsumi for permission of using a soil color meter. The landowners of the KYD-trench site are also thanked for allowing us to conduct our surveys on their properties. The editor Ken Ikehara and two anonymous reviewers provided constructive comments that improved the manuscript. This study was a part of "Geophysical and geological studies of earthquakes and tsunamis for off-Tohoku district, Japan" and supported by the Ministry of Education, Culture, Sports, Science, and Technology, Japan (MEXT). This work was supported by Intramural Research Grant for Special Project Researches from International Research Institute of Disaster Science, Tohoku University.

Author details

[1]Disaster Science Division, International Research Institute of Disaster Science, Tohoku University, 468-1 Aza-aoba, Aramaki, Aoba-ku, Sendai, Miyagi 980-0845, Japan. [2]Department of Earth Sciences, Graduate School of Science, Chiba University, 1-33 Yayoi-cho, Inage-ku, Chiba, Chiba 263-8522, Japan.

References

Aoki K, Machida H (2006) Major element composition of volcanic glass shards in the late Quaternary widespread tephras in Japan–Distribution of tephras using K2O-TiO2 diagrams. Bull Geol Surv Japan 57:239–258

Bos AJ, Busschers FS, Hoek WZ (2012) Organic-facies determination: a key for understanding facies distribution in the basal peat layer of the Holocene Rhine-Meuse delta, The Netherlands. Sedimentology 59:679–703

Danhara T, Yamashita T, Iwano H, Kasuya M (1992) An improved system for measuring refractive index using the thermal immersion method. Quat Int 13(14):89–91

Earthquake Research Committee Headquarters for Earthquake Research Promotion Prime Minister's Office, Government of Japan (1999) Seismic Activity in Japan -Regional perspectives on the characteristics of destructive earthquakes-. Association for the Development of Earthquake. Prediction, Tokyo

Earthquake Research Institute, the University of Tokyo (1934) Reports on the 3 March, Showa 8 Sanriku tsunami. Bull Earthq Res Inst Tokyo Univ, Suppl 1:1–250

Goff J, Chagué-Goff C, Nichol S, Jaffe B, Dominey-Howes D (2012) Progress in palaeotsunami research. Sedi Geol 243–244:70–88

Goldfinger C, Ikeda Y, Yeats RS, Ren J (2013) Superquakes and supercycles. Seis Res Lett 84:1–9

Hamilton S, Shennan I (2005) Late Holocene relative sea-level changes and the earthquake deformation cycle around upper Cook Inlet, Alaska. Quat Sci Rev 24:1479–1498

Haraguchi T, Goya K (2007) Tsunami deposits from Goishi-hama, of Ofunato City, Iwate Prefecture. In: Proceedings of Rekishi-Zishin Meeting 23. Ofunato Plaza Hotel, Ofunato, 15–17 September 2007

Haraguchi T, Ishibe T (2009) Large earthquake model in the central part of Off-Sanriku estimated from tsunami deposits and uplifted event layers. Chikyu Monthly 31:223–230

Haraguchi T, Iwamatsu A (2011) Detailed maps of the impacts of the 2011 Japan tsunami: Vol. 1: Aomori, Iwate and Miyagi prefectures. Kokon-Shoin Publishers Ltd, Tokyo

Haraguchi T, Fujiwara O, Shimazaki K, Imaizumi T (2006a) Tsunami deposits from Ohtsuchi Bay of Sanriku Coast, North Eastern Japan. In: Abstract of Japan Geoscience Union Meeting 2006. Makuhari Messe, Chiba, 14–18 May 2006

Haraguchi T, Torii K, Fujiwara O, Shimazaki K, Imaizumi T (2006b) Subaqueous tsunami deposits from Otsuchi Bay of Sanriku Coast, Northeastern Japan. Chikyu Monthly 28:539–545

Haraguchi T, Torii T, Imaizumi T, Miyauchi T, Shimazaki K (2007) Tsunami deposits from Sanriku coast NE Japan in late Holocene. In: Abstract of Japan Geoscience Union Meeting 2007. Makuhari Messe, Chiba, 19–24 May 2007

Hatori T (1975) Tsunami magnitude and wave source regions of historical Sanriku tsunamis in northeast Japan. Bull Earthq Res Inst 50:397–414

Hatori T (2000) Reexamination of magnitude of historical tsunamis off the Sanriku. Tsunami Eng Tech Rep Tohoku Univ 17:38–48

Hatori T (2009) Regional deviation of inundation heights due to the large Sanriku tsunamis. Histo Earthq 24:41–48

Hayward BW, Grenfell HG, Sabaa AT, Carter R, Cochran U, Lipps JH, Shane PR, Morley MS (2005) Micropaleontological evidence of large earthquakes in the past 7200 years in southern Hawke's Bay, New Zealand. Quat Sci Rev 25:1186–1207

Ikeda Y, Okada S, Tajikara M (2012) Long-term strain buildup in the Northeast Japan arc-trench system and its implications for gigantic strain-release events. J Geol Soc Japan 118:294–312

Imaizumi T, Miyauchi T, Haraguchi T, Shimazaki K, Kagohara K, Sasaki A, Kamal PABM, Goya K (2007) Tsunami deposits along the Sanriku coast and historical earthquakes. In: Proceedings of the 2007 General Meeting of the Association of Japanese Geographers. Toyo University, Tokyo, 20–22 March 2007

Imamura A (1934) Past tsunamis of the Sanriku District. Bull Earthq Res Inst Tokyo Univ, Suppl 1:1–16

Ishimura D, Yamada K, Miyauchi T, Hayase R (2014) Characteristics of tephras interbedded with the Holocene sediments in the Sanriku Coast, northeast Japan. J Geogra 123:671–697

Kawamura K, Sasaki T, Kanamatsu T, Sakaguchi A (2012) Large submarine landslides in the Japan Trench: a new scenario for additional tsunami generation. Geophys Res Lett 39, L05308

Kawamura K, Laberg JS, Kanamatsu T (2014) Potential tsunamigenic submarine landslides in active margins. Marine Geol 356:44–49

Kortekaas S, Dawson AG (2007) Distinguishing tsunami and storm deposits: an example from Martinhal, SW Portugal. Sedi Geol 200:208–221

Machida H, Arai F (2013) Atlas of Tephra in and around Japan [revised edition]. University of Tokyo Press, Tokyo

Minoura K, Nakaya S (1991) Traces of tsunami preserved in inter-tidal lacustrine and marsh deposits: some examples from northeast Japan. J Geol 99:265–287

Minoura K, Imamura F, Sugawara D, Kono Y, Iwashita T (2001) The 869 Jogan tsunami deposit and recurrence interval of large-scale tsunami on the Pacific coast of northeast Japan. J Nat Dis Sci 23:83–88

Morton RA, Gelfenbaum G, Jaffe BE (2007) Physical criteria for distinguishing sandy tsunami and storm deposits using modern examples. Sed Geol 200:184–207

Namegaya Y, Satake K (2014) Reexamination of the A.D. 869 Jogan earthquake size from tsunami deposit distribution, simulated flow depth, and velocity. Geophys Res Lett 41:2297–2303

Namegaya Y, Yata T (2014) Tsunamis which affected the Pacific coast of eastern Japan in medieval times inferred from historical documents. J Seis Soc Japan 2nd ser. 66:73–81

Ozawa S, Nishimura T, Suito H, Kobayashi T, Tobita M, Imakiire T (2011) Coseismic and postseismic slip of the 2011 magnitude-9 Tohoku-Oki earthquake. Nature 475:373–376

Phantuwongraj S, Choowong M (2012) Tsunamis versus storm deposits from Thailand. Nat Hazards 63:31–50

Prefecture I (1969) Record of recovery process from the 1960 Chile earthquake tsunami. Iwate Prefecture, Morioka

Rajendran K (2013) On the recurrence of great subduction zone earthquakes. Curr Sci 104:880–892

Ramsey BC (2009) Bayesian analysis of radiocarbon dates. Radiocarbon 51:337–360
Reimer PJ, Bard E, Bayliss A, Beck JW, Blackwell PG, Ramsey BC, Grootes PM, Guilderson TP, Haflidason H, Hajdas I, Hatte C, Heaton TJ, Hoffmann DL, Hogg AG, Hughen KA, Kaiser KF, Kromer B, Manning SW, Niu M, Reimer RW, Richards DA, Scott EM, Southon JR, Staff RA, Turney CSM, van der Plicht J (2013) IntCal13 and Marine13 radiocarbon Age calibration curves 0–50,000 years cal BP. Radiocarbon 55:1869–1887
Sawai Y, Nasu H (2005) A 4500-year record of emergence events at Onnetoh, Hokkaido, northern Japan, reconstructed using plant macrofossils. Mar Geol 217:49–65
Sawai Y, Shishikura M, Okamura Y, Takada K, Matsuura T, Aung TT, Komatsubara J, Fujii Y, Fujiwara O, Satake K, Kamataki T, Fujino S (2007) A study on paleotsunami using handy geoslicer in Sendai Plain (Sendai, Natori, Iwanuma, Watari, and Yamamoto), Miyagi, Japan. Ann Rep Act Fault Paleoearthquake Res 7:47–80
Sawai Y, Kamataki T, Shishikura M, Nasu H, Okamura Y, Satake K, Thomson KH, Matsumoto D, Fujii Y, Komatsubara J, Aung TT (2009) Aperiodic recurrence of geologically recorded tsunamis during the past 5500 years in eastern Hokkaido, Japan. J Geophys Res 114, B01319
Sawai Y, Namegaya Y, Okamura Y (2012) Challenges of anticipating the 2011 Tohoku earthquake and tsunami using coastal geology. Geophys Res Lett 39, L21309
Shennan I, Bruhn R, Barlow N, Good K, Hocking E (2014) Late Holocene great earthquakes in the eastern part of the Aleutian megathrust. Quat Sci Rev 84:86–97
Shishikura M, Sawai Y, Okamura Y, Komatsubara J, Aung TT, Ishiyama T, Fujiwara O, Fujino S (2007) Age and distribution of tsunami deposit in the Ishinomaki Plain, Northeastern Japan. Ann Rep Act Fault Paleoearthquake Res 7:31–46
Spiske M, Piepenbreier J, Benavente C, Bahlburg H (2013) Preservation potential of tsunami deposits on arid siliciclastic coasts. Earth-Sci Rev 126:58–73
Sugawara D, Goto K, Imamura F, Matsumoto H, Minoura K (2012) Assessing the magnitude of the 869 Jogan tsunami using sedimentary deposits: Prediction and consequence of the 2011 Tohoku-oki tsunami. Sed Geol 282:14–26
Switzer AD, Jones BG (2008) Large-scale washover sedimentation in a freshwater lagoon from the southeast Australian coast: sea-level change, tsunami or exceptionally large storm? The Holocene 18:787–803
Szczucinski W (2012) The post-depositional changes of the onshore 2004 tsunami deposits on the Andaman Sea coast of Thailand. Nat Hazards 60:115–133
Tappin DR, Grilli ST, Harris JC, Geller RJ, Masterlark T, Kirby JT, Shi F, Ma G, Thingbaijam KKS, Mai PM (2014) Did a submarine landslide contribute to the 2011 Tohoku tsunami? Marine Geol 357:344–361
The Headquarters for Earthquake Research Promotion (2006) Research report of high-priority observation and survey on miyagi-oki earthquakes, 2005 fiscal year. The Headquarters for Earthquake Research Promotion, Tokyo
The Headquarters for Earthquake Research Promotion (2007) Research report of high-priority observation and survey on miyagi-oki earthquakes, 2006 fiscal year. The Headquarters for Earthquake Research Promotion, Tokyo
The Headquarters for Earthquake Research Promotion (2008) Research report of high-priority observation and survey on miyagi-oki earthquakes, 2007 fiscal year. The Headquarters for Earthquake Research Promotion, Tokyo
The Headquarters for Earthquake Research Promotion (2009) Research report of high-priority observation and survey on miyagi-oki earthquakes, 2008 fiscal year. The Headquarters for Earthquake Research Promotion, Tokyo
The Headquarters for Earthquake Research Promotion (2010) Research report of high-priority observation and survey on miyagi-oki earthquakes, 2009 fiscal year. The Headquarters for Earthquake Research Promotion, Tokyo
Torii K, Haraguchi T, Imaizumi T, Miyauchi T, Shimazaki K (2007) Survey of tsunami deposits at Sanriku coast, North East Japan. In: Proceedings of Japan society of engineering geology 2007. Osaka City University, Osaka, 11–12 October 2007
Tsuji Y, Ueda K (1995) Evaluation of A.D. 1611 Keicho, A.D. 1677 Enpou, A.D. 1763 Houreki, A.D. 1793 Kansei, and A.D. 1856 Ansei Sanriku tsunamis. Histo Earthq 11:75–106
Unohana M, Ota T (1988) Disaster records of Meiji Sanriku tsunami by Soshin Yamana. Res Rep Tsunami Disaster Control Center Tohoku Univ 5:57–379
Utsu T (2004) Catalog of damaging earthquakes in the world (through 2002), Tokyo. http://iisee.kenken.go.jp/utsu/index_eng.html. Accessed 24 December 2014.
Watanabe H (1998) Comprehensive list of tsunamis to hit the Japanese Islands 2nd edn. University of Tokyo Press, Tokyo
Witter RC, Kelsey HM, Hemphill-Haley E (2003) Great Cascadia earthquakes and tsunamis of the past 6700 years, Coquille River estuary, southern coastal Oregon. Geol Soc Am Bull 115:1289–1306
Yoshida T, Ohsawa A, Katada M (1984) Geological Map of Japan 1:200,000. Geological Survey of Japan, Tsukuba, Morioka

8

Stratification of earth's outermost core inferred from SmKS array data

Satoshi Kaneshima[1*] and Takanori Matsuzawa[2]

Abstract

S*m*KS arrivals recorded by large-scale broadband seismometer arrays are analyzed to investigate the depth profile of P wave speed (V_p) in the outermost core. The V_p structure of the upper 700 km of the outer core has been determined using S*m*KS waves of Fiji-Tonga events recorded at stations in Europe. According to a recent outer core model (KHOMC), the V_p value is 0.45 % slower at the core mantle boundary (CMB) than produced by the Preliminary Reference Earth Model (PREM), and the slow anomaly gradually diminishes to insignificant values at ∼300 km below the CMB. In this study, after verifying these KHOMC features, we show that the differential travel times measured for S*m*KS waves that are recorded by other large-scale arrays sampling laterally different regions are well matched by KHOMC. We also show that KHOMC precisely fits the observed relative slowness values between S2KS, S3KS, and S4KS (S*m*KS waves with $m = 2, 3$, and 4). Based on these observations, we conclude that S*m*KS predominantly reflect the outer core structure. Then we evaluate biases of secondary importance which may be caused by mantle heterogeneity. The KHOMC V_p profile can be characterized by a significant difference in the radial V_p gradient between the shallower 300 km and the deeper part of the upper 700 km of the core. The shallower part has a V_p gradient of $-0.0018\ s^{-1}$, which is steeper by $0.0001\ s^{-1}$ when compared to the deeper core presented by PREM. The steeper V_p gradient anomaly of the uppermost core corresponds to a radial variation in the pressure derivative of the bulk modulus, $K' = dK/dP$. The K' value is 3.7, which is larger by about 0.2 than that of the deeper core. The radial variation in K' is too large to have a purely thermal origin, according to recent ab initio calculations on liquid iron alloys, and thus requires a thick and compositionally stratified layering at the outermost outer core.

Keywords: Outermost core; Compositional stratification; S*m*KS waves; Array processing

Background

Stratification of the outermost core has been long suspected (e.g., Lay and Young 1990; Buffett and Seagle 2010) and has been recently strongly advocated by Helffrich and Kaneshima (2010) and Kaneshima and Helffrich (2013) (respectively abbreviated as HK2010 and KH2013 hereafter). The present study consolidates the propositions of HK2010 and KH2013 by investigating additional S*m*KS data obtained for significantly more event-array pairs and by evaluating possible mantle effects on S*m*KS travel time and slowness observations. We analyze S*m*KS arrivals with integer *m* from 2 to 5 in order to better constrain the V_p profile of the topmost 700 km of the core, anticipating that the results will strengthen the evidence for a compositionally stratified layer. As shown in Fig. 1, the ray paths of S2KS, S3KS, and S4KS propagate quite a long distance through the outermost part of the core. Since each S*m*KS possesses a distinct depth sensitivity depending both on *m* and epicentral distance, when used in combination, S*m*KS waves can form an ideal data set for use in determining the V_p profile of the outermost core. In order to measure differential travel time and slowness differences between two S*m*KS waves, we used large-scale broadband seismometer arrays in Europe, Eastern Asia (including Japan), North and South America, Africa, and Australia.

Prior to HK2010 and KH2013, other studies also investigated the outer core structure by analyzing S*m*KS data. The proposed models show either slightly or rather strongly slower V_p anomaly relative to the Preliminary Reference Earth Model (PREM) (Dziewonski and Anderson 1981) near the top of the core. For instance,

*Correspondence: kane@geo.kyushu-u.ac.jp
[1] Department of Earth and Planetary Science, Kyushu University, Hakozaki, Higashi-ku, 812-8581 Fukuoka, Fukuoka, Japan
Full list of author information is available at the end of the article

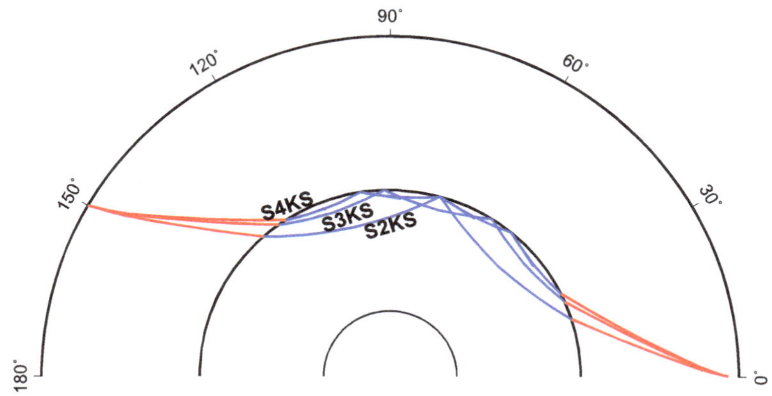

Fig. 1 *SmKS* ray paths. Ray paths of *SmKS* waves (S2KS, S3KS, and S4KS) propagating through the outermost part of the core to a receiver at a epicentral distance of 150°. Blue and red lines show P and S wave portions of the rays, respectively

Tanaka (2007) analyzed a composite record section of S2KS, S3KS, and S4KS, which were observed globally, and proposed a model with up to a maximum of 1.2 % slower V_p than PREM in the outermost 90 km of the core. His model (called Tanaka-1 hereafter) is similar to that presented in an earlier study by Garnero et al. (1993). Alexandrakis and Eaton (2010) investigated composite globally observed record sections of S2KS to S4KS for the distance range shorter than 140° and showed a permissible range of V_p profiles for the top 200 km of the outer core. The range of permissible models centers around slightly slower V_p values (from ~0.1 to 0.4 %) than PREM and falls between PREM and the IASP91 velocity model (Kennett and Engdahl 1991). In this study, we also include other global V_p models such as SP6 (Morelli and Dziewonski 2012), AK135 (Kennett et al. 1995), and another model proposed by Tanaka (2007) (called Tanaka-3 hereafter) with the models to be compared.

SmKS waveform data

SmKS data are analyzed for earthquakes that occurred in several different regions and were recorded at large-scale broadband stations prior to July 2014 (Table 1). The broadband seismograms in Europe are extracted from the Incorporated Research Institutions for Seismology (IRIS) and the Observatories and Research Facilities for European Seismology (ORFEUS) data management centers. Seismic stations in Japan belong to J-Array, F-net, and the tilt-meter network of Hi-net (Obara et al. 2005). Data of the seismic stations in Eastern Asia outside of Japan, the US, and Australia are extracted from the IRIS data management center (DMC). In this study, we loosely define a "large-scale array" as a group of 20 or more broadband seismometer stations with aperture lengths exceeding ~1000 km whose seismograms can be analyzed by standard array processing methods. As we shall show later, travel times of *SmKS* waves observed at individual stations may be significantly affected by the heterogeneous mantle structure, rather irrespective of the waveform quality. This makes it difficult to investigate a detailed V_p profile of the outer core based on *SmKS* measurements made for individual traces. However, HK2010 and KH2013 showed that stacking seismograms over an entire large-scale array substantially suppresses the mantle effects and thereby enables us to discern the fine structure of the outermost core. In this study, we shall also show that the processing of large-scale array data enables us to precisely retrieve differential slowness between two *SmKS* arrivals, which helps to better constrain the outer core structure. Figure 2 shows an example of observed and computed record sections for a Fiji-Tonga event recorded at the European stations. The synthetic seismograms shown in Fig. 2b are computed for PREM using reflectivity (Kind 1979) and a method similar to that of Wang et al. (2008) (HK2010). When *SmKS* touches an internal caustic and is underside reflected $m-1$ times at the core mantle boundary (CMB), it suffers from a phase delay by nearly $\frac{m-1}{2}\pi$ relative to SKS (Choy 1977). The arrivals of S2KS, S3KS, S4KS, S5KS, and S6KS are clearly visible on this section and are systematically delayed with respect to the theoretical PREM predictions throughout the array.

Methods and results

Array measurements

In this study, we measure differential travel times between two different *SmKS* waves on stacked waveforms; those between S3KS and S2KS (called dt^{3-2} hereafter), S4KS and S3KS (called dt^{4-3}), and S5KS and S3KS (called dt^{5-3}). Next, the differential time anomalies relative to the ray theoretical predictions based on PREM are computed. Since instrument corrections have negligible effects on the results, stacking and differential time measuring are performed directly on the observed broadband seismograms (HK2010). We also avoid unstable deconvolution of

Table 1 Event list

Event	Date	Time	Lat. N	Lon. E	h (km)	Mo (dyne cm)
FEV1	2007/10/05	7:17:52	−25.19	179.46	509	6.8×10^{25}
FEV2	2007/10/16	21: 5:43	−25.77	179.53	509	1.0×10^{26}
FEV3	2008/07/03	3: 2:37	−23.37	−179.78	581	2.8×10^{25}
FEV4	2011/07/29	7:42:24	−23.79	179.75	537.0	1.5×10^{26}
FEV5	2007/05/06	21:11:53	−19.40	−179.35	676	6.1×10^{25}
FEV6	2012/01/24	00:52: 5	−24.977	178.520	580.3	4.2×10^{25}
FEV7	2008/07/19	22:39:53	−17.337	−177.312	391.	5.2×10^{25}
FEV8	2011/08/19	3:54:27	−16.522	−177.004	407.9	2.9×10^{25}
FEV9	2008/10/19	5:10:34	−21.863	−173.819	29.0	2.9×10^{26}
FEV10	2009/03/19	18:17:41	−23.050	−174.660	34.0	3.4×10^{27}
FEV11	2009/11/24	12:47:16	−20.710	−174.040	18.0	1.7×10^{26}
FEV12	2014/07/21	14:54:41	−19.829	−178.464	616.	3.0×10^{26}
FEV13	1998/03/29	19:48:16	−17.552	−179.092	537.5	6.4×10^{26}
KEV1	2013/04/26	6:53:29	−28.68	−178.96	358.8	2.1×10^{25}
KEV2	2011/02/21	10:57:52	−26.142	178.394	558.0	7.9×10^{25}
KEV3	2007/09/25	5:16: 1	−30.965	179.998	417.	2.3×10^{25}
KEV4	2009/08/18	21:20:47	−26.064	−178.391	269.3	3.6×10^{25}
KEV5	2013/08/28	2:54:41	−27.75	179.62	480.3	2.9×10^{25}
KEV6	2014/06/23	19:19:16	−30.118	−177.67	20.	2.6×10^{26}
AEV1	2000/04/23	9:27:23	−28.31	−62.99	609	3.1×10^{26}
AEV2	2005/03/21	12:23:54	−24.98	−63.47	579	2.3×10^{26}
AEV3	2008/09/03	11:25:14	−26.74	−63.22	570	3.3×10^{25}
AEV4	2012/05/28	5: 7:24	−28.06	−63.11	589.3	1.3×10^{26}
AEV5	2011/09/02	13:47:11	−28.420	−63.150	592.8	1.4×10^{26}
AEV6	2014/04/03	1:58:30	−20.311	−70.576	24.1	9.6×10^{25}
NCEV	2014/04/11	20:29:12	11.642	−85.878	135.0	1.0×10^{25}
IEV1	2011/08/30	6:57:42	−6.36	126.76	469.0	2.7×10^{26}
IEV2	2010/03/20	14: 0:50	−3.36	152.24	414	9.2×10^{25}
IEV3	2010/07/23	22: 8:11	6.72	123.41	607.0	1.2×10^{27}
IEV4	2013/06/13	16:47:23	−10.0	107.24	9.0	1.1×10^{26}
IEV5	2011/03/10	17:08:37	−6.87	116.72	510.6	9.7×10^{25}
IEV6	2010/05/31	19:51:45	11.132	93.471	112.	6.4×10^{25}
MEV	2006/02/22	22:19: 8	−21.32	33.58	11	4.2×10^{26}
INEV	2008/05/31	4:37:56	−41.290	80.47	10.	5.1×10^{25}
SEV1	2014/06/29	7:52:56	−55.506	−28.451	16.5	2.4×10^{26}
PEV1	2013/05/14	0:32:26	18.728	145.287	602.3	2.1×10^{26}

the source time function, which may cause some biases to the dt^{m-n} (differential travel times between S*m*KS and S*n*KS waves) and their anomalies relative to ray theoretical predictions by PREM, which are usually positive. Those biases are mostly less than 0.2 s for the Fiji-Tonga events used in HK2010 and KH2013 but can be larger for other events. Therefore, we evaluate biases by using synthetic seismograms (Fig. 2b). To measure these synthetic seismograms dt^{m-n}, we use the same array method as the data and compare them with the theoretical ray calculations based on PREM. The difference, which is usually 0.2 to 0.3 s but sometimes could exceed 0.5 s (Table 2), is regarded as the bias. For each event, dt^{3-2} are measured both by picking peaks and cross-correlating the stacked waveforms after correcting for the $\frac{\pi}{2}$ phase shift due to the caustic (Choy 1977). Both S2KS and S3KS are

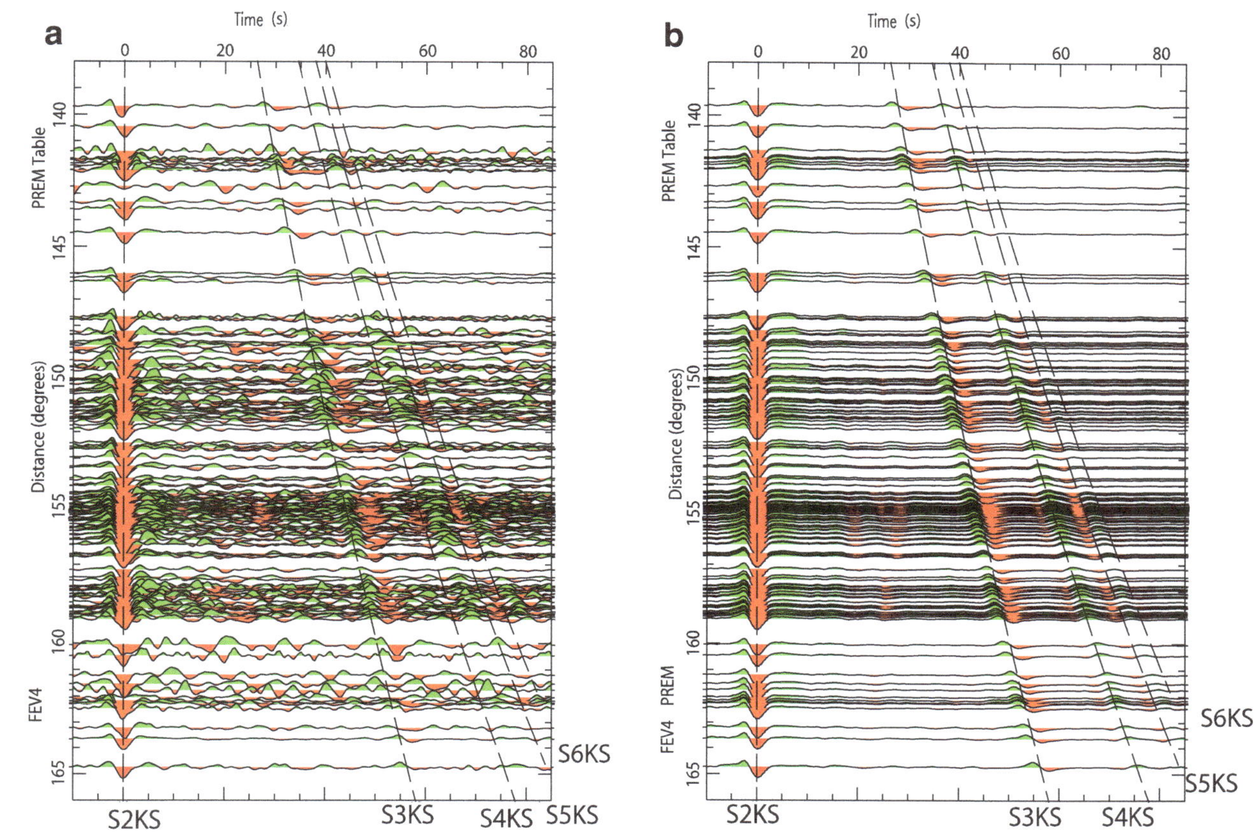

Fig. 2 Observed and synthetic record sections. The observed (**a**) and computed (**b**) record sections of a Fiji-Tonga event recorded at the European stations. The arrivals of S2KS, S3KS, S4KS, S5KS, and S6KS are clearly visible. When aligned on S3KS arrivals, S4KS and S5KS waves are systematically delayed relative to PREM across the entire array. In this example, dt^{3-2}, dt^{4-3}, and dt^{5-3} are anomalous by about +1.2 s

used as the alignment phase whenever the S3KS arrival is sufficiently clear, and the agreements between different alignment phases (S2KS or S3KS) and between different methods (peak-picking or cross correlating) are used for evaluating the error of each dt^{3-2} measurement. The measured differential times are classified into two categories, A (high quality) and B (moderate) (Table 2). When the values obtained by different methods agree with each other, the stacked synthetic and observed seismograms usually agree quite well, and the measurements can be regarded as category A, and have smaller errors.

In addition to the $\frac{\pi}{2}$ phase shift, waveforms of S4KS could be distorted by the interference by S5KS (Eaton and Kendall 2006) or by S3KS when epicentral distances are not large. The S5KS arrivals are not separated completely from S4KS, even when its peak, which is oppositely polarized to S3KS, can be clearly identified. Therefore, dt^{4-3} and dt^{5-3} are measured simply by identifying the corresponding peaks. Since alignment on S4KS arrivals is usually unstable, differential times between S4KS and S5KS (called dt^{5-4}) are calculated by subtracting dt^{4-3} from dt^{5-3} with error propagation. Other array measurement details are described in HK2010 and KH2013. We note that, especially for the Tonga-Fiji events (Fig. 2a), the anomalies of dt^{m-n} relative to PREM are observed more or less uniformly across the entire European array (KH2013). Except for the uniform delays, no systematic trends in the differential times with azimuth from the Tonga-Fiji events, which may be amenable to elaborate modeling, have been identified for the high quality events. This would suggest only minor influences of CMB structure on individual paths.

V_p model for the outermost core

$\tau - p$ inversion: effects of starting model

KH2013 built a V_p model of the topmost 700 km of the outer core using differential S*m*KS travel time anomalies of several Fiji-Tonga events recorded at Europe, dt^{3-2}, dt^{4-3}, and dt^{5-3}, to which a $\tau - p$ inversion method had been applied (Garmany et al. 1979; HK2010; KH2013), with PREM as the starting model. The resultant model, KHOMC, has a 0.45 % slower V_p than PREM at the CMB

Table 2 Differential travel times are measured on the waveforms which are obtained by linearly stacking the observed broadband seismograms with the relative slowness computed for PREM

Event	Array	Distance	dt^{3-2} (s)	Bias	dt^{4-3} (s)	Bias	dt^{5-3} (s)	Bias	Category
FEV1	EU	154.0	1.2 ± 0.2	0.1	1.0 ± 0.2	0.2	1.2 ± 0.3	0.2	A
FEV2	EU	154.0	1.1 ± 0.2	0.2	1.2 ± 0.2	0.2	1.2 ± 0.3	0.3	A
FEV3	EU	152.5	1.1 ± 0.3	0.1	1.3 ± 0.3	0.3	1.5 ± 0.4	0.2	A
FEV4	EU	152.	1.6 ± 0.2	0.2	1.3 ± 0.2	0.2	1.2 ± 0.3	0.3	A
FEV5	EU	147.3	1.5 ± 0.3	0.3	1.3 ± 0.4	0.2	1.6 ± 0.5	0.0	A
FEV6	EU	153.5	1.6 ± 0.4	0.2	0.9 ± 0.2	0.2	1.1 ± 0.4	0.1	A
FEV6	US	123.5	0.9 ± 0.3	0.3	-		-		B
FEV7	EU	148.5	1.5 ± 0.4	0.3	1.4 ± 0.3	0.4	1.3 ± 0.4	0.4	B
FEV8	EU	148.	1.1 ± 0.5	0.3	-		-		B
FEV9	EU	153.	1.2 ± 0.5	−0.1	-		-		B
FEV10	EU	151.9	0.9 ± 0.4	−0.3	-		-		B
FEV11	EU	151.9	1.1 ± 0.4	0.0	-		-		B
FEV12	EU	150.0	1.3 ± 0.2	0.3	1.2 ± 0.2	0.4	1.1 ± 0.3	0.3	A
FEV13	AF	131.	0.2 ± 0.3	−0.1	0.2 ± 0.2	−0.1	-		B
KEV1	EU	155.	1.2 ± 0.2	0.3	1.2 ± 0.3	0.4	1.0 ± 0.4	0.4	A
KEV2	EU	153.	1.9 ± 0.4	0.2	1.0 ± 0.3	0.2	-		B
KEV3	EU	155.	1.4 ± 0.4	0.2	-		-		B
KEV4	EU	155.	1.6 ± 0.5	0.4	1.5 ± -	0.4	-		B
KEV5	EU	155.5	1.2 ± 0.4	0.3	1.2 ± 0.4	0.2	1.5 ± 0.5	0.1	A
KEV6	EU	155.	1.4 ± 0.3	0.2	1.5 ± 0.5	0.3	-		B
AEV1	JP	162.4	0.5 ± 0.4	0.1	0.9 ± 0.4	0.2	-		B
AEV2	HI	157.8	0.9 ± 0.4	−0.1	1.3 ± 0.2	0.2	1.4 ± 0.3	−0.1	A
AEV3	HI	162.6	0.5 ± 0.5	0.0	0.9 ± 0.4	0.2	0.8 ± 0.5	0.2	B
AEV4	HI	160.0	0.6 ± 0.3	−0.1	0.6 ± 0.3	0.3	0.5 ± 0.3	0.2	A
AEV4	AU	124.8	0.7 ± 0.3	0.6	-		-		B
AEV4	AS	167.9	1.0 ± 0.4	0.0	-		-		B
AEV5	JP	163.4	1.0 ± 0.4	0.0	-		-		B
AEV5	AU	124.7	0.6 ± 0.4	0.4	-		-		B
AEV5	AS	167.8	1.0 ± 0.4	0.0	1.4 ± 0.5	0.6	-		B
AEV6	JP	153.9	0.6 ± 0.4	−0.1	-		-		B
NCEV	AU	140.1	2.0 ± 0.5	0.7	1.0 ± 0.5	0.3	-		B
IEV1	US	130.9	1.3 ± 0.4	0.2	-		-		B
IEV2	EU	122.8	0.9 ± 0.4	0.1	-		-		B
IEV2	US	119.5	0.9 ± 0.4	0.1	-		-		B
IEV3	US	120.9	0.8 ± 0.5	−0.4	-		-		B
IEV4	US	144.	1.0 ± 0.4	0.4	1.0 ± 0.5	0.4	0.9 ± 0.5	0.5	A
IEV5	US	141.	0.9 ± 0.3	0.1	0.3 ± 0.3	0.0	0.3 ± 0.5	0.2	A
IEV6	US	133.1	0.4 ± 0.2	0.2	0.1 ± 0.3	−0.2	-		B
MEV	US	149.2	1.0 ± 0.5	0.2	-		-		B
INEV	US	168.6	0.2 ± 0.5	−0.3	1.2 ± 0.4	0.5	-		B
SEV1	AK	154.1	0.4 ± 0.5	0.1	-		-		B
SEV1	JP	153.8	0.9 ± 0.3	0.2	-		-		A
PEV1	SA	149.5	1.9 ± 0.3	0.5	0.5 ± 0.5	0.6	-		B

Array distances correspond to approximate array centers, whose accurate value do not affect the results. The values of "Bias" are measured by applying the array method on the reflectivity seismograms computed for PREM and comparing the differential *SmKS* travel times with the ray theoretical predictions

(Fig. 3). The lower-than PREM V_p values persist down to about 300 km below the CMB. We note here that the $\tau - p$ method is a linear inversion that requires a good starting model. Although PREM is known to be a good global model, there may be other reference models as good as PREM. We perform the $\tau - p$ inversion using IASP91 as the starting model, as did Alexandrakis and Eaton (2010). By using only the Fiji-Tonga to European stations data, as in the case of KHOMC, we obtained a profile of V_p anomaly relative to IASP91 (called KOCTI). KOCTI has faster V_p values than IASP91 in the shallowest 300 km of the outer core (Fig. 3). KOCTI shows an improved fit to the data compared to IASP91 (Fig. 3), but its overall fit is less suitable than that of KHOMC, with the most significant disagreement for the dt^{3-2} data of larger distances, as will be shown later in this study. The V_p values of KOCTI for the upper 300 km are only about 0.01 km/s slower than those of KHOMC (the V_p at the CMB is −0.045 km/s relative to PREM, while that of KHOMC is −0.035 km/s). Therefore, it can be seen that the two models agree with each other within the uncertainty range (Fig. 3). We conclude that the V_p at CMB is constrained well by the $\tau - p$ inversion to 8.03 ± 0.01 km/s, irrespective of the starting model. The disagreement between KOCTI and KHOMC for the depth range from about 300 km to 700 km from the CMB is larger (by about −0.02 km/s), suggesting a poorer resolution than the shallower core.

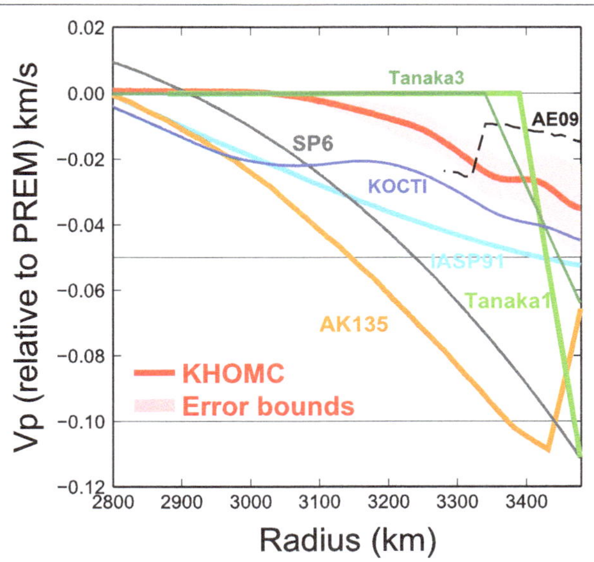

Fig. 3 V_p models of outermost core. V_p models of the upper 700 km of the outer core: KHOMC, AK135, SP6, Tanaka-1, Tanaka-3, IASP91, AE09 (Alexandrakis and Eaton 2010), and KOCTI. The V_p values (in km/s) as a function of depth relative to the PREM values are shown. The uncertainty range of KHOMC are shown in pink (KH2013)

Genetic algorithms

In order to check further the effects of the starting model on the result, we perform another V_p inversion using genetic algorithms (e.g., Yamanaka and Ishida 1995) and attempt to find the global minimum of dt^{m-n} misfit in a fashion that does not explicitly require a starting model. The V_p profile of the outer 700 km of the core is assumed to be continuous and consist of four layers with constant V_p gradients. For each layer, the V_p value at its top and its thickness are determined (see the caption of Fig. 4 for details).

We use the data set consisting of dt^{3-2}, dt^{4-3}, and dt^{5-4} data for four events from Fiji-Tonga to Europe (FEV1, FEV2, FEV3, FEV4) and three events from Argentina to

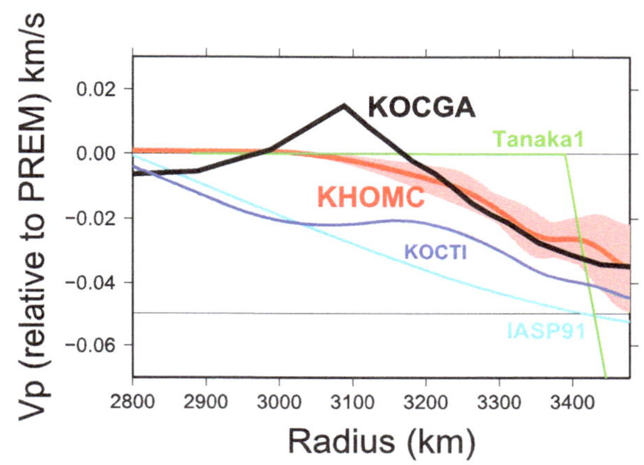

Fig. 4 Genetic algorithm. The V_p model obtained by the inversion using genetic algorithms. The thick black line labeled KOCGA is the obtained continuous V_p model of the outer 700 km of the core, which has four layers with uniform V_p gradients. Other details are the same as Fig. 3. The thickness of the deepest layer is fixed to 400 km and the V_p value at the bottom of the layer is fixed to that of PREM, so that there are seven parameters to be determined, each of which is described as a 6-bit binary number. Therefore, each gene has a length of 42 bits. The least bit for the V_p at the top of each layer corresponds to 0.003 km/s, while the least bit for the thickness corresponds to 2 km. The depth of the CMB is fixed to that of PREM, so the inversion is not entirely free from the reference model, but the effect is quite small insofar as the differential travel times of SmKS waves are concerned. The range of the V_p value sought by this parameterization covers about ±0.2 km/s relative to PREM, which is wide enough not to miss any successful models. First, the samples of V_p model of the first generation are constructed by randomly selecting the seven model parameters. Second, the misfit of the model predictions to the observations is computed, and each V_p model sample is selected with a probability proportional to the inverse of the squared sum of its misfit. After the selection, crossover of the genes between two randomly selected samples occurs with a given probability, followed by the mutation within each gene that occurs with a given probability, in order to generate the samples of the next generation. This process is iterated through the fixed number of generations. The total misfit over the entire samples usually converges rapidly enough. The parameters used in the inversion are as follows: both the mutation probability and the crossover probability are 0.2, and the numbers of sample and generation are 100 and 40, respectively

Japan (AEV1, AEV2, AEV3) without correcting for bias (Table 2). The inversion is repeated 40 times with different initial values, and the case with the least misfit, for which the total residual decreases from 0.5 to 0.15 s through the 40 generations, is chosen. As shown in Fig. 4, the obtained model (called "KOCGA" hereafter) is fairly close to KHOMC. The agreement between the model and KHOMC is remarkable especially for the shallower 300 km of modeled depth range. Based on this result, we conclude that the validity of KHOMC is not affected by the choice of the starting model.

S*m*KS slowness measurements

In addition to travel time differences, differences in slowness between two S*m*KS waves provide information on the relative arrival directions of the waves. In HK2010 and KH2013, it was reported that the relative slownesses for the Tonga and Argentina events are close to those predicted by PREM. This observation has been exploited in using the τ-p inversion. The slowness observations can also give constraints on the degree of large-scale heterogeneity in the receiver-side mantle, as shall be shown in the "Discussion" section. The differential slowness between two different S*m*KS waves changes gradually with distance. Therefore, the moveout of a S*m*SK wave, relative to the reference S*m*KS wave on a record section, is aligned on a slightly curved line rather than a straight line. Although the curvature is small, when the aperture of an array is as large as 20°, its effect on the differential travel times approaches a second, which could give rise to a small bias of the measured differential slowness between S*m*KS and S*n*KS relative to PREM (called "dp^{m-n}"). To ensure that dp^{m-n} is measured accurately, we correct for the curvature by assuming that the moveout around the array center (distance of Δ_0) is a quadratic function of epicentral distance when stacking seismograms. Accordingly, the differential slowness between S3KS and S2KS (and similarly between S4KS and S2KS) is expressed as

$$dp^{3-2}(\Delta) = dp^{3-2}(\Delta_0) + (\Delta - \Delta_0)\frac{d}{d\Delta}dp^{3-2}$$

The curvatures of the moveout curve are fixed to those of the reference model, so that $\frac{d}{d\Delta}dp$ is 0.014 and 0.021 s/°^{2} for dp^{3-2} and dp^{4-2}, respectively, which does not significantly affect the results. We obtain precise slowness measurements of S3KS and S4KS waves with S2KS as the reference wave for eight Fiji-Tonga and Kermadec events observed from Europe, and four Argentina events observed from Japan (Table 3). The dp^{3-2} and dp^{4-2} are then compared to the predictions of KHOMC (Table 3).

Figure 5a shows a vespagram example of the cube-root stack output as a function of slowness and arrival time relative to S2KS, with the small moveout curvature effects corrected. The observed dp^{3-2} and dp^{4-2} (red circles) closely agree with theoretical predictions by KHOMC (Table 3), which are shown by red stars in the figure. The RMS residuals computed for the Fiji-Tonga and Argentina events using the outer core models (PREM, IASP91, KOCTI, AK135, SP6, and Tanaka-1) show that the least residual is obtained for KHOMC (Fig. 5b). The RMS residual is less than 0.03 s/° for KHOMC (Fig. 5b), and dp^{3-2} for the majority of the events, which agree with KHOMC within 0.02 s/° (Table 3). The residuals for KHOMC of dp^{3-2} and dp^{4-2} are slightly improved in comparison to PREM, while no such dp^{m-n} improvement is seen for the other models. The small discrepancy of the observed dp^{3-2} and dp^{4-2}, with respect to PREM, justifies the usage of PREM as the reference model for stripping the mantle contributions from the observed distance ranges and travel times in the τ-p inversion that resulted in KHOMC (HK2010; KH2013). While the primary information about the core V_p structure is contained in the differential S*m*KS travel times, as shown in parameter tests (KH2013), the measured dp^{3-2} and dp^{4-2} data not only indicate the validity of our travel time analyses but also further support the validity of KHOMC.

S*m*KS differential times of other regions

Observations of dt^{3-2}

KHOMC was built by using the Tonga-Fiji data set that samples rather limited areas of the CMB, mostly beneath western Pacific and Siberia. KH2013 showed that the S*m*KS differential travel times of the Argentina to Japan data set are consistent with KHOMC, even though they sample entirely different parts of the CMB. In this study, we compile the dt^{3-2} measurements for additional pairs of event and large-scale array that sample a much larger portion of the CMB. We determine the anomalies of dt^{3-2} for several large-scale arrays (Japan, US, Australia, Alaska, Africa, South America, and Southeast Asia; Fig. 6a) using the same processing. The ray paths cover a reasonably large part of the outer core (Fig. 6a).

A striking feature of the result is that the dt^{3-2} observations based on the large-scale array processing are quite well matched by KHOMC (Fig. 6b). On the other hand, the dt^{3-2} values measured for individual stations (Souriau et al. 2003) show a much larger scatter than the array measurements (Fig. 6b), which suggests that measuring dt^{3-2} using large-scale array data significantly ameliorates the heterogeneous mantle structure effects. Another notable feature is that the array measurements fall approximately in the middle of the range of individual station measurements (Figure six b; Souriau et al. 2003), which also suggests that the array measurement of this study has successfully represented the overall core structure. The total residual of the array measurements is computed for KHOMC and for other outer core V_p models (IASP91, Tanaka-1, Tanaka-3, ak135, SP6, and PREM), and the least

Table 3 Differential slowness ($dp^{3\text{-}2}$ and $dp^{4\text{-}2}$,s/°)

Event	Distance Δ_0	dp^{3-2} (s/°)	KHOMC	Residual	dp^{4-2} (s/°)	KHOMC	Residual
FEV1	154.0	1.16	1.16	0.0	1.62	1.61	0.01
FEV2	154.0	1.20	1.16	0.04	1.65	1.61	0.04
FEV3	152.5	1.15	1.13	0.02	1.60	1.57	0.03
FEV4	152.0	1.14	1.13	0.01	1.59	1.57	0.02
FEV6	153.5	1.19	1.18	0.01	1.65	1.65	0.00
FEV12	150.0	1.12	1.11	0.01	1.56	1.53	0.03
KEV1	155.0	1.20	1.16	0.04	1.64	1.62	0.02
AEV2	157.8	1.19	1.21	−0.02	1.70	1.69	0.01
AEV3	162.6	1.20	1.18	0.02	1.65	1.65	0.00
AEV4	160.0	1.30	1.23	0.07	1.77	1.73	0.04
AEV5	167.8	1.30	1.28	0.02	1.80	1.77	0.02

The corresponding values calculated for KHOMC and the residuals of the observed values from the KHOMC predictions are also shown. Alignment is at the center distance of the array

misfit is obtained for KHOMC (Fig. 6c). As mentioned previously, we note that KHOMC definitely gives a better fit than the τ-p model based on IASP91 (KOCTI) for larger distances. These observations indicate that the mantle effects on dt^{3-2} are of secondary importance and that the essential features of KHOMC relative to PREM reflect the structure of the core.

Observations of dt^{4-3} and dt^{5-4}

The ray paths for which dt^{4-3} can be determined by using large-scale array data are geographically more restricted around Pacific than in the case of dt^{3-2} but still sample a reasonably broad portion of the CMB (Fig. 7a). The measured dt^{4-3} values fall in the middle of individual station measurements (Garnero et al. 1993) and are well consistent with KHOMC (Fig. 7b). Among the V_p models considered (Fig. 7d), the dt^{4-3} observations are best matched by KOCTI, but the data set (especially for category A) is explained almost equally well by KHOMC; the dt^{4-3} data set is not useful for discriminating between the two models. The other models (IASP91, Tanaka-1, Tanaka-3, AK135, and SP6) give significantly poorer fits when compared with KHOMC and KOCTI.

The dt^{5-4} data set is matched well by some of the models considered, KHOMC, PREM, IASP91, and KOCTI (Fig. 7c), so that it is not crucial to discriminate between the models. Nevertheless, it clearly refutes the class of models that have a strong V_p reduction in a thin layer at the top of the outer core, such as Tanaka-1, Tanaka-3, and that of Garnero et al. (1993). These models were built without using dt^{5-4} and were aimed to match mainly dt^{3-2}, which means that the amount of V_p anomaly relative to PREM across the top several hundred kilometers of the outer core can be well predicted by these models. The mismatch between these models with the observed dt^{5-4} means that the V_p anomaly relative to PREM needs to be distributed over a broader depth range than those in the models. This also indicates that the V_p gradient near the top of the core is not extremely anomalous compared to PREM (Fig. 3). The inference is further supported by the good match of the observed S6KS waveforms relative to S5KS by KHOMC (KH2013); the V_p gradient near the top of the core is tightly constrained by our data set.

We emphasize again that KHOMC was constructed by using the Fiji-Tonga to Europe data set alone, yet the observations of dt^{4-3} and dt^{5-4} for different regions can be matched by the same model quite well. The S*m*KS data we used, therefore, should primarily reflect the outer core structure.

Anomalous outermost core in terms of V_p gradient

The most important feature of our V_p models (KHOMC and KOCGA) is the presence of marked radial change in the V_p gradient, $V_p' = dVp/dr$; the outermost core of KHOMC and KOCGA is essentially characterized by two distinctive layers with different V_p'. We parameterize the V_p structure of the outermost 700 km of the core with two layers that have constant V_p' and compute the misfits of dt^{3-2}, dt^{4-3}, and dt^{5-4} for the Fiji-Tonga and Argentina data sets (KH2013).

The minimum misfit in this modeling is obtained for the V_p value at the CMB near that of KHOMC (−0.35 km/s with respect to PREM). For this case, the obtained V_p' value of the shallower layer approximately agrees with the mean values of V_p' of the shallower 300 km for KHOMC and KOCGA (Fig. 8), which indicates that the mean V_p gradient of the shallower layer is well constrained (Fig. 8) by the presence of additional different S*m*KS waves passing through it. Although the lower layer gradient is rather poorly constrained, the significant difference in the values of V_p' anomalies relative to PREM between the two layers is robust; the anomaly of V_p' with respect to PREM

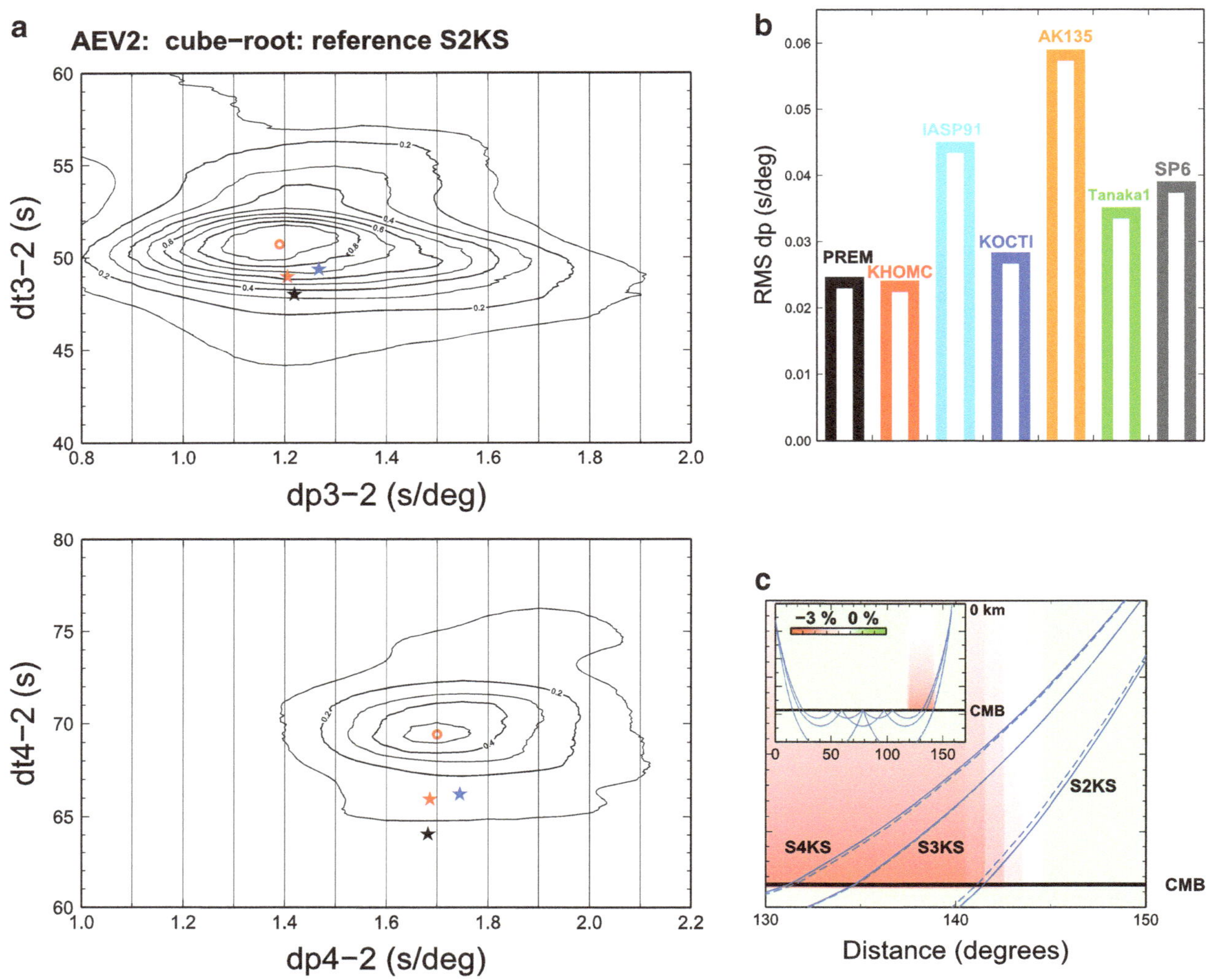

Fig. 5 Slowness measurements. (**a**) The observed differential slownesses, dp^{3-2} (top) and dp^{4-2} (bottom). Horizontal axis is slowness (in $s/°$) as a function of time after the arrival of S2KS (vertical axis). Red, blue, and black stars indicate the predictions from KHOMC, IASP91, and PREM, respectively. (**b**) RMS misfits of dp^{3-2} and dp^{4-2} in $s/°$ for different outer core models. The difference in the misfit between Takana-1 and Tanaka-3 is small. (**c**) An artificial model of 2D V_S heterogeneity localized near the bottom of the mantle receiver-side structure and the ray paths of S2KS, S3KS, and S4KS, after the Earth-flattening transformation. Dashed and solid blue rays are those for the unperturbed and perturbed Earth models, respectively. Inset shows the entire paths, while the rays near the receiver side CMB are focused below. The V_S anomaly is shown with color scale

($dV_p'=dVp/dr_{profile}$ - dVp/dr_{PREM}) is much larger in the shallower part of the depth range considered than below it (Fig. 8). Accordingly, the V_p' of the shallower 300 km of the outer core is steeper than PREM by about 0.0001 (1/s), while that of the deeper part is closer to that of PREM. The number anomalies of V_p' (relative to PREM) obtained for the Genetic algorithms (KOCGA) is 0.00012 (1/s) for the upper ~300 km of the core, while that is −0.00002 (1/s) for the deeper part (300 to 700 km from the CMB).

The V_p gradient and the pressure derivative of bulk modulus ($K' = dK_s/dP$) are interrelated, and the principal feature of KHOMC indicates a substantial radial variation in K' within the uppermost 700 km of the outer core. By using the equation, $V_p' = -g(2V_p)^{-1}(K' - 1)$, K' can be computed from V_p. We find that the K' value of the outermost 300 km of the core is nearly 3.7, which is larger than that of the deeper core by about 0.2 (Butler and Anderson 1978). The estimated anomaly of K' for the upper 300 to 400 km of the outer core amounts to a nearly 5 % radial anomaly, which is more than an order of magnitude larger than the V_p anomaly itself.

Discussion

Effects of receiver-side mantle

Our previous studies have shown that large-scale array analyses adequately extract differential $SmKS$ travel time

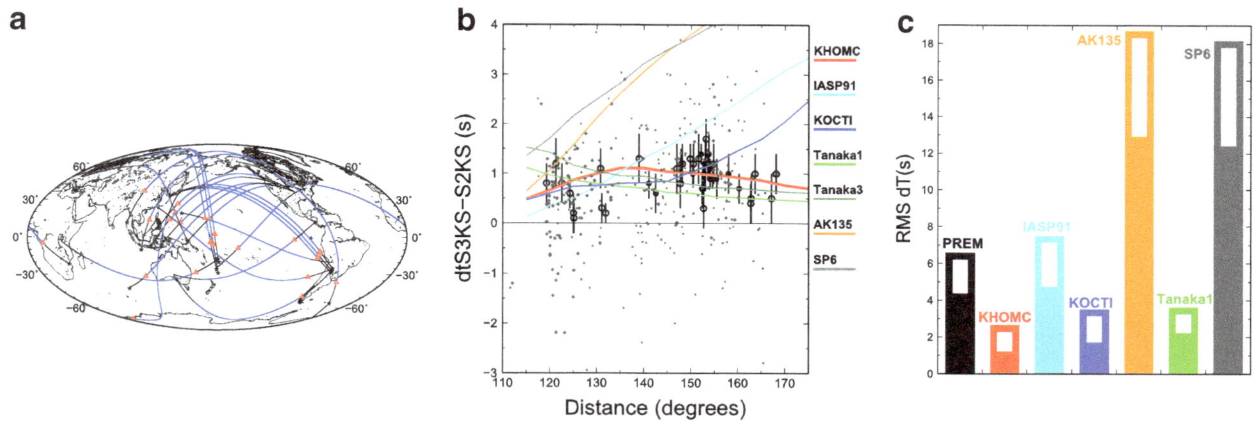

Fig. 6 Globally measured dt^{3-2} values. (**a**) Ray paths corresponding to the measured dt^{3-2}. Stars and red triangles show the epicenters and source side CMB entry points of S3KS. Blue portions of the ray paths indicate the segments within the outer core. For the events at Tonga-Fiji and Kermadec, only selected events and ray paths are plotted in order to prevent the figure from becoming excessively cluttered. (**b**) Measured dt^{3-2} for other large-scale arrays converted to a hypothetical focal depth of 500 km. The dt^{3-2} values predicted by global outer core models are shown with lines. The measured dt^{3-2} are shown by circles with error bars. Filled symbols show the data of category A, while open symbols are of category B (Table 2). The observed dt^{3-2} values are explained quite well by KHOMC (red line). Superimposed are the individual station-based dt^{3-2} data converted to 500 km focal depth (large diamonds) (courtesy of A. Souriau). Smaller squares are other dt^{3-2} measurements made for individual stations taken from Souriau et al. (2003) without depth corrections. (**c**) Total residuals, defined as the square root of the squared sum of the residuals of dt^{3-2} for different outer core models. Filled portions are the misfits for the category A data alone. The colors used for the models correspond to those in (**b**)

anomalies representative to the array as a whole (HK2010; KH2013). The receiver-side CMB piercing points for the Tonga-Fiji events scatter widely beneath Europe (HK2010; KH2013), making it difficult to envisage a receiver-side mantle heterogeneity which causes a systematic anomaly of S*m*KS differential travel times across the entire array.

Nevertheless, we attempt to conservatively evaluate the effects of mantle heterogeneity on the dt^{3-2} and dt^{4-3} measurements, by focusing on the observation of very

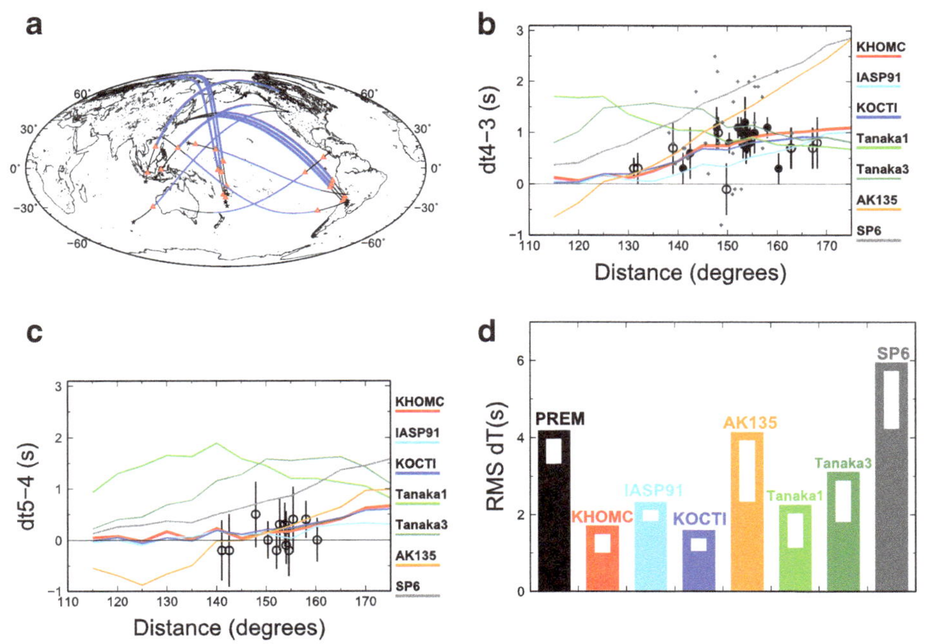

Fig. 7 Globally measured dt^{4-3} and dt^{5-4} values. (**a**) Ray paths for the dt^{4-3} and dt^{4-3} measurements. The CMB piercing points are of S4KS. Thick lines show the ray paths for which both dt^{4-3} and dt^{5-4} are measured, while thin lines show the ray paths for which only dt^{4-3} is measured. Other details are the same as Fig. 6a. (**b**) Measured dt^{4-3}. Details are the same as in Fig. 6b. (**c**) Measured dt^{5-4}. Details are the same as in Fig. 6b. (**d**) Total residuals for dt^{4-3}. Details are the same as in Fig. 6c

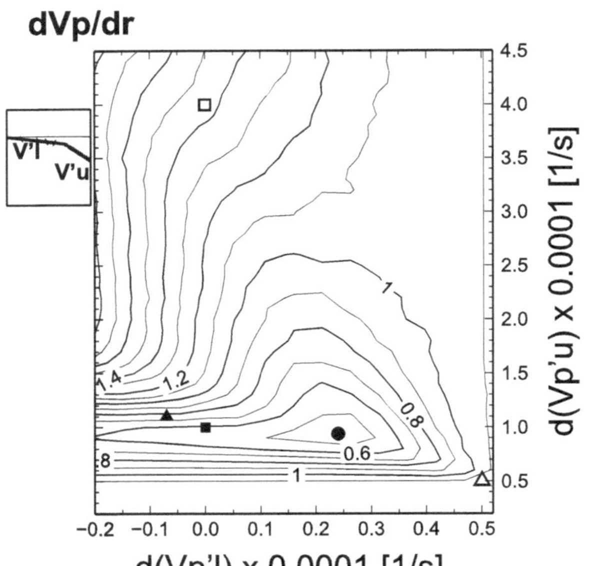

Fig. 8 V_p gradient with respect to PREM for a two-layer outermost core. Contour plot of total residual (in seconds) of dt^{3-2}, dt^{4-3}, and dt^{5-3} for the two-layered outer core model for the case of V_p at the CMB of 8.03 km/s, which give the least residual. The input data are as follows: $dt^{3-2} = 1.0$, $dt^{4-3} = 1.0$, and $dt^{5-3} = 0.9$ for 154° and h = 500 km representing typical values of the Fiji-Tonga data set; and $dt^{3-2} = 0.7$, $dt^{4-3} = 0.9$, and $dt^{5-3} = 0.9$, 162° and h = 590 km, representative of the Argentina data set. The deviation of V_p' from PREM ($dVp/dr_{profile}$ - dVp/dr_{PREM}) for the shallower part of the depth range (vertical axis) and the that of the deeper layer (horizontal axis). The points approximately representing KHOMC, KOCGA, IASP91, and Tanaka-1 are shown by a solid square, solid triangle, open triangle, and open square, respectively

small residuals of differential slownesses dp^{3-2} and dp^{4-2} described above from KHOMC (Fig. 5b). The observations indicate that the relative arrival angles of the different S*m*KS waves are barely anomalous and the rays of the S*m*KS waves are not substantially bent with respect to each other. A dp^{3-2} anomaly of 0.02 s/° or less for the majority of the events (Table 3) corresponds to the anomaly in the separation of S3KS and S2KS piercing points at the CMB of the receiver side by less than 5 km.

We argue that a differential travel time anomaly (dt^{3-2} and dt^{4-3}) as large as those observed would need to be accompanied by a large anomaly in the relative direction of ray arrivals at the receivers when the anomaly is caused by the mantle heterogeneity beneath the receiver (Fig. 5c). For a low V_s heterogeneity to cause a dt^{3-2} anomaly exceeding 1 s across the array, S2KS waves would need to more effectively avoid the heterogeneous body compared to S3KS (e.g., Garnero and Helmberger 1995). This effect on the ray angle deviations was evaluated by ray tracing experiments. The typical dominant period of S2KS and S3KS is nearly 10 s, and dt^{3-2} values measured for 3 s high-pass filtered seismograms of three Tonga-Fiji events (FEV2, FEV3, and FEV4) and two Argentina events (AEV2 and AEV4) agree with those for the original broadband seismograms within 0.3 s. They agree within 0.1 s for two of the Fiji events. This result confirms the utility of conducting theoretical ray estimations on dt^{3-2} to identify possible mantle heterogeneity effects. Therefore, we will next consider a test case involving an artificially strong and sharp-edged 2D low-V_s anomaly that extends about 1000 km from the CMB with a V_s anomaly that is a maximum 3 % slower in the receiver side of the lowermost mantle (Fig. 5c). The dt^{3-2} and dt^{4-3} anomalies at approximately 150° are 1.8 s and 0.8 s, respectively, which are comparable to the observations. The relative slownesses, dp^{3-2} and dp^{4-2}, are 0.10 and 0.14 s/°, respectively, which are nearly five times larger than the observations. Accordingly, the rays of S3KS and S2KS, as well as those of S4KS and S2KS, bend relatively by approximately 25 km at the CMB. The observed minute anomalies in dp^{3-2} and dp^{4-2} indicate that the receiver-side piercing points are much less significantly bent than is required by this model. The maximum V_s anomaly needs to be as low as 0.6 % in order to match the observed dp^{3-2} and dp^{4-2}, which sets the upper bounds on the allowable biases of dt^{3-2} and dt^{4-3} due to the receiver-side heterogeneity to much less than 0.4 s and 0.2 s, respectively.

Effects of source-side mantle

A source-side lower mantle structure that is capable of causing a dt^{3-2} anomaly of ∼1 s across the entire European array would need to be laterally much larger than 200 km (KH2013). For the source-side mantle sampled by the Fiji-Tonga data set, the V_s structure in the D″ of very large scale (≥3000km) beneath the north of Vanuatu seems to have been resolved moderately well by global seismic tomography (e.g., Lekic et al. 2012). Therefore, it would appear worthwhile to check whether V_s heterogeneity of a larger scale in the source side deep mantle accounts for a significant portion of the S*m*KS differential travel time anomalies.

We focus on five Fiji-Tonga events that are located from 300 to 1000 km laterally separately from each other (Fig. 9a) and for which high quality S2KS, S3KS, and S4KS have been observed (Table 2). The five events sample the CMB regions that are shifted systematically by 200 to 600 km; each event covers a CMB area of ∼1000 km by 300 km (Fig. 9b). Effects of D″ should be most significant for dt^{3-2}, since S2KS and S3KS have larger CMB piercing point separations; dt^{4-3} should be less sensitive to heterogeneity in D″ as the separations of the CMB entry points between S3KS and S4KS are less than half of dt^{3-2}.

We find that the dt^{3-2} anomalies for the five events are nearly uniform and are excellently matched by KHOMC (Fig. 10a), which is also the case for dt^{4-3} (Fig. 10b). The

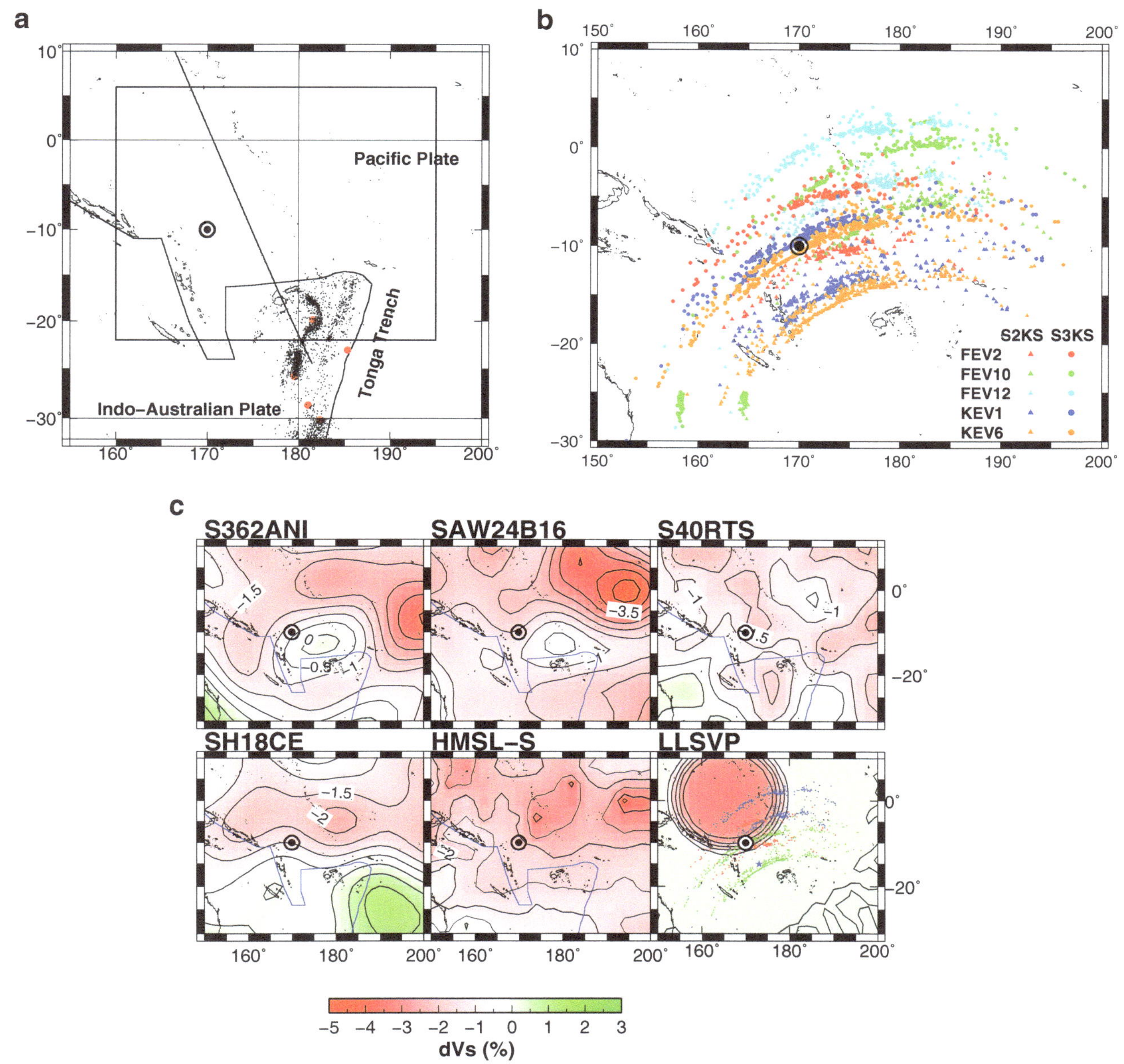

Fig. 9 Tonga-Fiji events. (**a**) Locations of the five events, FEV2, FEV10, FEV12, KEV1, and KEV6. (**b**) CMB piercing points for S2KS and S3KS for the four events. Triangles and circles show the core entry points of S2KS and S3KS for each event, respectively. The large black circles are drawn to mark a reference point referred as below. (**c**) The V_S anomalies in the D″ beneath Vanuatu around the CMB piercing points of the Fiji-Tonga data set for the five tomography models: SAW24B16 (Megnin and Romanowicz 2000), S362ANI (Kustowski et al. 2008), HMSL (Houser et al. 2008), SH18CE (Takeuchi 2007), and S40RTS (Ritsema et al. 2011). The V_S anomalies are shown with color code and contours. The large black circles show the reference point. Bottom right panel shows a model with an extreme and sharp-edged anomaly, like LLSVP. The piercing points of S2KS and S3KS for three of the five events are shown in this panel only

V_s models produced by global tomography appear to provide moderately reliable images of the deepest ~200 km of the mantle (essentially corresponding to D″) of this region, at least for wavelength features exceeding 2000 km. According to the V_s models for the 2800 km depth of five different tomography studies (Fig. 9c), despite considerable differences in detail, the presence of 1 to 3 % low V_s in the D″ of the northern half of the study area seems to be commonly resolved. On the other hand, the depth extent of such anomalies are very poorly constrained, and 2500 km depth tomography images show only insignificant anomalies that differ substantially from model to model. Therefore, we only deem the heterogeneity in the D″ to be significant and evaluate its effects on S*m*KS travel

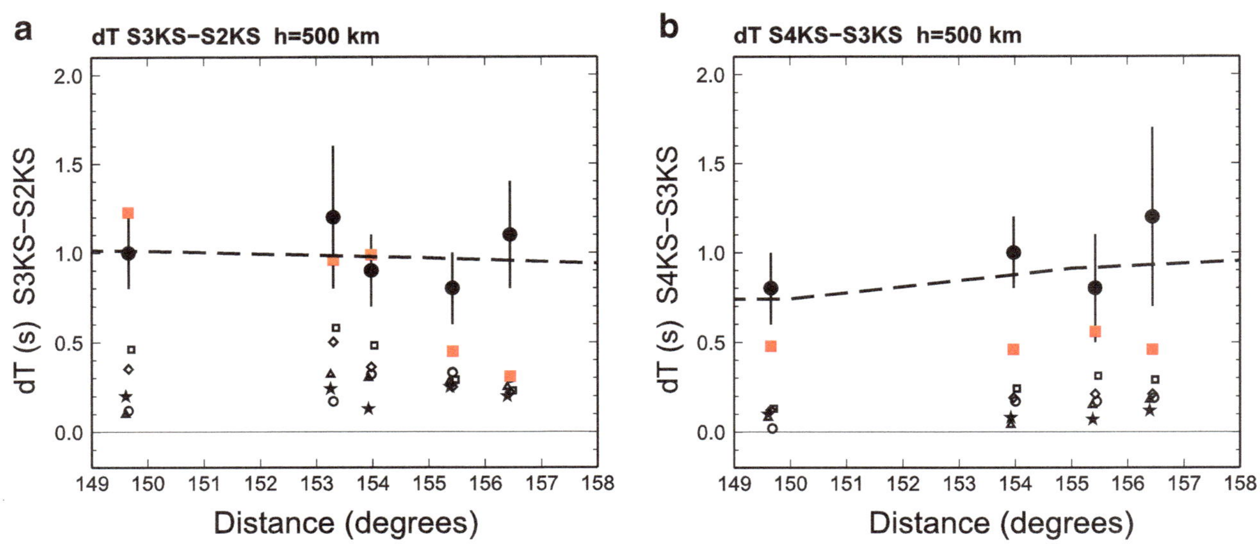

Fig. 10 Effects of source side D″. (**a**) Symbols with error bars show the anomalies with respect to PREM of dt^{3-2} measured for the five events in Fig. 9a as a function of array center distance from the epicenters. Small symbols are the predicted effects of the D″ heterogeneity for five different tomography models (those in Fig. 9c). Large red squares show the anomalies of dt^{3-2} predicted by the artificial lower mantle heterogeneity model shown in the bottom right panel of Fig. 9c. (**b**) Same as (**a**) for dt^{4-3}

times, rather than attempting to make corrections on the S*m*KS travel times for the effects of mantle heterogeneity. Ray theoretical travel times computations show that the tomography-derived structures of D″ could systematically affect the anomalies of dt^{3-2} values by 0.3 s on average (Fig. 10a), while the dt^{4-3} anomalies by 0.2 s (Fig. 10b). These effects may cause small biases to the core models but do not alter them significantly.

If an extreme but currently unresolved heterogeneity that is analogous to LLSVP with a sharp edge exists near the source side CMB entry points of S*m*KS, it might cause dt^{3-2} and dt^{4-3} anomalies of the observed magnitude. As an example of such a scenario, we will next consider a simplified but significantly exaggerated model that has a qualitative resemblance with the tomographic V_s anomalies of the D″ (Fig. 9c) and evaluate the effects of the extreme lower mantle heterogeneity. The model that we will consider has an axisymmetrical tabular-shaped low V_s heterogeneity that has a maximum anomaly of 3.5 % at the CMB and that exponentially decays upward with a scale height of 500 km (Fig. 9c, bottom right). Theoretical dt^{3-2} and dt^{4-3} for this model are computed by ray theory (Fig. 10a, b). The values of dt^{3-2} can be as large as the observed values depending on the epicentral distance. However, there should be a clear trend in dt^{3-2} with the epicentral distance by about 1 s, which is entirely different from the observed dt^{3-2} trend. The relative magnitudes of dt^{3-2} and dt^{4-3} are also grossly inconsistent with the observation. The heterogeneity model significantly underestimates as a dt^{4-3} value that is less than half of the observations, mostly because of the smaller separation of the CMB piercing points (Fig. 10a). Although this demonstrates only just one example, the basic feature of V_s structures that potentially cause dt^{3-2} anomalies as large as the observed values should be more or less the same. S2KS more effectively avoids the low V_s body than S3KS. Simplified mantle heterogeneity models resembling tomography images, no matter how pronounced and sharp they are, have difficulty matching the observed dt^{3-2} and dt^{4-3} of the Tonga-Fiji data set. Therefore, we conclude that an unresolved lower mantle heterogeneity is unlikely to be the predominant cause of the observed S*m*KS anomalies, and estimate its effects on dt^{3-2} and dt^{4-3} values based on the current tomography models to be less than 0.3 s and 0.2 s, respectively.

Interpretations in terms of the composition of the core

As suggested by high estimates of liquid iron thermal conductivity, a thermally stratified layer in the neighborhood of 100 km thick might exist in the outermost core (Pozzo et al. 2012). However, a more recent numerical estimate of the electrical conductivity of iron under the Earth's core conditions has negated the thermal stratification requirement (Zhang et al. 2015). If the stratified layer is nearly 300 km thick, as estimated in this study, the maximum temperature excess over the adiabat at the CMB would be about 300 K. Recent ab initio calculations of liquid and solid iron show that the bulk sound speeds of iron are mostly independent of temperature (Ichikawa et al. 2014; Vocadlo et al. 2009). Therefore, the observed V_p and V_p' anomalies in the uppermost 300 km of the outer core cannot be of thermal origin and should primarily reflect compositional heterogeneity. Since the effective dK/dP in the layer is larger than the bulk of the core, if the light elements diffuse downward from the CMB (and thus have concentrations decreasing with depth), the addition of the light elements must decrease not only the density of the liquid iron alloy but also its bulk modulus. According to recent ab initio calculations of liquid iron-alloy under the core conditions (Badro et al. 2014), these requirements are satisfied. However, the same calculations show that including light elements increases V_p . Thus, it seems that matching the observed V_p value at the CMB cannot be done by simply by adding light elements, even though the effects of non-ideal mixing in the iron alloy (which might not be adequately modeled in the simulations) might still play a role in reducing V_p (Helffrich 2012).

The estimated thickness of the compositionally stratified layer (~300 km) cannot be interpreted via a straightforward process. If the stratified layer evolved from the CMB through the diffusion of light elements, the thickness of the layer is essentially determined by the diffusion coefficient of the core liquid. The mass diffusivity of liquid iron under core conditions is thought to be reasonably well constrained (Koci et al. 2007; Pozzo et al. 2012; Helffrich 2014), and the expected thickness is no more than 80 km (Buffett and Seagle 2010; Helffrich and Kaneshima 2013). Helffrich (2014) suggests that the presence of a thick layer is a feature of the Earth's core that was formed at the time of the putative giant impact.

While the V_p profile of the top 700 km of the core is adequately represented by two layers with nearly constant radial V_p profile gradients, there is certainly room for the profile to be optimized with some physically plausible constraints, such as the diffusion profile of light elements (Helffrich 2014). However, the revelation of detailed features of the V_p profile is somewhat more difficult due to the presence of mantle effects that have secondary importance. Based on the lack of corresponding anomalies in the waveforms, a sharp interface with a large V_p jump at the bottom of the shallower layer near the depth of 300 km is unlikely to exist, but the presence of a weak jump cannot be ruled out. If light elements diffuse from the CMB, and if double diffusion takes place to form the stratified layer, a succession of thin homogeneous layers might occur near the bottom of the stratified layer (e.g., Buffett and Seagle 2010). In such cases, scattering of seismic energy might occur near the bottom of the layer, depending on the contrasts in the elastic properties between the materials enriched and depleted in light elements. A search for such scattering waves might reveal further details about the enigmatic region of deep Earth. On the other hand, the very top of the core is obviously another locality where an anomalous structure is possible. The existence of a thin and anomalously high V_p and low density layer at the top of the core (Helffrich and Kaneshima 2004) is not supported, if the layer thickness exceeds 10 km or so, by a good fit of the waveforms S6KS to KHOMC (KH2013). Nevertheless, a thinner layer might exist.

Conclusions

The differential travels between *Sm*KS measured by analyzing large-scale broadband seismometer arrays are shown to predominantly reflect the V_p structure of the outermost outer core. The combination of dt^{3-2}, dt^{4-3}, and dt^{5-4} anomalies restrict permissible V_p models within a narrow range. There is a significant radial change in gradient of V_p at the depth about 300 km below the CMB. The gradient of the shallower layer corresponds to an effective change in dK_s/dP by about 0.2, which is too large to be attributed to thermal effects alone, and requires compositional stratification.

Abbreviations

LLSVP, Large low shear velocity province; HK2010, Helffrich and Kaneshima (2010); KH2013, Kaneshima and Helffrich (2013).

Competing interests

The authors declare that they have no competing interests.

Authors' contributions

SK performed the analyses and wrote the manuscript. TM prepared for the Hi-net tilt-meter data. Both authors read and approved the final manuscript.

Acknowledgements

This study owes a great deal to the management of waveform data by IRIS DMC, the NDIC's F-net in Japan, the J-Array Data Center, the Taiwan Data Center, and the ORPHEUS Data Center in Europe. Generic Mapping Tools (GMT) (Wessel and Smith 1995) were used for drawing all of the figures. Appreciation is extended to A. Souriau and G. Helffrich for providing individual S3KS-S2KS travel time measurement data. Thanks are also extended to George Helffrich who kindly checked our manuscript, and J. Ritsema who generously provided data from his tomography model. T. Tsuchiya is thanked for his enlightening discussions, and the comments and suggestions of Satoru Tanaka and the two anonymous reviewers were very helpful for improving the manuscript.

Author details

[1]Department of Earth and Planetary Science, Kyushu University, Hakozaki, Higashi-ku, 812-8581 Fukuoka, Fukuoka, Japan. [2]National Research Institute for Earth Science and Disaster Prevention, 3-1 Tennodai, 305-0006 Tsukuba, Japan.

References

Alexandrakis C, Eaton DW (2010) Precise seismic-wave velocity atop Earth's core: No evidence for outer-core stratification. Phys Earth Planet Inter 180:59–65

Badro J, Cote AS, Brodholt JP (2014) A seismologically consistent compositional model of Earth's core. Proc Nat Acad Sci 111:7542–7545

Buffett BA, Seagle CT (2010) Stratification of the top of the core due to chemical interactions with the mantle. J Geophys Res 115. doi:10.1029/2009JB006751

Butler R, Anderson DL (1978) Equation of state fits to the lower mantle and outer core. Phys Earth Planet Inter 17:147–162

Choy G (1977) Theoretical seismograms of core phases calculated by a frequency-dependent full wave theory, and their interpretation. Geophys J Astr Soc 51:275–311

Dziewonski A, Anderson DL (1981) Preliminary reference Earth model. Phys Earth Planet Inter 25:297–356

Eaton DW, Kendall J-M (2006) Improving seismic resolution of outermost core structure by multichannel analysis and deconvolution of broadband SmKS phases. Phys Earth Planet Inter 155:104–119
Garmany J, Orcutt JA, Parker RL (1979) Travel time inversion: a geometrical approach. J Geophys Res 84:3615–3622
Garnero EJ, Helmberger DV (1995) On seismic resolution of lateral heterogeneity in the Earth's outemost core. Phys Earth Planet Inter 88:117–130
Garnero EJ, Helmberger DV, Grand SP (1993) Constraining outermost core velocity with SmKS waves. Geophys. Res. Lett. 20:2463–2466
Helffrich G (2012) How light element addition can lower core liquid wave speeds. Geophys J Int 118:1065–1070
Helffrich, G (2014) Outer core compositional layering and constraints on core liquid transport properties. Earth Planet Sci Lett 391:256–262
Helffrich, G, Kaneshima S (2004) Seismological constraints on core composition from Fe-O-S liquid immiscibility. Science 306:2239–2242
Helffrich G, Kaneshima S (2010) Outer-core compositional stratification from observed core wave speed profiles. Nature 468:807–810
Helffrich G, Kaneshima S (2013) Cause and consequence of outer-core stratification. Phys Earth Planet Inter 223:2–7
Houser C, Masters G, Shearer P, Laske G (2008) Shear and compressional velocity models of the mantle from cluster analysis of long-period waveforms. Geophys J Int 174:195–212
Ichikawa H, Tsuchiya T, Tange Y (2014) The P-V-T equation of state and thermodynamic properties of liquid iron. J Geophys Res 119. doi:10.1002/2013JB010732
Kaneshima S, Helffrich G (2013) Vp structure of the outermost core derived from analyzing large-scale array data of SmKS waves. Geophys J Int 193:1537–1555
Kennett BLN, Engdahl ER (1991) Traveltimes for global earthquake location and phase identification. Geophys J Int 105:429–465
Kennett BLN, Engdahl ER, Buland R (1995) Constraints on seismic velocities in the Earth from traveltimes. Geophys J Int 126:108–124
Kind R (1979) Extensions of the reflectivity method for a buried source. J Geophys 45:373–380
Koci L, Belonoshko AB, Ahuja R (2007) Molecular dynamics calculation of liquid iron properties and adibatic temperature gradient in the Earth's outer core. Geophys J Int 168:890–894
Kustowski B, Ekstrom G, Dziewonski AM (2008) Anisotropic shear-wave velocity structure of the Earth's mantle. J Geophys Res 113:B06306
Lay T, Young C (1990) The stably-stratified outermost core revisited. Geophys. Res. Lett. 17:2001–2004
Lekic V, Cottaar S, Dziewonski AM, Romanowicz B (2012) Cluster analysis of global lower mantle tomography: a new structure and implications for chemical heterogeneity. Earth Planet Sci Lett 357-358:68–77
Mégnin C, Romanowicz B (2000) The shear velocity structure of the mantle from inversion of body, surface and higher modes waveforms. Geophys J Int 143:709–728
Morelli A, Dziewonski AM (1993) Body wave traveltimes and a spherically symmetric P- and S-wave velocity model. Geophys J Int 112:178–194
Obara K, Kasahara K, Hori S, Okada Y (2005) A densely distributed high sensitivity seismograph network in Japan: Hi-net by National Research Institute for Earth Science and Disaster Prevention. Rev Sci Intrum 76:021301
Pozzo M, Davies C, Gubbins D, Alfe D (2012) Thermal and electrical conductivity of iron at Earth's core conditions. Nature 485:355–358
Ritsema J, Deuss A, Heijst H, Woodhouse J (2011) S40RTS: a degree-40 shear-velocity model for the mantle from new Rayleigh wave dispersion, teleseismic traveltime and normal-mode splitting function measurements. Geophys J Int 184:1223–1236
Souriau A, Teste A, Chevrot S (2003) Is there any structure inside the liquid outer core? Goephys. Res. Lett. 30. doi:10.1029/2003GL017008
Takeuchi N (2007) Whole mantle SH velocity model constrained by waveform inversion based on 3D Born kernels. Geophys J Int 169:1153–1163
Tanaka S (2007) Possibility of a low P-wave velocity layer in the outermost core from global SmKS waveforms. Earth Planet Sci Lett 259:486–499
Vocadlo L, Dobson DP, Wood IG (2009) Ab initio calculations of the elasticity of hcp-Fe as a function of temperature at inner-core pressure. Earth Planet Sci Lett 288:534–538
Wang P, de Hoop MV, van der Hilst R (2008) Imaging the lowermost mantle (D") and the core-mantle boundary with SKKS coda waves. Geophys J Int 175:103–115
Wessel P, Smith WH (1995) New version of generic mapping tools. EOS Trans. AGU Electron. Suppl. Aug. 15
Yamanaka H, Ishida H (1995) Phase velocity inversion using genetic algorithms. J. Struct. Constr. Eng. AIJ 468:9–17
Zhang P, Cohen RE, Haule K (2015) Effects of electron correlations on transport properties of iron at Earth's core conditions. Nature 517:605–607

Water and magmas: insights about the water solution mechanisms in alkali silicate melts from infrared, Raman, and ^{29}Si solid-state NMR spectroscopies

Charles Le Losq*, Bjorn O. Mysen and George D. Cody

Abstract

Degassing of water during the ascent of hydrous magma in a volcanic edifice produces dramatic changes in the magma density and viscosity. This can profoundly affect the dynamics of volcanic eruptions. The water exsolution history, in turn, is driven by the water solubility and solution mechanisms in the silicate melt. Previous studies pointed to dissolved water in silicate glasses and melts existing as molecules (H_2O_{mol} species) and hydroxyl groups, OH. These latter OH groups commonly are considered bonded to Si^{4+} but may form other bonds, such as with alkali or alkaline-earth cations, for instance. Those forms of bonding influence the structure of hydrous melts in different ways and, therefore, their properties. As a result, exsolution of water from magmas may have different eruptive consequences depending on the initial bonding mechanisms of the dissolved water. However, despite their importance, the solution mechanisms of water in silicate melts are not clear. In particular, how chemical composition of melts affects water solubility and solution mechanism is not well understood. In the present experimental study, components of such information are reported via determination of how water interacts with the cationic network of alkali (Li, Na, and K) silicate quenched melts. Results from ^{29}Si single-pulse magic-angle spinning nuclear magnetic resonance (^{29}Si SP MAS NMR), infrared, and Raman spectroscopies show that decreasing the ionic radius of alkali metal cation in silicate melts results in decreasing fraction of water dissolved as OH groups. The nature of OH bonding also changes as the alkali ionic radius changes. Therefore, as the speciation and bonding of water controls the degree of polymerization of melts, water will have different effects on the transport properties of silicate melts depending on their chemical composition. This conclusion, in turn, may affect volcanic phenomena related to the viscous relaxation of hydrous magmas, such as for instance the fragmentation process that occurs during explosive eruptions.

Keywords: Melt; Glass; Structure; Water bonding; Water speciation; NMR spectroscopy; Raman spectroscopy; Viscosity; Magma; Fragmentation

Background

Water is typically the major volatile component in magmas and can reach concentrations up to ~6–7 wt. % in silicic magma chambers (e.g., see Blake 1984; Clemens 1984). When magmas migrate from their source up towards the Earth surface, water exsolution can result in large (up to several orders of magnitude) increases in magma viscosity because of the strong effect of water on the viscosity of silicate melts (Leonteva 1940; Saucier 1952; Friedman et al. 1963; Shaw 1963; Dingwell et al. 1996; Richet et al. 1996). Water exsolution is also accompanied by the formation of vapor bubbles that participate in decreasing the bulk density of the magma, reinforcing, in this way, the buoyant force that pushes the magma out of the Earth crust. In silica-poor magmas with viscosities lower than 10^6 Pa s at eruptive temperature, such as with basalts, gas bubbles can coalesce and escape from the magma (Thomas et al. 1994; Gardner et al. 1996). This situation usually results in gentle (lava flows) to slightly explosive

* Correspondence: clelosq@carnegiescience.edu
Geophysical Laboratory, Carnegie Institution of Washington, 5251 Broad Branch Road NW, Washington, DC 20015, USA

(lava fountain, strombolian explosion) eruptions. In contrast, for silica-rich compositions, the magma viscosity can be considerably higher than 10^6 Pa s at eruptive temperature, so that bubbles are trapped in the melts (Thomas et al. 1994; Gardner et al. 1996). Exsolution of water from such magmatic liquids often results in foaming. In such case, brittle fragmentation of the magmatic foam can occur, for example, when the elongation strain rate of the foam ascending in the conduit exceeds its viscous relaxation time (Papale 1999). This situation leads to explosive eruptions.

The effects of water on magma properties, together with their possible consequences as described above, result from how water affects the disordered tetrahedral silicate network of melts. A simple and commonly proposed solution mechanism of water in such melts involves the presence of H_2O molecules (H_2O_{mol}), which can react with the oxygens O of the silicate network to form OH groups:

$$H_2O_{mol(melt)} + O_{(melt)} <=> 2\ OH_{(melt)} \quad (1)$$

The solution mechanism illustrated in Eq. (1) is documented by numerous nuclear magnetic resonance (NMR) and infrared (IR) spectroscopic studies (e.g., Scholtze 1960; Moulson and Roberts 1961; Bartholomew et al. 1980; Stolper 1982; Kohn et al. 1989a, b; Zotov and Keppler 1998), which also indicated that OH groups seem bonded predominantly to Si (see also Farnan et al. 1987; Cody et al. 2005; Mysen and Cody 2005). Therefore, in this simple model, H_2O_{mol} reacts predominantly with bridging oxygens (BO), breaking T-O-T bonds (with T = Si, Al), so that Eq. (1) can be rewritten as:

$$H_2O_{mol(melt)} + T{-}O{-}T_{(melt)} <=> 2\ T{-}OH_{(melt)} \quad (2)$$

Equation (2) implies formation of non-bridging oxygens (NBO) when the reaction shifts to the right. However, reaction (2) does not always proceed to completion. Indeed, both H_2O_{mol} and OH groups coexist in silicate melts, and hence glasses, in proportions defined by temperature (Nowak and Behrens 1995; Sowerby and Keppler 1999; Behrens and Yamashita 2008) and apparently water concentration (e.g., Stolper 1982).

However, Eq. (2) is just one reaction describing the reaction of H_2O with the silicate network of silicate melts. Other dissolution mechanisms have been proposed, such as the formation of MOH bonds (with $M^{+,2+}$ a metallic ion) (Xue and Kanzaki 2004, 2006, 2008; Moretti et al. 2014) or the protonation of T-O-T bonds (H^+ does not break T-O-T bonds but is attached to it; see for example Kohn et al. 1989b, 1992, 1998 and references therein). Moreover, Eq. (2) does not consider the Q^n speciation often employed to describe the silicate structure (where n denotes the number of bridging oxygen in a silicate complex, see Schramm et al. 1984). For example, water can react with different Q^n species in sodium silicate melts depending on their initial water content and Na/Si ratio (Zotov and Keppler 1998; Mysen and Cody 2005; Mysen 2007). In addition, it appears likely that the electronic properties of the alkali and alkaline earth metal cations will affect the reaction mechanisms of water with the silicate network (Xue and Kanzaki 2004).

To understand, quantitatively, how water interacts with the silicate structure of glasses and melts, we examined the structure of alkali (Li^+, Na^+, K^+) tetrasilicate glasses using Raman and ^{29}Si SP MAS NMR spectroscopy. Existing infrared spectroscopic information (Le Losq et al. 2015a, b) was also used. By studying such compositionally simple systems, it is possible to isolate effects of individual components on the water solution behavior. Furthermore, the chosen glasses have anhydrous glass transition temperature, Tg, within a ~30 K range: the Tg of the sodium tetrasilicate glass is 751 K (Neuville 2006), that of the potassium tetrasilicate glass to 762 K (Neuville 1992), and that of the LS4 glass to 740 K (Voigt et al. 2005). They also have very similar distributions of silicate Q^n species (e.g., Maekawa et al. 1991). Those two characteristics (comparable anhydrous T_gs and Q^n distributions) allow us to emphasize differences that characterize the hydrous quenched melts rather than differences related to the starting anhydrous glasses.

Methods

Sample preparation

The starting materials as well as the hydrous glasses used for this study are the same as those reported in Le Losq et al. (2015a, b). We refer readers to these references for details on sample synthesis. The chemical compositions of glass starting materials are reported in Table 1.

Table 1 Chemical composition of LS4, NS4, and KS4 anhydrous glasses

		SiO_2	Li_2O	Na_2O	K_2O
	nom. mol %	80.00	20.00	0.00	0.00
LS4	nom. wt. %	88.94	11.08	0.00	0.00
	meas. wt. %	88.77 (94)	11.23 (75)	*n.a.*	*n.a.*
	nom. mol %	80.0	0.00	20.00	0.00
NS4	nom. wt. %	79.50	0.00	20.50	0.00
	meas. wt. %	80.38 (36)	*n.a.*	19.62 (31)	*n.a.*
	nom. mol %	80.0	0.00	0.00	20.0
KS4	nom. wt. %	71.84	0.00	0.00	28.16
	meas. wt. %	71.16 (49)	*n.a.*	*n.a.*	28.84 (28)

nom. mol %, nom. wt. %, and meas. wt. % refer respectively to the nominal compositions in mol and wt. % and to the measured composition in wt. %. Chemical compositions are from Le Losq et al. (2015a). Errors are given at the 1σ confidence interval

The ^{29}Si NMR signal from the anhydrous glasses suffers from a very long T_1 relaxation time. To reduce this T_1 relaxation time, a portion of the starting glasses was synthesized with a few hundred ppm Fe_2O_3, as previously done by Maekawa et al. (1991) to study similar alkali silicate glasses. To ensure homogeneous distribution of this small amount of iron oxide in the materials, the glass synthesis was done in two steps. The first step consisted in enriching a proportion of the starting glasses to 15 wt. % Fe_2O_3. During the second step, the iron-rich glasses were crushed and then mixed with crushed iron-free starting materials, in order to produce iron-enrichment in the 100–1000 ppm concentration range.

Quantitative analysis of the glasses by energy-dispersive x-ray spectrometry (EDS) was performed with a JEOL field-emission scanning electron microscope (FE-SEM) equipped with an Oxford silicon drift detector and using a 15 kV accelerating voltage and a 1.00 nA current. Measurements were performed scanning 25 by 25 micron areas. The EDS results were calibrated using the following standards: pyrope for O and Si, scapolite for Na, K1597 GI for K, and cossyrite for Fe. Calibration was based on a set of glass standards. The EDS results are within error of standards analyzed as unknowns.

The compositions of the Fe-doped glasses are essentially the same as those of the Fe-free starting materials. The iron contents are below detection limit (~200 ppm) for Fe-doped LS4, 1500 ± 300 ppm for Fe-doped NS4, and 700 ± 400 ppm for Fe-doped KS4 (mean values and errors from >10 individual measurements made at different spots). These Fe-doped glasses were used only to perform the ^{29}Si NMR measurements of anhydrous end-members. Optically, they are clear and colorless with the same Raman signal as iron-free starting glasses, indicating a homogeneous distribution of Fe in their structure. This is supported by SEM observations that show no clustering of iron at the micrometer scale. The ^{29}Si NMR data of the hydrous glasses were obtained on iron-free glasses. Indeed, in those water-bearing glasses, the ^{1}H nuclei that have a strong magnetic moment serve to aid the relaxation of ^{29}Si nuclei, as Fe at the ‰ level does in the anhydrous glasses.

Solid-state NMR spectroscopy

Silicon-29 solid-state NMR experiments were carried out with a Chemagnetics CMX Infinity 300 solid-state NMR spectrometer. The static field strength of the magnet is ~7.05 T, and the Larmor frequency of ^{29}Si is ~59 MHz at this magnetic field. ^{29}Si spectra were referenced to tetramethylsilane (TMS), defined as 0 ppm.

Single-pulse ^{29}Si MAS spectra of each glass were obtained by spinning the samples at 11 kHz. ^{29}Si excitation pulse widths corresponded to 30° tip angles. One thousand acquisitions separated by a 100 s recycling time were co-added. Proton decoupling was performed with continuous RF irradiation at 62.5 kHz. As indicated in the "Sample preparation" section, the anhydrous starting materials had a long ^{29}Si spin lattice T_1 relaxation time. For the KS4 glass, this long relaxation time resulted in a signal from ^{29}Si in Q^4 polymerized units much too low compared to a prediction of a mixture of 50 % Q^3:50 % Q^4 units based on glass chemical composition. Therefore, a portion of the anhydrous samples was doped with iron (see "Sample preparation" section) in order to decrease of the ^{29}Si T_1. The resulting signal from KS4 glass was much more intense than without the iron dopant and the spectra yielding Q^3–Q^4 proportion near the nominal 50:50. In hydrous glasses, which were iron-free, the protons played a similar role as iron, thus reducing the ^{29}Si spin–lattice relaxation time, so that spectra are directly quantitative. More details on the ^{29}Si single-pulse NMR experiments can be found in Cody et al. (2005).

Raman spectroscopy

Raman spectra were recorded with a Jasco NRS 3100 spectrometer, equipped with holographic gratings, a single monochromator, and a 1024 × 128 Andor DV401-F1 CCD Peltier-cooled detector operating at −70 °C. The 490 nm line of a coherent solid-state laser was used for sample excitation, with a power of ~44 mW on sample. The laser was focused on sample through a 50× Olympus lens. The groove density of the gratings was 2400 lines/mm, which results in an accuracy of ±3 cm^{-1}. Precision was ±1 cm^{-1}. Alignment of the Raman spectrometer was tested against the 520.7 cm^{-1} Raman peak of pure Si. All spectra reported in this study are unpolarized. The spectra were not corrected from temperature and excitation line effects (e.g., see Neuville et al. 2014, for further discussion about the relevance of such corrections) because thermal effects mostly affect the Raman intensity at frequency below 500 cm^{-1}, which does not contain much information relevant to the focus of the present study, and because excitation line effects are very small and in all cases will not affect our results since we used the same laser line for recording all spectra.

The Raman spectra were fitted using Gaussian peaks. A two-step fitting procedure was employed. As both the number of peaks and the frequency range of the fitted region are large, a first fitting step was carried out by using an equal peak width for all peaks. This fit was accomplished by using the *nlinfit* non-linear least-square optimization function of Matlab®. Then, in order to refine the models produced by *nlinfit*, the parameters and associated errors from those fits were injected in the pseudo-Newtonian algorithm described in Tarantola (2005) and used in Le Losq et al. (2014). This algorithm has the advantage of accounting of the a priori errors on the initial models (e.g., the errors on the estimations provided by the user or by *nlinfit* in our case). Errors in peak areas were calculated by propagating the peak parameter errors.

Water concentration

Water concentrations [H_2O^{TOT}] of hydrous glasses were determined from IR absorption spectroscopy data of the NS4 compositions (for which molar absorption coefficients are available; Yamashita et al. 2008) and an FTIR-based calibration of the relative area of the OH fundamental stretch Raman band for the LS4 and KS4 glasses as described by Le Losq et al. (2015a). The resulting water contents together with glass synthesis conditions are reported in the Table 2.

The molar absorption coefficients of OH groups and H_2O_{mol} species for LS4 glasses have been estimated by using the IR data reported in Le Losq et al. (2015b) and the data treatment reported in Le Losq et al. (2015a), which follows the protocol described in Yamashita et al. (2008). A six-order polynomial baseline was fitted on spectra to derive a baseline under the 4500 and 5200 cm^{-1} IR absorption peaks (see Fig. 1, top left) and subtracted from spectra. Then, areas of the 4500 (A4500) and 5200 (A5200) cm^{-1} IR peaks, assigned to OH groups and H_2O_{mol} species respectively (Scholtze 1960; Moulson and Roberts 1961; Stolper 1982), have been measured (Table 2). The absorption coefficients of these two peaks (ε_{5200} and ε_{4500}, respectively) were calculated from the following relationship (Yamashita et al. 2008):

$$\frac{1820 \times A5200}{d \times \rho \times [H_2O^{TOT}]} = \varepsilon_{5200} - \frac{\varepsilon_{5200}}{\varepsilon_{4500}} \times \left[\frac{1820 \times A4500}{d \times \rho \times [H_2O^{TOT}]}\right] \quad (3)$$

with d, the thickness of the double-polished glass piece, and ρ, its density. The density of anhydrous LS4 glass was taken from Tischendorf et al. (1998). The density of hydrous glasses was then estimated by assuming a linear mixing between the molar volume of water and that of the glasses. In the water content range of the present study, such an assumption seems valid in light of the results of Yamashita et al. (2008) for sodium silicate glasses, and, on a larger dataset, with the results of Richet et al. (2000). According to results from those two studies, a partial molar volume of 12 cm^3 mol^{-1} for dissolved water was chosen. Such a value may overestimate slightly the actual molar volume of water in lithium silicate glasses, as inferred from results from ^1H NMR spectroscopy on these same glasses (Le Losq et al. 2015a). However, a change of the partial molar volume of water of 4 cm^3 mol^{-1} for instance can result in a maximum change of 4 L mol^{-1} cm^{-2} of the absorption coefficients. Such variations are incorporated in the reported error bars. By using Eq. (3), $\varepsilon_{5200} = 56 \pm 8\ (1\sigma)\ L\ mol^{-1}\ cm^{-2}$ and $\varepsilon_{4500} = 70 \pm 7\ (1\sigma)$ L mol^{-1} cm^{-2} for the LS4 glass.

Use of the IR combination peaks at 4500 and 5200 cm^{-1} to determine water content of glasses is only possible when those peaks are well defined. In water-bearing alkali silicate glasses, an increase of the ionic radius of alkali (decreasing their ionic field strength) produces a decrease of the O-O distances around protons dissolved as H_2O_{mol} species and OH groups (Le Losq et al. 2015a). This perturbs significantly the frequency and intensity of the signal assigned to O-H stretching vibrations in IR and Raman spectra of hydrous alkali silicate glasses (Le Losq et al. 2015b). For example, in IR spectra of KS4 glasses, this signal extends to frequencies much lower than 3000 cm^{-1}, presenting several intensity maxima near ~2350, ~2800, and ~3600 cm^{-1}. This extended frequency range results in a poorly defined 4500 cm^{-1} combination band (Le Losq et al. 2015b) so that this

Table 2 Water concentration of glasses

Glass	nom. mol % H_2O	Conditions of synthesis	FTIR values					RAMAN
			A_{4200}	A_{5200}	[OH]	[H_2O_{mol}]	[H_2O^{TOT}]	[H_2O^{TOT}]
LS4	3.28	1650 °C	151 (12)	93 (9)	2.11 (35)	1.04 (22)	3.15 (41)	3.08 (60)
	9.40	1.5 GPa	325 (18)	452 (21)	4.56 (70)	5.06 (42)	9.61 (81)	10.09 (60)
	17.64		447 (21)	1109 (33)	6.26 (94)	12.42 (55)	18.68 (1.09)	18.74 (60)
NS4	3.28	1500 °C	155 (12)	3 (2)	3.85 (40)	0.07 (3)	3.92 (40)	3.44 (67)
	9.40	1.5 GPa	301 (17)	86 (9)	7.35 (63)	1.89 (20)	9.24 (66)	9.46 (67)
	17.64		403 (20)	440 (21)	9.52 (80)	9.35 (60)	18.87 (1.00)	17.54 (67)
KS4	3.28	1550 °C	-	-	3.98 (70)[a]	0.00 (0)[a]	-	3.98 (74)
	9.40	1.5 GPa	-	-	8.89 (70)[a]	0.73 (7)[a]	-	9.62 (74)
	17.64		-	-	13.65 (70)[a]	2.30 (24)[a]	-	15.95 (74)

Nom. refers to the nominal water concentrations. [H_2O_{mol}], [OH], and [H_2O^{TOT}] concentrations are in mol %. A_{4500} and A_{5200} are the integrated intensities of the IR peaks at ~4500 and ~5200 cm^{-1} that have been used to quantify the content of [OH] and [H_2O_{mol}] for the LS4 and NS4 glasses. FTIR and Raman values are from Le Losq et al. (2015a). Errors are given at the 1σ confidence interval

[a]Calculated with the calibration linking ^1H MAS NMR sideband area to H_2O_{mol} content shown in Fig. 1; the concentration of OH is estimated by difference to [H_2O_{mol}] measured with using Raman spectroscopy

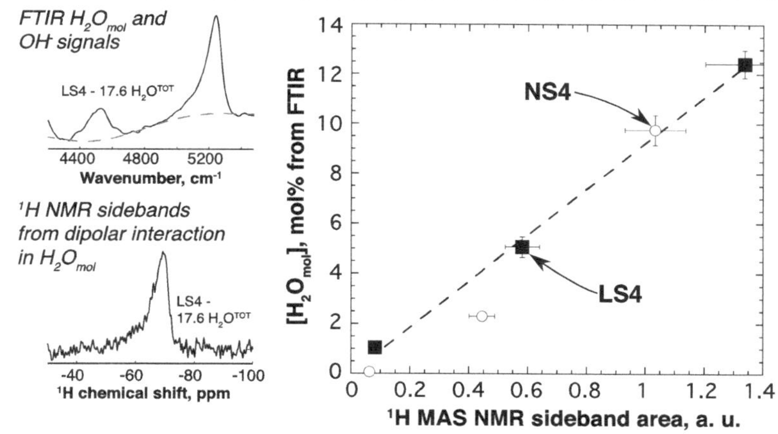

Fig. 1 Relationship between the area of ^{1}H MAS NMR sidebands, which arise from ^{1}H-^{1}H dipolar interaction in H_2O_{mol} species, and the concentration of the latters as measured by IR spectroscopy in the LS4 (plain squares) and NS4 (open circles) glasses. On the *left*, examples of the 4500 and 5200 cm^{-1} IR combination bands, assigned to OH and H_2O_{mol} species, respectively, and of one ^{1}H NMR sideband are shown for the LS4 glass containing 17.6 nom. mol% water. On the FTIR example, the *plain line* represents the experimental spectrum and the *dashed line* is a polynomial baseline defining the background (see text)

combination band cannot be used for measuring the concentration of OH groups. In such cases, an alternative to derive H_2O_{mol}/OH is to use the sidebands in ^{1}H MAS NMR spectra. These sidebands are residual signals from ^{1}H-^{1}H dipolar coupling, this interaction being strongly averaged by the fast magic angle spinning in the present experiments. The ^{1}H-^{1}H dipolar coupling is very strong in the case of H_2O_{mol} species and negligible in the case of isolated OH groups (Eckert et al. 1988). Therefore, in normalized ^{1}H NMR spectra acquired under similar conditions, sideband area would express the fraction of H_2O_{mol} species. Following this idea, Fig. 1 shows that the area of the sidebands ($A_{sidebands}$) of the ^{1}H MAS NMR spectra of LS4 and NS4 glasses, from Le7#146;Losq et al. (2015a), seem to be a linear function of the glasses H_2O_{mol} concentrations measured by IR. Particularly, the relationship seems nearly perfect for LS4 glasses. Slight deviations are observed for NS4 glasses. This can be rationalized with considering that most of the signal from O-H stretching vibrations is located above 3000 cm^{-1} in the IR spectra of LS4 glasses whereas it slightly extends toward lower frequency in the IR spectra of NS4 glasses (Le Losq et al. 2015b). This results in better defined ~ 4500 and ~ 5200 cm^{-1} combination bands in IR spectra of LS4 glasses than in those of NS4 glasses (Le Losq et al. 2015b), and hence, in higher accuracy on the calculated [H_2O_{mol}] concentrations. Because of that, we can take the IR-based H_2O_{mol} concentrations and $A_{sidebands}$ from LS4 glasses as standards. Doing so, we obtain the relationship: mol % $H_2O_{mol} = 9.21 \pm 0.21 \times A_{sidebands}$. This equation allows us to determine the concentration of H_2O_{mol} species and of OH groups (by difference to the total water content) in the KS4 glasses (Table 2).

Results

Raman spectroscopy

The Raman-active vibrational modes of the silicate tetrahedral network give rise to numerous bands between 400 and 1400 cm^{-1} in the Raman spectra of silicate glasses (Fig. 2). A strong band near ~500 cm^{-1} is always observed. Smaller peaks and shoulders near 550 and 590–600 cm^{-1} are also present. Their intensity and frequency depend on glass composition. The respective intensity and frequencies of the contributions located between 450 and 650 cm^{-1} depend on the water content of the glasses and on the type of alkali metal (Fig. 2). Those signals arise from the bending/breathing modes of Si-O-Si bonds that link Q^n units (noted "Q^n-Q^n modes" in Fig. 2; see for instance Bell et al. 1968; Sen and Thorpe 1977; Furukawa et al. 1981 or McMillan 1984).

A band near 800 cm^{-1} is also observed in all of the Raman spectra (Fig. 2). For a given alkali, the intensity of this peak changes with water content, but not its shape. At given water content, increasing ionic radius of alkali results in decreasing the intensity of this peak. This band may arise from vibrations involving the motion of the Si in its oxygen cage (McMillan 1984), this being possibly related to the threefold—degenerate "rigid cage" vibrational mode of SiO_2 tetrahedra (Sen and Thorpe 1977; Galeener 1979). However, its origin is not firmly established.

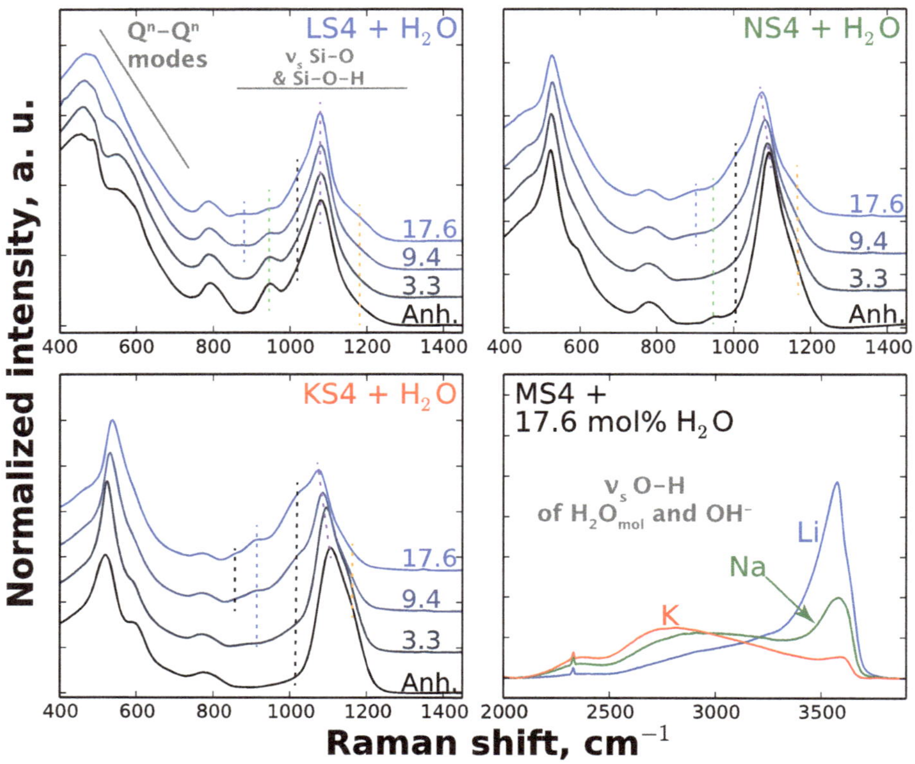

Fig. 2 Raman spectra of the lithium (LS4), sodium (NS4), and potassium (KS4) tetrasilicate glasses with 0.0 (Anh.), 3.3, 9.4, and 17.6 nom. mol % [H_2O^{TOT}]. Raman intensities are offset vertically for a better visibility, with the exception of the figure in *bottom right*. *Dashed lines* are used as visual markers of the signals contributing in the 820–1350 cm^{-1} portion of the spectra (see text). The small sharp peak located near 2330 cm^{-1} in the *lower right* figure is a signal from atmospheric N_2

Signals assigned to Si-O stretch vibrations in various Q^n units (ν_s Si-O in Fig. 2) form the broad envelope of bands between 850 and 1300 cm^{-1} (Brawer and White 1975, 1977; Virgo et al. 1980; Mysen et al. 1982; McMillan 1984; Neuville et al. 2014 and references therein). In hydrous alkali silicate glasses, a band often assigned to Si-O-H stretch vibrations contributes to the intensity in the 900–970 cm^{-1} frequency range (Stolen and Walrafen 1976; McMillan et al. 1993; Zotov and Keppler 1998; Malfait 2009). At frequencies higher than 1300 cm^{-1}, bands are observed between 2000 and 4000 cm^{-1} (Fig. 2, bottom right). These are assigned to O-H stretching (ν_s O-H) in H_2O_{mol} and OH groups (Scholtze 1960; Bartholomew et al. 1980; Wu 1980; Zotov and Keppler 1998). Their intensity and frequency depend on the chemical composition of the alkali silicate glasses. The larger the alkali ionic radius *r*, the lower the O-H stretch frequency (Fig. 2; see also Le Losq et al. 2015a; Le Losq et al. 2015b).

The topology of the 850–1300 cm^{-1} frequency region of the Raman spectra systematically varies as a function of both water content and ionic field strength of alkali in the glass (Fig. 2). This variation likely reflects changes of the Q^n species fractions. To quantify those changes, this portion of Raman spectra was fitted with Gaussian peaks. Such peak shape was used because it returns the best fitting results compared to other peak shapes (e.g., Lorentzian or Pseudo-Voigt shapes). This conclusion also is coherent with an expected pseudo Boltzmann distribution of Si-O vibrators in silicate melts in regards of their high thermal agitation and structural disorder at the glass transition temperature (Mysen et al. 1982). Table 3 summarizes the peaks used to perform the fits as well as their assignments and the references that substantiate those assignments.

The peak near 1100 cm^{-1} is assigned to Si-O stretching in Q^3 species. It is the most intense signal in the Raman spectra of all anhydrous glasses, whatever the alkali metal (Figs. 3 and 4). Areas of the peaks assigned to Si-O stretching in Q^2 and Q^4 species (near 950 and 1150–1200 cm^{-1}) respectively decrease and increase with increasing ionic radius of the alkali cation. These intensity variations indicate variations in the Q^n distribution as a function of the alkali ionic radius/field strength, similar

Table 3 Assignment of the peaks from curve-fitted Raman spectra

Raman shift, cm^{-1}	Attribution	Source
~850 cm^{-1}	ν_s Si-O in Q^1	Furukawa et al. (1981), Virgo et al. (1980), Mysen et al. (1982), McMillan (1984)
~880–910 cm^{-1}	ν_s Si-O-H	McMillan et al. (1993), Zotov and Keppler (1998), Malfait (2009), Spiekermann et al. (2012)
~960 cm^{-1}	ν_s Si-O in Q^2	Brawer and White (1975), Brawer and White (1977), Virgo et al. (1980), Furukawa et al. (1981), Mysen et al. (1982), McMillan (1984)
~1020 cm^{-1}	ν_s Si-BO in Q^n	Mysen et al. (1982), Spiekermann et al. (2012), Le Losq and Neuville (2013)
~1090 cm^{-1}	ν_s Si-O in Q^3	Brawer and White (1975), Virgo et al. (1980), Furukawa et al. (1981), Mysen et al. (1982)
~1140 cm^{-1}	ν_s Si-O in $Q^{4,II}$ (a)	Mysen et al. (1982), Seifert et al. (1982), Mysen (1990), Neuville and Mysen (1996), Le Losq and Neuville (2013), Le Losq et al. (2014)
~1190 cm^{-1}	ν_s Si-O in $Q^{4,I}$ (a)	Mysen et al. (1982), Seifert et al. (1982), Mysen (1990), Neuville and Mysen (1996), Le Losq and Neuville (2013), Le Losq et al. (2014)

[a] Two Gaussians named $Q^{4,II}$ and $Q^{4,I}$ are used for fitting the Si-O stretch of Q^4 because of its non-Gaussian shape. See Mysen et al. (1982), Seifert et al. (1982), Neuville and Mysen (1996), and Le Losq and Neuville (2013) for further discussion on this topic. In more depolymerized glasses, the signal of Q^4 becomes weak and only one band is needed (see in Fig. 2 the spectrum of KS4 + 17.6 mol % H_2O for instance)

to the conclusions drawn from previous ^{29}Si NMR (e.g., Maekawa et al. 1991) and Raman (e.g., Matson et al. 1983; Mysen 1990) spectra of binary alkali silicate glasses.

Solution of water in the melts quenched to glasses results in the appearance in the Raman spectra of a new peak near 900 cm^{-1} (Fig. 3). Its intensity increases with increasing water content, but the extent of this change depends on ionic radius of alkali (Fig. 3). The greater the alkali ionic radius, the larger the intensity of the ~900 cm^{-1} peak. Moreover, it appears that the type of alkali in the glass affects the frequency of this peak. For example, in the Raman spectra of glasses with 17.6 mol % H_2O, it shifts from 882 cm^{-1} for LS4 to 902 cm^{-1} for NS4 and 918 cm^{-1} for KS4. According to the study of Zotov and Keppler (1998) on sodium silicate glasses, this ~900 cm^{-1} peak may be assigned to Si-O-H stretching vibrations in Q^n species with Si-O-H bonds, regardless their number of bridging oxygens n. In pure hydrous silica, a peak near 976 cm^{-1} has been assigned to the signal of Si-O-H stretching vibrations (Stolen and Walrafen 1976). In hydrous aluminosilicate glasses, a peak located near 900 cm^{-1} has been assigned to a mixture of Si-O-H and Al-O-H vibrations, with the assumption that the Al-O-H stretching frequency is lower than that of Si-O-H (McMillan et al. 1993; Malfait 2009). However, recent results from molecular dynamic simulations (Spiekermann et al. 2012) questioned the attribution of the 976 cm^{-1} signal in pure hydrous silica and suggested that this peak may actually arise from the asymmetric Si-O stretching vibration of Q^2 units and that the Si-O-H stretching signal may form a shoulder near 920 cm^{-1} on the side of the 976 cm^{-1} peak. The fact that a peak near 900 cm^{-1} is detected and grows with solution of water in silicate glasses (Figs. 2 and 3, see also Zotov and Keppler 1998) may support such an interpretation, at least for silicate glasses.

The area of the ~950 cm^{-1} peak, assigned to Si-O stretching vibrations in Q^2, increases with solution of water in KS4 and NS4 glasses but decreases slightly in spectra of LS4 glasses (Fig. 4). The area of the ~1100 cm^{-1} peak, assigned to Si-O stretching vibrations in Q^3, seems to increase slightly with solution of water, but this effect is small (≤6 % with addition of 17.6 mol% H_2O, Fig. 4). Finally, the area of the two peaks at ~1150 and 1200 cm^{-1}, assigned to Si-O stretching vibrations in Q^4, varies differently as a function of water concentration and alkali ionic radius (Fig. 4). In Raman spectra of the NS4 and KS4 glasses, the area of these two peaks decreases with increasing water content (~10–20 % with addition of 17.6 mol % H_2O). In contrast, their area increases slightly (~3 %) with increasing water content from 0.0 up to 17.6 mol % in LS4 glasses.

There is another peak, near 850 cm^{-1}, in spectra of the KS4 glasses at high water concentration (Figs. 2 and 3). Its origin could be related to the presence of Q^1 species (e.g., see Mysen et al. 1982 or McMillan 1984), but Q^1 species are not detected by ^{29}Si NMR spectroscopy (see below and Fig. 5). This could be explained by an enhanced Raman signal of Q^1 species compared to other more polymerized Q^2, Q^3, or Q^4 species because of the increase of the Raman cross section of Q^n units that results from their depolymerization (decreasing n; see for instance Furukawa et al. 1981). Such effect has been observed for Q^2 species in sodium silicate glasses (Le Losq et al. 2014). Therefore, a very small proportion of Q^1 species (<1 %) may be present in the KS4 + 17.6 mol % H_2O and undetected by ^{29}Si NMR spectroscopy. This interpretation is speculative but does not affect the global observations because the 850 cm^{-1} signal intensity is negligible compared with those of the other signals.

^{29}Si single-pulse (SP) MAS NMR spectra

The ^{29}Si SP MAS NMR spectra of anhydrous glasses comprise two main peaks near −95 and −105 ppm

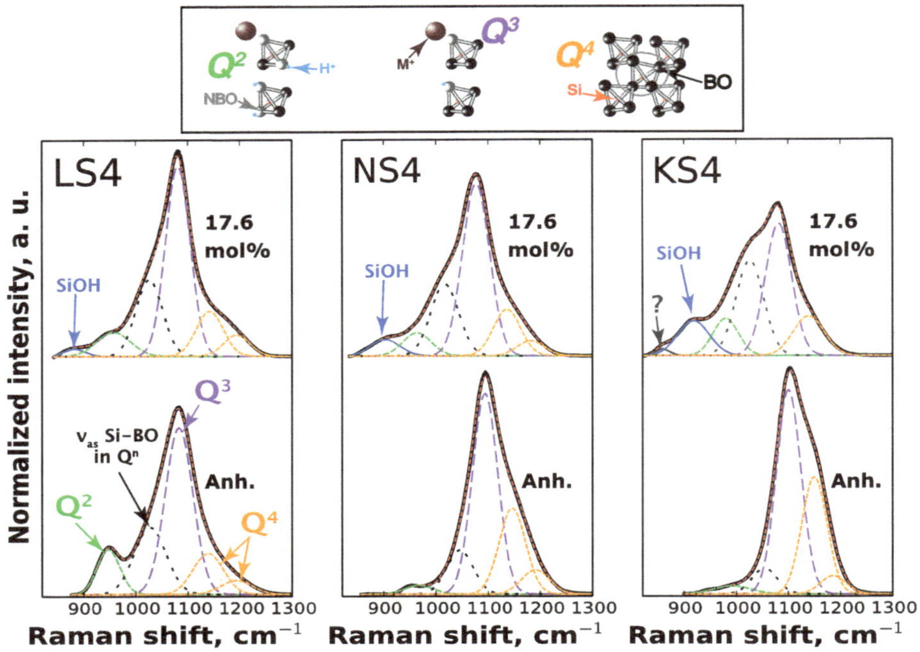

Fig. 3 Examples of curve-fitted Raman spectra of anhydrous and 17.6 mol% H_2O LS4 (*left*), NS4 (*middle*), and KS4 (*right*) glasses. Addition of water in the glasses clearly produces an increasing intensity near 900 cm^{-1}. It could be related to the stretching signal of Si-O-H bonds regardless of the Q^n specie carrying those bonds. Signal near 850 cm^{-1} in the Raman spectrum of KS4 + 17.6 mol % H_2O glass may be related to Q^1 species in the glass in negligible quantities. The various Q^n species are schematically represented at the top of the figure (ionic size is not at scale). *BO* bridging oxygen (Si-O-Si bonds), *NBO* non-bridging oxygen (Si-O-M or Si-O-H bonds, with M^+ an alkali)

(Fig. 5). A third contribution, located near −85 ppm, forms a shoulder in the spectrum of the anhydrous LS4 glass. The ~ −105, −95, and −85 ppm contributions are assigned to the signals of ^{29}Si in Q^4, Q^3, and Q^2 species, respectively (Lippmaa et al. 1980; Murdoch et al. 1985; Maekawa et al. 1991). The Q^3 and Q^4 contributions dominate the ^{29}Si SP MAS NMR spectra of anhydrous glasses (Fig. 5), in agreement with interpretation of Raman data (Figs. 3 and 4). The signal of Q^2 species is visible only in the spectrum of the LS4 anhydrous glass (Fig. 5). The relative intensities of Q^3 and $Q^{4\,29}Si$ NMR signals vary slightly with ionic radius of alkali in anhydrous glasses. The global width of the ^{29}Si NMR signal also decreases with increasing ionic radius of alkali metal, and the frequency difference between Q^3 and Q^4 signals narrows (Fig. 5). The Q^4 frequency is the lowest in the spectrum of LS4 glass compared with those in the spectra of the NS4 and KS4 glasses. Such ^{29}Si NMR frequency changes can result from a modification of the shielding of the ^{29}Si nuclei with changing ionic radius of alkali. Maekawa et al. (1991) suggested that such shielding evolution may reflect changes in the third neighbor environments (e.g., Olivier et al. 2001), bond length, or bridging Si-O-Si angle as a function of ionic radius of alkali in silicate glasses.

In order to highlight the perturbations induced by the water addition in the glasses networks, the ^{29}Si SP MAS NMR spectra of the anhydrous glasses have been subtracted from those of hydrous glasses, giving residual ^{29}Si NMR spectra where positive and negative peaks can be observed (Fig. 5, bottom). Those peaks represent the changes of the proportions in the various Q^n species upon water addition. In all glasses, solution of water results in a reduction in intensity of the Q^4 peak, whereas that of the Q^3 peak increases (Fig. 5). However, the extent of this intensity change depends on ionic radius of alkali metal. The Q^3/Q^4 ratio changes only slightly for LS4 glass, whereas it increases strongly for KS4 glass. In the NMR spectra of NS4 and KS4 glasses, the intensity near −80/−85 ppm also increases upon water solution but not in spectra of LS4 glasses. It follows, therefore, that upon solution of water in the silicate glasses, Q^4 units are converted to Q^3 and Q^2 units, and that this effect gets more pronounced with increasing ionic radius of alkali.

Quantification of the effect of water on glass polymerization

The relative areas, A_{Q^n} $\left(\sum_{n=0}^{4} A_{Q^n} = 1\right)$, of the Raman peaks assigned to Si-O stretching in Q^n units (Fig. 4) can be related to the fraction of Q^n units through (Mysen 1990; Mysen and Frantz 1993):

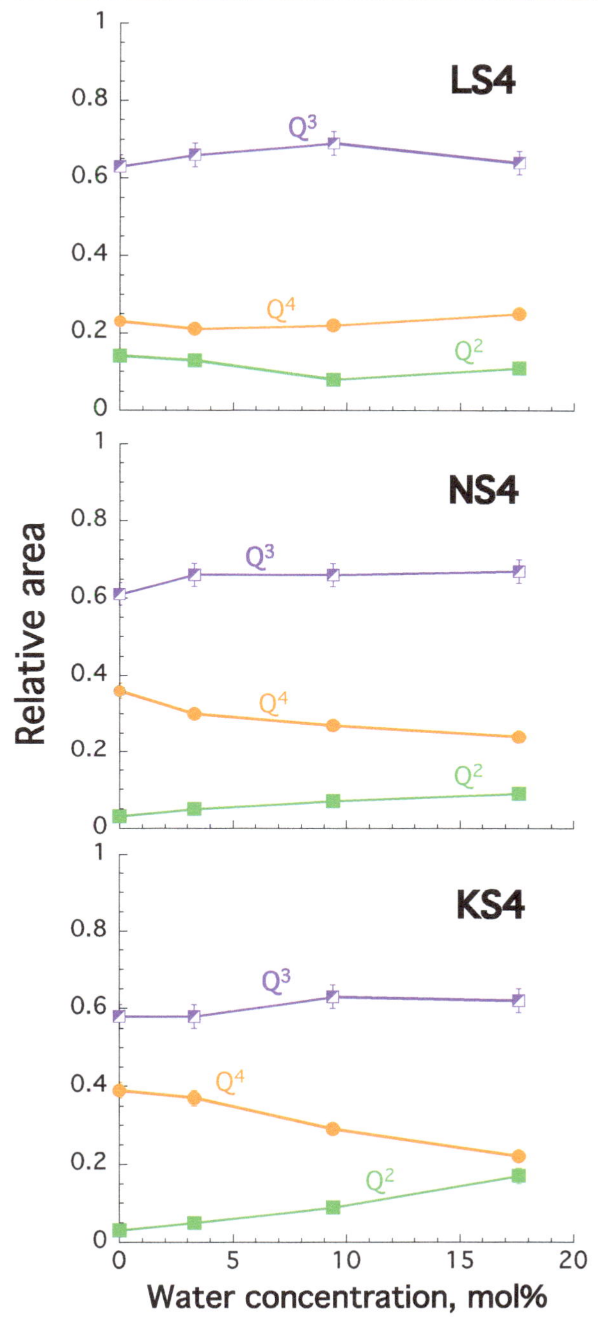

Fig. 4 Relative area of the Raman signals assigned to Q^2, Q^3, and Q^4 units composing the LS4 (*top*), NS4 (*middle*), and KS4 (*bottom*) glasses as a function of their nominal water concentration. Errors are given at the 1σ confidence interval. If not visible, they are smaller than the symbol sizes

$$[Q^n] = \theta^n A_{Q^n} \tag{4}$$

with θ^n a conversion factor that directly depends on the Raman cross section of the Si-O stretching vibrations in the Q^n units. The polymerization of a glass that contains a mixture of Q^2, Q^3 and Q^4units can be quantified by calculating the ratio of the mean number of NBOs other tetrahedral BOs as:

$$\mathrm{NBO/T} = 2[Q^2] + [Q^3] = 2 \times \theta^2 A_{Q^2} + \theta^3 A_{Q^3} \tag{5}$$

Between the anhydrous and hydrous glasses, we have the following equation:

$$\Delta \mathrm{NBO/T} = \mathrm{NBO/T}^{\mathrm{hyd}} - \mathrm{NBO/T}^{anh} = 2[\Delta \mathrm{Q}^2] + [\Delta \mathrm{Q}^3], \tag{6}$$

where ΔQ^2 and ΔQ^3 are the variations of the Q^2 and Q^3 fractions between the anhydrous and hydrous glasses. The θ^2 and θ^3 values are near unity for sodium silicates and the effect of water solution on those values is negligible (Mysen 2007). Therefore, as a first approximation, we can neglect the θ^2 and θ^3 factors in Eqs. (5) and (6). It follows from the Raman data that addition of 17.6 mol % water in the LS4, NS4, and KS4 glasses results in *ΔNBO/T* of −0.04 ± 0.05, 0.18 ± 0.05, and 0.30 ± 0.05, respectively (Table 4).

Fitting ^{29}Si SP MAS NMR spectra with a set of Gaussian lines usually allows quantification of the Q^n fractions in silicate glasses (e.g., Maekawa et al. 1991). However, fitting the ^{29}Si SP MAS NMR spectra of the hydrous glasses is difficult because of the poor resolution of the shoulder associated to the signal of ^{29}Si in Q^2 units. Several solutions can be returned by the fitting procedure, and it is not clear how to discriminate between them. Fortunately, spectral fitting is not a requirement for quantifying the polymerization change induced by solution of water in glasses. Indeed, the areas of the peaks assigned to Q^2, Q^3, and Q^4 units in the residual ^{29}Si SP MAS NMR spectra (Fig. 5, bottom) are equivalent to the fractions of transformed Q^n units (ΔQ^n) upon solution of water. For the LS4 glasses, the ΔQ^3 signal area was measured between −87 and −102 ppm and the ΔQ^4 signal area between −102 and −122 ppm. In the spectra of the NS4 glasses, areas of the ΔQ^2, ΔQ^3, and ΔQ^4 signals were measured from −70 to −88, −88 to −101, and −101 to −120, respectively. For the spectra of KS4 glasses, the areas were measured from −70 to −91, −91 to −100, and −100 to −120, respectively. In order to evaluate the errors affecting the area determinations, the boundaries reported above were varied in a ±2–3 ppm uncertainty range. We have chosen such an uncertainty range because it is similar to the variations that affect the frequency of the peaks observed on the ^{29}Si MAS NMR spectra of the various glasses, and because it produces residual intensity variations higher than the noise level. As a result, such uncertainty range appears to be the best estimate for estimating the errors affecting the areas of residual signals.

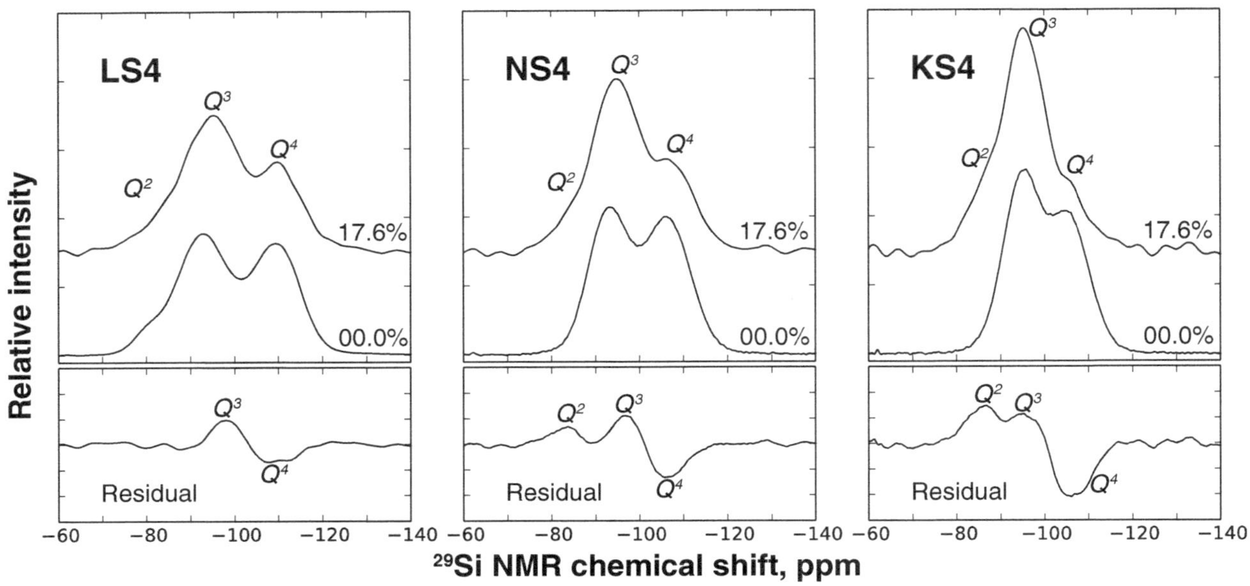

Fig. 5 *Top*: ^{29}Si single-pulse MAS (5 kHz) NMR spectra of the anhydrous and 17.6 nom. mol % H_2O LS4 (*left*), NS4 (*middle*), and KS4 (*right*) glasses. Intensities on spectra are normalized such that the total areas of the spectra are equal to unity. Changes in relative intensity and areas are thus directly proportional to the changes in the chemical concentrations of ^{29}Si in particular Q^n species. *Bottom*: residuals from the subtraction of the ^{29}Si NMR signals from hydrous glasses minus those from anhydrous glasses. Areas under the positive and negative peaks are quantitative and represent the fraction of transformed Q^n species upon hydration

The ΔQ^2 and ΔQ^3 values thus obtained can be used to compute $\Delta NBO/T$ with Eq. (6). The $\Delta NBO/T$ calculated from Raman data and from ^{29}Si SP MAS NMR difference spectra are in good agreement (Fig. 6 and Table 4) except for LS4 glass. Indeed, for the latter, the $\Delta NBO/T$ values obtained with the two different spectroscopic methods are different: -0.04 ± 0.05 (Raman) and 0.07 ± 0.01 (NMR). The Q^n speciation derived from the Raman spectra of LS4 glasses suggests almost no depolymerization of those glasses upon water solution. This conclusion differs somewhat from that drawn from the ^{29}Si NMR data, which indicate that there is a weak, but real, effect of water on the polymerization of LS4 glasses (Fig. 5). The errors affecting the fitting of Raman spectra might be too large to allow detection of the weak effect of water solution on the structure of LS4 glasses. Indeed, the Raman-derived $\Delta NBO/T$ values are affected by errors inherent to the fitting procedure (optimization error) as well as by errors inherent to the initial assumptions made for fitting (e.g., perfect Gaussian shape for the peaks). In this regards, NMR-derived $\Delta NBO/T$ values are more robust than the Raman-derived $\Delta NBO/T$ values because they do not rely on any fitting procedure but directly on calculations from raw data. This notwithstanding, both sets of values indicate a small effect of water addition on the degree of polymerization of the LS4 composition.

Table 4 ΔNBO/T, concentrations of total OH groups, SiOH, and *unknown* OH groups

	H_2O^{TOT} (Raman)	[OH]	ΔNBO/T (Raman)	[SiOH] (Raman)	[Unknown OH] (Raman)	ΔNBO/T (NMR)	[SiOH] (NMR)	[Unknown OH] (NMR)
LS4	3.08 (60)	2.11 (35)	0.02 (5)	0.63 (1.25)	1.47 (3.41)	-	-	-
	10.09 (60)	4.56 (70)	−0.05 (5)	0.00 (–)	4.56 (1.27)	-	-	-
	18.74 (60)	6.26 (94)	−0.04 (5)	0.00 (–)	6.26 (1.57)	0.07 (1)	2.77 (16)	3.49 (95)
NS4	3.44 (67)	3.85 (40)	0.08 (5)	3.24 (66)	0.61 (77)	-	-	-
	9.46 (67)	7.35 (63)	0.13 (5)	5.02 (43)	2.33 (76)	-	-	-
	17.54 (67)	9.52 (80)	0.18 (5)	7.28 (30)	2.24 (85)	0.18 (1)	7.17 (5)	2.35 (80)
KS4	3.98 (74)	3.98 (70)	0.04 (4)	1.42 (1.22)	2.56 (40)	-	-	-
	9.46 (74)	8.89 (70)	0.15 (5)	5.88 (32)	3.01 (77)	-	-	-
	15.95 (74)	13.65 (70)	0.30 (5)	12.08 (19)	1.57 (72)	0.34 (2)	13.63 (7)	0.02 (70)

Concentrations of OH, H_2O^{TOT}, SiOH, and *unknown* OH are in mol %

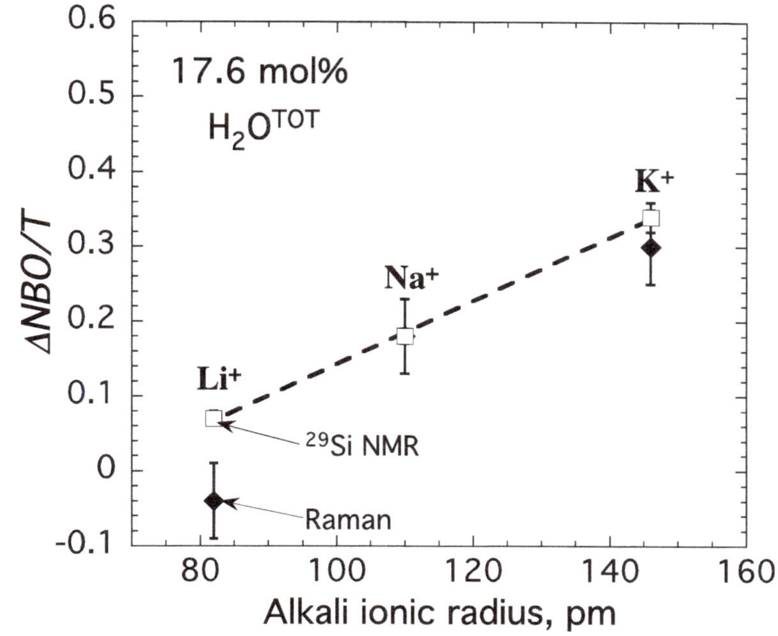

Fig. 6 Variation of the polymerization of the glasses (ΔNBO/T) upon addition of 17.6 nom. mol % water represented as a function of the ionic radius of alkali. ΔNBO/T (= $NBO/T_{anhydrous} - NBO/T_{17.6\ mol\%\ water}$) is calculated from the variations of the Q^n species as observed in Raman spectra (*plain symbols*) or in ^{29}Si NMR spectra (*open symbols*). Ionic radii of alkalis are from Whittaker and Muntus (1970), assuming sixfold oxygen-coordinated alkalis. The *black dotted line* is a linear fit of the ΔNBO/T vs ionic radius relationship for NMR data. For the Na-bearing silicate glass, Raman and ^{29}Si NMR give the same value, so that symbols are difficult to distinguish. *Error bars* are given at the 1σ confidence interval

Discussion

The dissolution of water into alkali silicate glasses results in their depolymerization, as shown by ^{29}Si NMR and Raman data (Figs. 4, 5, and 6). The Q^4 units transform to Q^3 and Q^2 units according to the reaction.

$$Q^4 \ <=> \ \phi Q^2 \ + (1-\phi) Q^3, \tag{7}$$

where ϕ is a constant reflecting the partitioning of the NBOs (formed upon H_2O addition) between Q^3 and Q^2. The equilibrium constant K_7 of reaction 7 depends on ionic radius of alkali. The larger the radius, the greater the *ΔNBO/T* and, thus, the higher the K_7 (Fig. 6). This dependence likely is linked to the influence of ionic radius of alkali metal on the OH/H_2O_{mol} ratio. At given water content, the larger the ionic size of alkali cation, the higher this ratio (Fig. 7). The variations of ϕ seem primarily related to the extent of depolymerization of the silicate network because the ratio $\Delta Q^2/\Delta Q^3$ is correlated with *ΔNBO/T*. At low *ΔNBO/T*, and hence for *NBO/*T close to 0.5, ϕ is low and Q^3 units are produced by reaction (7). At high *ΔNBO/T*, and hence for *NBO/T* > 0.5, ϕ is high and Q^2 units formation may dominate.

The concentration of OH groups that form SiOH bonds upon hydration of the glasses can be determined by using the *ΔNBO/T* values derived from the Q^n-species abundances. This is possible because oxygen in the

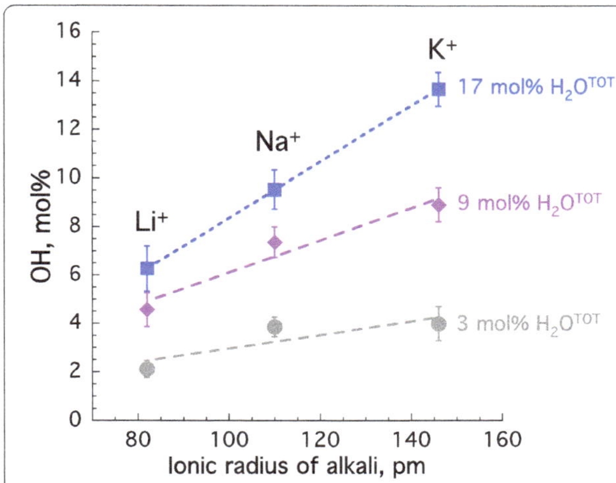

Fig. 7 Total OH group contents (mol %) of the alkali silicate glasses represented as a function of the alkali ionic radius (in picometers) for different water concentrations. Total OH group contents in KS4 glasses were determined by subtracting the H_2O_{mol} species contents determined with using ^1H MAS NMR spectra (see text and Fig. 1) to the total water content measured by Raman spectroscopy (see values in Table 2). The *dotted lines* are linear fits of the data. *Error bars* are given at the 1σ confidence interval

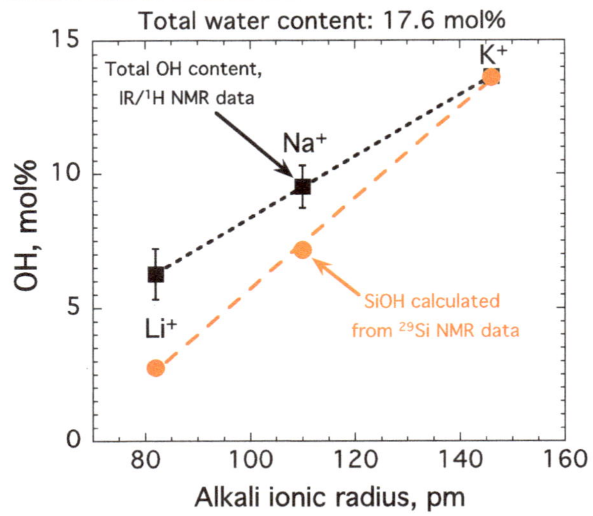

Fig. 8 Comparison between the concentration of OH groups measured with IR (LS4 and NS4 glasses) and ^{1}H NMR (KS4 glass) data and that of OH groups bonded to Si determined with using ^{29}Si NMR data for glasses with 17.6 mol % total H_2O. The *dotted lines* are linear fits of the data. *Error bars* are given at the 1σ confidence interval

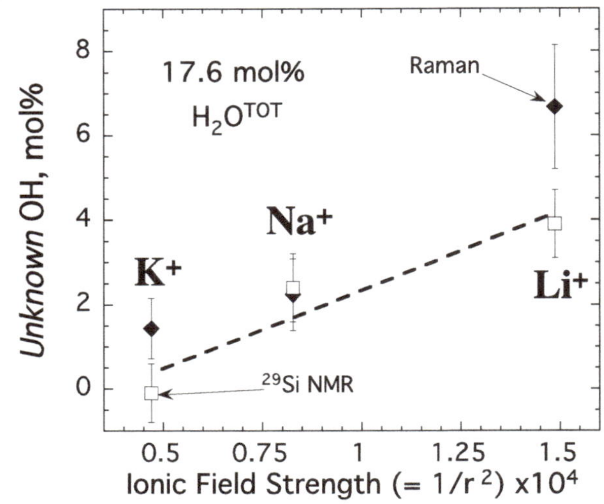

Fig. 9 Concentrations of *unknown* OH groups that do not participate in the depolymerization of the silicate network in the 17.6 mol % water glasses, represented as a function of the ionic field strength (IFS) of alkali. IFS = Z/r^2 with Z the electric charge and r the ionic radii of the alkalis (from Whittaker and Muntus 1970 assuming sixfold oxygen-coordinated alkalis). IFS was multiplied by 10^4 for clarity purpose. The *black dashed line* is a linear fit of the mol% unknown OH vs alkali IFS relationship calculated using NMR data. *Error bars* are given at the 1σ confidence interval

SiOH groups is non-bridging, so that SiOH group formation testifies to the depolymerization of the silicate network upon water solution (Eq. (2)). Figure 8 presents a comparison between the SiOH content determined with using *ΔNBO/T* derived from ^{29}Si SP MAS NMR data and the concentration of OH groups determined by using IR and ^{1}H MAS NMR data for glasses with 17.6 mol % water. At this water content, it appears that there are significant differences between these two sets of values for hydrous NS4 and LS4 glasses. For those glasses, the concentrations of SiOH groups are lower than those of OH groups detected by IR and ^{1}H MAS NMR spectroscopy. For the KS4 glass with 17.6 mol % water, the SiOH and the OH values obtained by the different methods agree. The SiOH fractions calculated from Raman-derived *ΔNBO/T* values confirm the above observations by showing a very similar trend following changes in ionic radius of alkali metal for glasses with 17.6 nom. mol % [H_2O^{TOT}] (Table 4).

The comparison in Fig. 8 and Table 4 between the SiOH concentrations retrieved through the ^{29}Si NMR data, validated by Raman results, and the total OH concentration from IR spectroscopy indicates that there is a significant fraction of OH groups in the NS4 and LS4 glasses that do not reflect depolymerization of the silicate network and for which we do not know the nature of their bonding. At 17.6 nom. mol % water, the concentration of those "unknown" OH groups increases with decreasing ionic radius, and hence, increasing ionic field strength of alkali metal (Fig. 9). It is negligible for the KS4 glass whereas it is maximum for the LS4 glass, which also has a negligible *ΔNBO/T*. It follows, therefore, that the chemical composition of the melt, in terms of ionic radius of the alkali metal, not only affects the speciation of water between OH groups and H_2O_{mol} species (Fig. 7), but also affects the bonding of the OH groups with the quenched melt structure (Figs. 8 and 9). The effect of the water concentration is less clear. From the Raman-derived SiOH contents, the fraction of *unknown* OH groups may increase in LS4 and NS4 with addition of water, whereas it may slightly decrease in the case of the KS4 glasses (Table 4). However, those values are difficult to interpret with confidence because they have not been validated by ^{29}Si NMR spectroscopy. This validation seems necessary in regards of the errors that can affect the *ΔNBO/T* determined from Raman spectroscopy, as shown by the data for LS4 (see Fig. 6).

Previous studies of hydrous sodium silicate glasses indicated that, at 80 mol% SiO_2, Eq. (2) appears to describe the mechanism of water solution in glasses (Zotov and Keppler 1998; Cody et al. 2005; Mysen and Cody 2005). The content of *unknown* OH is low in the NS4 glasses, near 2 mol % for a nominal total water content of 17.4 mol % (Table 4 and Fig. 9). The errors affecting the Q^n fractions determined by Raman or ^{29}Si NMR spectroscopy combined with those affecting the determination of the concentrations of total OH groups render difficult the detection of such a low amount. This may explain why previous studies of sodium silicate glasses containing high silica contents did not see such small concentrations of OH groups not bonded to Si. However, in alkaline-earth and alkali silicate glasses with low silica contents (NBO/T > ~ 1), presence of OH groups not bonded to Si has been inferred (e.g., Xue and Kanzaki 2004; Cody et al. 2005; Mysen and Cody 2005).

Present data indicate that Eq. (2) alone cannot explain the solution mechanism of water in silicate glasses containing small alkali metals, such as Li and Na. Another mechanism implying a different bonding of OH group must be considered. The concentration of total OH groups measured by IR in LS4 glasses is much higher than that of OH bonded to Si (Figs. 8 and 9). This indicates that the *unknown* OH groups contribute to the 4100–4500 cm^{-1} portion of IR spectra, where the peak assigned to the combination of the OH groups vibrational modes is usually observed (e.g., see Fig. 1). To our knowledge, three hypotheses can explain observations: (1) the formation of MOH groups, where M is an alkali; (2) the protonation of bridging Si-O-Si bonds; or (3) the protonation of Si-O-M bonds.

Infrared spectra of crystalline alkali hydroxide show absorption bands in the 4000–4400 cm^{-1} portion of IR spectra. It has been proposed that they arise from the combination of the O-H fundamental stretching vibrations with the lattice modes (Buchanan 1959; Wickersheim 1959; Snyder et al. 1960). Therefore, contribution from MOH species in the 4100–4500 cm^{-1} range of the IR spectra of the present glasses seems possible, so that the observed *unknown* OH could be MOH species. In this line of thinking, we note that the IR peak near 4500–4600 cm^{-1} is also observed in hydrous soda-lime-borate glasses and assigned to the combination of the OH stretching with lattice vibrational modes (Bauer et al. 2015). Therefore, the 4500 cm^{-1} IR peak in hydrous glasses not only contains contributions of OH groups bonded to Si, as it has been previously inferred (e.g., Malfait 2009 and references therein), but also possible contributions from other OH species bonded to metallic elements or even boron, as shown by the data of Bauer et al. (2015).

Hypothesis (2) involves the protonation of Si-O-Si bonds and thus the formation of Si-(OH)-Si protonated bonds. A similar hypothesis has been proposed to account for water solution in glasses along the joined SiO_2-$NaAlO_2$ because of the apparently negligible changes of ^{29}Si and ^{27}Al NMR signals upon water addition to these glasses (Kohn et al. 1989b, 1992, 1998). Molecular orbital calculations indicate that both Si-OH and Si-(OH)-Si bonds give signals near 900 cm^{-1} (Kubicki and Sykes 2004), so that the OH groups involved in Si-(OH)-Si bonds will contribute to the IR combination peak near 4500 cm^{-1}. Therefore, the *unknown* OH groups may be involved in Si-(OH)-Si bonds.

Hypothesis (3) involves the protonation of Si-O-M bonds and the formation of Si-(OH)-M protonated bonds. In this hypothesis, it must be inferred that the Si-(OH)-M bonds give a signal similar to that of Si-(OH)-Si protonated bonds near 900 cm^{-1}, such that those bonds will contribute to the IR combination peak near 4500 cm^{-1}.

To discriminate between those hypotheses, several points have to be considered. First, the link between the size/ionic field strength (abbreviated IFS in the following) of the alkali and the concentration of *unknown* OH (Fig. 9) indicates that alkalis that form short (and thus strong) M-O bonds promote *unknown* OH groups. Considering hypothesis (3), it is unlikely that protonation of Si-O-M bonds occurs when the M-O bond strength is high, such as for Li-O bonds, because protonation of Si-O-M bonds may be more likely to occur for stabilizing Si-O-M bonds with low M-O bond strengths. Considering now hypothesis (2), protonation of Si-O-Si bonds would result in the appearance of positively charged $Si-(OH)^{+}-Si$ bonds that would require charge balance by a negatively charged ion, i.e., by NBO oxygens or "free" oxygens O^{2-}. Such situation would require specific structural positions for the $Si-(OH)^{+}-Si$ bonds and the compensating oxygens and thus seems unlikely to occur at high temperature in silicate melts that lack long- and medium-range orders. It should be noted that protonation of Si-O-Al bonds has been inferred in aluminosilicate compositions because it may help the electrical charge compensation of AlO_2 tetrahedral units, but it also has been strongly questioned (see introduction of Malfait and Xue 2010 and references cited therein). Indeed, the most recent data point to formation of SiOH and AlOH groups in aluminosilicate glasses and not to protonation of bridging oxygen (e.g., see Xue and Kanzaki 2009, Malfait and Xue 2010 and references cited therein).

Therefore, the most plausible hypothesis to explain the observations of a fraction of OH groups not bonded to Si in the alkali silicate glasses is the formation of MOH groups. In particular, this hypothesis explains the link between the alkali IFS and the concentration of *unknown* OH groups (Fig. 9): metal cations with strong M-O bonds may help stabilizing MOH complexes, in agreement with interpretations of Xue and Kanzaki (2004). In addition, this hypothesis is supported by ^{1}H MAS and ^{1}H->^{29}Si->^{1}H CP MAS NMR data on Ca and Mg silicate glasses that indicate the present of MOH groups in such compositions (Xue 2009). Therefore, from the above discussion, the *unknown* OH groups likely are bonded to alkali elements, forming MOH molecular complexes.

We propose the following framework to explain the spectroscopic observations. A condition for the reaction of H_2O_{mol} species with the network of silicate melts is the self-ionization of H_2O_{mol} species (Moretti et al. 2014):

$$H_2O_{mol\,(melt)} <=> H^+{}_{(melt)} + OH^-{}_{(melt)} \tag{8}$$

When H_2O_{mol} reacts with Si-O-Si bonds, H^+ and OH^- are consumed to form SiOH bonds, leading to the process described by Eq. (2). In the present silica-rich glasses, H^+ and OH preferentially react with Q^4 bonds to form Q^2 and Q^3 species (Fig. 5), so that the depolymerization mechanism can be described by Eq. (7), confirming previous observations in similar composition (Zotov and Keppler 1998; Cody et al. 2005; Xue and Kanzaki 2004; Mysen and Cody 2005). Decreasing the ionic radius, and hence, increasing the IFS of alkali metal in silicate melts inhibits the self-ionization reaction of water (Fig. 7) but also promotes reaction of H^+ and OH^- with the non-bridging oxygen NBO and the alkalis of the Si-O-M bonds (Fig. 9) following:

$$H_2O_{mol} + SiOM <=> SiOH + MOH \tag{9}$$

Such a mechanism was proposed by Xue and Kanzaki (2004), who suggested the formation of MOH complexes in highly depolymerized Ca, Mg, and Na silicate melts (NBO/T > 1) on the basis of molecular dynamic calculations and ^1H MAS NMR observations. They also proposed the following reaction (see also Mysen and Cody 2005):

$$2\ SiOM + H_2O_{mol} <=> SiOSi + 2\ MOH, \tag{10}$$

Formation of MOH species can result in no change (Eq. (9)) or even in an increase (Eq. (10)) of the melt polymerization. Xue and Kanzaki (2004) indicated that such mechanisms mostly concerned depolymerized melts (NBO/T > ~1) (see also the discussion in Xue and Kanzaki 2009), in agreement with the findings of Cody et al. (2005) and Mysen and Cody (2005). However, the water solution mechanism for the LS4 glass is best described by the combination of Eqs. (8) and (9). Therefore, the present data support the solution mechanism described with Eq. (9) in melts with relatively high silica contents (NBO/T ~ 0.5) that contain network modifier elements with high IFS (Fig. 9).

According to the present data (Figs. 7, 8, and 9), IFS of network modifying alkali metals plays an important role in defining the equilibrium constants of Eqs. (8) (self-ionization of water), (2) (SiOH formation), (9) and (10) (MOH formation). The origin of such role could arise from electronic/steric considerations, hydration shell formation, or volume effects around alkalis governed by their size/IFS (Le Losq et al. 2015a). For instance, electronic effects may influence the hydrolysis energy of Si-O-Si and Si-O-M bonds, and hence, the equilibrium constants of reactions (2), (9), and (10) (i.e., see discussion in Xue and Kanzaki 2004). Volume effects cannot be excluded. For instance, large alkalis may inhibit the presence of cavities in which H_2O_{mol} species might be contained, and hence, will result in a large equilibrium constant for reaction (8). Formation of hydration shell around small alkalis may also help explaining the speciation of water (Fig. 7). Finally, IFS of metal cations may also affect the stability of MOH molecular complexes (Xue and Kanzaki 2004), explaining the link between the IFS of metal cations and the concentration of observed MOH groups (Fig. 9). However, the present data do not allow us to discriminate between those hypotheses. In addition, it cannot be excluded that several mechanisms play roles at the same time. Despite that, it is clear that the influence of the alkali IFS on the speciation, environment, and bonding of water in silicate quenched melts results in differences in water solubility (Mysen and Acton 1999; Mysen and Cody 2004) and partial molar volume (Mysen and Cody 2004). Beyond such concerns, the differences of bonding and speciation of water in hydrous glasses as a function of their chemical composition probably also result in variations in their physical properties such as their glass transition temperature and density.

Conclusions

The water speciation in glasses is that of the melts at their glass transition temperature, *Tg* (see for instance Nowak and Behrens 1995). Present results show that IFS of alkali affects the speciation of dissolved water between H_2O_{mol} species and OH groups, as well as the OH bonding mechanisms in glasses, and hence, in melts at their *Tg*. Therefore, the chemical composition of silicate melts affects the solution mechanisms of water (Figs. 7, 8, and 9), resulting in different effects on melt structures at *Tg* (Figs. 3, 4, 5, and 6). In addition, variations in the solubility of water with ionic radius of the alkali in tetrasilicate melts (see data for NS4 and KS4 in Mysen and Cody 2004) at given temperature and pressure indicate that such interplay between the melt composition, water solution mechanism, and the melt structure affects the solubility of water. As a result, such interplay probably drives the water exsolution history of magmas.

In addition to such effect, this complex interplay implies that water will affect differently the rheological properties of magmas depending on their composition. Indeed, the strong changes in melt structure (Figs. 4, 5, and 6) that accompanies variations in water solution mechanism (Figs. 7, 8, and 9) probably result in different glass transition temperatures for the hydrous alkali tetrasilicate glasses at given [H_2O^{TOT}], and, as a corollary, in different viscosity vs temperature paths. The present data do not allow us to make inferences on the extent of those differences, but they are expected to occur in view of the intimate link between the structure and the viscosity of alkali silicate melts (e.g., see Mysen and Richet 2005 for further details). Therefore, magmas with different chemical compositions will not only present different degassing paths, but also they will present different changes in their viscosity as a function of water content and temperature. Such consideration might be important for assessing about volcanic processes intimately linked to the viscous relaxation of magmas, such as the fragmentation that occurs in explosive eruptions (Papale 1999). Indeed, following the above ideas, the complex interplay between the magma composition, molecular structure, viscous properties, and the water solution mechanisms and solubility may influence the fragmentation threshold of magmas.

Abbreviations

EDS: energy-dispersive x-ray spectroscopy; SP: single pulse; MAS: magic angle spinning; NMR: nuclear magnetic resonance; IR: infrared; IFS: ionic field strength.

Competing interests

The authors declare that they have no competing interests.

Authors' contributions

BOM and GDC proposed the study. CLL made the experimental study, treated and interpreted data, and drafted the manuscript. BOM helped CLL for the experiments and lead the study. GDC performed the NMR experiments with the help of CLL. BOM and GDC collaborated with CLL in the construction of the manuscript. All authors read and approve the final manuscript.

Authors' information

CLL performed this study as a Postdoctoral Fellow at the Geophysical Laboratory, Carnegie Institution of Science, Washington DC, USA, under the mentorship of BOM and GDC, respectively Senior Scientist at, and Acting Director of the Geophysical Laboratory.

Acknowledgements

This research was partially supported by grants EAR1212754 and EAR1250449 to BOM and the NASA Astrobiology Institute. All NMR experiments were performed at the W. M. Keck Solid State NMR Facility at the Geophysical Laboratory, supported in part by the W. M. Keck Foundation, the NSF, and the Carnegie Institution of Washington.
This paper belongs to the Solid Earth Sciences section.

References

Bartholomew RF, Butler BL, Hoover HL, Wu CK (1980) Infrared spectra of a water-containing glass. J Am Ceram Soc 63:481–485

Bauer U, Behrens H, Fechtelkord M, Reinsch S, Deubener J (2015) Water- and boron speciation in hydrous soda-lime-borate glasses. J Non Cryst Solids 423–424:58–67

Behrens H, Yamashita S (2008) Water speciation in hydrous sodium tetrasilicate and hexasilicate melts: constraint from high temperature NIR spectroscopy. Chem Geol 256:306–315

Bell RJ, Bird NF, Dean P (1968) The vibrational spectra of vitreous silica, germania and beryllium fluoride. J Physics C (Proc phys soc) 1:299–303

Blake S (1984) Volatile oversaturation during the evolution of silicic magma chambers as an eruption trigger. J Geophys Res 89:8237–8244

Brawer SA, White WB (1975) Raman spectroscopic investigation of the structure of silicate glasses. I. The binary alkali silicates*. J Chem Phys 63:2421–2432

Brawer SA, White WB (1977) Raman spectroscopic investigation of the structure of silicate glasses (II). Soda-alkaline earth-alumina ternary and quaternary glasses. J Non Cryst Solids 23:261–278

Buchanan RA (1959) Near infrared spectra of crystalline alkali hydroxides. J Chem Phys 31:870–874

Clemens JD (1984) Water contents of silicic to intermediate magmas. Lithos 17:273–28

Cody GD, Mysen BO, Lee SK (2005) Structure vs. composition: a solid-state ^1H and ^{29}Si NMR study of quenched glasses along the Na_2O-SiO_2-H_2O join. Geochim Cosmochim Acta 69:2373–2384

Dingwell DB, Romano C, Hess KU (1996) The effect of water on the viscosity of a haplogranitic melt under P-T-X conditions relevant to silicic volcanism. Contribut Mineral Petrol 124:19–28

Eckert H, Yesinowski JP, Silver LA, Stolper EM (1988) Water in silicate glasses: quantification and structural studies by ^1H solid echo and MAS-NMR methods. J Phys Chem 92:2055–2064

Farnan I, Kohn SC, Dupree R (1987) A study of the structural role of water in hydrous silica glass using cross-polarisation magic angle spinning NMR. Geochim Cosmochim Acta 51:2869–2873

Friedman I, Long W, Smith RL (1963) Viscosity and water content of rhyolite glass. J Geophys Res 68:6524–6535

Furukawa T, Fox KE, White WB (1981) Raman spectroscopic investigation of the structure of silicate glasses. III. Raman intensities and structural units in sodium silicate glasses. J Chem Phys 75:3226–3237

Galeener FL (1979) Band limits and the vibrational spectra of tetrahedral glasses. Phys Rev B 19:4292–4397

Gardner JE, Thomas RME, Jaupart C, Tait S (1996) Fragmentation of magma during Plinian volcanic eruptions. Bull Volcanol 58:144–162

Kohn SC, Dupree R, Mortuza MG (1992) The interaction between water and aluminosilicate magmas. Chem Geol 96:399–409

Kohn SC, Dupree R, Smith ME (1989a) Proton environments and hydrogen-bonding in hydrous silicate glasses from proton NMR. Nature 337:539–541

Kohn SC, Dupree R, Smith ME (1989b) A multinuclear magnetic resonance study of the structure of hydrous albite glass. Geochim Cosmochim Acta 53:2925–2935

Kohn SC, Smith ME, Dirken PJ, Van Eck ERH, Kentgens APM, Dupree R (1998) Sodium environment in dry and hydrous albite glasses: Improved ^{23}Na solid state NMR data and their implications for water dissolution mechanisms. Geochim Cosmochim Acta 62:79–87

Kubicki JD, Sykes DG (2004) Ab initio calculation of ^1H, ^{17}O, ^{27}Al and ^{29}Si NMR parameters, vibrational frequencies and bonding energetics in hydrous silicate and Na-aluminosilicate glasses. Geochim Cosmochim Acta 68:3909–3918

Le Losq C, Neuville DR (2013) Effect of the Na/K mixing on the structure and the rheology of tectosilicate silica-rich melts. Chem Geol 346:57–71

Le Losq C, Cody GD, Mysen BO (2015a) Alkali influence on the water speciation and the environment of protons in silicate glasses revealed by ^1H MAS NMR spectroscopy. Am Mineral 100:466–473

Le Losq C, Cody GD, Mysen BO (2015b) Complex stretching signal of OH groups in silicate glasses: implication for the use of the 4500 cm^{-1} IR peak as a marker of OH^- groups concentration. Am Mineral 100:945–950

Le Losq C, Neuville DR, Florian P, Henderson GS, Massiot D (2014) The role of Al^{3+} on rheology and structural changes of sodium silicate and aluminosilicate glasses and melts. Geochim Cosmochim Acta 126:495–517

Leonteva A (1940) Measurements of the viscosity of obsidian and of hydrated glasses. *Izvestiya Akademii Nauk SSSR*. Ser Geol 2:44–54

Lippmaa E, Mägi M, Samoson A, Engelhardt G, Grimmer A-R (1980) Structural studies of silicates by solid-state high-resolution ^{29}Si NMR. J Am Chem Soc 102:4889–4893

Maekawa H, Maekawa T, Kawamura K, Yokohama T (1991) The structural groups of alkali silicate glasses determined from ^{29}Si MAS-NMR. J Non Cryst Solids 127:53–64

Malfait WJ (2009) The 4500 cm−1 infrared absorption band in hydrous aluminosilicate glasses is a combination band of the fundamental {Si, Al}-OH and O-H vibrations. Am Mineral 94:849–852

Malfait WJ, Xue X (2010) The nature of hydroxyl groups in aluminosilicate glasses: Quantifying Si-OH and Al-OH abundances along the SiO_2-$NaAlSiO_4$ join by ^1H, ^{27}Al-^1H and ^{29}Si-^1H NMR spectroscopy. Geochim Cosmochim Acta 74:719–737

Matson DW, Sharma SK, Philpotts JA (1983) The structure of high-silica alkali-silicate glasses. A Raman spectroscopic investigation. J Non Cryst Solids 58:323–352

McMillan PF (1984) Structural studies of silicate glasses and melts—applications and limitations of Raman spectroscopy. Am Mineral 69:622–644

McMillan PF, Poe BT, Stanton TR, Remmele RL (1993) A Raman spectroscopic study of H/D isotopically substituted hydrous aluminosilicate glasses. Phys Chem Min 19:454–459

Moretti R, Le Losq C, Neuville DR (2014) The amphoteric behavior of water in silicate melts from the point of view of their ionic-polymeric constitution. Chem Geol 367:23–33

Moulson AJ, Roberts JP (1961) Water in silica glass. Trans Faraday Soc 57:1208–1216

Murdoch JB, Stebbins JF, Carmichael ISE (1985) High-resolution ^{29}Si NMR study of silicate and aluminosilicate glasses: the effect of network-modifying cations. Am Mineral 70:332–343

Mysen BO (1990) Role of Al in depolymerized, peralkaline aluminosilicate melts in the systems Li_2O-Al_2O_3-SiO_2, Na_2O-Al_2O_3-SiO_2, and K_2O-Al_2O_3-SiO_2. Am Mineral 75:120–134

Mysen BO (2007) The solution behavior of H_2O in peralkaline aluminosilicate melts at high pressure with implications for properties of hydrous melts. Geochim Cosmochim Acta 71:1820–1834

Mysen BO, Acton M (1999) Water in H_2O-saturated magma-fluid systems: solubility behavior in K_2O-Al_2O_3-SiO_2-H_2O to 2.0 GPa and 1300 °C. Geochim Cosmochim Acta 63:3799–3815

Mysen BO, Cody G (2005) Solution mechanisms of H_2O in depolymerized peralkaline melts. Geochim Cosmochim Acta 69:5557–5566

Mysen BO, Cody G (2004) Solubility and solution mechanism of H_2O in alkali silicate melts and glasses at high pressure and temperature. Geochim Cosmochim Acta 68:5113–5126

Mysen BO, Frantz JD (1993) Structure and properties of alkali silicate melts at magmatic temperatures. Eur J Mineral 5:393–407

Mysen BO, Richet P (2005) Silicate glasses and melts—properties and structure. Developments in Geochemistry 10. Elsevier B.V, Amsterdam

Mysen BO, Finger LW, Virgo D, Seifert FA (1982) Curve-fitting of Raman spectra of silicate glasses. Am Mineral 67:686–695

Neuville, D. R. (1992) Etude des propriétés thermodynamiques et rheologiques des silicates fondus. Ph.D. Thesis of the Université Paris 7, 274.

Neuville DR (2006) Viscosity, structure and mixing in (Ca, Na) silicate melts. Chem Geol 229:28–41

Neuville DR, Mysen BO (1996) Role of aluminium in the silicate network: in situ, high-temperature study of glasses and melts on the join SiO_2-$NaAlO_2$. Geochim Cosmochim Acta 60:1727–1737

Neuville DR, de Ligny D, Henderson GS (2014) Advances in Raman spectroscopy applied to Earth and Material Sciences. Rev Mineral Geochem 78:509–541

Nowak M, Behrens H (1995) The speciation of water in haplogranitic glasses and melts determined by in situ near-infrared spectroscopy. Geochim Cosmochim Acta 59:3445–3450

Olivier L, Yuan X, Cormack AN, Jäger C (2001) Combined ^{29}Si double quantum NMR and MD simulation studies of network connectivities of binary $Na_2O.SiO_2$ glasses: new prospects and problems. J Non Cryst Solids 293–295:53–66

Papale P (1999) Strain-induced magma fragmentation in explosive eruptions. Nature 397:425–428

Richet P, Lejeune A-M, Holtz F, Roux J (1996) Water and the viscosity of andesite melts. Chem Geol 128:185–197

Richet P, Whittington AG, Holtz F, Behrens H, Ohlhorst S, Wilke M (2000) Water and the density of silicate glasses. Contribut Mineral Petrol 138:337–347

Saucier H (1952) Quelques expériences sur la viscosité à haute température de verres ayant la composition d'un granite. Influence de la vapeur d'eau sous pression. Bull de la Société Française de Minéralogie et de Cristallographie 75:1–45

Scholtze H (1960) Zur Frage der unterscheidung zwischen H_2O-Molekeln und OH-gruppen in gläsern und mineralen. Naturwissenschaften 47:226–227

Schramm CM, de Jong BHWS, Parziale VE (1984) ^{29}Si magic angle spinning NMR study on local silicon environments in amorphous and crystalline lithium silicates. J Am Chem Soc 106:4396–4402

Seifert F, Mysen BO, Virgo D (1982) Three-dimensional network structure of quenched melts (glass) in the systems SiO_2-$NaAlO_2$, SiO_2-$CaAl_2O_4$ and SiO_2-$MgAl_2O_4$. Am Mineral 67:696–717

Sen PN, Thorpe MF (1977) Phonons in AX_2 glasses: from molecular to band-like modes. Phys Rev B 15:4030–4038

Shaw HR (1963) Obsidian-H_2O viscosities at 1000 and 2000 bars in the temperature range 700° to 900 °C. J Geophys Res 68:6337–6343

Snyder RG, Kumamoto J, Ibers JA (1960) Vibrational spectrum of crystalline potassium hydroxide. J Chem Phys 33:1171–1177

Sowerby JR, Keppler H (1999) Water speciation in rhyolitic melt determined by in-situ infrared spectroscopy. Am Mineral 84:1843–1849

Spiekermann G, Steel-McInnis M, Schmidt C, Jahn S (2012) Vibrational mode frequencies of silica species in SiO_2-H_2O liquids and glasses from ab initio molecular dynamics. J Chem Phys 136:154501

Stolen RH, Walrafen GE (1976) Water and its relation to broken bond defects in fused silica. J Chem Phys 64:2623–2631

Stolper A (1982) Water in silicate glasses: an infrared spectroscopic study. Contribut Mineral Petrol 81:1–17

Tarantola, A. (2005) Inverse problem theory and methods for model parameter estimation. Society of Industrial and Applied Mathematics

Thomas N, Jaupart C, Vergniolle S (1994) On the vesicularity of pumice. J Geophys Res 99:15633–15644

Tischendorf B, Ma C, Hammersten E, Venhuizen P, Peters M, Affatigato M, Feller S (1998) The density of alkali silicate glasses over wide compositional ranges. J Non Cryst Solids 239:197–202

Virgo D, Mysen BO, Kushiro I (1980) Anionic constitution of silicate melts quenched at 1 atm from Raman spectroscopy: implications for the structure of igneous melts. Science 208:1371–1373

Voigt U, Lammert H, Eckert H, Heuer A (2005) Cation clustering in lithium silicate glasses: quantitative description by solid-state NMR and molecular dynamics simulations. Phy Rev B 72:064207-1–11

Whittaker EJW, Muntus R (1970) Ionic radii for use in geochemistry. Geochim Cosmochim Acta 34:945–956

Wickersheim KA (1959) Infrared absorption spectra of lithium hydroxide. J Chem Phys 31:863–869

Wu C-K (1980) Nature of incorporated water in hydrated silicate glasses. J Am Ceram Soc 63:453–457

Xue X (2009) Water speciation in hydrous silicate and aluminosilicate glasses: direct evidence from ^{29}Si-^{1}H and ^{27}Al-^{1}H double-resonance NMR. Am Mineral 94:395–398

Xue X, Kanzaki M (2004) Dissolution mechanisms of water in depolymerized silicate melts: constraints from ^{1}H and ^{29}Si NMR spectroscopy and *ab initio* calculations. Geochim Cosmochim Acta 68:5027–5057

Xue X, Kanzaki M (2006) Depolymerization effect of water in aluminosilicate glasses: direct evidence from ^{1}H-^{27}Al heteronuclear correlation NMR. Am Mineral 91:1922–1926

Xue X, Kanzaki M (2008) Structure of hydrous aluminosilicate glasses along the diopside—anorthite join: a comprehensive one- and two-dimensional ^{1}H and ^{27}Al NMR study. Geochim Cosmochim Acta 72:2331–2348

Xue X, Kanzaki M (2009) Proton distributions and hydrogen bonding in crystalline and glassy hydrous silicates and related inorganic materials: insights from high-resolution solid-state nuclear magnetic resonance spectroscopy. J Am Ceram Soc 92:2803–2830

Yamashita S, Behrens H, Schmidt BC, Dupree R (2008) Water speciation in sodium silicate glasses based on NIR and NMR spectroscopy. Chem Geol 256:231–241

Zotov N, Keppler H (1998) The influence of water on the structure of hydrous sodium tetrasilicate glasses. Am Mineral 83:823–834

Rainfall intensity–duration conditions for mass movements in Taiwan

Chi-Wen Chen[1*], Hitoshi Saito[2,3] and Takashi Oguchi[1,3]

Abstract

Mass movements caused by rainfall events in Taiwan are analyzed during a 7-year period from 2006 to 2012. Data from the Taiwan Soil and Water Conservation Bureau reports were compiled for 263 mass movement events, including 156 landslides, 91 debris flows, and 16 events with both landslides and debris flows. Rainfall totals for each site location were obtained from interpolated rain gauge data. The rainfall intensity–duration (*I–D*) relationship was examined to establish a rainfall threshold for mass movements using random sampling: $I = 18.10(\pm 2.67)D^{-0.17(\pm 0.04)}$, where *I* is mean rainfall intensity (mm/h) and *D* is the time (h) between the beginning of a rainfall event and the resulting mass movement. Significant differences were found between rainfall intensities and thresholds for landslides and debris flows. For short-duration rainfall events, higher mean rainfall intensities were required to trigger debris flows. In contrast, for long-duration rainfall events, similar mean rainfall intensities triggered both landslides and debris flows. Mean rainfall intensity was rescaled by mean annual precipitation (MAP) to define a new threshold: $I_{MAP} = 0.0060(\pm 0.0009)D^{-0.17(\pm 0.04)}$, where I_{MAP} is rescaled rainfall intensity and MAP is the minimum for mountainous areas in Taiwan (3000 mm). Although the *I–D* threshold for Taiwan is high, the I_{MAP}–*D* threshold for Taiwan tends to be low relative to other areas around the world. Our results indicate that Taiwan is highly prone to rainfall-induced mass movements. This study also shows that most mass movements occur in high rainfall-intensity periods, but some events occur before or after the rainfall peak. Both antecedent and peak rainfall play important roles in triggering landslides, whereas debris flow occurrence is more related to peak rainfall than antecedent rainfall.

Keywords: Mass movements; Landslides; Debris flows; *I–D* thresholds; Rainfall; Typhoons

Background

Mass movements such as landslides and debris flows are common natural hazards in mountainous areas (Aleotti and Chowdhury 1999; Guzzetti et al. 1999; Dai et al. 2002; Dou et al. 2014). These hazards not only cause considerable loss of property and life but also ecological and environmental problems such as increased soil erosion and sediment discharge (Hovius et al. 1997; Claessens et al. 2007). Mass movements, especially shallow landslides and debris flows, are often triggered by heavy rainfall. Therefore, many studies have investigated rainfall thresholds for the initiation of mass movements across the globe using both physical and empirical models. Although the simplicity of the empirical approach neglects important hydrological controls, it offers a straightforward means for issuing regional-scale mass movement warnings based solely on rainfall data.

In order to derive empirical critical thresholds for mass movements, many mass movement studies have focused on rainfall parameters such as rainfall intensity, duration, cumulative rainfall, and antecedent rainfall (see Guzzetti et al. 2007, 2008 for detailed reviews). Among these, the relationship between rainfall intensity (*I*) and duration (*D*) is most commonly used to estimate rainfall thresholds (e.g., Caine 1980; Guzzetti et al. 2007, 2008; Brunetti et al. 2010; Saito et al. 2010a, 2014). Several studies have also found that peak rainfall intensity and the initiation of mass movements are often concurrent (Aleotti 2004; Chien-Yuan et al. 2005; Guzzetti et al. 2007, 2008; Dahal and Hasegawa 2008; Saito et al. 2010a, 2010b). Although antecedent rainfall also plays an important role in the initiation of mass movement (Kim et al. 1992; Glade 1997; Crozier 1999; Glade et al. 2000; Guzzetti et al. 2007, 2008; Dahal and Hasegawa 2008), intensity–duration (*I–D*) thresholds and/or the peak

* Correspondence: kevin4919@gmail.com
[1]Department of Natural Environmental Studies, Graduate School of Frontier Science, The University of Tokyo, 5-1-5 Kashiwanoha, Kashiwa Chiba 277-8568, Japan
Full list of author information is available at the end of the article

rainfall intensity are often used to predict mass movements and warn the appropriate authorities of potential mass movement hazards (Onodera et al. 1974; Keefer et al. 1987; Aleotti 2004; Hong et al. 2005; Cannon et al. 2008; Coe et al. 2008; Dahal and Hasegawa 2008; Guzzetti et al. 2008).

Taiwan is extremely susceptible to mass movements because of steep mountainous topography and frequent heavy rainfall and earthquakes. Thus, many studies have focused on understanding the underlying triggers of mass movements in Taiwan. For example, Chang and Chiang (2009) combined slope stability and statistical models to establish a landslide probability model based on geographical and hydrological data for landslides triggered by typhoon events. Lin and Chen (2012) used a rainfall kinetic energy conversion formula and rainfall intensity data from typhoon events between 1994 and 2008 to determine the landslide rainfall threshold for mountainous areas in central Taiwan (2000 J/m^2). They then used rainfall, sediment discharge, and flow discharge data from another typhoon event to validate this threshold as a predictor of landslides and resultant sediment discharge. In addition, Jan and Chen (2005) examined the *I–D* thresholds of debris flows before and after the 1999 Chi-Chi earthquake for the Chenyoulan River catchment in central Taiwan. They used the debris flows caused by Typhoon Herb in 1996 for the *I–D* threshold before the Chi-Chi earthquake and the debris flows caused by Typhoon Toraji and Nari in 2001 for the *I–D* threshold after the Chi-Chi earthquake. They found that the *I–D* threshold decreased after the Chi-Chi earthquake.

Although Taiwanese mass movement events have been widely investigated, studies have typically examined specific events or locations. Surprisingly, few studies have explored rainfall conditions associated with mass movements in Taiwan as a whole. Although Chien-Yuan et al. (2005) investigated the *I–D* threshold of debris flows in Taiwan, their data were generally sourced from specific areas in northern and central Taiwan and only spanned the time period from 1989 to 2001, thus excluding more recent data. Since mass movements are densely distributed throughout Taiwan, it is important to develop a general relationship between the initiation of mass movements and rainfall conditions. Therefore, the objectives of this study are to analyze the mean intensity, peak intensity, and duration of rainfall associated with mass movements across Taiwan and to establish the *I–D* threshold using recent data.

Study area

Taiwan is located on a convergent plate boundary between the Eurasian Continental and the Philippine Sea plates, with the Philippine Sea plate moving towards the Eurasian Continental plate at a rate of 80 mm/year (Yu et al. 1997). The subduction of the Philippine Sea plate beneath the Eurasian Continental plate has resulted in the formation of an active mountain belt with over 200 peaks higher than 3000 m a.s.l. (Ho 1986; Teng 1990) and is responsible for frequent large earthquakes and an orogenic uplift rate of about 5–7 mm/year (Li 1976; Willett et al. 2003). The geological map provided by the Central Geological Survey of Taiwan (Ho 1988) illustrates the seven major geological zones of Taiwan (Fig. 1). Tertiary marine sedimentary rocks mainly underlie the western region and part of the northeastern coastal region. Central Taiwan is composed of metamorphic rocks. The pre-Tertiary metamorphic complex is exposed mainly in the eastern Central Range, while the majority of the western Central Range consists of Tertiary indurated metamorphosed argillaceous sedimentary rocks (Ho 1988). The eastern coastal region is underlain by volcanic-arc sediments, orogenic sediments, and a subduction-collision complex formed in the Pliocene to Pleistocene. Northern Taiwan consists mainly of agglomerates-masses of igneous rock fragments, and, finally, lowlands such as alluvial fans are mostly composed of Quaternary sediments.

Taiwan is located between 120°E and 122°E and between 22°N and 25°N. According to the elevation histogram derived from a 10 m digital terrain model of Taiwan, about 68 % of the country lies below 1000 m above sea level. The slope of mountainous areas is mostly between 30° and 50° (Fig. 1). Taiwan experiences tropical and subtropical-monsoon climates, with the boundary between the two climate regimes located in southern Taiwan (Wang and Ho 2002). The average temperature over the Taiwanese lowlands during the wet season (May–October) is above 20 °C, while the average temperature during the dry season (November–April) is between 14 and 20 °C. In the mountain regions of central and southern Taiwan, the average temperature is lower than in other regions of the country.

On average, four typhoons strike Taiwan every year (Wu and Kuo 1999), causing heavy and concentrated rainfall. Average annual rainfall over Taiwan is 2500 mm, but annual rainfall in mountainous regions can surpass 3000 mm (Shieh 2000). Approximately 60–80 % of rainfall falls during the wet season (Fig. 2). Heavy rainfall and the steep topography of Taiwan have contributed to an erosion rate of 3–7 mm/year (Dadson et al. 2003). Landslides and debris flows represent the primary mechanisms for this erosion and are important for maintaining a balance between erosion and uplift (Dadson et al. 2003).

Taiwan is characterized by a fragile geological environment, and areas that display a high density of geological discontinuity are more prone to mass movements (Chen et al. 1999; Chen and Su 2001; Chuang et al. 2009). Interestingly, the number and magnitude of rainfall-induced mass movements increased after the 1999 Chi-Chi earthquake because slopes became more vulnerable (Chang and Slaymaker

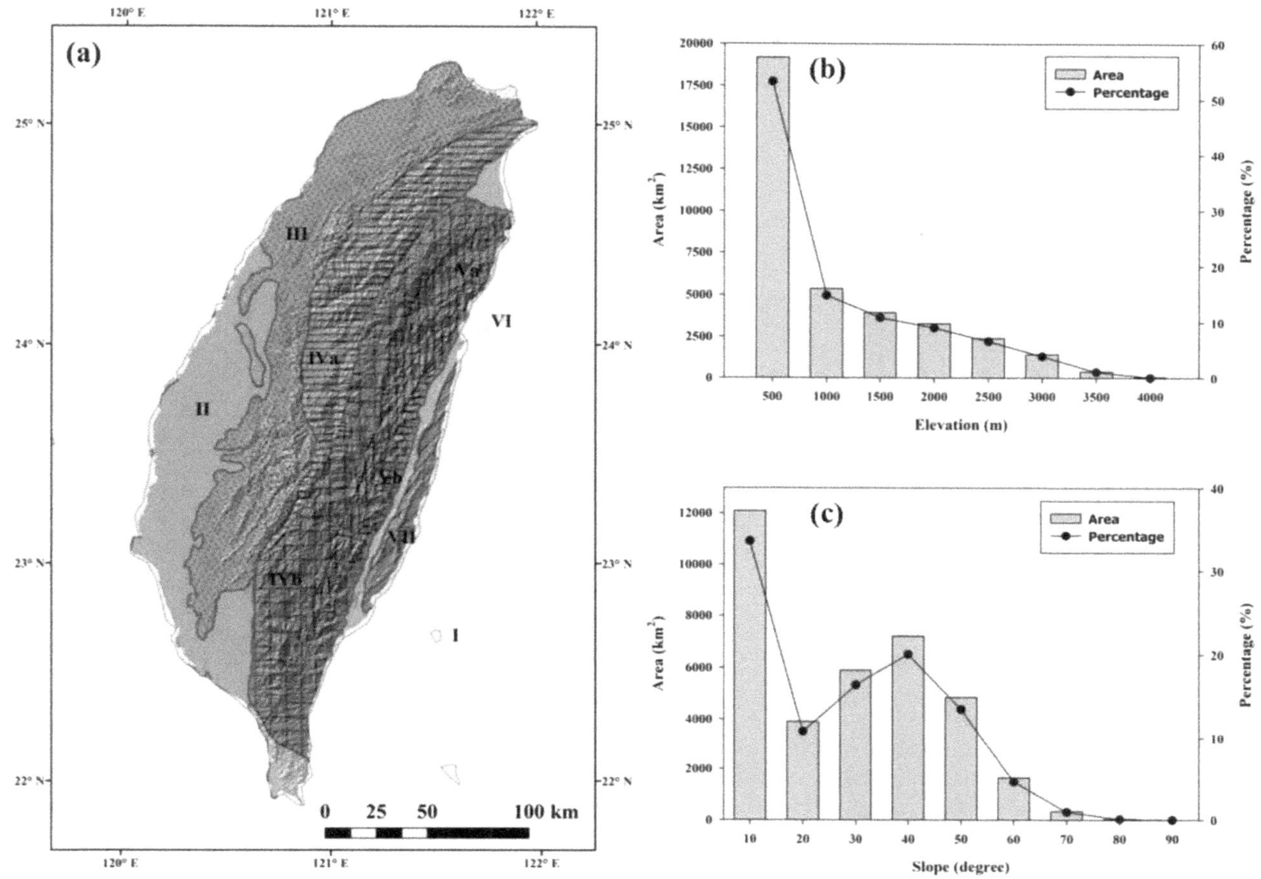

Fig. 1 **a** Location, geology, **b** elevation, and **c** slope of Taiwan. *I* Pleistocene basalt, *II* Quaternary alluvial deposit, *III* Neocene clastic rock, *IV* Tertiary sub-metamorphic rock (*a* argillite and slate; *b* slate and phyllite), *V* Pre-Tertiary metamorphic complex (*a* schist, marble, and granite;, *b* schist and serpentine), *VI* plate suture belt, *VII* Neogene volcanic rock

2002; Dadson et al. 2004; Cheng et al. 2005), which reduced both the maximum rainfall intensity and critical cumulative rainfall required to trigger mass movements (Lin et al. 2003; Chang et al. 2009; Shieh et al. 2009; Chen 2011).

Methods

Mass movement data

A total of 263 mass movements caused by rainfall events were analyzed during the 7-year period from 2006 to 2012 (Fig. 3). Mass movement data were compiled from the Soil and Water Conservation Bureau (SWCB) of Taiwan reports for 156 landslides, 91 debris flows, and 16 events with both landslides and debris flows (http://117.56.222.245/disasterInfo/ImpDisasterReport.aspx). The 16 events that included both types of mass movement were initiated as landslides, but these were subsequently completely or partially mobilized to form debris flows during the same rainfall event. Over the study period, mass movements across Taiwan occurred only during the wet season, with 1 in May, 26 in June, 33 in July, 141 in August, 24 in September, and 38 in October (Fig. 3).

To prevent secondary disasters and enable quick recovery, SWCB has conducted detailed field surveys after rainfall-induced mass movements that cause damage to houses and roads. The resultant reports contain information on the type, location, and approximate time (to the closest hour) of each event. The time of the event is estimated from real-time videos taken at observation stations, the time people informed authorities of the disaster, and interviews with residents. The field surveys include investigations of the extent of affected areas and the damage caused to residences and infrastructures. These reports classify disasters into three types: landslides, debris flows, and floods. As landslides and debris flows are characterized by different processes, they are separated in the analysis to help identify the triggers associated with each type of failure. Only a few cases of gravity-driven falls or topples were reported, and all such events occurred during a dry period with almost no rainfall. Therefore, we carefully checked all reports ($n = 292$) and excluded events not caused by rainfall (almost no rainfall during mass movements).

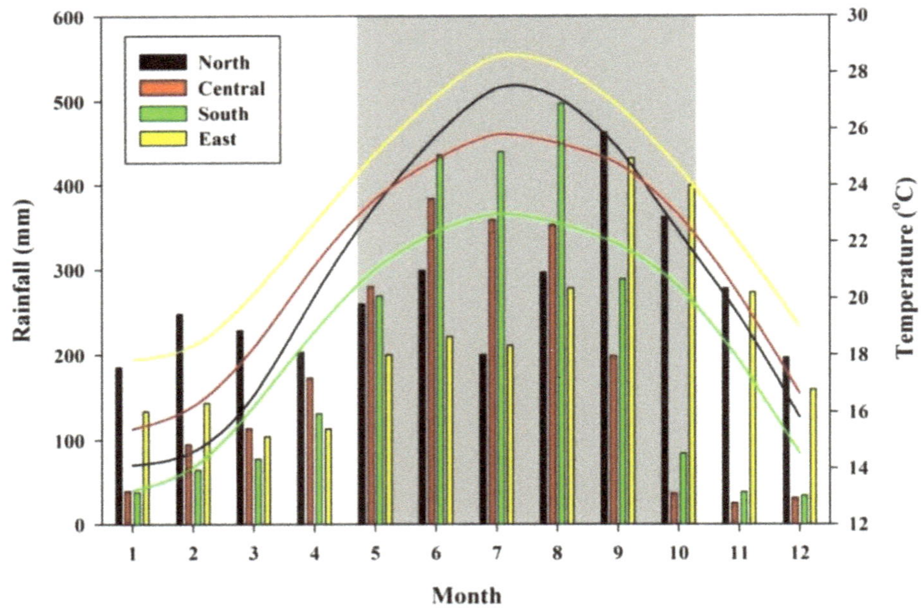

Fig. 2 Average monthly temperature (*curved lines*) and rainfall (*bars*) in different regions of Taiwan (averaged from 1981 to 2010). *North* Taipei, Keelung, Taoyuan, Hsinchu, and Miaoli; *Central* Taichung, Changhua, Nantou, and Yunlin; *South* Chiayi, Tainan, Kaohsiung, and Pingtung; *East* Ilan, Hualien, and Taitung

Although mass movements occur frequently in Taiwan, there is currently no complete inventory that can provide detailed information about these events. As SWCB data have a relatively high level of integrity and consistency, it is the best source of information to compile an inventory and analyze the relationships between mass movements and rainfall conditions in Taiwan.

Rainfall data

Taiwan's Central Weather Bureau has installed more than 400 rain gauges to record hourly data with a density of approximately one gauge every 76 km^2 (Fig. 3). Since rain gauges are not always located close to mass movement sites, we selected the nearest five rain gauges for each mass movement and conducted a kriging interpolation to estimate rainfall at each mass movement location. This technique uses ordinary kriging and a spherical semivariogram model with a variable search radius to estimate hourly rainfall data at each site over the duration of a rainfall event. A continuous rainfall event is considered to begin when hourly rainfall surpasses 4 mm and ends when hourly rainfall decreases below 4 mm over the next six consecutive hours (e.g., Chang et al. 2011).

We calculated the mean rainfall intensity (I, mm/h), rainfall duration (D, h), and peak rainfall intensity (mm/h) from the beginning of each rainfall event to the time of mass movement occurrence. Some previous studies have used 1 week of antecedent rainfall for analyzing the antecedent soil moisture and found that it is an important factor for triggering landslides (Crosta 1998; Jakob 2006). Therefore, we also examined rainfall totals for 168 h (1 week) before the beginning of rainfall events that triggered mass movement events.

I–D thresholds

The I–D threshold for mass movements is identified on an I–D plot as the minimum rainfall total for which a mass movement could occur (e.g., Guzzetti et al. 2007). In early years, the lower rainfall boundary (threshold) was determined manually (e.g., Caine 1980; Larsen and Simon 1993; Chien-Yuan et al. 2005). However, more recently, mathematical and/or statistical criteria have been used to objectively determine the lower boundary. Guzzetti et al. (2007) were the first to propose the Bayesian inference method for determining I–D thresholds; Brunetti et al. (2010) proposed a frequentist approach; and Saito et al. (2010a) adopted the quantile-regression method.

Although many different methods have been proposed, the general form of the relationship between rainfall intensity and duration is:

$$I = \alpha D^{-\beta} \quad (1)$$

where α is a scaling constant (intercept), and β is the shape parameter (slope).

Like all landslide inventories, data analyzed in this study contain inherent uncertainties, including measurement errors, erosional or geographical censoring, and reporting biases (Kirschbaum et al. 2012; Korup et al. 2012; Petley

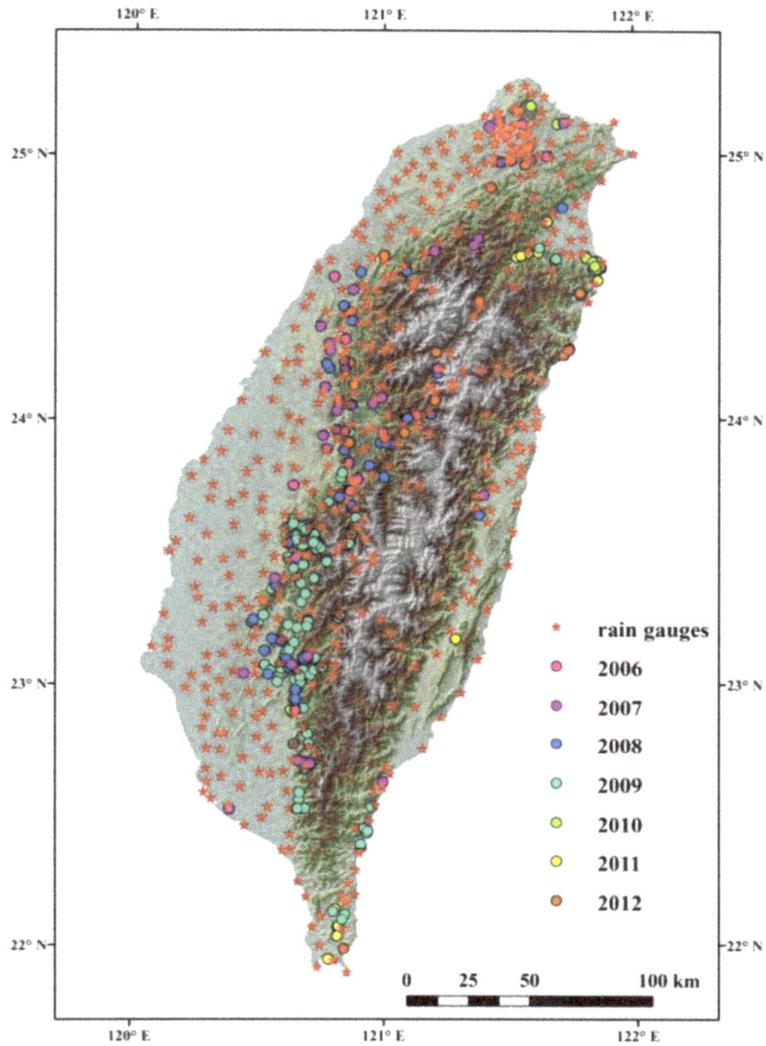

Fig. 3 Distribution of rain gauges and mass movements that occurred between 2006 and 2012

2012). We therefore determined thresholds by the frequentist statistical method and the random sampling method, which are statistically robust approaches. Brunetti et al. (2010) suggested the frequentist statistical method for the definition of objective empirical-rainfall thresholds. Following this method, the standard *I*–*D* threshold was defined as 5 %. We also used the robust random sampling method due to its resistance to errors and outliers. To establish the threshold equation, two-thirds of the data were randomly sampled, and this was repeated 100 times, allowing us to calculate the mean and standard deviation of α and β in Eq. (1).

Regional *I*–*D* thresholds are limited by their specificity to a certain region. They cannot be easily applied to other areas because topographic, lithological, meteorological, and climatological characteristics may vary between regions (Jakob and Weatherly 2003; Guzzetti et al. 2007, 2008). For example, limited surface vegetation and the accumulation of regolith on slopes during long dry periods allow a relatively short and weak rainfall event to initiate mass movements in arid areas. Therefore, rainfall intensity (I) was divided by mean annual precipitation (MAP) (Aleotti 2004; Guzzetti et al. 2007, 2008; Dahal and Hasegawa 2008; Saito et al. 2010a) to offset regional effects and obtain comparable rainfall thresholds for different areas. The relationship between rescaled rainfall intensity (I_{MAP}) and rainfall duration (I_{MAP}–D condition) was analyzed using the same procedure described for raw values.

The MAP of mountainous areas in Taiwan is greater than 3000 mm but varies spatially (Shieh 2000). Therefore, for mountainous regions, I_{MAP} is conservatively estimated by dividing I by 3000 mm. The *I*–*D* and I_{MAP}–D thresholds for Taiwan were then compared with previously reported threshold values compiled by Saito et al. (2010a).

Time between mass movements and peak rainfall intensity

To account for the effect of temporal rainfall intensity variations on mass movements, we calculated the time difference (T_d) between mass movements and peak rainfall intensity (maximum hourly rainfall) as follows:

$$T_d = \text{time of a mass movement} - \text{time of peak rainfall intensity}. \quad (2)$$

Positive T_d values indicate that mass movements occurred after peak rainfall intensity, and negative T_d values indicate that mass movements occurred before peak rainfall intensity.

Results

Rainfall conditions and *I–D* thresholds

Mean rainfall intensity for all mass movements ranged from 8.9 to 64.8 mm/h with an average of 21.7 mm/h, and rainfall duration ranged between 2 and 71 h with an average of 33.2 h. Cumulative rainfall was as low as 60.3 mm and as high as 1916.5 mm with an average of 668.8 mm. Antecedent rainfall was between 1.8 and 707.7 mm with an average of 124.9 mm (Table 1). The *I–D* correlation (Fig. 4) shows that an increase in rainfall duration results in a reduction in the rainfall intensity required to trigger a mass movement.

Average rainfall intensity, duration, cumulative rainfall, and antecedent rainfall for debris flows were 23.2 mm/h, 33.5 h, 692.6 mm, and 101.1 mm, respectively, and those for landslides were 20.7 mm/h, 32.5 h, 652.4 mm, and 142.2 mm, respectively (Table 1). A *t*-test and resultant *p*-values indicate that mean rainfall intensity and antecedent rainfall are significantly different for landslides and debris flows (Table 1). Figure 4 also shows that the *I–D* regression line for debris flows is higher than the *I–D* regression line for landslides.

Figure 5 shows the time difference (T_d) between mass movements and peak rainfall intensity. Approximately 16 % of mass movements occurred very close to the time of peak rainfall intensity (within ±30 min), and most mass movements (75 %) occurred within 7 h of peak rainfall intensity.

As previously stated, we utilized the frequentist method with random sampling to delineate the *I–D* thresholds for all mass movements, landslides only, and debris flows only (Fig. 4). The resultant equations for these three conditions are as follows:

$$I = 18.10(\pm 2.67)D^{-0.17(\pm 0.04)}(2 \le D \le 71\text{ h; all mass movements}), \quad (3)$$

$$I = 13.25(\pm 2.76)D^{-0.08(\pm 0.06)}(2 \le D \le 71\text{ h; landslides only}), \text{ and} \quad (4)$$

$$I = 26.92(\pm 6.67)D^{-0.26(\pm 0.07)}(2 \le D \le 68\text{ h; debris flows only}). \quad (5)$$

The equation for all mass movements (Eq. 3) indicates that short-duration (e.g., <2 h) and high-intensity (e.g., >16.1 mm/h) rainfall events can potentially trigger mass movements. However, long-duration (e.g., >71 h) and low-intensity (e.g., <8.8 mm/h) rainfall events may also trigger mass movements. Figure 6 shows *I–D* thresholds for Taiwan and other global areas from Saito et al. (2010a). This shows that the *I–D* threshold for Taiwan is generally higher than that of other areas.

I_{MAP}–*D* thresholds

Figure 7 shows the I_{MAP}–D thresholds for Taiwan. The new thresholds for all mass movements, landslides only,

Table 1 Rainfall conditions

Rainfall conditions	Mean intensity (mm/h)		Duration (h)		Cumulative rainfall (mm)		Antecedent rainfall (mm)	
	Mass movements		Mass movements		Mass movements		Mass movements	
	Landslides	Debris flows	Landslides	Debris flows	Landslides	Debris flows	Landslides	Debris flows
Average	21.7		33.2		668.8		124.9	
	20.7	23.2	32.5	33.5	652.4	692.6	142.2	101.1
	$p < 0.05$		$p > 0.05$		$p > 0.05$		$p < 0.05$	
Max.	64.8		71.0		1916.5		707.7	
	64.8	58.8	71.0	68.0	1733.5	1916.5	707.3	707.7
Min.	8.9		2.0		60.3		1.8	
	8.9	9.9	2.0	2.0	60.3	64.0	1.8	3.5

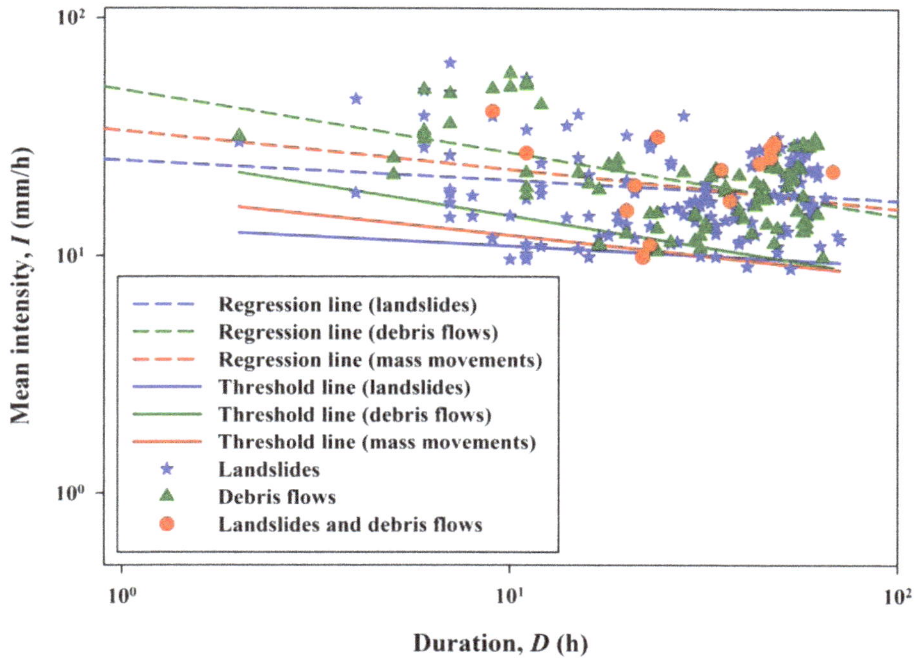

Fig. 4 *I–D* correlations and thresholds for all mass movements, landslides only, and debris flows only

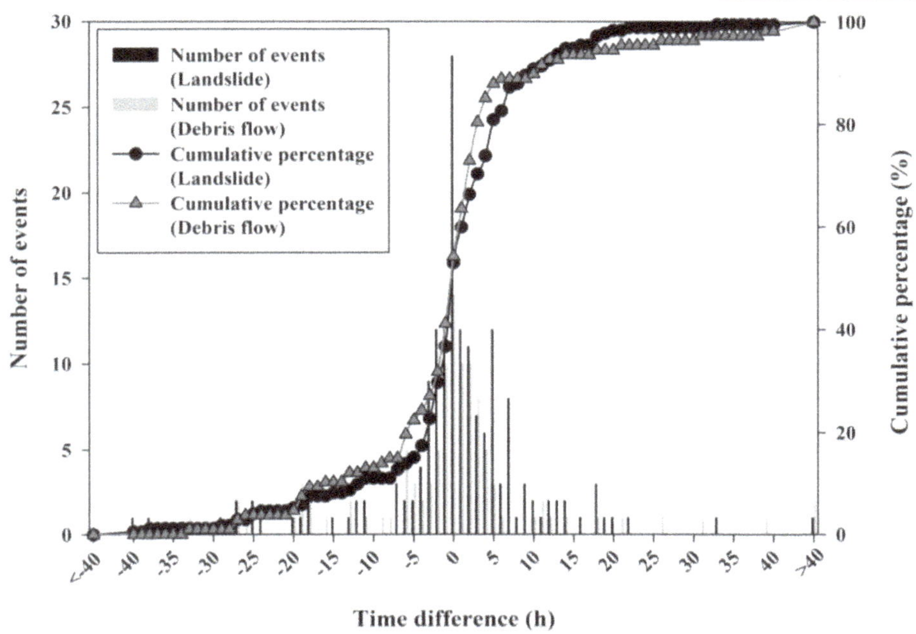

Fig. 5 Time difference between mass movement occurrence and peak rainfall intensity

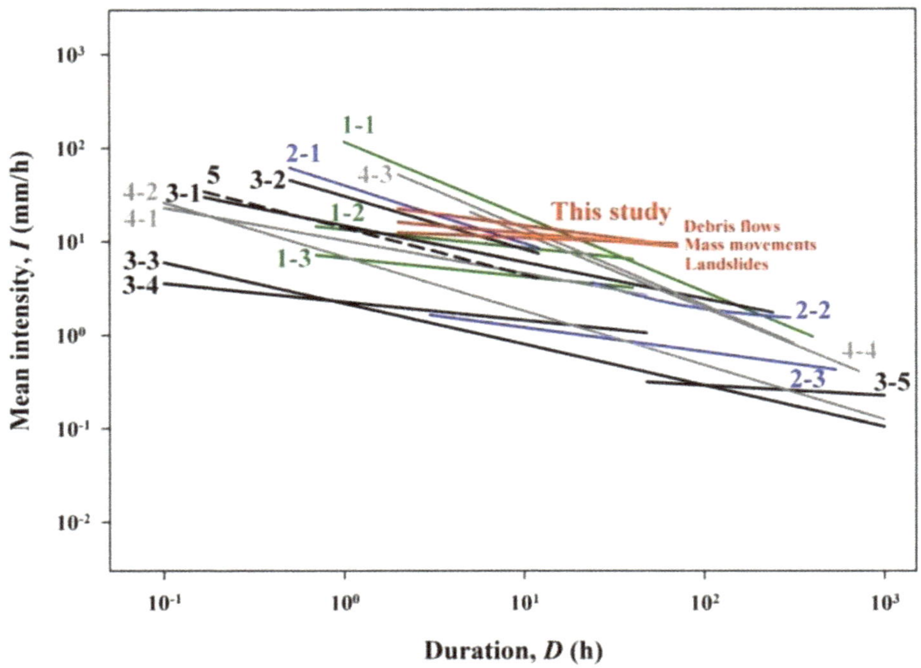

Fig. 6 Comparison of *I–D* thresholds. *Green lines* thresholds for Taiwan. *Blue lines* thresholds for Japan. *Black lines* global thresholds. *Gray lines* thresholds for humid (sub)tropics or Asian monsoon regions. *Dashed line* other regional thresholds. *1-1* Chien-Yuan et al. (2005); *1–2* and *1–3* Jan and Chen (2005); *2–1* and *3–2* Jibson (1989); *2–2* Hong et al. (2005); *2–3* Saito et al. (2010a); *3–1* Caine (1980); *3–3*, *3–4*, and *3–5* Guzzetti et al. (2008); *4–1* and *4–2* Guzzetti et al. (2008), *Cfa* climate of humid subtropical east coast in Köppen's system; *4–3* Larsen and Simon (1993), Puerto Rico; *4–4* Dahal and Hasegawa (2008), Nepal Himalaya; *5* Cannon et al. (2008), Southern California

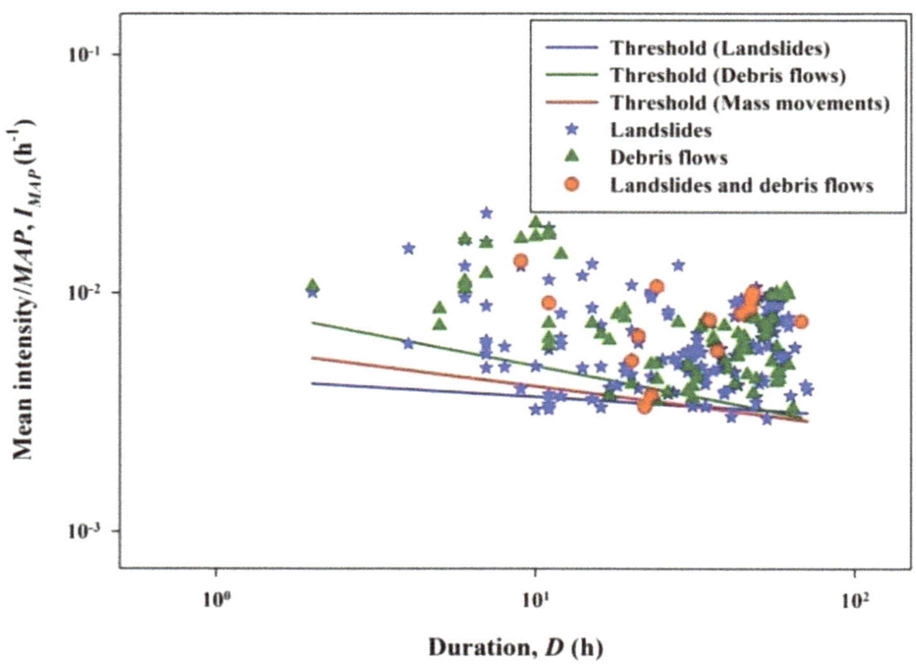

Fig. 7 I_{MAP}–*D* thresholds for all mass movements, landslides only, and debris flows only

and debris flows only are described by the following equations:

$$I_{\mathrm{MAP}} = 0.0060(\pm 0.0009)D^{-0.17(\pm 0.04)}(2 \le D \le 71 \text{ h; all mass movements}), \quad (6)$$

$$I_{\mathrm{MAP}} = 0.0044(\pm 0.0009)D^{-0.08(\pm 0.06)}(2 \le D \le 71 \text{ h; landslides only}), \text{ and} \quad (7)$$

$$I_{\mathrm{MAP}} = 0.0090(\pm 0.0022)D^{-0.26(\pm 0.07)}(2 \le D \le 68 \text{ h; debris flows only}). \quad (8)$$

Figure 8 shows the I_{MAP}–D threshold for mass movements in Taiwan alongside data from Saito et al. (2010a). After rescaling, the I_{MAP}–D threshold for Taiwan is generally lower than that of other areas (Fig. 8).

Differences in thresholds for landslides and debris flows

I–*D* and I_{MAP}–D thresholds for landslides and debris flows show that debris flows tend to occur during more severe (high intensity) rainfall conditions than landslides. In addition, *t*-tests show that the α and β of *I*–*D* and I_{MAP}–D thresholds for landslides (Eqs. 4 and 6) and debris flows (Eqs. 5 and 7) are statistically different at the 5 % significance level. However, Figs. 4 and 7 also demonstrate that the *I*–*D* correlations and thresholds for landslides and debris flows become closer as *D* increases, and their difference becomes negligible once *D* reaches approximately 50 h.

Discussion

I–*D* and I_{MAP}–*D* conditions and thresholds for landslides and debris flows

The difference in *I*–*D* and I_{MAP}–D thresholds between landslides and debris flows, as well as data in Table 1, indicate that a higher mean rainfall intensity is required to flush surface materials and cause a debris flow for short-duration rainfall events. We rule out reporting biases from our mass movement inventory based on unaltered trends obtained using the frequentist statistical and random sampling methods. Our results reflect the fact that high water levels in channels and streams are required to trigger a debris flow. A long-duration rainfall event leads to a gradual increase in groundwater level, soil moisture, and pore water pressure (Wieczorek and Glade 2005), and this leads to the concurrent increase in surface flow in channels and streams. Therefore, similar rainfall intensity levels can trigger both landslides and debris flows if rainfall duration is sufficiently long. In addition, landslides may completely or partially mobilize to form debris flows (Iverson et al. 1997), as was the case in the 16 events in which a landslide and debris flow both occurred (Figs. 4 and 7).

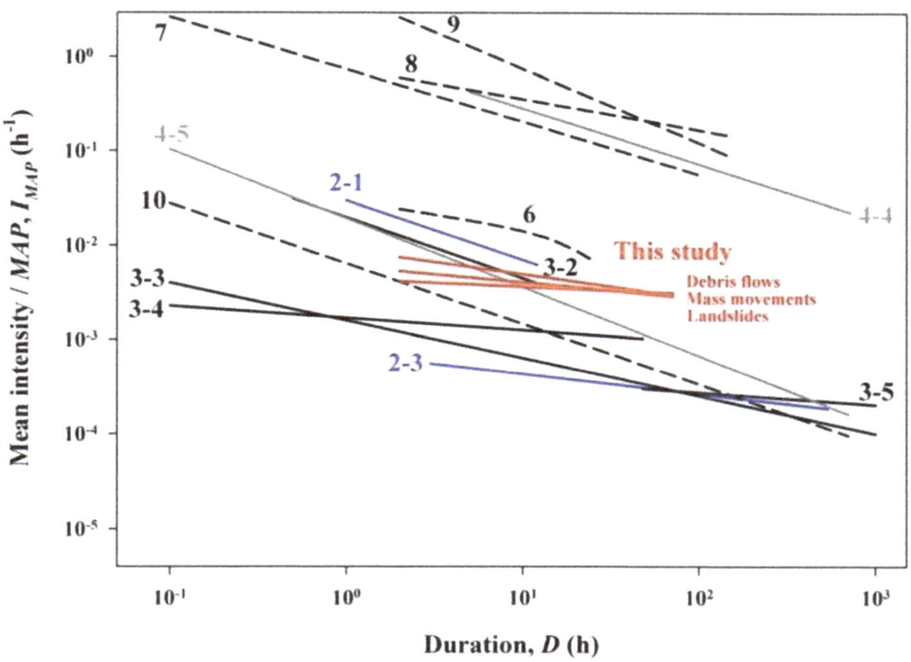

Fig. 8 Comparison of I_{MAP}–D thresholds. *Blue lines* thresholds for Japan. *Black lines* global thresholds. *Gray lines* thresholds for humid (sub)tropics or Asian monsoon regions. *Dashed lines* other regional thresholds. *2-1* and *3-2* Jibson (1989); *2-3* Saito et al. (2010a); *3-3*, *3-4*, and *3-5*: Guzzetti et al. (2008); *4-4* Dahal and Hasegawa (2008), Nepal Himalaya; *4-5* Guzzetti et al. (2007), mild mid-latitude climates; *6* Cannon (1988), San Francisco; *7* Bacchini and Zannoni (2003), Cancia, Dolomites, Italy; *8* and *9* Aleotti (2004), Piedmont, Italy; *10* Guzzetti et al. (2007), Central and Southern Europe

Comparing I–D and I_{MAP}–D thresholds for Taiwan with those of other areas

Jan and Chen (2005) established thresholds for debris flows in the Chenyoulan river catchment in central Taiwan before and after the 1999 Chi-Chi earthquake. The threshold determined in this study for the whole of Taiwan is similar to the threshold for the Chenyoulan river catchment before the Chi-Chi earthquake (1–2 in Fig. 6), lower than the threshold for Taiwan from 1989 to 2001 (1–1 in Fig. 6), and higher than the threshold for the Chenyoulan river catchment just after the Chi-Chi earthquake (1–3 in Fig. 6). These results suggest that the rainfall threshold for mass movements in Taiwan decreased in 1999 when the Chi-Chi earthquake generated an enormous quantity of loose material, which resulted in a rapid decrease of the mass movement rainfall threshold for this region. Our results indicate that the threshold recovered to pre-earthquake values during 2006–2012. Many previous studies reported similar decreases in both the maximum rainfall intensity and critical cumulative rainfall required to trigger mass movements after the Chi-Chi earthquake. As observed in this study, these values gradually returned to pre-earthquake conditions after several years (Lin et al. 2003; Chang et al. 2009; Shieh et al. 2009; Chen 2011).

The slope of the rainfall threshold line for debris flows determined in this study is similar to the value calculated by Jan and Chen (2005) but quite different from the I–D threshold value reported in Chien-Yuan et al. (2005). The later study utilized data from 61 debris flows between 1989 and 2001, obtained from newspapers and interviews with local residents, and the I–D threshold was determined visually rather than statistically. The differences in threshold line slope between our study and Chien-Yuan et al. (2005) are likely attributable to their relatively small sample size and subjective method for establishing I–D thresholds. Comparatively, we determined the I–D threshold using 107 debris flow events and employed statistical methods to ensure the validity of our results. Nevertheless, our study along with results from Chien-Yuan et al. (2005) and Jan and Chen (2005) all show debris flow thresholds larger than landslide thresholds.

Comparing our I–D thresholds with those from other areas of the world shows that the I–D threshold for Taiwan is relatively high, particularly for long-duration rainfall events (Fig. 6). As Taiwan is characterized by high-relief topography and complex geology (Fig. 1), both of which facilitate mass movements, this observation seems contradictory. In Taiwan, high cumulative precipitation amounts and frequent heavy storms, particularly those associated with typhoons (Wu and Kuo 1999), have created landforms that are well adjusted to the extreme climatic conditions of the region. Therefore, since unconsolidated hillslope materials are limited, strong rainfall is needed to trigger mass movements.

Several studies on mass movements in Japan also found that hillslopes in regions with high cumulative precipitation displayed greater resistance to rainfall than hillslopes in regions with low cumulative precipitation (Omura 1980, 1982). Saito and Matsuyama (2012) confirmed this for Japan using MAP data. They found that because the MAP is extremely high in the eastern part of the Kii Peninsula, Japan, it is not prone to landslides even with the high cumulative rainfall and maximum hourly rainfall intensity due to typhoon events. The resistance of the landscape in Taiwan to mass movements therefore explains the higher I–D threshold relative to other areas of the world. In addition, the gentle slope of the threshold in Taiwan suggests that rainfall intensity plays a much more important role in triggering mass movements, This observation further confirms the above statement that landforms are well adjusted to the extreme climatic conditions.

As indicated above, the I_{MAP}–D threshold is useful for eliminating biases caused by variable rainfall conditions. Even though we used a relatively conservative MAP value of 3000 mm to rescale the data, the I_{MAP}–D threshold for Taiwan was generally lower than in other areas. Therefore, the steep topography and highly deformed underlying bedrock of Taiwan likely contribute to frequent mass movements in the region. In summary, analysis of both I–D and I_{MAP}–D thresholds is useful for examining the effects of topography, geology, and hillslope adjustment to rainfall conditions on mass movements.

Mass movements in relation to rainfall peaks and antecedent rainfall

I–D plots represent average rainfall event conditions and do not necessarily reflect high rainfall intensities at the time of mass movement occurrence. Therefore, it is difficult to predict the exact time of a mass movement using only I–D thresholds. Many studies have found that peak rainfall intensity occurs very close to the time of mass movement occurrence (Aleotti 2004; Chien-Yuan et al. 2005; Guzzetti et al. 2007, 2008; Dahal and Hasegawa 2008; Saito et al. 2010a, 2010b). Thus, analyzing peak rainfall intensity in conjunction with I–D thresholds is useful for characterizing mass movements.

Most mass movement occurrences investigated were concentrated around the time of peak rainfall intensity; approximately 75 % of mass movements occurred within 7 h of peak rainfall (Fig. 5). To focus on the remaining 25 % of mass movements that occurred more than 7 h before or after peak rainfall, we classified mass movements into three groups: $T_d < -7$ (group 1 = G1), $T_d = -7$ to 7 (G2), and $T_d > 7$ (G3). Table 2 shows average peak intensity, mean

Table 2 Average rainfall conditions for different T_d groups. Italic values show groups with significantly higher values using a *t*-test ($p < 0.05$)

Rainfall condition groups	Peak intensity (mm/h)		Mean intensity (mm/h)		Duration (h)		Antecedent rainfall (mm)	
	Landslides	Debris flows	Landslides	Debris flows	Landslides	Debris flows	Landslides	Debris flows
G1: $T_d < -7$ ($n = 19, 16$)	71.5	65.6	16.8	13.7	32.6	29.9	79.4	102.0
G2: $T_d = -7$ to 7 ($n = 131, 79$)	80.7	*90.3*	*21.9*	*25.9*	30.8	32.2	*159.3*	90.1
G3: $T_d > 7$ ($n = 22, 12$)	60.4	56.6	18.7	18.1	*45.6*	*46.9*	94.7	172.0

intensity, duration, and antecedent rainfall for these three groups.

G2 has the highest peak and mean rainfall intensities for both landslides and debris flows, and rainfall values tend to be significantly different from the other groups. Antecedent rainfall for landslides in G2 is also significantly higher than other groups. This indicates that G2 is composed primarily of landslides, with release mechanisms directly related to hillslope hydrology. Soil moisture content is directly related to landslide occurrence (Terlien 1998; Wu and Chen 2009), and high antecedent rainfall elevates moisture content (Glade 1997; Crosta 1998; Crozier 1999; Glade et al. 2000; Jakob 2006). Therefore, abundant antecedent rainfall combined with high-intensity rainfall can easily trigger landslides, especially around the time of peak rainfall. In addition, during or just after peak rainfall, high-intensity rainfall flushes large amounts of debris into rivers, triggering debris flows. In this case, abundant antecedent rainfall is not a requirement because debris flows can be triggered by riverbed material saturated with surface water (Berti et al. 1999; Coe et al. 2008), not necessarily hillslope material saturated with underground soil water.

G3 has the longest rainfall duration for both landslides and debris flows. Although G3 events are characterized by low peak and mean rainfall intensities, rainfall durations are relatively long, resulting in mass movements long after peak rainfall. Terlien (1998) also showed that when there is no significant antecedent rainfall and rainfall intensity is low, a long-duration rainfall event is required to trigger landslides with high soil moisture content.

According to Table 2, G1 events are associated with relatively short-duration rainfall, with low peak and mean rainfall intensities, and low antecedent rainfall, meaning that mass movements occurred in spite of these conditions. This is an unexpected result; our first assumption was that mass movements well before the rainfall peak must reflect abundant antecedent rainfall, which was found to be invalid. For G1, the role of factors other than rainfall, such as local topography or geology highly susceptible to mass movements, seems to be more important. However, the result may also be due to the definition of antecedent rainfall. Although antecedent rainfall was calculated over a period of 168 h, even earlier rainfall might have played a key role in triggering the G1 mass movements. Indeed, the definition of antecedent rainfall is a difficult task (Crozier 1999; Glade et al. 2000; Dahal and Hasegawa 2008), and future research is needed to refine this definition and further investigate the effect of topographical and geological factors in triggering mass movements.

Conclusions

This study established empirical I–D thresholds for mass movements in Taiwan from landslide and debris flow data, and these thresholds were compared with those for other areas of the world. For short-duration rainfall events, higher mean rainfall intensities were required to trigger debris flows, while long-duration rainfall events can trigger both landslides and debris flows with almost the same rainfall intensity. I–D thresholds for mass movements in Taiwan tended to be higher than those for other areas, but the rescaled I_{MAP}–D thresholds (in which rainfall intensity was rescaled using a MAP of 3000 mm) tended to be lower. These results reflect the rainfall characteristics of the region along with the topographic and geological conditions of Taiwan.

This study also demonstrated the importance of peak rainfall intensity. Most mass movements (74.9 %) occurred within 7 h of peak rainfall. Antecedent rainfall and saturated hillslope material are also important for triggering landslides, but antecedent rainfall is less important in triggering debris flows from river bed material. Other events that occurred significantly before or after peak rainfall may be related to the rainfall duration and non-meteorological factors such as local topography and geology. Further studies are necessary to investigate I–D thresholds and related issues in more detail.

Abbreviations

I: rainfall intensity; D: rainfall duration; SWCB: Soil and Water Conservation Bureau; MAP: mean annual precipitation; I_{MAP}: rescaled rainfall intensity using mean annual precipitation; T_d: time difference between mass movements and peak rainfall intensity.

Competing interests

The authors declare that they have no competing interests.

Authors' contributions

CC collected the data, performed the statistical analysis, and drafted the manuscript. HS participated in the design of the study, provided statistical methods, and helped to draft the manuscript. TO conceived of the study, participated in its design and coordination, and helped to draft the manuscript. All authors read and approved the final manuscript.

Acknowledgements

We would like to thank the Typhoon and Flood Research Institute (TTFRI) of Taiwan for providing rainfall data. We would also like to thank the Soil and Water Conservation Bureau (SWCB) of Taiwan for providing mass movement data.

Author details

[1]Department of Natural Environmental Studies, Graduate School of Frontier Science, The University of Tokyo, 5-1-5 Kashiwanoha, Kashiwa Chiba 277-8568, Japan. [2]College of Economics, Kanto Gakuin University, 1-5-1 Mutsuura-higashi, Yokohama, Kanagawa 236-8501, Japan. [3]Center for Spatial Information Science, University of Tokyo, 5-1-5 Kashiwanoha, Kashiwa Chiba 277-8568, Japan.

References

Aleotti P (2004) A warning system for rainfall-induced shallow failures. Eng Geol 73:247–265. doi:10.1016/j.enggeo.2004.01.007

Aleotti P, Chowdhury R (1999) Landslide hazard assessment: summary review and new perspectives. Bull Eng Geol Env 58:21–44. doi:10.1007/s100640050066

Bacchini M, Zannoni A (2003) Relations between rainfall and triggering of debris-flow: case study of Cancia (Dolomites, Northeastern Italy). Natural Hazard Earth Sys 3:71–79. doi:10.5194/nhess-3-71-2003

Berti M, Genevois R, Simoni A, Tecca PR (1999) Field observations of a debris flow event in the Dolomites. Geomorphology 29:265–274. doi:10.1016/S0169-555X(99)00018-5

Brunetti MT, Peruccacci S, Rossi M, Luciani S, Valigi D, Guzzetti F (2010) Rainfall thresholds for the possible occurrence of landslides in Italy. Nat Hazards Earth Syst Sci 10:447–458. doi:10.5194/nhess-10-447-2010

Caine N (1980) The rainfall intensity–duration control of shallow landslides and debris flows. Geogr Ann A 62:23–27. doi:10.2307/520449

Cannon S (1988) Regional rainfall–threshold conditions for abundant debris-flow activity. In: Ellen SD, Wieczorek GF (eds) Landslides, floods, and marine effects of the storm of January 3–5, 1982, in the San Francisco Bay Region, California. US Geological Survey Professional Paper, vol 1434:35–42

Cannon S, Gartner J, Wilson R, Bowers J, Laber J (2008) Storm rainfall conditions for floods and debris flows from recently burned areas in southwestern Colorado and southern California. Geomorphology 96:250–269. doi:10.1016/j.geomorph.2007.03.019

Chang CW, Lin PS, Tsai CL (2011) Estimation of sediment volume of debris flow caused by extreme rainfall in Taiwan. Eng Geol 123:83–90. doi:10.1016/j.enggeo.2011.07.004

Chang FJ, Chiang YM, Lee WS (2009) Investigating the impact of the Chi-Chi earthquake on the occurrence of debris flows using artificial neural networks. Hydrol Process 23:2728–2736. doi:10.1002/hyp.7369

Chang JC, Slaymaker O (2002) Frequency and spatial distribution of landslides in a mountainous drainage basin: Western Foothills, Taiwan. Catena 46:285–307. doi:10.1016/S0341-8162(01)00157-6

Chang K, Chiang S (2009) An integrated model for predicting rainfall-induced landslides. Geomorphology 105:366–373. doi:10.1016/j.geomorph.2008.10.012

Chen H, Chen RH, Lin ML (1999) Initiative anatomy of Tungmen debris flow, eastern Taiwan. Environ Eng Geosci 5:459–473. doi:10.2113/gseegeosci.V.4.459

Chen H, Su DI (2001) Geological factors for hazardous debris flow in Hoser, central Taiwan. Environ Geol 40:1114–1124. doi:10.1007/s002540100312

Chen JC (2011) Variability of impact of earthquake on debris-flow triggering conditions: case study of Chen-Yu-Lan watershed, Taiwan. Environ Earth Sci 64:1787–1794. doi:10.1007/s12665-011-0981-4

Cheng JD, Huang YC, Wu HL, Yeh JL, Chang CH (2005) Hydrometeorological and land use attributes of debris flows and debris floods during typhoon Toraji, July 29–30, 2001 in central Taiwan. J Hydrol 306:161–173. doi:10.1016/j.jhydrol.2004.09.007

Chien-Yuan C, Tien-Chien C, Fan-Chieh Y, Wen-Hui Y, Chun-Chieh T (2005) Rainfall duration and debris-flow initiated studies for real-time monitoring. Environ Geol 47:715–724. doi:10.1007/s00254-004-1203-0

Chuang SC, Chen H, Lin GW, Lin CW, Chang CP (2009) Increase in basin sediment yield from landslides in storms following major seismic disturbance. Eng Geol 103:59–65. doi:10.1016/j.enggeo.2008.08.001

Claessens L, Knapen A, Kitutu MG, Poesen J, Deckers JA (2007) Modelling landslide hazard, soil redistribution and sediment yield of landslides on the Ugandan footslopes of Mount Elgon. Geomorphology 90:23–35. doi:10.1016/j.geomorph.2007.01.007

Coe J, Kinner D, Godt J (2008) Initiation conditions for debris flows generated by runoff at Chalk Cliffs, central Colorado. Geomorphology 96:270–297. doi:10.1016/j.geomorph.2007.03.017

Crosta G (1998) Regionalization of rainfall thresholds: an aid to landslide hazard evaluation. Environ Geol 35:131–145. doi:10.1007/s002540050300

Crozier MJ (1999) Prediction of rainfall-triggered landslides: a test of the antecedent water status model. Earth Surf Proc Land 24:825–833. doi:10.1002/(SICI)1096-9837(199908)24:9<825::AID-ESP14>3.3.CO;2-D

Dadson SJ, Hovius N, Chen H, Dade WB, Hsieh ML, Willett SD et al (2003) Links between erosion, runoff variability and seismicity in the Taiwan orogeny. Nature 426:648–651. doi:10.1038/nature02150

Dadson SJ, Hovius N, Chen H, Dade WB, Lin JC, Hsu ML et al (2004) Earthquake-triggered increase in sediment delivery from an active mountain belt. Geology 32:733–736. doi:10.1130/G20639.1

Dahal R, Hasegawa S (2008) Representative rainfall thresholds for landslides in the Nepal Himalaya. Geomorphology 100:429–443. doi:10.1016/j.geomorph.2008.01.014

Dai FC, Lee CF, Ngai YY (2002) Landslide risk assessment and management and management: an overview. Eng Geol 64:65–87. doi:10.1016/S0013-7952(01)00093-X

Dou J, Oguchi T, Hayakawa YS, Uchiyama S, Saito H, Paudel U (2014) GIS-based landslide susceptibility mapping using a certainty factor model and its validation in the Chuetsu area, central Japan. Landslide Science for a Safer Geoenvironment, Springer International Publishing:419–424. doi:10.1007/978-3-319-05050-8_65

Glade T (1997) The temporal and spatial occurrence of rainstorm-triggered landslide events in New Zealand. Victoria University of Wellington, Dissertation

Glade T, Crozier M, Smith P (2000) Applying probability determination to refine landslide-triggering rainfall thresholds using an empirical "antecedent daily rainfall model". Pure Appl Geophys 157:1059–1079. doi:10.1007/s000240050017

Guzzetti F, Carrara A, Cardinali M, Reichenbach P (1999) Landslide hazard evaluation: a review of current techniques and their application in a multi-scale study, central Italy. Geomorphology 31:181–216. doi:10.1016/S0169-555X(99)00078-1

Guzzetti F, Peruccacci S, Rossi M, Stark C (2007) Rainfall thresholds for the initiation of landslides in central and southern Europe. Meteorol Atmos Phys 98:239–267. doi:10.1007/s00703-007-0262-7

Guzzetti F, Peruccacci S, Rossi M, Stark C (2008) The rainfall intensity–duration control of shallow landslides and debris flows: an update. Landslides 5:3–17. doi:10.1007/s10346-007-0112-1

Ho CS (1986) A synthesis of the geologic evolution of Taiwan. Tectonophysics 125:1–16. doi:10.1016/0040-1951(86)90004-1

Ho CS (1988) An introduction to the geology of Taiwan, explanatory text of the geologic map of Taiwan. Central Geological Survey, the Ministry of Economic Affairs, Taipei, Taiwan

Hong Y, Hiura H, Shino K, Sassa K, Suemine A, Fukuoka H et al (2005) The influence of intense rainfall on the activity of large-scale crystalline schist landslides in Shikoku Island, Japan. Landslides 2:97–105. doi:10.1007/s10346-004-0043-z

Hovius N, Stark CP, Allen PA (1997) Sediment flux from a mountain belt derived by landslide mapping. Geology 25:231–234. doi:10.1130/0091-7613(1997)025<0231:SFFAMB>2.3.CO;2

Iverson RM, Reid ME, LaHusen RG (1997) Debris-flow mobilization from landslides. Annu Rev Earth Pl Sc 25:85–138. doi:10.1146/annurev.earth.25.1.85

Jakob M (2006) Hydrometeorological thresholds for landslide initiation and forest operation shutdowns on the north coast of British Columbia. Landslides 3:228–238. doi:10.1007/s10346-006-0044-1

Jakob M, Weatherly H (2003) A hydroclimatic threshold for landslide initiation on the North Shore Mountains of Vancouver, British Columbia. Geomorphology 54:137–156. doi:10.1016/S0169-555X(02)00339-2

Jan CD, Chen CL (2005) Debris flows caused by typhoon Herb in Taiwan. In: Jakob M, Hungr O (eds) Debris flow hazards and related phenomena. Springer, Berlin Heidelberg, pp 539–563. doi:10.1007/3-540-27129-5_21

Jibson R (1989) Debris flow in southern Puerto Rico. Geological Society of America, Special Paper, vol 236:29–55. doi:10.1130/SPE236-p29

Keefer D, Wilson R, Mark R, Brabb E, Brown W, Ellen S et al (1987) Real-time landslide warning during heavy rainfall. Science 238:921–925. doi:10.1126/science.238.4829.921

Kim SK, Hong WP, Kim YM (1992) Prediction of rainfall-triggered landslides in Korea. In: Bell DH (ed) Landslides, Proceedings of the Sixth International Symposium on Landslides, Christchurch, 10–14 February 1992, vol 2. Balkema, Rotterdam, pp 989–994

Kirschbaum D, Adler R, Adler D, Peters-Lidard C, Huffman G (2012) Global distribution of extreme precipitation and high-impact landslides in 2010 relative to previous years. J Hydrometeorol 13:1536–1551. doi:10.1175/JHM-D-12-02.1
Korup O, Gorum T, Hayakawa Y (2012) Without power? Landslide inventories in the face of climate change. Earth Surf Proc Land 37:92–99. doi:10.1002/esp.2248
Larsen M, Simon A (1993) A rainfall intensity–duration threshold for landslides in a humid-tropical environment: Puerto Rico. Geogr Ann A 75:13–23. doi:10.2307/521049
Li YH (1976) Denudation of Taiwan island since the Pliocene epoch. Geology 4:105–107. doi:10.1130/0091-7613(1976)4<105:DOTIST>2.0.CO;2
Lin CW, Shieh CL, Yuan BD, Shieh YC, Liu SH, Lee SY (2003) Impact of Chi-Chi earthquake on the occurrence of landslides and debris flows: example from the Chenyulan River watershed, Nantou, Taiwan. Eng Geol 71:49–61. doi:10.1016/S0013-7952(03)00125-X
Lin GW, Chen H (2012) The relationship of rainfall energy with landslides and sediment delivery. Eng Geol 125:108–118. doi:10.1016/j.enggeo.2011.11.010
Omura H (1980) Fundamental study on forecasting of landslide area (I) – relation between total precipitation, resistance index and landslide area ratio –. Journal of the Japan Society of Erosion Control Engineering 33:15–25 (in Japanese with English abstract). doi:10.11475/sabo1973.33.2_15
Omura H (1982) Forecast method of landslide area ratio by gamma distribution model. Journal of the Japan Society of Erosion Control Engineering 35:31–37 (in Japanese with English abstract). doi:10.11475/sabo1973.35.3
Onodera T, Yoshinaka R, Kazama H (1974) Slope failures caused by heavy rainfall in Japan. Proc. 2nd Int. Congress of the Int Ass Eng Geol, v. 11. San Paulo 11:1–10. doi:10.5110/jjseg.15.191
Petley D (2012) Global patterns of loss of life from landslides. Geology 40:927–930. doi:10.1130/G33217.1
Saito H, Nakayama D, Matsuyama H (2010a) Relationship between the initiation of a shallow landslide and rainfall intensity–duration thresholds in Japan. Geomorphology 118:167–175. doi:10.1016/j.geomorph.2009.12.016
Saito H, Nakayama D, Matsuyama H (2010b) Two types of rainfall conditions associated with shallow landslide initiation in Japan as revealed by normalized soil water index. SOLA 6:57–60. doi:10.2151/sola.2010-015
Saito H, Matsuyama H (2012) Catastrophic landslide disasters triggered by record-breaking rainfall in Japan: their accurate detection with normalized soil water index. SOLA 8:81–84. doi:10.2151/sola.2012-021
Saito H, Korup O, Uchida T, Hayashi S, Oguchi T (2014) Rainfall conditions, typhoon frequency, and contemporary landslide erosion in Japan. Geology 42:999–1002. doi:10.1130/G35680.1
Shieh CL, Chen YS, Tsai YJ, Wu JH (2009) Variability in rainfall threshold for debris flow after the Chi-Chi earthquake in central Taiwan, China. Int J Sed Res 24:177–188. doi:10.1016/S1001-6279(09)60025-1
Shieh SL (2000) User's Guide for Typhoon Forecasting in the Taiwan Area (VIII). Central Weather Bureau, Taipei
Teng LS (1990) Geotectonic evolution of late Cenozoic arc-continent collision in Taiwan. Tectonophysics 183:57–76. doi:10.1016/0040-1951(90)90188-E
Terlien MT (1998) The determination of statistical and deterministic hydrological landslide-triggering thresholds. Environ Geol 35:124–130. doi:10.1007/s002540050299
Wang B, Ho L (2002) Rainy season of the Asian-Pacific summer monsoon. J Climate 15:386–398. doi:10.1175/1520-0442(2002)015<0386:RSOTAP>2.0.CO;2
Wieczorek G, Glade T (2005) Climatic factors influencing occurrence of debris flows. In: Jakob M, Hunger O (eds) Debris-flow Hazards and Related Phenomena. Springer, Berlin, pp 325–362. doi:10.1007/3-540-27129-5_14
Willett SD, Fisher D, Fuller C, Yeh EC, Lu CY (2003) Erosion rates and orogenic wedge kinematics in Taiwan inferred from apatite fission track thermochronometry. Geology 31:945–948. doi:10.1130/G19702.1
Wu CC, Kuo YH (1999) Typhoons affecting Taiwan: current understanding and future challenges. B Am Meteorol Soc 80:67–80. doi:10.1175/1520-0477(1999)080<0067:TATCUA>2.0.CO;2
Wu CH, Chen SC (2009) Determining landslide susceptibility in central Taiwan from rainfall and six site factors using the analytical hierarchy process method. Geomorphology 112:190–204. doi:10.1016/j.geomorph.2009.06.002
Yu SB, Chen HY, Kuo LC (1997) Velocity field of GPS stations in the Taiwan area. Tectonophysics 274:41–59. doi:10.1016/S0040-1951(96)00297-1

11

Towards scaling laws for subduction initiation on terrestrial planets: constraints from two-dimensional steady-state convection simulations

Teresa Wong* and Viatcheslav S Solomatov

Abstract

The strongly temperature-dependent viscosity of rocks leads to the formation of nearly rigid lithospheric plates. Previous studies showed that a very low yield stress might be necessary to weaken and mobilize the plates, for example, due to water. However, the magnitude of the yield stress remains poorly understood. While the convective stresses below the lithosphere are relatively small, sublithospheric convection can induce large stresses in the lithosphere indirectly, through thermal thinning of the lithosphere. The magnitude of the thermal thinning, the stresses associated with it, and the critical yield stress to initiate subduction depend on several factors including the viscosity law, the Rayleigh number, and the aspect ratio of the convective cells. We conduct a systematic numerical analysis of lithospheric stresses and other convective parameters for single steady-state convection cells. Such cells can be considered as part of a multi-cell, time-dependent convective system. This allows us a better control of convective solutions and a relatively simple scaling analysis. We find that subduction initiation depends much stronger on the aspect ratio than in previous studies and speculate that plate tectonics initiation may not necessarily require significant weakening and can, at least in principle, start if a sufficiently long cell develops during planetary evolution.

Keywords: Convection; Stress; Lithosphere; Subduction; Plate tectonics

Background

Plate tectonics is central to many aspects of the geology and evolution of terrestrial planets. While Earth is the only planet where plate tectonics is observed, its driving mechanism and timing of initiation are still poorly understood. Subduction is thought to be the fundamental process for plate tectonics initiation because the slab pull of subducting slab contributes most to the forces that drive plate movements. On the Earth, initiation of subduction is greatly facilitated by tectonic forces associated with plate motions *already occurring elsewhere* (Mueller and Phillips 1991; Hall et al. 2003). Various models for subduction initiation has been proposed (e.g., McKenzie 1977; Turcotte 1977; Ogawa 1990; Mueller and Phillips 1991; Kemp and Stevenson 1996; Toth and Gurnis 1998; Stern 2004; Solomatov 2004a; Ueda et al. 2008; Nikolaeva et al. 2010), many of which involve existing plate boundaries or weak zones. Incipient subduction zones are often found near transform faults or fracture zones because of their physical weakness (e.g., Mueller and Phillips 1991; Gurnis et al. 2004).

On one-plate planets such as Venus and Mars, the absence of plate tectonics is likely to be due to the difficulty of subduction initiation in the absence of forces due to plate motions. In other words, the problem of plate tectonics initiation can be viewed as the problem of the very first occurrence of subduction. Due to the high sensitivity of viscosity to temperature, the lithosphere acts as the cold rigid thermal boundary layer that has a very high strength. On these planets, mantle convection is likely to be in the stagnant lid regime (e.g., Morris and Canright 1984; Fowler 1985; Solomatov 1995). One possible mechanism for the very first episode of subduction is due to the lithospheric stresses generated by mantle convection (Ogawa

*Correspondence: twong@levee.wustl.edu
Department of Earth and Planetary Sciences, Washington University, 1 Brookings Drive, 63130-4899 St. Louis, MO, USA

1990; Fowler and O'Brien 2003; Solomatov 2004a). The magnitude of these stresses is relatively small compared to the lithospheric strength suggested by laboratory and field observations (e.g., Kohlstedt et al. 1995; Gurnis et al. 2004), and thus, it is usually believed that to initiate subduction, some weakening mechanisms must be present in the lithosphere.

Much effort has been devoted to understand the weakening mechanisms of the lithosphere. Several studies showed that the frictional shear stress resisting subduction at transform faults and fracture zones have to be less than 10 MPa for subduction to occur (Toth and Gurnis 1998; Hall et al. 2003; Gurnis et al. 2004). Stress drop estimates from earthquakes also indicate that fault strength may be approximately 10 MPa (Kanamori 1994; Kanamori and Brodsky 2004). Models are able to describe global reduction in the lithospheric strength, as well as localized weak zones such that plate-like features can be generated from mantle convection in a self-consistent manner (e.g., Trompert and Hansen 1998; Moresi and Solomatov 1998; Tackley 2000a; Bercovici 2001; Branlund et al. 2001; Regenauer-Lieb et al. 2001; Regenauer-Lieb and Kohl 2003; Regenauer-Lieb et al. 2006; Korenaga 2007; Landuyt et al. 2008). Various approaches have been used to deal with the creation of weak zones (Bercovici 2001; Bercovici and Ricard 2005; Landuyt et al. 2008; Branlund et al. 2001; Regenauer-Lieb and Kohl 2003). The two-phase damage theory with a grain-size dependent rheology was developed to explain the formation of weak plate boundaries and track the evolution of deformation (e.g., Bercovici and Ricard 2005; Landuyt et al. 2008; Bercovici 2012). Some studies suggested that water might play an important role in the localization of deformation (Regenauer-Lieb et al. 2001; Regenauer-Lieb and Kohl 2003; Regenauer-Lieb et al. 2006). Water also weakens the lithosphere by lowering the activation energy (Regenauer-Lieb et al. 2001; Regenauer-Lieb and Yuen 2004; Regenauer-Lieb et al. 2006) and increasing the pore fluid pressure (Kohlstedt et al. 1995).

One approach to quantify the weakening of lithosphere is to set a yield value to the rheology of the lithosphere to simulate brittle behavior (Fowler 1993; Trompert and Hansen 1998; Moresi and Solomatov 1998; Richards et al. 2001; Tackley 2000a,b; Fowler and O'Brien 2003; Solomatov 2004a; Stein et al. 2004; O'Neill et al. 2007; Stein and Hansen 2008). The yield stress can be regarded as a simplification of mechanisms that describe the strength of the lithosphere. Convection with yield stress is usually categorized into three regimes: mobile lid regime, transitional regime with some episodic failure, and stagnant lid regime (Moresi and Solomatov 1998; Tackley 2000b; Stein et al. 2004). Stein and Hansen (2008) further subdivided the transitional regime into episodically mobile and stable plate mobilization regimes. To access the conditions of a planet to have plate tectonics, some researchers presented regime diagrams in terms of Rayleigh number, viscosity contrast, and yield stress (e.g., Stein et al. 2004; O'Neill and Lenardic 2007).

A number of studies attempted to derive scaling relations for convective stresses and yield stress to extrapolate to planetary conditions (e.g., Moresi and Solomatov 1998; Fowler and O'Brien 2003; Solomatov 2004a,b; O'Neill et al. 2007; Valencia and O'Connell 2009; Korenaga 2010b; Van Heck and Tackley 2011; Stamenkovic and Breuer 2014). Yet the accurate description of these convection-induced stresses inside the lithosphere and thus the yield stress is lacking.

This study seeks to understand the stress distribution of the steady-state convecting cell with respect to various convective parameters using the pseudoplastic rheology as a first step. The goal of this study is to find a scaling law for the lithospheric stress (hereafter referred as lid stress) and the critical yield stress, which is the highest yield value at which the stagnant lid could be mobilized. Note that an alternative and, perhaps, more intuitive approach would be to assume that the yield stress is known and to try to figure out under what dynamic conditions it can be reached. However, (a) the 'normal' yield stress is so high that it is nearly impossible to reach and (b) given the uncertainties in the weakening mechanisms and thus the actual magnitude of the yield stress, it should be treated as an unknown.

In this study, we first examine the stress structure in steady-state stagnant lid convection and explore scaling relationships between convective parameters especially in relation to aspect ratio to develop a scaling theory for lid stress and critical yield stress. We then compare the theoretical scaling laws with numerical results. In addition, we investigate the accuracy of the Frank-Kamenetskii approximation for modeling the initiation of plate tectonics.

Rheology

Viscous creep governs the flow in the mantle as it has high temperatures and low stresses. It can be described by a constitutive relation (Hirth and Kohlstedt 2003), which is an Arrhenius function of temperature T, activation energy E, pressure P, and activation volume V with power law dependences on stress τ (second invariant of stress tensor), grain size d, water fugacity f_{H_2O}, and an exponential function of melt fraction ϕ:

$$\eta = A\tau^{1-n}d^m f_{H_2O}^r \exp(-\alpha\phi)\exp\left(\frac{E+PV}{RT}\right), \quad (1)$$

where A and α are constants, R is the gas constant, and m, n, and r are exponents for grain size, stress, and water fugacity, respectively. Depending on the temperature, grain size, stress, pressure, and composition, the dominating deformation mechanism in the mantle would be different (Karato and Wu 1993; Karato et al. 1995; Hirth and Kohlstedt 2003). In the lithosphere, the major factor controlling the viscosity is temperature. Thus, the viscosity function to investigate subduction initiation is often written as:

$$\eta = A' \exp\left(\frac{E}{RT}\right). \quad (2)$$

Frank-Kamenetskii approximation

Many numerical studies used use a relatively low viscosity contrast to observe plate behavior, which has limited applications to realistic planetary situations. Moresi and Solomatov (1998) investigated the convective regimes with viscosity contrast ranging from 3×10^4 to 3×10^7, and in Tackley (2000a), the viscosity contrast was limited to 10^4, whereas Richards et al. (2001) and Stein and Hansen (2008) used viscosity contrast on the order of 10^5. The viscosity contrast across the terrestrial lithosphere is many orders of magnitude higher.

The low-viscosity contrast is used because high-viscosity contrasts are difficult to treat in numerical calculations (Moresi and Solomatov 1995). Thus, the Arrhenius function is often approximated by the Frank-Kamenetskii function, which reduces the viscosity contrast by many orders of magnitude compared to Arrhenius viscosity function. This makes the problem of convection with strongly temperature-dependent viscosity more computationally tractable.

Frank-Kamenetskii approximation originated from the combustion theory. Frank-Kamenetskii pointed out that since the activation energy E is large, we can consider the rate of reaction only in a narrow range of temperature around the combustion temperature (Frank-Kamenetskii 1969). The equation for the rate of reaction is similar to the strain rate in the constitutive relations, which also has an Arrhenius form $\exp(-E/RT)$. Since convection mostly takes place in the interior of the cell where the temperature is close to the interior temperature T_i, we use the same approximation by expanding the exponent E/RT in the Arrhenius form so that the viscosity can be expressed as an exponential function of temperature only:

$$\eta = B \exp(-\gamma T), \quad (3)$$

where B and γ are constants. In the interior:

$$\eta_{i,\mathrm{Arr}} = \eta_{i,\exp}, \quad (4)$$

$$\left(\frac{d\eta_{i,\mathrm{Arr}}}{dT}\right)_{T=T_i} = \left(\frac{d\eta_{i,\exp}}{dT}\right)_{T=T_i}, \quad (5)$$

where η_{Arr} and $\eta_{\exp}$ are the interior viscosities of the Arrhenius function and that of Equation 3 (hereafter referred as exponential viscosity), respectively. Equations 4 and 5 give γ in terms of activation energy and interior temperature:

$$\gamma = \frac{E}{RT_i^2}. \quad (6)$$

This method of expanding the terms in the exponent preserves the interior viscosity and the change of viscosity with temperature close to T_i, where convection actively takes place. Some studies expanded the terms inside the exponents differently (e.g., King 2009). However, it is important to use Equations 4 and 5 to ensure the asymptotic accuracy of Frank-Kamenetskii approximation (Morris 1982; Morris and Canright 1984; Fowler 1985; Frank-Kamenetskii 1969).

Frank-Kamenetskii approximation was shown to be sufficiently accurate for the interior of the convective layer with large viscosity contrast (Solomatov and Moresi 1996; Ratcliff et al. 1997; Reese et al. 1999). Recent studies have examined convection with Arrhenius rheology and suggested slightly different scaling laws compared to convection with Frank-Kamenetskii viscosity (Korenaga 2009; Stein and Hansen 2013). Here, we assess the accuracy of the Frank-Kamenetskii approximation in predicting the values of critical yield stress.

Pseudoplastic rheology and plastic yielding

The brittle behavior of the lithosphere can be simplified with a viscoplastic rheology that causes yielding when the convective stresses exceed a plastic yield stress τ_y (Moresi and Solomatov 1998; Trompert and Hansen 1998; Tackley 2000b; Fowler and O'Brien 2003). The yield stress can be defined by Byerlee's law (Byerlee 1978):

$$\tau_y = \tau_0 + \mu \rho g z, \quad (7)$$

where τ_0 is the yield stress at the zero hydrostatic pressure, μ is the frictional coefficient, and $\rho g z$ is the hydrostatic pressure. Viscous deformation occurs according to Equation 3 when stresses are less than the yield stress. Above the yield stress, the deformation follows a plastic flow law defined by a non-linear effective viscosity:

$$\eta_{\mathrm{eff}} = \frac{\tau_y}{\dot{e}}, \quad (8)$$

where $\dot{e}$ is the second invariant of the strain rate tensor. The yield stress defines a change on deformation mechanism based on the second invariant of the deviatoric stress tensor, which corresponds to the Von Mises yield criterion. In this study, we consider two types of yield stress: a constant yield stress τ_y or a depth-dependent yield stress with a constant gradient τ'_y.

Methods

Formulation of the problem

Equations of thermal convection

The equations of thermal convection of an incompressible fluid in Boussinesq approximation and infinite Prandtl number are as follows:

$$\frac{\partial u_i}{\partial x_i} = 0, \tag{9}$$

$$\alpha\rho g_i T' - \frac{\partial p'}{\partial x_i} + \frac{\partial \tau_{ij}}{\partial x_j} = 0, \tag{10}$$

$$\frac{\partial T'}{\partial t} + u_i \frac{\partial T'}{\partial x_i} = \kappa \frac{\partial^2 T'}{\partial x_i^2}, \tag{11}$$

where ρ is density, p' and T' are pressure and temperature perturbations, g_i is the gravity vector, α is the thermal expansivity, $\kappa = \frac{k}{\rho c_p}$ is the thermal diffusivity, k is the thermal conductivity, and c_p is the isobaric specific heat. τ represents the elements of the stress tensor according to the following equation:

$$\tau_{ij} = \eta \left(\frac{\partial u_i}{\partial x_j} + \frac{\partial u_j}{\partial x_i} \right), \tag{12}$$

where i and j, are indices of the coordinate axes.

The boundary conditions are as follows. For a cell with only base heating, the top and bottom surfaces are isothermal. The temperature of the top surface as T_0 and that of the bottom surface as T_1. The temperature difference is $\Delta T = T_1 - T_0$. The vertical boundaries are thermally insulated. All surfaces are free-slip. The velocity normal to a cell boundary is zero.

Non-dimensionalization

The above equations are non-dimensionalized as follows:

$$\bar{x}_i = \frac{x_i}{d}, \ \bar{u}_i = u_i \frac{d}{\kappa}, \ \bar{\eta} = \frac{\eta}{\eta_1}, \ \bar{\tau} = \tau \frac{d^2}{\kappa \eta_1}, \ \bar{t} = t \frac{\kappa}{d^2}, \ \bar{T} = \frac{T}{\Delta T}, \tag{13}$$

where d is the layer depth, t is time, η_1 is the reference viscosity (at the bottom of the convective layer), and ΔT is the temperature drop across the layer. The Rayleigh number can then be used to characterize the system:

$$Ra = \frac{\alpha\rho g \Delta T d^3}{\kappa \eta_1}. \tag{14}$$

The Arrhenius viscosity is non-dimensionalized as:

$$\bar{E} = \frac{E}{R\Delta T}, \tag{15}$$

$$\bar{\eta} = \eta_{r,\mathrm{Arr}} \exp \frac{\bar{E}}{\bar{T}_0 + \bar{T}'}, \tag{16}$$

while the dimensionless exponential viscosity is as follows:

$$\bar{\eta} = \eta_{r,\exp} \exp(-\theta \bar{T}'), \tag{17}$$

where the pre-factors $\eta_{r,\mathrm{Arr}}$ and $\eta_{r,\exp}$ are chosen to ensure that the viscosity is equal to unity at $\bar{T}' = 1$ and the Frank-Kamenetskii parameter θ is the non-dimensionalized form of the constant γ (Equation 6):

$$\theta = \gamma \Delta T. \tag{18}$$

In this case, the viscosity contrast is characterized by only one parameter (θ):

$$\Delta\eta = e^{\theta}, \tag{19}$$

The non-dimensional yield stress is as follows:

$$\bar{\tau_y} = \bar{\tau_0} + \bar{\tau'}_y \bar{z}, \tag{20}$$

where

$$\bar{\tau_0} = \frac{d^2}{\kappa \eta_1} \tau_0 \tag{21}$$

is the non-dimensional yield stress at the surface and

$$\bar{\tau'}_y = \frac{\rho g d^3}{\kappa \eta_1} \mu \tag{22}$$

is the non-dimensional yield stress gradient. The dimensionless pseudoplastic viscosity is as follows:

$$\bar{\eta}_{\mathrm{eff}} = \frac{\bar{\tau}_y}{\bar{\dot{e}}}, \ \bar{\dot{e}} = \frac{d^2}{\kappa} \dot{e}. \tag{23}$$

In the following discussion, all parameters are assumed to be non-dimensionalized and the bar sign will be dropped. The non-dimensional forms of Equations 9 to 11 are as follows:

$$\frac{\partial u_i}{\partial x_i} = 0, \tag{24}$$

$$RaT'\mathbf{e}_i - \frac{\partial p'}{\partial x_i} + \frac{\partial \tau_{ij}}{\partial x_j} = 0, \tag{25}$$

$$\frac{\partial T'}{\partial t} + u_i \frac{\partial T'}{\partial x_i} = \frac{\partial^2 T'}{\partial x_i^2}. \tag{26}$$

where $\mathbf{e}_i$ is a unit vector in the direction of gravity.

Matching Arrhenius viscosity and exponential viscosity in non-dimensional form

To compare the two viscosity laws, the Arrhenius viscosity and the exponential viscosity are matched according to Equations 4 and 5:

$$\eta_{\mathrm{r,Arr}} \exp \frac{E}{T_i + T_0} = \eta_{\mathrm{r,exp}} \exp(-\theta T_i), \tag{27}$$

$$\frac{E}{(T_i + T_0)^2} \eta_{\mathrm{r,Arr}} \exp \frac{E}{T_i + T_0} = \theta \eta_{\mathrm{r,exp}} \exp(-\theta T_i). \tag{28}$$

Equations 27 and 28 would yield:

$$\theta = \frac{E}{(T_i + T_0)^2}. \tag{29}$$

T_i differs from the bottom temperature T_1 by a rheological temperature difference ΔT_{rh}, which is on the order of θ^{-1}.

Equation 29 shows that there are various combinations of E and T_0 that would give the same θ, and they would result in different Arrhenius viscosity contrasts:

$$\Delta\eta_{\mathrm{Arr}} = \exp\frac{ET_i}{T_0(T_i + T_0)}. \tag{30}$$

Thus the ratio of Arrhenius viscosity contrast to exponential viscosity contrast $\exp\theta$ is as follows:

$$\frac{\Delta\eta_{\mathrm{Arr}}}{\exp(\theta)} = \exp\frac{ET_i}{T_0(T_i + T_0)^2}. \tag{31}$$

Results

Steady-state convection

We use the finite element code CITCOM (Moresi and Solomatov 1995) to simulate convection in a $64a \times 64$ box, where a is the aspect ratio. Several high viscosity cases were ran with $128a \times 128$ resolution for more accurate results. All cases were run until they reached a steady state. We consider the range of parameters in which convection is in the stagnant lid regime (Solomatov 1995).

The structure of a steady-state convection cell is shown in Figure 1. Due to the temperature-dependent viscosity, the top part of a convective cell forms a stagnant lid and convection only penetrates into the lid by length of δ_{rh} - the rheological layer thickness (e.g., Solomatov 1995). A cold rigid lid, which is often defined by an isotherm, is naturally developed in the top part of the cell sloping downward to the downwelling end. The stress field (Figure 1, right) shows a stress boundary layer near the surface. This is consistent with the analytical solutions of Fowler (1985).

To compare the stresses in exponential and Arrhenius viscosity, we choose a range of T_0 and calculate their corresponding E that gives the same θ according to Equation 29 with $T_i \approx 1$.

Aspect ratio

The horizontally averaged profiles (Figure 2) show that the lid thickness slightly depend on the aspect ratio of the convective cell but the bulk of the temperature and viscosity profile does not vary much with the aspect ratio. However, the stress profile in small aspect ratio cells ($a = 0.25$) is distinctly different from that in larger cells ($a = 0.5$ to 1).

This difference is more apparent in the 2-D plots (Figure 3). In wider cells, the layer with highest stress (red) is approximately symmetric along the half-width of the cell, increasing in depth towards both edges and greater towards the downwelling edge. Below the surface stress boundary layer, the stresses in the middle of the lid are highest as they are not affected by the free-slip boundary conditions at the vertical edges. Although surface stress boundary layer is obvious in horizontally averaged stress profiles, the 2-D stress fields reveal that the surface stress are not always greater than that at depth. Figure 3 shows that at mid-width, it is possible that the surface stresses are lower than the interior. The high-stress region (orange to lime) roughly correspond to the cold lid shown in the temperature distribution, both having slopes towards the downwelling edge. In the narrowest cell however, this high-stress slope deviates from the thermal lid slope. There is a high-stress 'core' within the lid where the magnitude of stresses is close to that of the surface stress boundary layer. This implies that steady-state convection in small aspect ratios may be in another

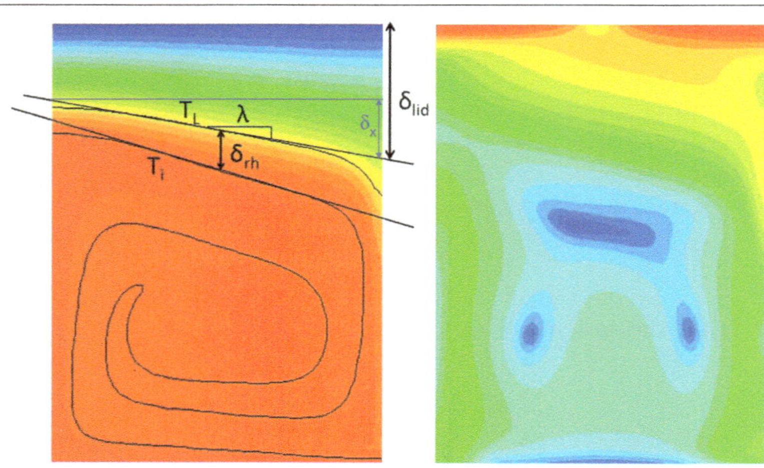

Figure 1 Temperature (left) and stress fields (right) of a steady-state convective cell. Color scale goes from high (red) to low (blue). The lid is defined by an isotherm T_L, and the interior temperature T_i is found by averaging the temperature of the convecting interior excluding boundary effects. For scaling purposes, the lid slope λ and rheological sublayer thickness δ_{rh} is taken at mid-width, whereas lid thickness δ_{lid} is extrapolated to the edge from the lid slope in the middle.

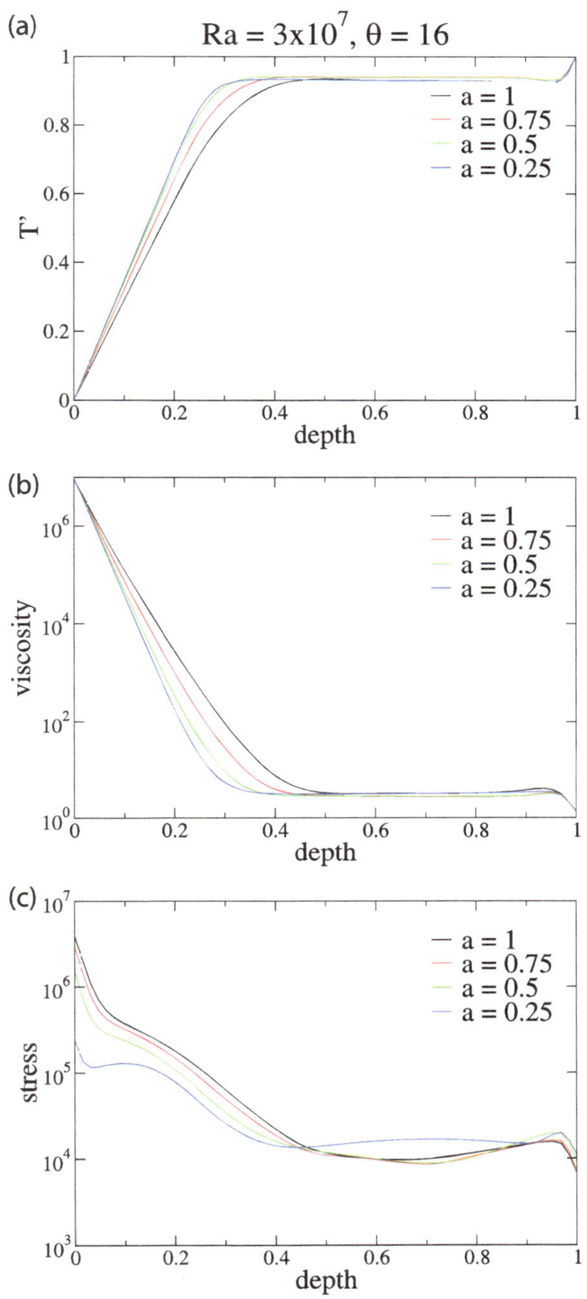

Figure 2 Comparison of **(a)** viscosity, **(b)** temperature, **(c)** stress profiles of exponential viscosities for $\theta = 16$, $Ra = 3 \times 10^7$, and varying a. The scmewhat different stress profile of $a = 0.25$ suggests that the surface stress boundary layer is not the region with highest stress.

regime of plastic failure where the interior stresses reach the yield stress first, such that the plastic zone could initiate at depth while the surface may or may not be plastic, depending on the yield stress.

For cases with high-viscosity contrasts and larger aspect ratios, convection is localized to a small part of the cell, or there could be multiple convective cells (Figure 4). These cases might belong to the subcritical convective regime (Solomatov 2012), which has a different behavior from supercritical convection and therefore not in our scope of study. To stabilize one-cell flow, smaller aspect ratio cases are considered to investigate the dependence of critical yield stress on aspect ratio.

Rayleigh number

Increasing Ra reduces the lid thickness as well as the lid slope (Figure 5). The stresses are larger with higher Ra, as is expected with more vigorous convection.

Viscosity contrast

The effects of viscosity contrast on the interior profile are illustrated in the plots in Figure 6. The conductive lid becomes thicker and the interior temperature is closer with the bottom temperature with increasing θ. In the stagnant lid regime ($\theta > 10$), the stress boundary layer near the surface is more pronounced than in the transitional regime.

The interior viscosity, temperature, and stress for Arrhenius viscosity and exponential viscosity are close. We investigate a range of different Arrhenius viscosity values by varying T_0, noting that at $T_0 = 2.0$ is a rather high surface temperature. As T_0 decreases, the difference in viscosity contrasts calculated by the two viscosity laws becomes larger. However, the temperature and the stress profiles are similar, as shown in the horizontally averaged profiles in Figure 7. There is only a slight decrease in thickness of the stress boundary layer as the Arrhenius viscosity contrast increases. The 2-D stress distributions in Figures 8 and 9 reflect the small differences in the stress distribution as viscosity contrast increases, both in exponential viscosity and Arrhenius viscosity. This is contrary to the findings of Stein and Hansen (2013), which observed a distinctly thinner lid in Arrhenius viscosity and slightly different temperature and viscosity profiles. The small difference that we observe may affect the scaling laws for subduction. The vertical variation in stresses at the downwelling edge may be particularly important. Although the lid thickness is about the same, the contrast in stresses at $x = a$ seems to be greater at low T_0 or high-viscosity contrast (Figure 9). Since subduction occurs at the downwelling edge, this may influence the scaling of yield stress.

Lid stress scaling theory

Fowler (1985) obtained a polynomial expression for the stress in the lid below the surface stress boundary layer in large lid slope approximation, which allows a comparatively simple scaling relation for stress. In order to solve the equations of convection, Fowler took Ra and θ to be asymptotically large and assumed the magnitude of lid

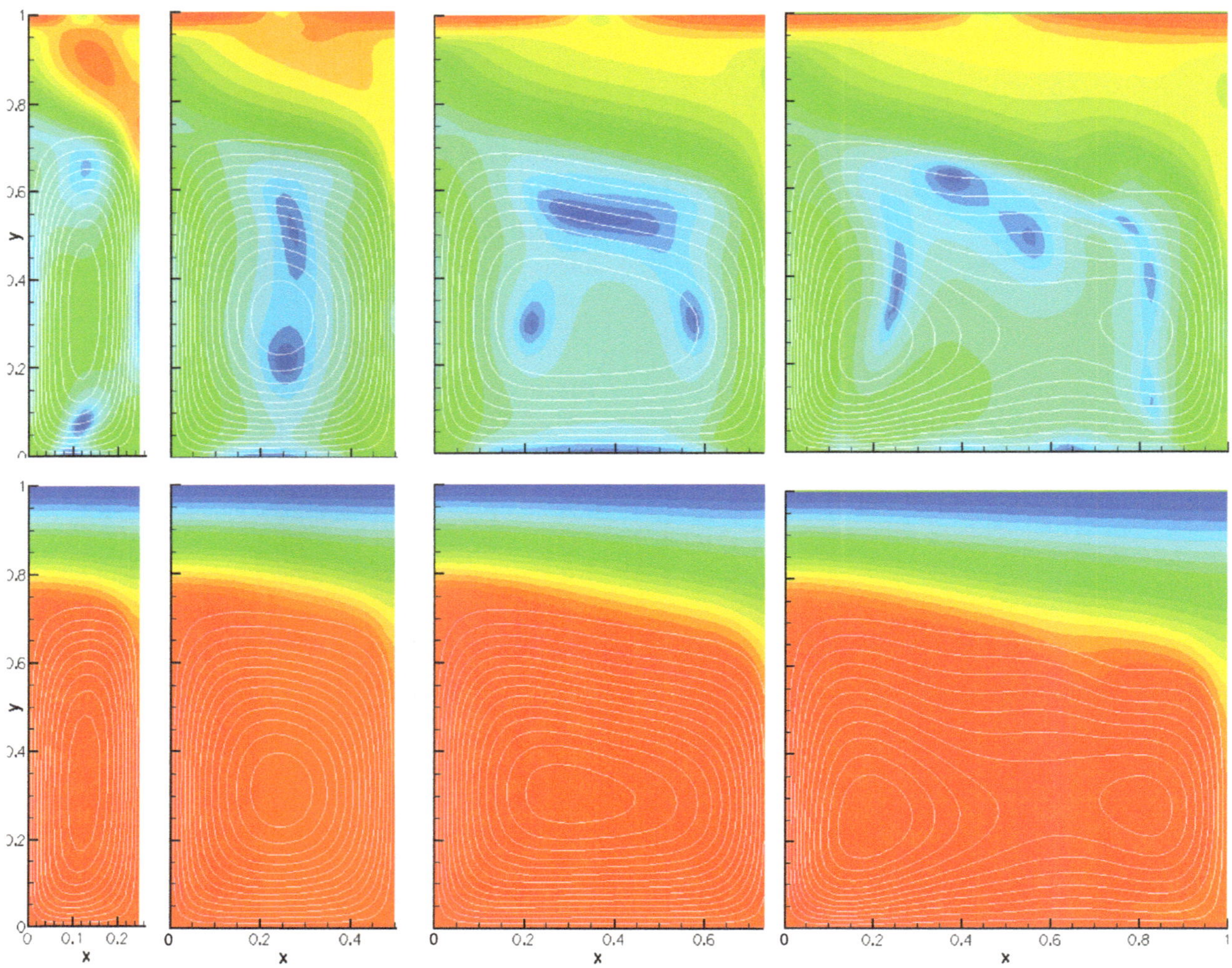

Figure 3 Stress (top row) and temperature fields (bottom row) of convecting cells with $Ra = 3 \times 10^7$, $\theta = 16$, and various a.

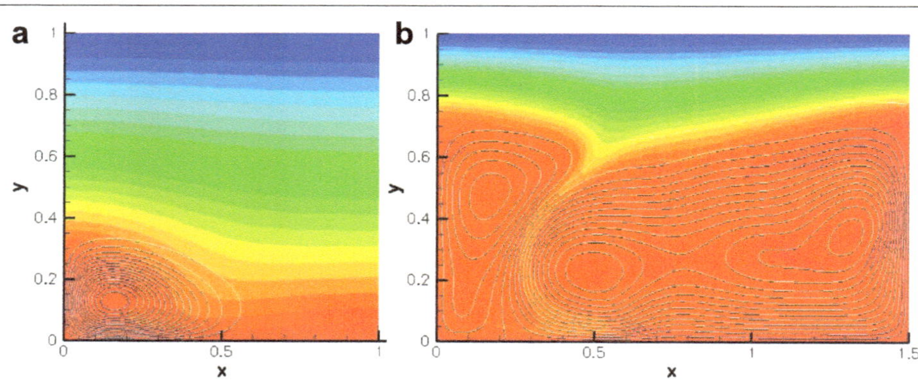

Figure 4 Localized convection and multiple subcells in steady-state convection. **(a)** $\theta = 19, T_0 = 0.6, a = 1, Ra = 1 \times 10^7$. **(b)** $\theta = 16, T_0 = 0.6, Ra = 3 \times 10^7$.

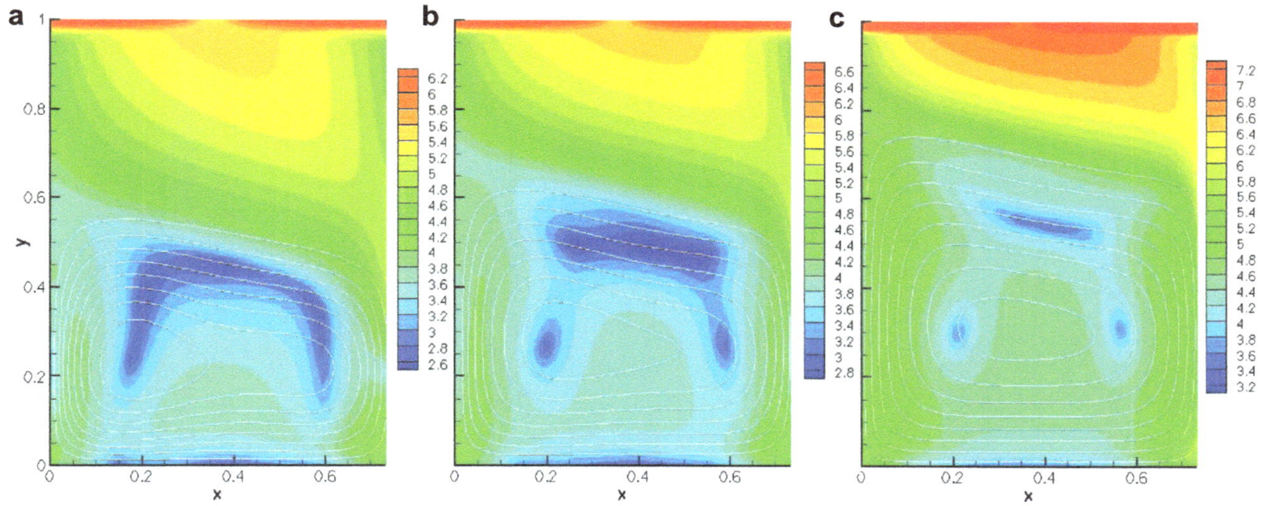

Figure 5 Stress structure of steady state cases with $\theta = 16, a = 0.75, T_0 = 0.8$. **(a)** $Ra = 10^7$. **(b)** $Ra = 3 \times 10^7$. **(c)** $Ra = 10^8$. The area of convecting interior becomes larger as Ra increases and the thickness of the stagnant lid decreases. Scale shows the values of log τ.

slope to be either on the order of lid thickness or rheological sublayer thickness. For the Earth and some smaller terrestrial planets as well as most numerical simulations, Ra may not be as high as the asymptotic theory require, and thus, some other theory may be needed for lid stress scaling. Moreover, as we find in this study, the lid slope does not follow either of these two end-member cases and thus needs to be scaled based on numerical simulations.

Fowler also found that the interior flow can be uncoupled from the rheological sublayer, which makes the problem setup akin to a viscous lid gravitationally sliding along a slope. We can therefore estimate the shear stress in the rigid lid τ_{lid} by considering the force balance on the lid (Figure 1):

$$\frac{d\tau}{dy} = \Delta\rho g_x. \tag{32}$$

For density changes due to temperature variations, the lid stress can be integrated from Equation 32:

$$\tau_{\text{lid}} = -\alpha\rho_0 \frac{dT}{dy} g \sin\lambda \frac{y^2}{2} + \tau_i, \tag{33}$$

where τ_i is the stress at the interior temperature T_i.

There are two assumptions that allow us to simplify Equation 33. The first is that τ_i is negligible since $\tau_i \ll \tau_{\text{lid}}$. The second is the small lid slope approximation. For small λ, $\sin(\lambda)$ is approximately equal to λ. The non-dimensional form of Equation 33 becomes:

$$\tau_{\text{lid}} = -\text{Ra}\frac{dT}{dy}\frac{y^2}{2}\lambda. \tag{34}$$

Thus, the lid stress is determined by Ra, dT/dy, λ, and y, which will be defined in the following discussion. This scaling is similar to that obtained from the analytical solutions of Fowler (1985).

Lid base temperature

The lid base is often defined by an isotherm:

$$T_L = T_i - C\theta^{-1} \tag{35}$$

where C is a constant. We determine the lid base from velocity profile (Solomatov and Moresi 2000) to find the constant C. We first find the greatest velocity gradient at a specific distance x from the upwelling edge of the cell. This velocity gradient is then extended to the depth at which velocity is zero, as shown in Figure 10. This depth defines the lid thickness. This process is repeated for all x (from 0 to a) to obtain the shape of the lid base across the convecting cell. This velocity gradient-defined lid base is then used to find the thermal lid base defined by Equation 35. The interior temperature T_i is found by averaging the temperature in the middle part of the interior to exclude the boundary effects. The constant C is determined by matching the lid thickness at mid-width ($x = 0.5a$) given by T_L and that defined by the velocity gradient (Figure 11). We choose the value at mid-width because in most cases, the two lid bases are closest around the middle of the cell for a large lateral extent. In some cases, especially for narrower cells, the temperature lid base and the velocity lid base may not match, so the mid-width serves as a reference point for consistency in defining the lid base.

Equation 35 allows us to determine the rheological temperature difference $\Delta T_{\text{rh}} = T_i - T_L = C\theta^{-1}$. It varies with Ra, θ, and aspect ratio, and the scaling exponents are summarized in Table 1.

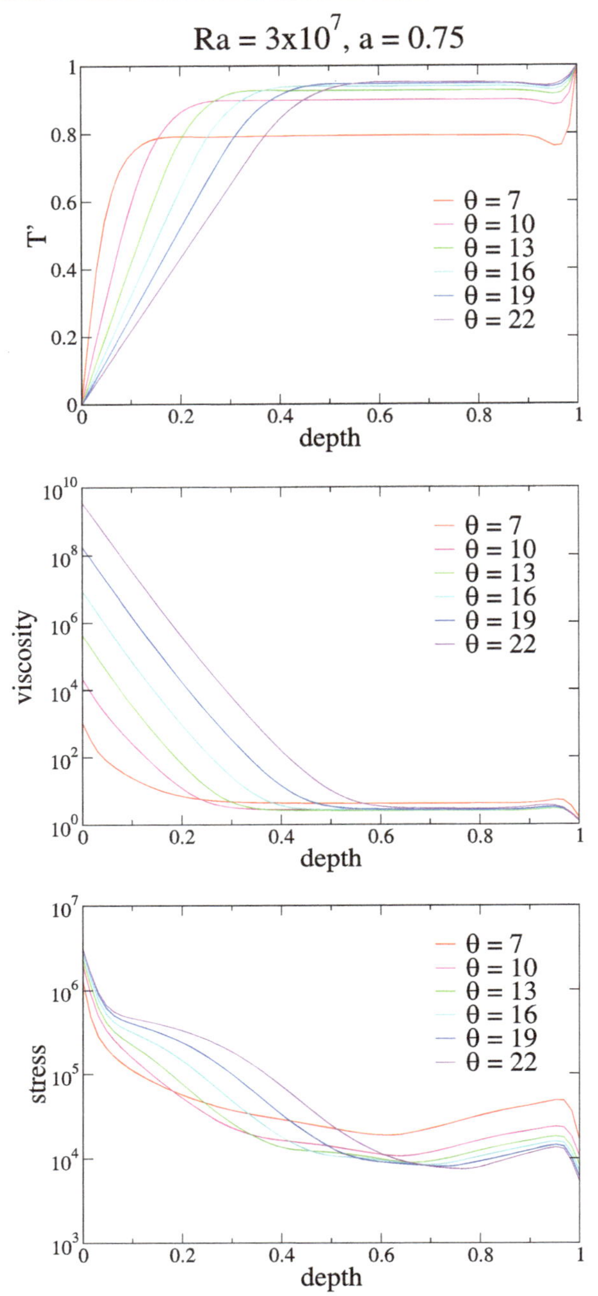

Figure 6 Temperature, viscosity, and stress profiles of $Ra = 3 \times 10^7$, $a = 0.75$, and varying Frank-Kamenetskii parameter θ.

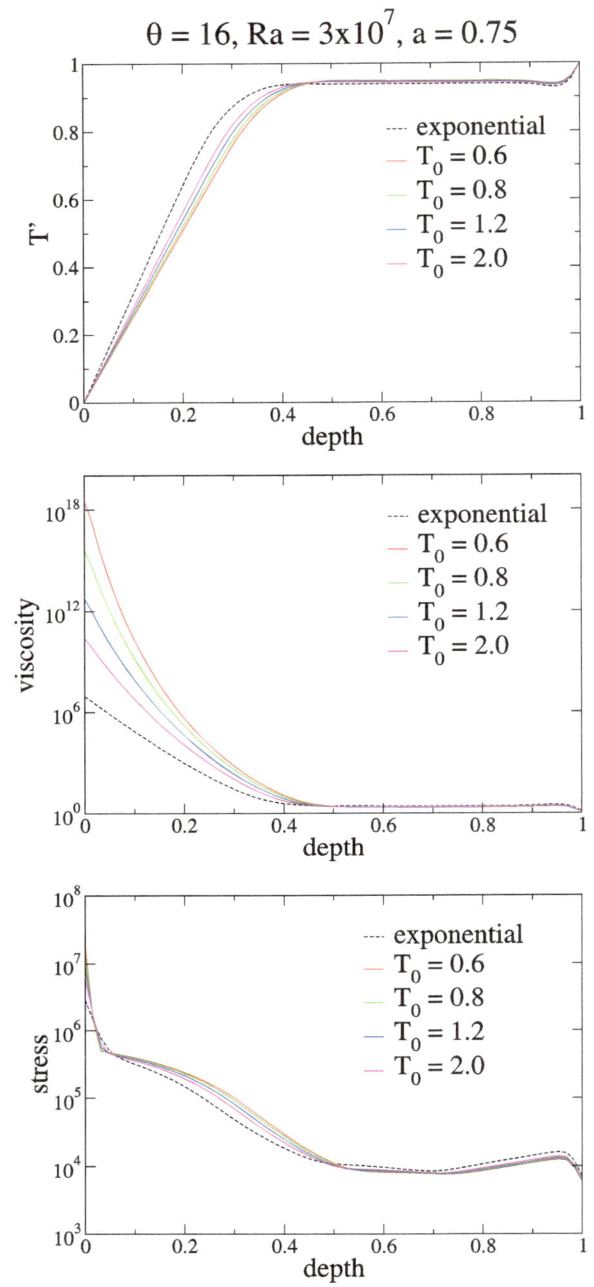

Figure 7 Comparison of temperature, viscosity, and stress profiles of Arrhenius and exponential viscosities for $\theta = 16$, $Ra = 3 \times 10^7$, $a = 0.75$, and various T_0.

Lid slope and lid thickness

Gravitational sliding requires a downward dipping slope λ as indicated in Equation 34. In larger aspect ratio cells, although the lid thickness varies horizontally, the lid slope is approximately constant in the middle portion of the cell. This is different for smaller cells, where the lid base could be some function of x instead of a straight line (Figure 12). For example, Fowler suggests that the lid base varies with $x^{0.4}$ (Fowler 1985). For consistency in our scaling analysis, the lid slope is taken to be the slope of the thermal lid base at mid-width.

To check whether the lid slope scales with lid thickness or rheological sublayer thickness (Fowler 1985), we look at the vertical drop δ_x (Figure 1) and define the lid slope as δ_x/a. If the lid slope scales with the lid thickness, then $\delta_x \sim \delta_{lid} \sim Nu^{-1}$, and thus

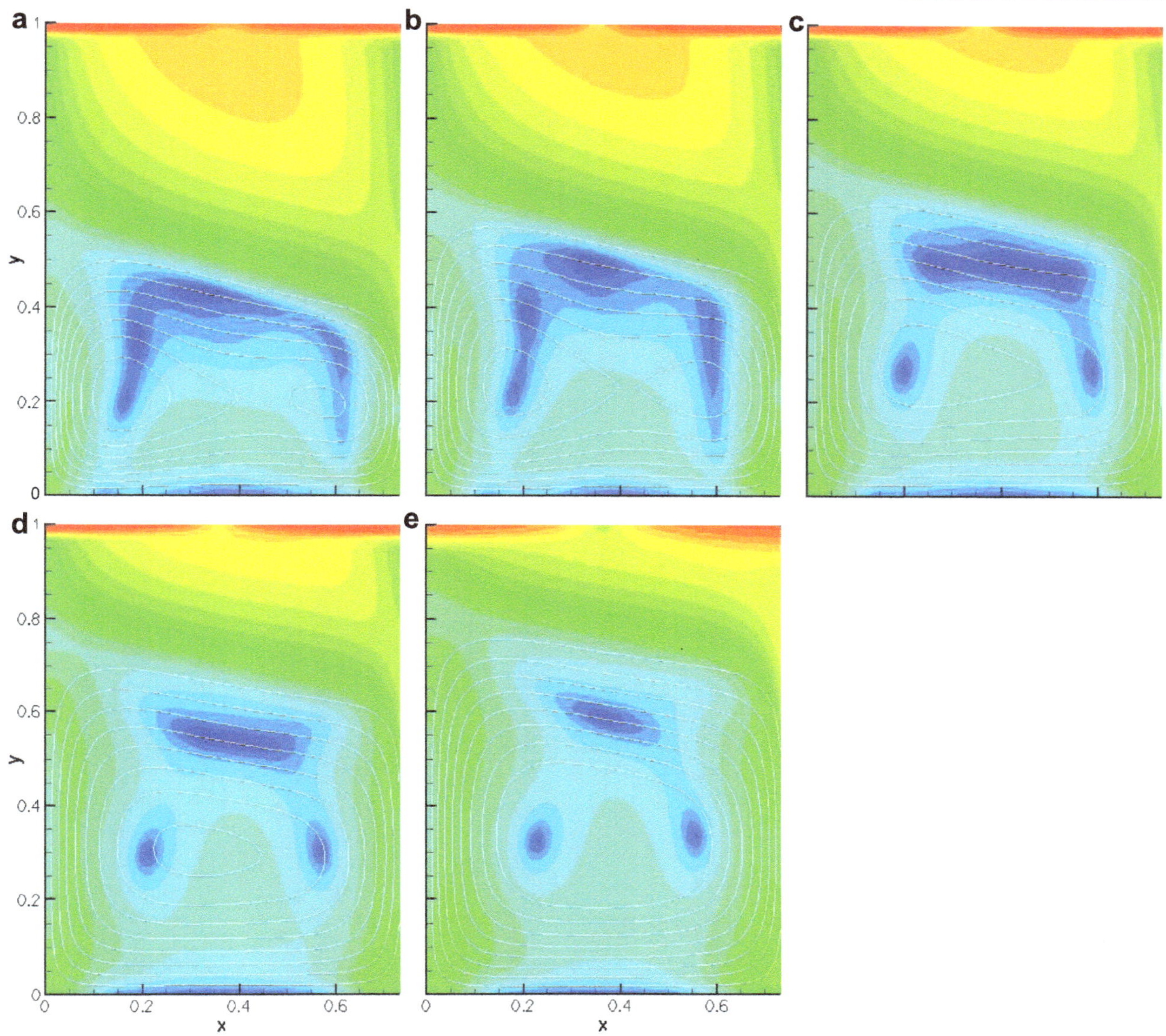

Figure 8 Stress structure of steady state cases with $Ra = 3 \times 10^7$, $T_0 = 0.8$, $a = 0.75$. **(a-d)** Arrhenius viscosity with **(a)** $\theta = 22$, **(b)** $\theta = 19$, **(c)** $\theta = 16$, **(d)** $\theta = 13$, and **(e)** exponential viscosity with $\theta = 13$. Color scale shows the values of $\log(\tau)$. White lines represent streamlines.

$$\delta_x/\delta_{lid} \sim \mathrm{Nu}\delta_x \sim \text{constant}. \tag{36}$$

If the lid slope scales with δ_{rh}, then $\delta_x \sim \delta_{rh}$. As $\delta_{rh}/\delta_{lid} \sim \theta^{-1}$, thus $\delta_x/\delta_{lid} \sim \theta^{-1}$, or

$$\theta\delta_x/\delta_{lid} \sim \mathrm{Nu}\delta_x\theta \sim \text{constant}. \tag{37}$$

We plot $\mathrm{Nu}\delta_x$ and $\mathrm{Nu}\delta_x\theta$ in Figure 13. Neither combination remains constant with θ, with $\mathrm{Nu}\delta_x$ increases with θ and $\mathrm{Nu}\delta_x\theta$ decreases with θ. This suggests that the lid slope is somewhere in between these two extreme cases. Therefore in deriving the scaling laws for stresses, we need to determine the dependence of lid slope on various convective parameters (Table 1).

Thermal gradient

The temperature is approximately a linear function of depth and the thermal gradient in the lid is about constant with depth. At mid-width ($x = 0.5a$), it is approximately equal to the Nusselt number, which is the non-dimensional horizontally averaged surface temperature gradient. Since we are looking at temperature changes from the interior to the bottom of the lid which includes the rheological sublayer, we also check the thermal gradient in the rheological sublayer $\Delta T_{rh}/\delta_{rh}$ to note any difference in the scaling relations. As before, we choose the values $\Delta T_{rh}/\delta_{rh}$ at the mid-width to exclude boundary effects for scaling purposes.

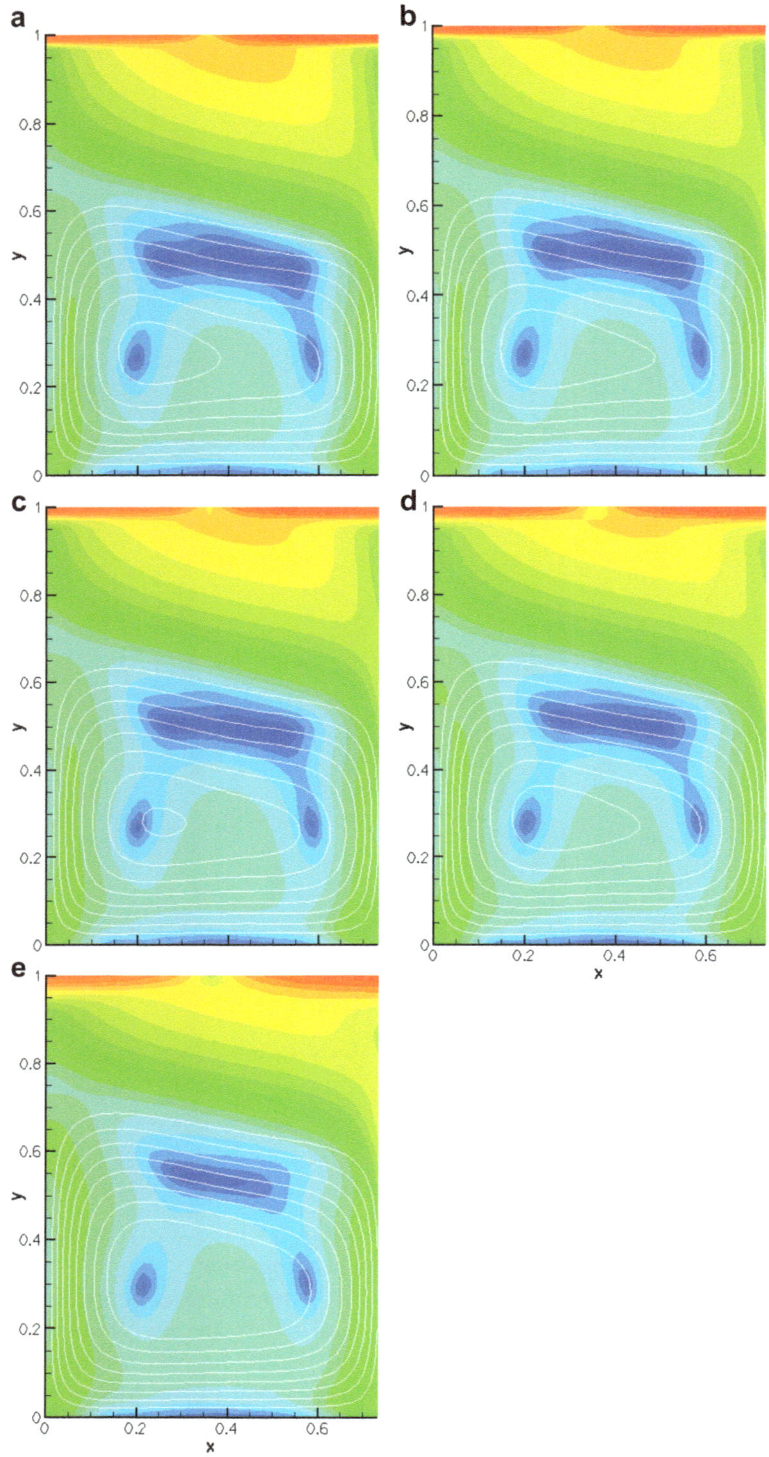

Figure 9 Stress structure of steady state cases with $Ra = 3 \times 10^7$, $\theta = 16$, $a = 0.75$. Cases **(a-d)** use Arrhenius viscosity with **(a)** $T_0 = 0.6$, **(b)** $T_0 = 0.8$, **(c)** $T_0 = 1.2$, and **(d)** $T_0 = 2.0$. Case **(e)** uses exponential viscosity. Color scale shows the values of $\log(\tau)$. White lines represent streamlines.

In previous theories, $\Delta T_{\text{rh}}/\Delta T \sim \theta^{-1}$ and $\delta_{\text{rh}}/\delta_{\text{lid}} \sim \theta^{-1}$. The determination of C follows the description in the previous section on lid base temperature, and it is found to be dependent on aspect ratio and θ. Therefore, $\Delta T_{\text{rh}}/\Delta T$ and $\delta_{\text{rh}}/\delta_{\text{lid}}$ will also have a dependence on a and θ, and their scaling relations are summarized in Table 1.

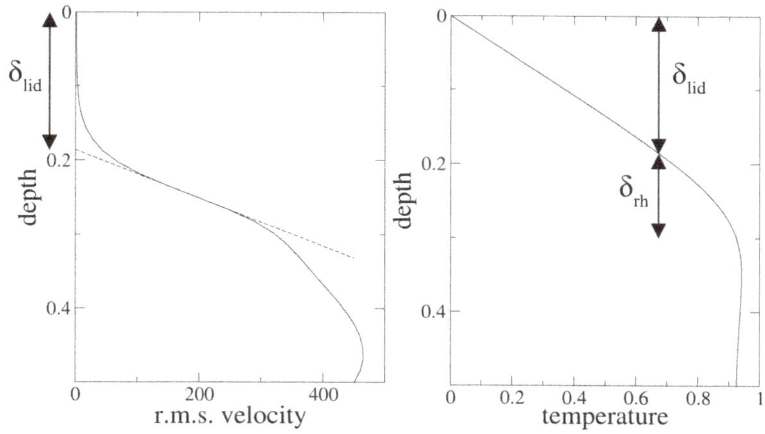

Figure 10 Velocity profile taken at mid-width of the cell ($x = 0.5a$). Dotted line is the linear extrapolation of the maximum velocity gradient, and the lid base is marked at the depth at which this line intersect with the vertical axis.

Stress scaling

Equation 34 can now be expressed in a non-dimensional form with the definitions of various parameters in the previous discussion. At the bottom of the lid at δ_{rh} from the interior, where the temperature difference is ΔT_{rh}, the stresses are as follows:

$$\tau_{\mathrm{lid}} \sim \mathrm{Ra}\frac{\Delta T_{\mathrm{rh}}}{\delta_{\mathrm{rh}}}\lambda y^2. \tag{38}$$

Further into the lid, the non-dimensional temperature gradient is the Nusselt number. Therefore Equation 34 can be alternatively scaled as:

$$\tau_{\mathrm{lid}} \sim \mathrm{RaNu}\lambda y^2. \tag{39}$$

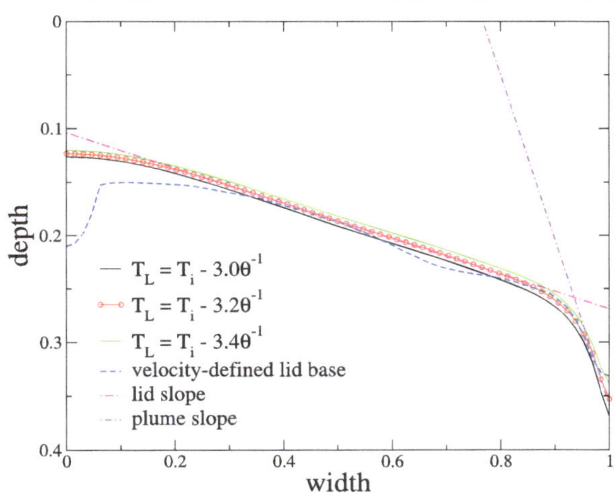

Figure 11 Top part of a convective cell showing various definitions of the lid base. The lid base defined by the velocity gradient is shown in dashed line, and it matches the temperature lid base defined at $T_L = T_i - 3.2\theta^{-1}$ at around the mid-point. The lid slope is estimated at the mid-point.

As shown in Figure 10, there is a slight difference in the thermal gradient in the lid and in the rheological sublayer; therefore, Equations 38 and 39 may result in slightly different scaling exponents. We check both scaling relations to see whether the stresses at the lid base and those in the lid can be scaled similarly.

We plot the stress profile according to Equations 38 and 39 and compare with that from numerical solutions (Figure 14). The prefactor of the stress as a function of y calculated from the thermal gradient in the rheological sublayer ($T_{\mathrm{rh}}/\delta_{\mathrm{rh}}$) is 5.9×10^6, whereas that from the Nusselt number is 9.6×10^6. This demonstrates that the theoretical stress profiles match fairly closely with the numerical one, and the best fit can be obtained with some small adjustments in the coefficient.

All the above parameters depend on Ra, aspect ratio a, and θ; thus, they can be expressed as $\mathrm{Ra}^{\beta}a^{\zeta}\theta^{\alpha}$ where β, ζ,

Table 1 Numerical results of power law coefficients in scalings of different parameters with Ra, aspect ration *a*, and Frank-Kamenetskii parameter θ

Parameter	c in 10^c	Ra	a	θ
Nu	0.23 ± 0.07	0.24 ± 0.01	-0.17 ± 0.03	-1.17 ± 0.05
$\Delta T_{rh}/\Delta T$	0.25 ± 0.17	0.010 ± 0.02	0.39 ± 0.08	-0.84 ± 0.12
δ_0/d	-0.26 ± 0.10	-0.22 ± 0.01	0.27 ± 0.04	1.12 ± 0.06
δ_{lid}/d	-0.74 ± 0.06	-0.23 ± 0.01	0.12 ± 0.03	1.44 ± 0.04
$\delta_{\mathrm{lid,max}}/d$	-0.62 ± 0.07	-0.19 ± 0.01	0.33 ± 0.03	1.26 ± 0.05
δ_{rh}/d	0.02 ± 0.33	-0.20 ± 0.04	0.60 ± 0.15	0.34 ± 0.21
Lid slope	-0.98 ± 0.13	-0.07 ± 0.02	0.14 ± 0.06	0.63 ± 0.09
τ_{lid}	0.36 ± 0.21	0.68 ± 0.03	0.92 ± 0.09	-0.15 ± 0.13
$\tau_{y,cr}$	0.40 ± 0.22	1.09 ± 0.03	1.67 ± 0.10	-1.49 ± 0.14
$\tau'_{y,cr}$	2.34 ± 0.32	1.34 ± 0.04	1.45 ± 0.14	-3.47 ± 0.21

$\Delta T_{rh}/\Delta T$ and τ_{lid} taken at mid-width.

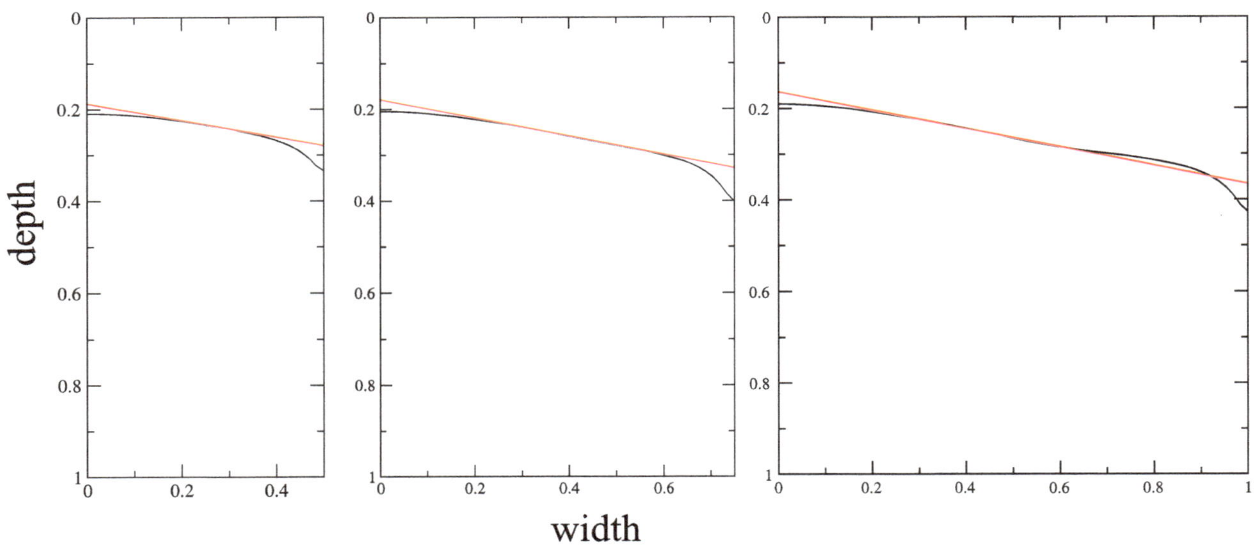

Figure 12 The slope of the lid base at various aspect ratios. The lid slope deviates from the linear approximation as the cell aspect ratio and viscosity contrast increase; therefore, a non-constant value of the lid slope may affect the scalings.

and α are scaling exponents. The results are summarized in Table 2.

Convection with yield stress

We use the steady-state solutions as the starting point before imposing a yield stress to simulate plastic yielding. For the yield stress gradient, a small cohesion term (surface yield stress) was introduced to stabilize the solution.

Regimes of convection with constant yield stress or constant yield stress gradient

When a yield stress is present, the regions with stresses that reach the yield value would have an approximately constant stress close to τ_y. These plastic zones develop first at the corners of the cell where the stresses are highest. As τ_y decreases, the plastic zones extend both in depth and horizontally, narrowing the width of the high-viscosity part of the lid (Figure 15). If yield stress is too high, the depth of the plastic zone is small or the plastic zone is entirely absent; thus, the stagnant lid does not fail. If the yield stress is sufficiently low, the plastic zone extends suffcieintly deep so that the stagnant lid is mobilized.

We examine the stress and viscosity profiles at various locations x to see how they change in the presence of a yield stress (Figures 16 and 17). At high yield stress or high yield stress gradient, the plastic zone only occurs at shallow depths and the bulk of the stress and viscosity profiles are unaltered from the stagnant lid state. The plastic zone extends deeper as the yield stress or yield stress gradient decreases. As the yield stress approaches the critical value, a small change in yield stress induces a change in plastic depth that is comparable to the change caused by an order of magnitude change in yield

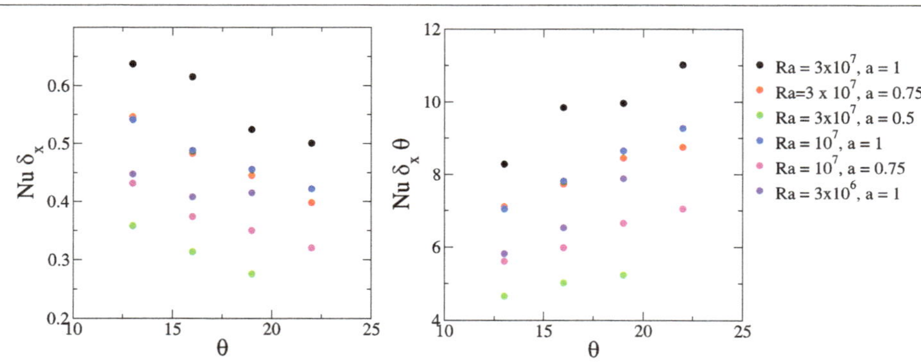

Figure 13 Plots of $Nu\delta_x$ (or δ_x/δ_{lid}) and $Nu\delta_x\theta$ (or $\theta\delta_x/\delta_{lid}$) as a function of θ.

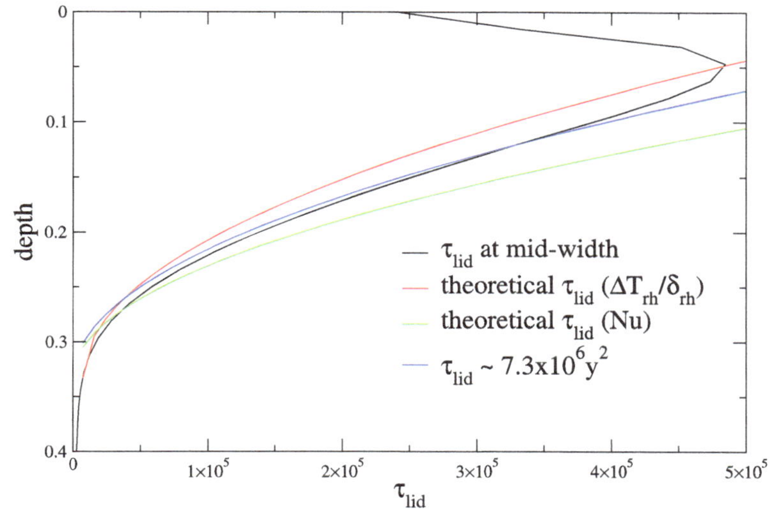

Figure 14 Comparison of stress profiles at mid-width obtained from numerical calculations and theory. Blue line represents the best fit to numerical solutions of stresses. The stress profile is taken at mid-width, where the stresses in the surface boundary layer is lower than the interior as shown in the 2-D stress fields.

stress when it is far from critical. Although subduction does not occur, this implies a change in convection regime from stagnant lid to some sort of transitional regime.

In this transitional regime, the yield stress slightly changes the interior dynamics as can be seen in various convective parameters of the interior region of the cell. The yield stress increases the lid slope whereas the lid thickness remains approximately the same. When the yield stress is slightly above the critical value, these changes caused by yield stress are negligible and convection remains in the stagnant lid regime. Therefore in deriving the critical yield stress scalings, we refer to the steady-state structure that has a yield stress just above the critical value, so that the scaling relations for various convective parameters (Table 1) from steady-state stagnant lid convection can still be used.

Time evolution of lid weakening and failure

When the lid fails, the surface velocity continuously increases and overturn occurs (Figure 18). Figure 19 shows the time sequence of stress structures before and during failure. When the surface velocity is still low compared to the bottom velocity (Figure 18, left), the variation in stress structure is not obvious. It is not until the velocity begins to increase drastically that the plastic yield zones from the two corners start to connect in the middle of the cell to form a plastic lid. The weak lid then becomes unstable and starts to subduct.

Table 2 Power law coefficients in scaling laws for stresses: numerical results vs theory

Parameter	Ra	a	θ	Method
τ_{lid}	0.68 ± 0.03	0.92 ± 0.09	-0.15 ± 0.13	Numerical
	0.73 ± 0.09	1.14 ± 0.29	0.146 ± 0.42	Theory (in terms of $\Delta T_{rh}/\delta_{rh}$)
	0.77 ± 0.12	1.18 ± 0.39	0.14 ± 0.56	Theory (in terms of Nu)
$\tau_{y,cr}$	1.09 ± 0.03	1.67 ± 0.10	-1.49 ± 0.14	Numerical
	0.94 ± 0.10	1.81 ± 0.32	-1.13 ± 0.47	Theory (in terms of $\Delta T_{rh}/\delta_{rh}$)
	0.97 ± 0.13	1.85 ± 0.42	-1.16 ± 0.61	Theory (in terms of Nu)
$\tau'_{y,cr}$	1.34 ± 0.04	1.45 ± 0.14	-3.47 ± 0.21	Numerical
	1.13 ± 0.11	1.48 ± 0.35	-2.39 ± 0.51	Theory (in terms of Nu)
	1.16 ± 0.14	1.52 ± 0.45	-2.38 ± 0.65	Theory (in terms of $\Delta T_{rh}/\delta_{rh}$)

$\Delta T_{rh}/\Delta T$ and τ_{lid} taken at mid-width.

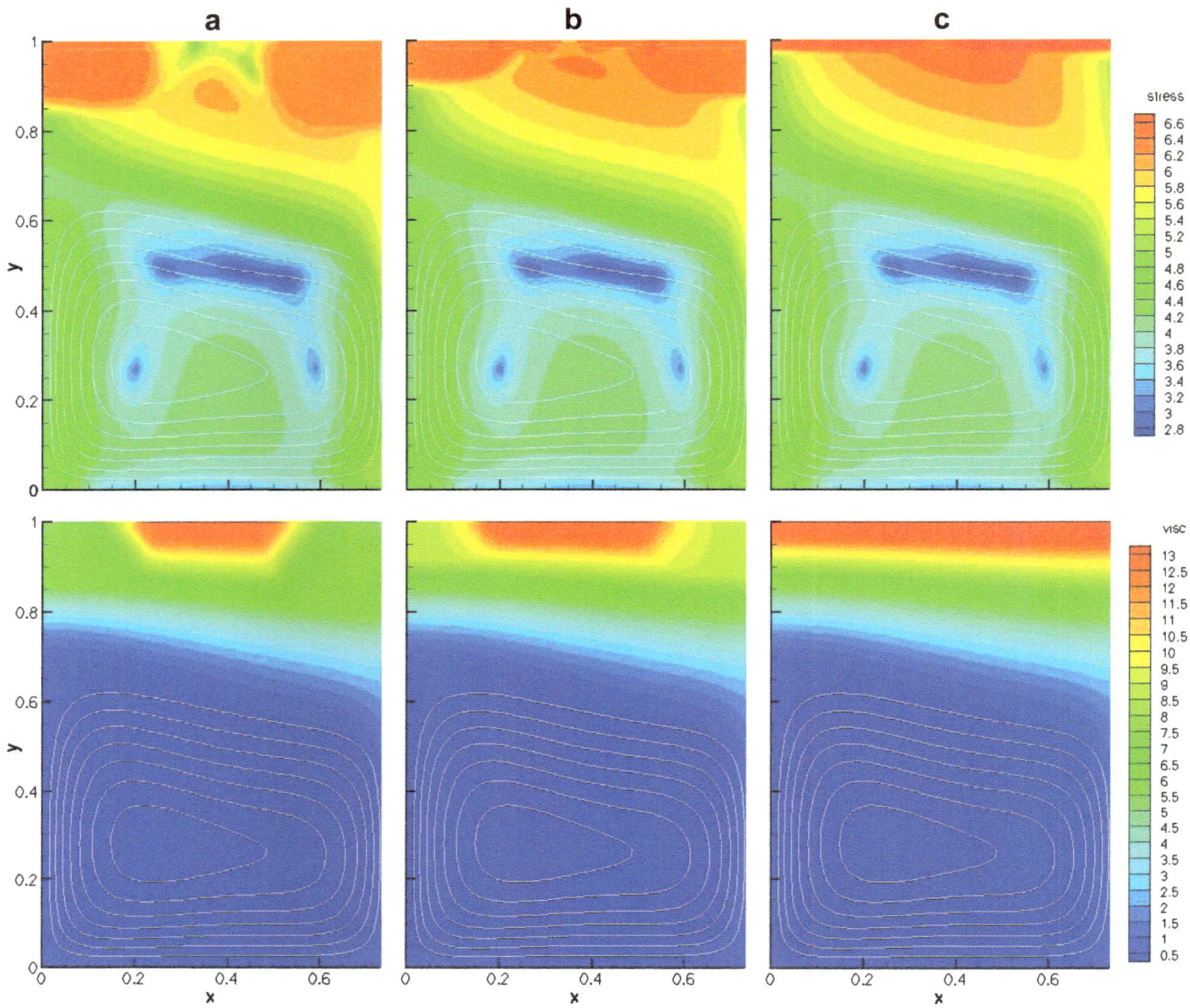

Figure 15 Stress (top) and viscosity (bottom) fields of case $Ra = 3 \times 10^7, \theta = 16, a = 0.75, T_0 = 0.8$. **(a)** $\tau_y = 6.7 \times 10^5$, **(b)** $\tau_y = 7 \times 10^5$, **(c)** $\tau_y = \infty$. Failure occurs at $\tau_y = 6.6 \times 10^5$, while the stagnant lid remains at higher τ_y. Color bars show $\log\tau$ values.

A possible contribution to the uncertainty in determining the critical yield stress is that at the vicinity of the critical value, the behavior may be difficult to interpret. In some cases that as the yield stress gets close to a critical value, the surface velocity increases slowly and it may take more than 10^5 timesteps to reach a point of overturn, whereas typically, it takes less than 10^4 timesteps to a drastic increase in surface velocity (Figure 18 right). This may be due to the behavior of dynamic system near a critical point.

Critical depth of plastic failure zone

For subduction to occur, the lithosphere has to be sufficiently weak so that it can be mobilized by stresses arisen from mantle convection. As seen in Figures 16 and 17, the lid remains stagnant if the plastic yield zone is small. Therefore, the question is how deep does this plastic zone have to penetrate for the lid to be mobilized? Previous theories proposed that the plastic zone has to penetrate through some critical depth $\delta_{\rm pl}$ defined by a critical temperature T_c:

$$\frac{\delta_{\rm pl}}{\delta_0} = \frac{T_c - T_0}{T_i - T_0}, \tag{40}$$

where $\delta_0 = \delta_{\rm lid} + \delta_{\rm rh}$. The model of Fowler and O'Brien (2003) predicts that δ_{pl} is defined by the temperature that gives the interior viscosity. For Newtonian rheology, this means that the plastic zone has to extend through the base of the lid. Solomatov (1995,2004a) suggested that δ_{pl} only has to penetrate to the isotherm at which the viscosity

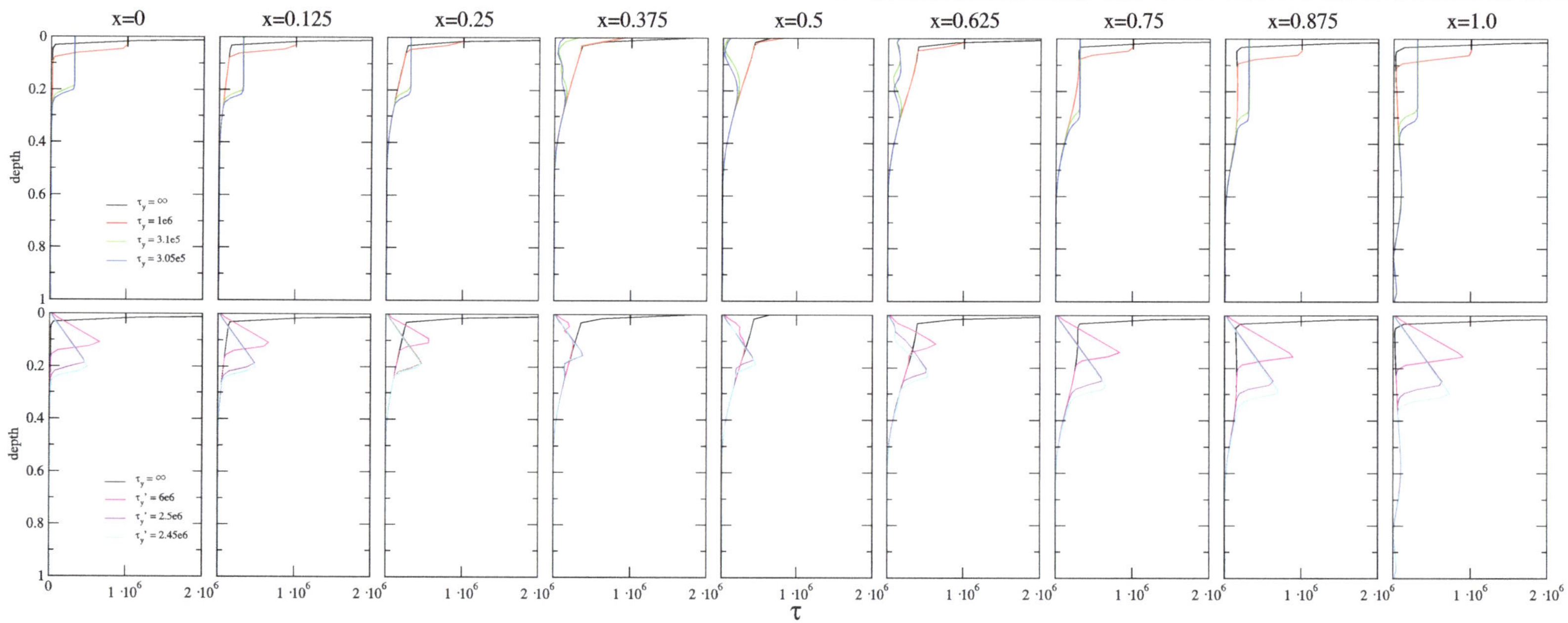

Figure 16 Stress profiles at different widths at various yield stress or yield stress gradient.

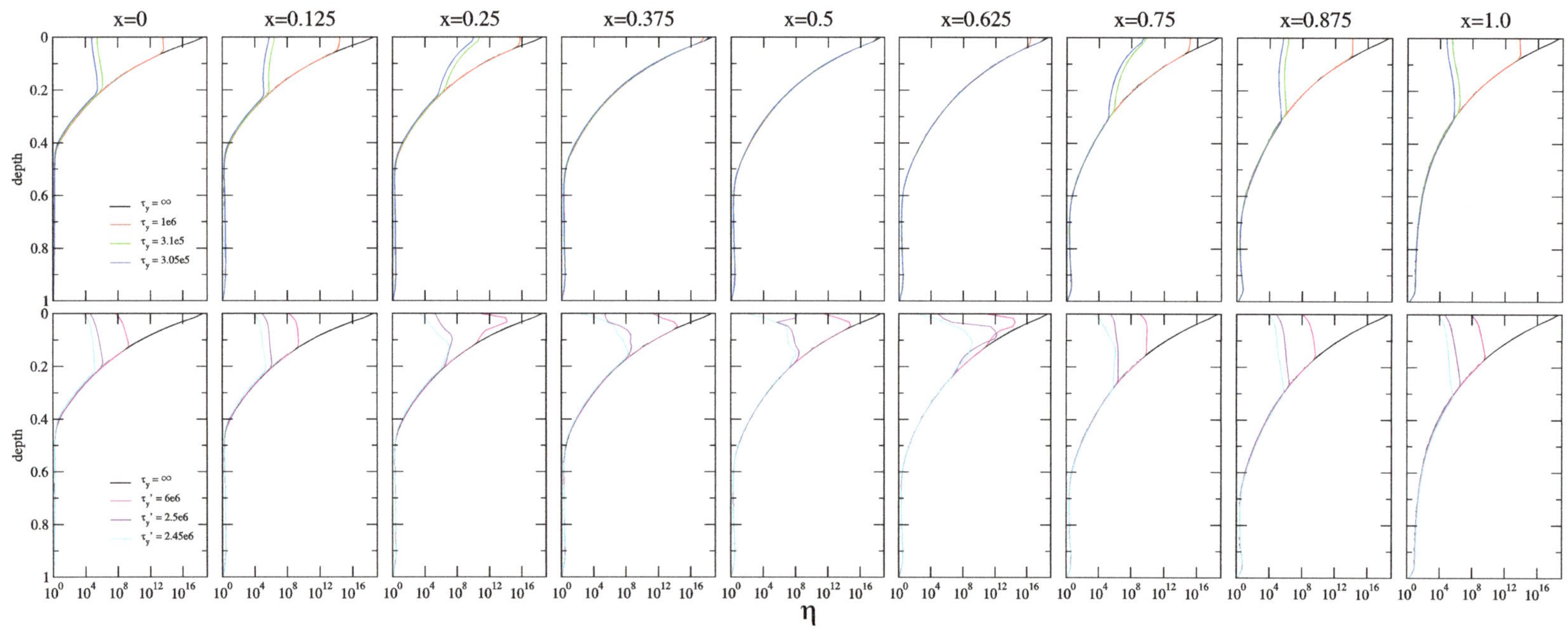

Figure 17 Viscosity profiles at different widths at various yield stress or yield stress gradient.

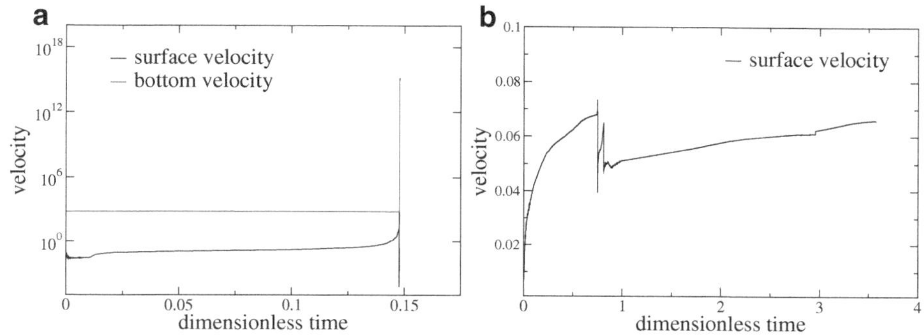

Figure 18 Surface velocities over time for $Ra = 3 \times 10^7, \theta = 16, a = 0.75$, **(a)** $T_0 = 0.8, \tau_y = 6.6 \times 10^5$, **(b)** $T_0 = 0.4, \tau_y = 6.1 \times 10^5$, bottom velocity ≈ 650. The surface velocity in (b) increases slowly. Although it has evolved for approximately 25 times as much as the period for (a) to fail, it is far from reaching the bottom velocity.

contrast with the interior viscosity is $e^{4(n+1)}$, where n is the stress exponent for non-Newtonian viscosity.

To examine these hypotheses, we look at the stress profile at the downwelling edge ($x = a$) to determine the depth of the plastic zone, as the stresses at this edge are the highest and this is where subduction starts (Figure 20). The depth of the plastic zone is defined by extent of the stress modified by the plastic flow law in Equation 7. The exact value of δ_{pl} and the stress at this depth are found by the intercept of the stress calculated from linear extrapolation from the top (where stress is determined by the yield stress) and exponential extrapolation from the creep regime just below the plastic depth. The values of δ_{pl} for constant yield stress cases and that for constant yield stress gradient cases are close. It is noted that the stress at depth δ_{pl} in constant yield stress gradient cases is about twice as much as that in constant yield stress cases. The force on the lid is $\delta_{pl}\tau_{y,cr}$ for constant yield stress and $0.5\delta_{pl}\tau'_{y,cr}$ for constant yield stress gradient. This implies that the force on the lid is about the same for both constant yield stress and yield stress gradient.

As we see from the effects of aspect ratio on steady-state stress distribution, the stresses in the interior become comparable or even exceed the surface stresses as the aspect ratio and Ra decrease and viscosity contrast increases. This means that the plastic zone may not propagate from the top but also develop at depths in the lid, and the plastic zone does not span the whole top part of the cell. (Figure 21). This may represent another regime of lid failure. These cases are thus excluded from our scaling analysis.

The depth of the plastic zone is determined from drop in stress by the yield stress. We note that the zone of reduced viscosity due to the yield stress may correspond to the zone of reduced stress as shown in Figure 20. However, this does not always hold, especially for cases with higher θ. Figure 22 shows that the transition of the plastic zone to creep flow may not correspond to the sharp change in the viscosity. This means that the viscosity at the plastic depth and the maximum viscosity are different, since the reduction in viscosity is not only determined by the yield stress but also by the strain rate.

To find out whether there is a critical viscosity contrast in lithospheric failure, we examine both the viscosity contrast at plastic depth $\Delta\eta_{\mathrm{pl}}$ and the maximum viscosity contrast $\Delta\eta_{\max}$. The maximum viscosity needs to be determined by extrapolation as the resolution near the point where the viscosity is maximum is not high enough to resolve sharp changes in stress and viscosity. The point of maximum viscosity is found by extrapolating the values from both above and below the maximum point (e.g., Figure 22 right). The viscosity is extrapolated linearly from the two points above the maximum point, and below the maximum point, the viscosity is calculated from temperature. The intersection of these two curves determines the maximum viscosity and its depth. We found that these two viscosity contrasts are mostly within the same order of magnitude. Figure 23 shows that the maximum viscosity seems to depend on the original non-yielding viscosity contrast $\Delta\eta$. For exponential viscosity and Arrhenius viscosity cases at low non-yielding viscosity contrast, the maximum viscosity contrast $\Delta\eta_{\max}$ increases with non-yielding viscosity contrast. This may be because they are close to transitional regime. At higher $\Delta\eta$, the increase in $\Delta\eta_{\max}$ seems to decline with increasing $\Delta\eta$, but the spread of data prevents us from concluding that $\Delta\eta_{\max}$ converges towards higher viscosities. As with the maximum viscosity contrast, $\Delta\eta_{\mathrm{pl}}$ also does not display linearity or convergence clearly with either the non-yielding viscosity contrast or θ (Figure 24).

Since the critical viscosity contrast is neither a constant or a function of θ, we look at the depth of the plastic zone to derive scaling relations for the yield stress. We investigate the plastic depth δ_{pl} as a fraction of the lid thickness. For scaling purposes, the lid thickness

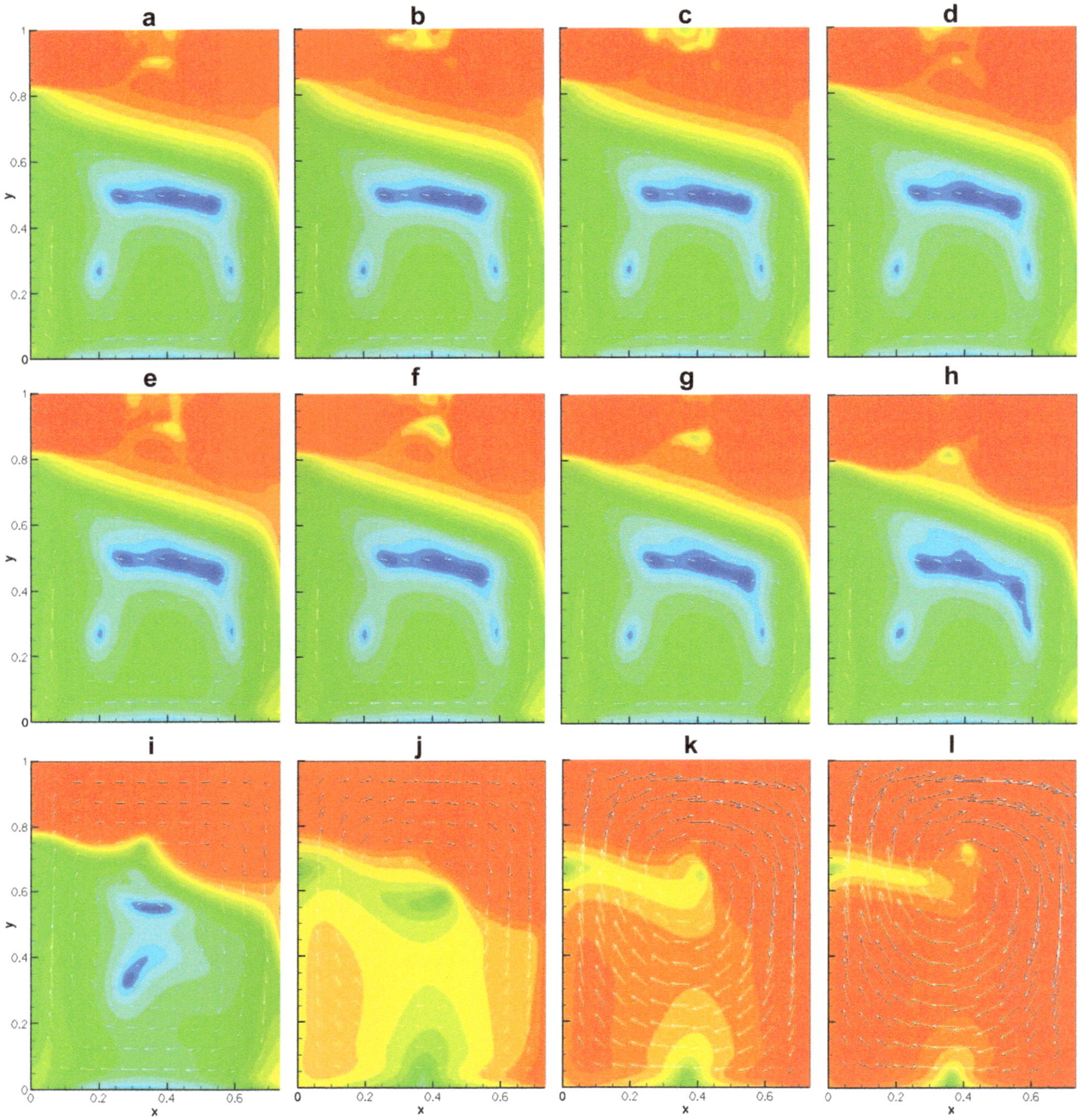

Figure 19 Snapshots of stress fields before and at the point of failure for case $Ra = 3 \times 10^7, \theta = 16, a = 0.75, T_0 = 0.8, \tau_y = 6.6 \times 10^5$. Time sequence goes from left to right and top to bottom. White arrows show velocities. **(a)** and **(b)** are close to beginning of simulation, **(c)** is at the mid-point between the start and failure, and **(d-i)** are right before overturning (depicted in **(j-l)**) occurs.

was previously defined at the middle of the convecting cell. Here, since δ_{pl} is defined at the downwelling edge, we have to determine a lid thickness at the edge $\delta_{lid,max}$. This is done by extrapolating the mid-width lid slope to the downwelling edge (Figure 10). As shown is Figure 24, the plastic depth is approximately 0.3 to 0.5 of the lid thickness. We take approximate values for our scaling relations rather than scaling these properties with convective parameters because the trends observed in Figure 24 maybe due to insufficient viscosity contrasts which place convection on the boundary of transitional regime, especially for $\theta = 13$ in which the viscosity is reduced to approximately 10^4 by the critical yield stress.

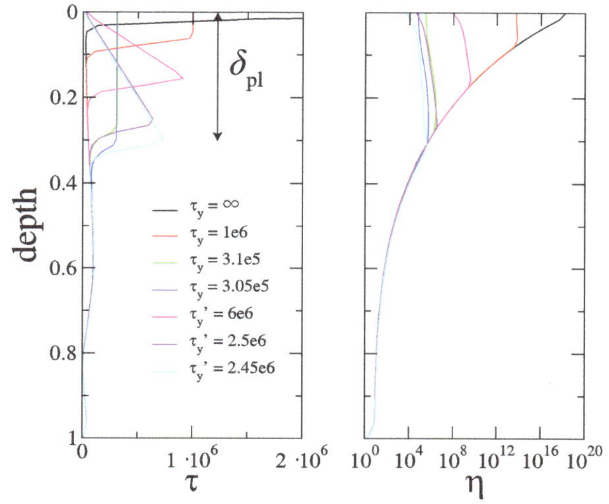

Figure 20 Depth of plastic zone determined by the yield stress (right) and the drop in viscosity due to the yield stress (right). Profiles taken at the downwelling edge ($x = a$) of the convecting cell.

To find the lid stress at δ_{pl} using Equation 34, we also need to determine the distance of the base of the plastic zone from the convective interior y_{pl}. The lid base is at δ_{rh} from the interior, so we express y_{pl} in terms of δ_{rh} to give a sense of distance in relation to rheological sublayer thickness. While there is a general trend of increasing of $y_{\mathrm{pl}}/\delta_{\mathrm{rh}}$ with θ, it is difficult to discern a correlation as $y_{\mathrm{pl}}/\delta_{\mathrm{rh}}$ fluctuates; thus, we take $y_{\mathrm{pl}} \approx 3\delta_{\mathrm{rh}}$ for our scaling relations.

The dependence of δ_{pl}, y_{pl}, and $\Delta\eta_{\mathrm{pl}}$ on Ra and aspect ratio are very weak and therefore assumed negligible.

Scaling for critical yield stress and critical yield stress gradient

Critical yield stress scaling theory

In deriving a theoretical scaling for the critical yield stress, we assume an approximate balance between the force generated by the shear stresses acting at the base of the lid and the normal stress acting on the side of the lid and we assume that the latter are largely dominated by the stresses in the plastic zone δ_y (Solomatov 2004a and Figure 25). Thus, we can express the yield stress as:

$$\begin{aligned} \tau_y &\sim \tau_{\mathrm{lid}} \frac{a}{\delta_y} \\ &= \mathrm{Ra} \frac{dT}{dy} \lambda \frac{y_{\mathrm{pl}}^2}{2} \frac{a}{\delta_y}. \end{aligned} \tag{41}$$

In the case of a constant yield stress, gradient ($\tau_y = \tau_y' z$), τ_y' can be scaled as

$$\begin{aligned} \tau_y' &\sim \tau_{\mathrm{lid}} \frac{a}{\delta_y^2} \\ &= \mathrm{Ra} \frac{dT}{dy} \lambda \frac{y_{\mathrm{pl}}^2}{2} \frac{a}{\delta_y^2}. \end{aligned} \tag{42}$$

In Equations 41 and 42, the yield stress is treated as the normal stress whereas the lithosphere stresses are shear stress. However, the magnitude of stresses is expressed in second invariant, and the yield stress in von Mises

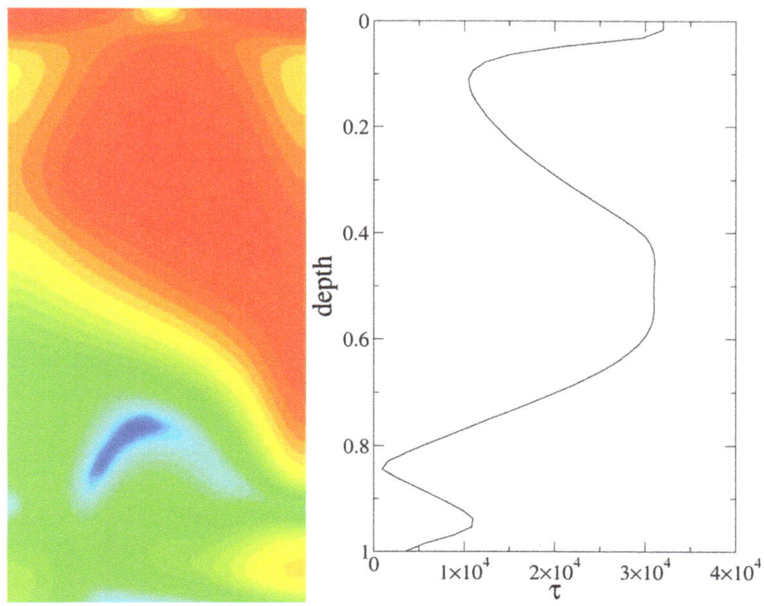

Figure 21 Stress field (left) and stress profile at the downwelling edge. $Ra = 3 \times 10^6$, $a = 0.sx5$, $\theta = 19$, $\tau_y = 3.2 \times 10^4$, exponential viscosity.

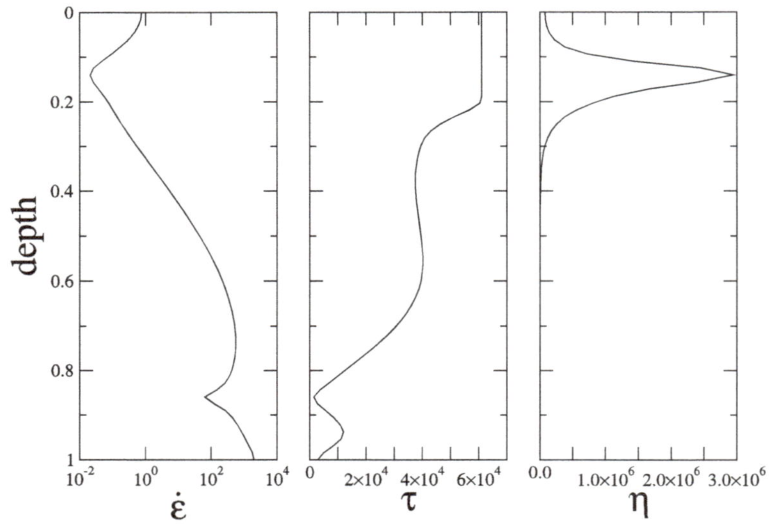

Figure 22 Strain rate, stress, and viscosity profile at the edge with $Ra = 3 \times 10^6$, $a = 0.75$, $\theta = 19$, exponential viscosity. The point of maximum viscosity may not correspond to the brittle-plastic transition.

criterion also put a limit the second stress invariant. To see which stress component contributes to the second invariant, we plot the normal stress and shear stress profiles in Figure 26. Inside the lid where plastic failure occurs, normal stress dominates, whereas shear stress exceeds normal stress below the plastic depth.

The critical yield stress is the stress at $\delta_y = \delta_{pl}$. The critical plastic depth δ_{pl} is taken to be 0.3 to $0.5\delta_{\text{lid}}$, and y_{pl} is about 2 to $4\delta_{rh}$.

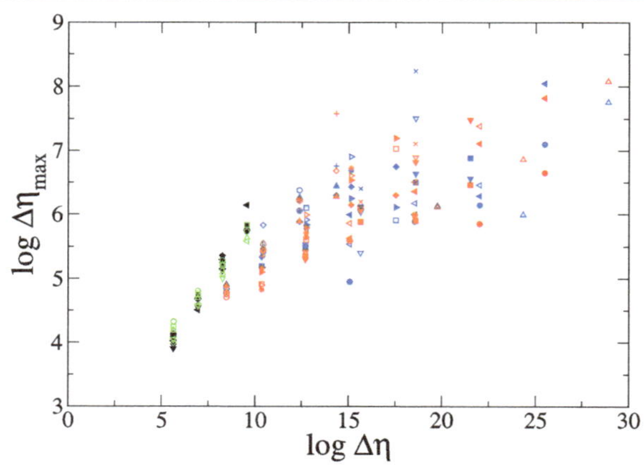

Figure 23 Maximum apparent viscosity in the convective cell at the point of lithospheric failure as a function of the viscosity contrast in the absence of yield stress. Black, exponential viscosity with constant $\tau_{y,cr}$; red, Arrhenius viscosity with constant $\tau_{y,cr}$; green, exponential viscosity with constant $\tau'_{y,cr}$; blue, Arrhenius viscosity with constant $\tau'_{y,cr}$.

$$\tau_{\text{lid}} \sim C_1 \text{Ra} \Delta T_{rh} \lambda \delta_{rh}, \tag{43}$$

$$\tau_{y,cr} \sim C_2 \text{Ra} \Delta T_{rh} \lambda a \frac{\delta_{rh}}{\delta_{\text{lid}}}, \tag{44}$$

$$\tau'_{y,cr} \sim C_3 \text{Ra} \Delta T_{rh} \lambda a \frac{\delta_{rh}}{\delta_{\text{lid}}^2}. \tag{45}$$

For τ_{lid} expressed in terms of Nu (Equation 39), noting that $\text{Nu} \sim dT/dy \sim \delta_{\text{lid}}^{-1}$,

$$\tau_{\text{lid}} \sim C_1 \text{Ra} \text{Nu} \lambda \delta_{rh}^2, \tag{46}$$

$$\tau_{y,cr} \sim C_2 \text{Ra} \text{Nu}^2 \lambda a \delta_{rh}^2, \tag{47}$$

$$\tau'_{y,cr} \sim C_3 \text{Ra} \text{Nu}^3 \lambda a \delta_{rh}^2. \tag{48}$$

where C_1, C_2, C_3 ranges from 4 to 16, 8 to 53, and 16 to 178, respectively.

From the previous sections, since $\Delta T_{\text{rh}}/\Delta T$, δ_{rh}/d, δ_{lid}/d, and λ are all scaled in terms of Ra, a, and θ with scaling exponents summarized in Table 1. τ_{lid}, $\tau_{y,cr}$, and $\tau'_{y,cr}$ can be scaled in terms of Ra, a, and θ. The results are listed in Table 2.

Numerical results for critical yield stress and yield stress gradient: Arrhenius vs. exponential viscosity

Figures 27 and 28 show that both $\tau_{y,cr}$ and $\tau'_{y,cr}$ decrease with increasing θ and total viscosity contrast and might converge to asymptotic values at high viscosity contrasts, although it is difficult to tell from our limited data. To estimate the accuracy of Frank-Kamenetskii approximation in the prediction of critical yield stress, we express the ratio of yield stress for Arrhenius viscosity to that for exponential viscosity $R_\tau = \tau_{y,cr,\text{Arr}}/\tau_{y,cr,\text{exp}}$ and similarly

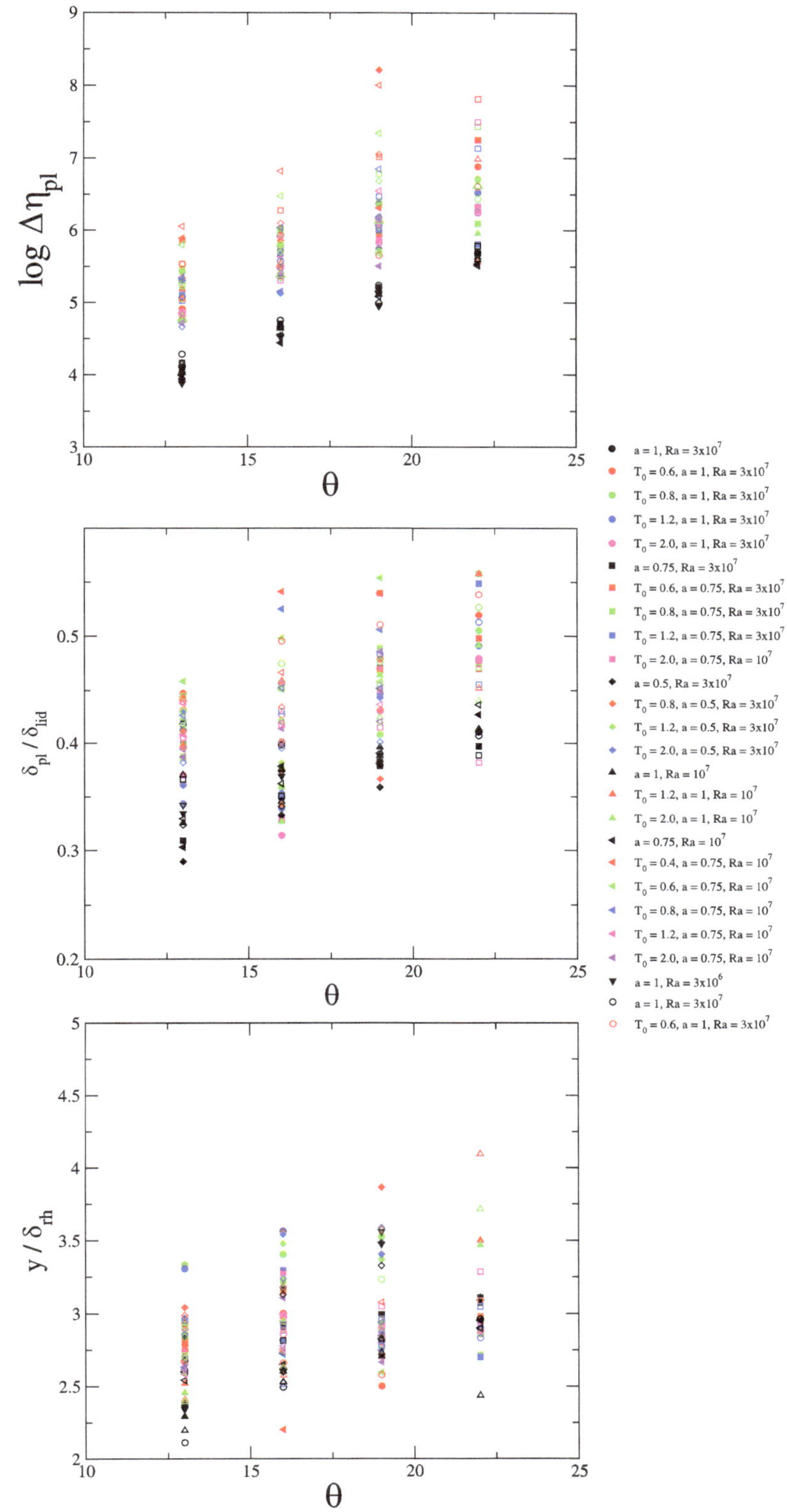

Figure 24 Viscosity contrast at plastic depth, fraction of lid that experience plastic failure, and the distance between the interior and plastic depth as a function of θ. Solid symbols represent constant $\tau_{y,cr}$ cases, while open symbols are constant $\tau'_{y,cr}$.

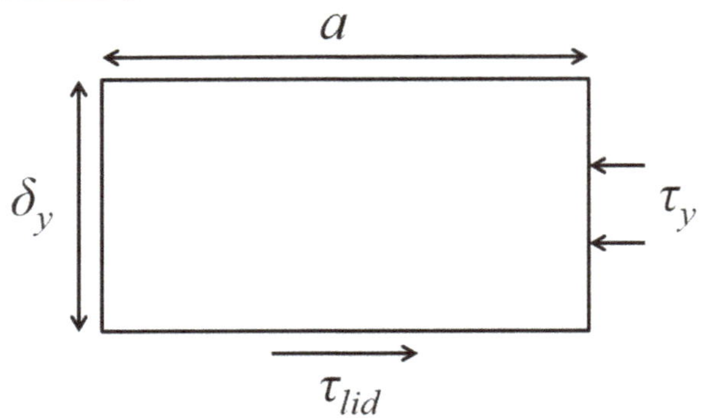

Figure 25 Schematic diagram of surface stresses on the plastic zone in the lid. The shear stress τ_{lid} acting on the base of the lid of horizontal length a is balanced by the normal stress τ_y acting on the side with depth δ_y, developed under free-slip boundary conditions.

for critical yield stress gradient with $R_{\tau'} = \tau'_{y,cr,\mathrm{Arr}}/\tau'_{y,cr,\exp}$ to look at the dependence of these ratios on θ (Figures 29 and 30). Both R_τ and $R_{\tau'}$ increase with θ and the values of yield stress for Arrhenius viscosity and that for Frank-Kamenetskii approximation get closer as the viscosity contrast increases, assuming that $R_\tau < 1$ and $R'_\tau < 1$ at all θ.

For the cases tested at resolution that is doubled, we find that the values for R_τ and $R_{\tau'}$ are within 5% difference. Therefore, $64 \times 64a$ resolution is sufficient for our single-cell steady-state convection analysis.

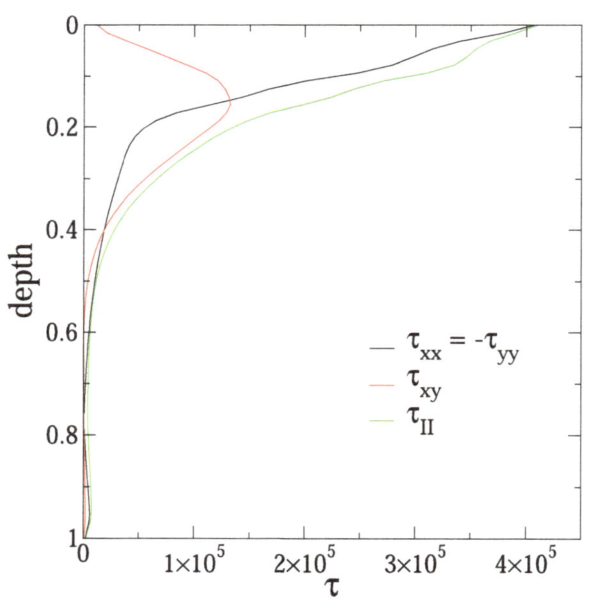

Figure 26 Horizontally averaged stress components and second invariant of stress as a function of depth.

To find the dependence of R_τ and $R_{\tau'}$ on the Arrhenius viscosity contrast $\Delta\eta_{\mathrm{Arr}}$, we consider the ratio of Arrhenius viscosity contrast to exponential viscosity contrast ($\Delta\eta_{\mathrm{Arr}}/\exp\theta$). This ratio reflects the difference between Arrhenius viscosity contrast and exponential viscosity contrast: the larger the ratio, the greater the difference. Figure 31 shows that values of the two $\tau_{y,cr}$ approach each other as the Arrhenius viscosity gets closer to the exponential viscosity. It also suggests that R_τ and $R_{\tau'}$ are mainly determined by the Frank-Kamenetskii parameter θ but less sensitive to Ra and a. As the difference between the Arrhenius viscosity and exponential viscosity increases, both R_τ and $R_{\tau'}$ appear to approach some asymptotic value.

We also investigate the dependence of critical yield stress and yield stress gradient on aspect ratio (Figures 32 and 33). The positive power law coefficient (see Table 1) implies that with smaller aspect ratio, the critical yield value is lower and thus more difficult to reach the yielding criterion. Therefore, smaller cells are more stable. This may explain the phenomenon where overturning of the cold lid is observed once or a few times but then stabilized after reconfiguring into smaller cells.

Figures 34 and 35 and Table 1 show that $\tau_{y,cr}$ and $\tau'_{y,cr}$ are approximately proportional with Ra.

The plots in Figures 27, 28, 32, 33, 34, and 35 show that the scaling exponents could have a range of values, and the cases using Arrhenius viscosity may have a bit different values from those with exponential viscosities. Figure 36 shows that while the scaling exponent for Ra and a for exponential viscosity cases lie between the range obtained from Arrhenius viscosity cases, the exponent for θ varies with surface temperature T_0, Ra, and a. In general, the scaling exponent of θ increases with the viscosity contrast (i.e., lower T_0), a, and moderately with Ra. This may be related to the difference in the stress distribution in the lid in Arrhenius cases, especially towards the downwelling edge, as discussed previously. In the future, it will be worthwhile to look in more detail at the stress variations for the Arrhenius viscosity and to see if there are any relationships between R_τ, $R_{\tau'}$, and T_0 at even lower T_0 (i.e., at higher viscosity contrasts.)

Discussion and conclusions

Comparison with other studies of stress scaling laws

The studies dealing with convective stress scaling often aim to provide an expression for stress in terms of radius and mass of a planet to predict the likelihood of plate tectonics. They reached various conclusions (e.g., O'Neill et al. 2007; O'Neill and Lenardic 2007; Valencia and O'Connell 2007;2009; Korenaga 2010a; Van Heck and Tackley 2011; Stamenkovic and Breuer 2014). One of the main difficulties in deriving convincing scaling laws for

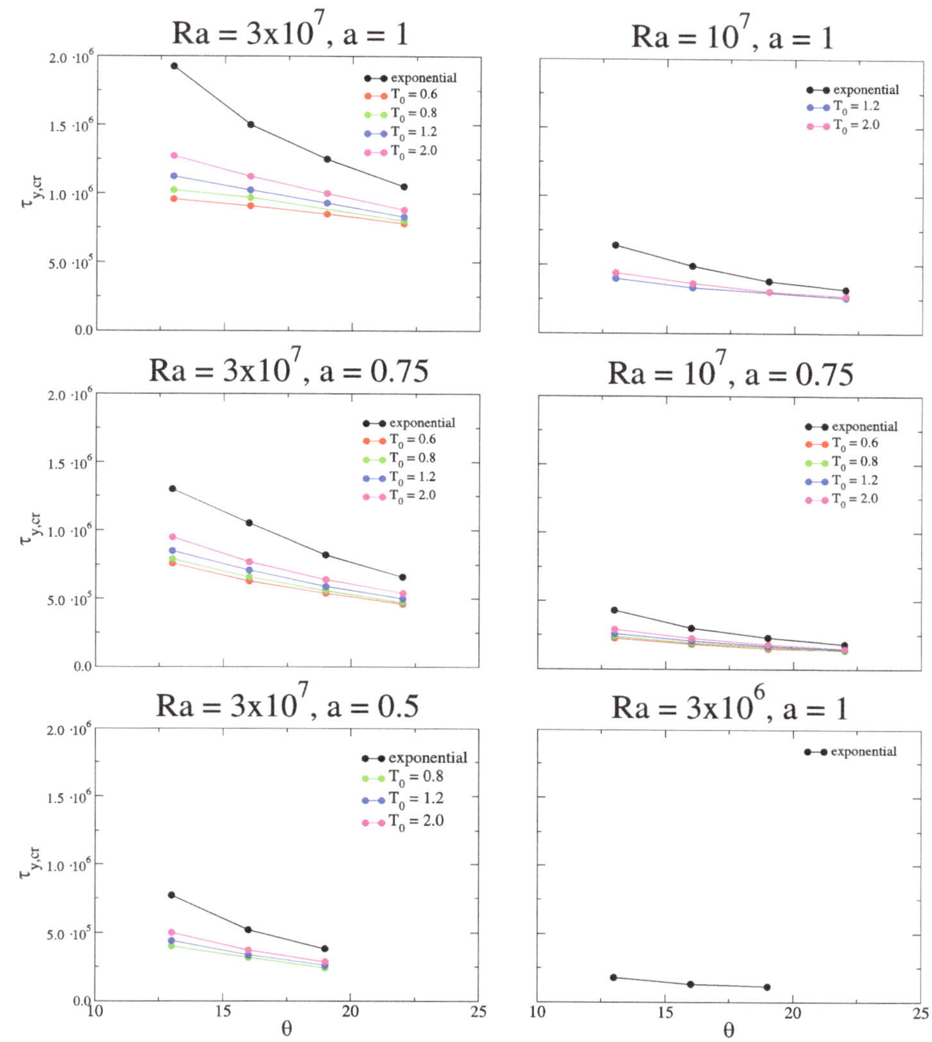

Figure 27 Critical yield stress $\tau_{y,cr}$ as a function of Frank-Kamenetskii parameter θ. The Arrhenius viscosity is calculated with the non-dimensional surface temperature T_0 and the activation energy E that gives the corresponding θ. Lower T_0 gives a higher viscosity contrast $\Delta\eta_{Arr}$.

plate tectonics initiation was a poor understanding of lid stresses and how they are related to lid failure. In the present study, we have addressed these issues using two-dimensional steady-state convective cell simulations. This is the simplest system to analyze, and yet even for this system, the derivation of scaling laws proved to be complicated and not well described by the existing asymptotic theories. Below, we discuss some differences between our study and previous studies and summarize our scaling laws in a dimensional form.

In some studies (e.g., Moresi and Solomatov 1998; Trompert and Hansen 1998; Tackley 2000b; Fowler and O'Brien 2003), the authors assumed that subduction occurs when the stresses in the convective interior exceed the yield stress. This means that subduction begins when not only the lid but also the interior of the convective cell fails. However, subduction initiation may not necessarily require the failure of interiors but instead may only require failure of just a small portion of the lid. The stresses in the lid are several orders of magnitude higher than the stresses in the interior, and also, they scale differently. Thus, the assumption regarding what part of the convective cell must fail in order for subduction to begin is critically important. In this study, we have investigated this assumption quantitatively, based on a detailed analysis of stresses and other parameters in the convective cell, and then formulated the critical conditions for subduction initiation.

In agreement with Fowler (1985), we have shown that the lid slope is a key factor in determining the stresses

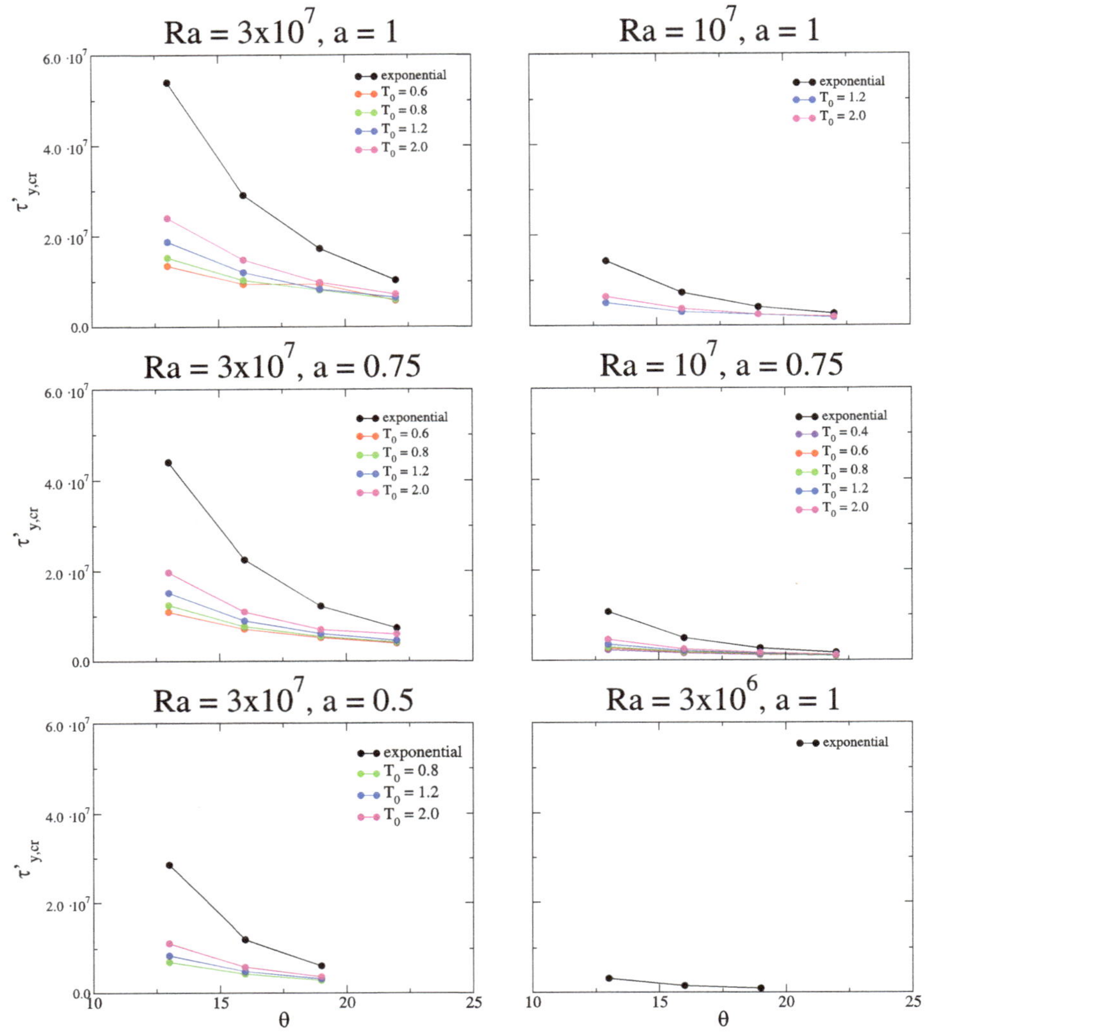

Figure 28 Critical yield stress gradient $\tau'_{y,cr}$ as a function of θ.

in the lid. However, our model has several important differences from Fowler (1985). The theoretical solution in Fowler (1985) is a similarity solution and does not take into the finite horizontal extent of the lid. Our model has vertical boundaries, and thus, the structure of the lid in our model is more complex. Also, the solution in Fowler (1985) is an asymptotic solution requiring very high values of parameters, such as Ra and θ, and a satisfaction of certain asymptotic conditions, which are not reached in our simulations and may not necessarily be reached on planets. Thus, our scaling laws are not asymptotic in this sense. Also, solutions in Fowler (1985) are obtained for two end-member cases, the large lid slope case, and the small lid slope case. We find that the lid slope behaves in a more complex way and is between these two end-member cases. We have determined a scaling law for the lid slope numerically and used it to derive the scaling law for the stresses in the lid.

Our analysis suggests that the stresses in the lid increase approximately as a square of the distance from the bottom of the lid (Equation 34 and Figure 14). This agrees with the asymptotic analysis of Fowler (1985) but is different from the stress distribution in Solomatov (2004a). In Solomatov (2004a), the stress distribution was more complex because the convective cell was heated from within rather than from the bottom and the internal heating affected the temperature-induced density distribution in the lid. At Rayleigh numbers higher than those reached in Solomatov (2004a), the lid is expected to become sufficiently thin so that the heat production inside the lid would be negligible compared to the heat flux at the base of the lid. Thus, we expect that for convection with internal heating, the

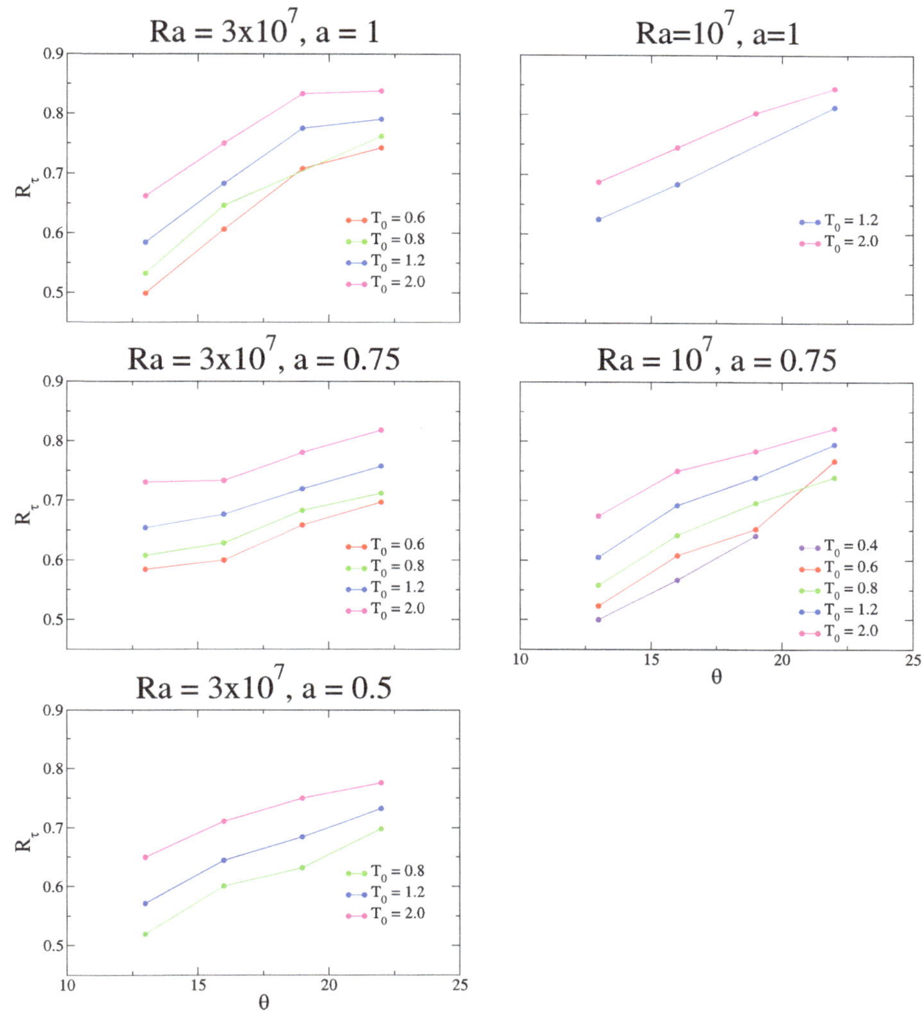

Figure 29 The ratio of Arrhenius yield stress to exponential yield stress R_τ as a function of Frank-Kamenetskii parameter θ.

stress distribution in the lid should approach the quadratic distribution that we observe for convection with bottom heating.

We find that subduction initiation requires that only a part of the lid undergoes plastic failure, roughly 0.3 to 0.5 of the total lid thickness. This generally agrees with the analysis in Solomatov (2004a) and confirms that the plastic failure does not have to extend all the way to the bottom of the lid as was assumed in Fowler and O'Brien (2003). However, unlike Solomatov (2004a), we determine the distance to the boundary of the plastic failure zone by measuring it from the base of the lid and scaling it in terms of the rheological boundary layer thickness. We find that such an approach is more appropriate because the mobility of the lid is largely controlled by the viscosity contrast between the zone of failure and the convective interior of the cell, which in turn is scaled with the rheological boundary layer thickness.

Estimates for the Earth

To compare our results with those obtained in Solomatov (2004a,b), we convert the critical yield stress $\tau_{y,cr}$ and critical yield stress gradient $\tau'_{y,cr}$ into their dimensional forms (Equations 20 and 22) and estimate the critical yield strength and the critical coefficient of friction μ for subduction initiation on the Earth.

The interior viscosity cannot be reliably estimated from the viscosity law alone and is usually determined from better constrained properties such as lithospheric thickness. Therefore, following Solomatov (2004a), we use the scaling law for δ_0 (Table 1) and present

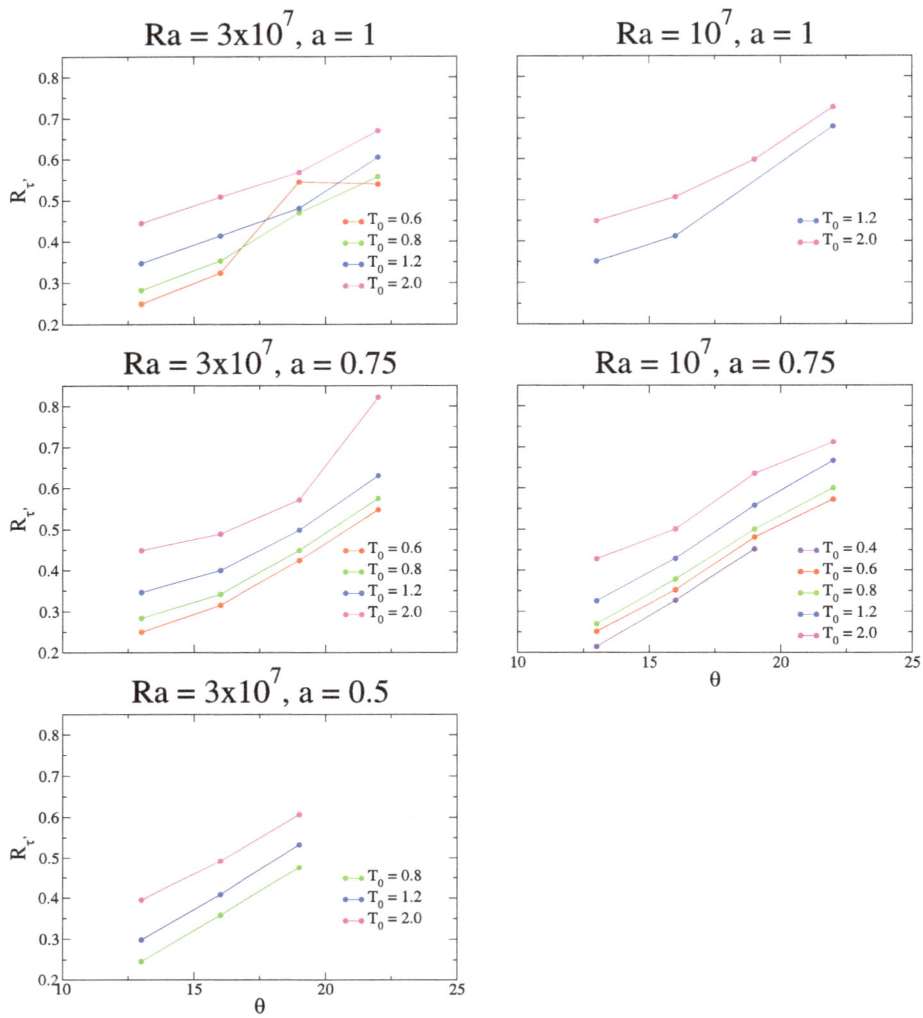

Figure 30 The ratio of Arrhenius yield stress gradient to exponential yield stress gradient $R_{\tau'}$ as a function of θ.

the results in terms of the thickness of the thermal boundary layer $\delta_0 \sim 100$ km instead of the mantle viscosity η_1.

The scaling law for the critical yield stress depends strongly on aspect ratio a. Previous studies have scaled the aspect ratio from half-space cooling of lithosphere (Korenaga 2010b; Stamenkovic and Breuer 2014) or estimated from numerical simulations (Solomatov 2004a,b), whereas it was assumed to be on the order of 1 in Valencia and O'Connell (2009). We use the horizontal width of the convective cells as $l_{\text{hor}} = ad$ ~100 km as a very rough value to compare our estimates with those in Solomatov (2004a,b). This value was inferred from observational constraints on the present-day horizontal scale of sublithospheric convective structures (Solomatov 2004a).

Using the results in Table 1, we obtain that the dimensional critical yield stress for subduction initiation is as follows:

$$\tau_{y,cr} \sim 1.95\alpha\rho g \left(\frac{E}{RT_i^2}\right)^{-1.03} \Delta T^{-0.03}\delta_0^{-0.41} l_{\text{hor}}^{1.78} d^{-0.37}. \quad (49)$$

For comparison, Solomatov (2004a,b) give:

$$\tau_{y,cr} \sim 13\alpha\rho g \left(\frac{E}{RT_i^2}\right)^{-2} \Delta T^{-1} l_{\text{hor}}. \quad (50)$$

The critical coefficient of friction μ for subduction initiation is as follows:

$$\mu \sim 89\alpha \left(\frac{E}{RT_i^2}\right)^{-1.74} \Delta T^{-0.74}\delta_0^{-1.55} l_{\text{hor}}^{1.87} d^{-0.32}, \quad (51)$$

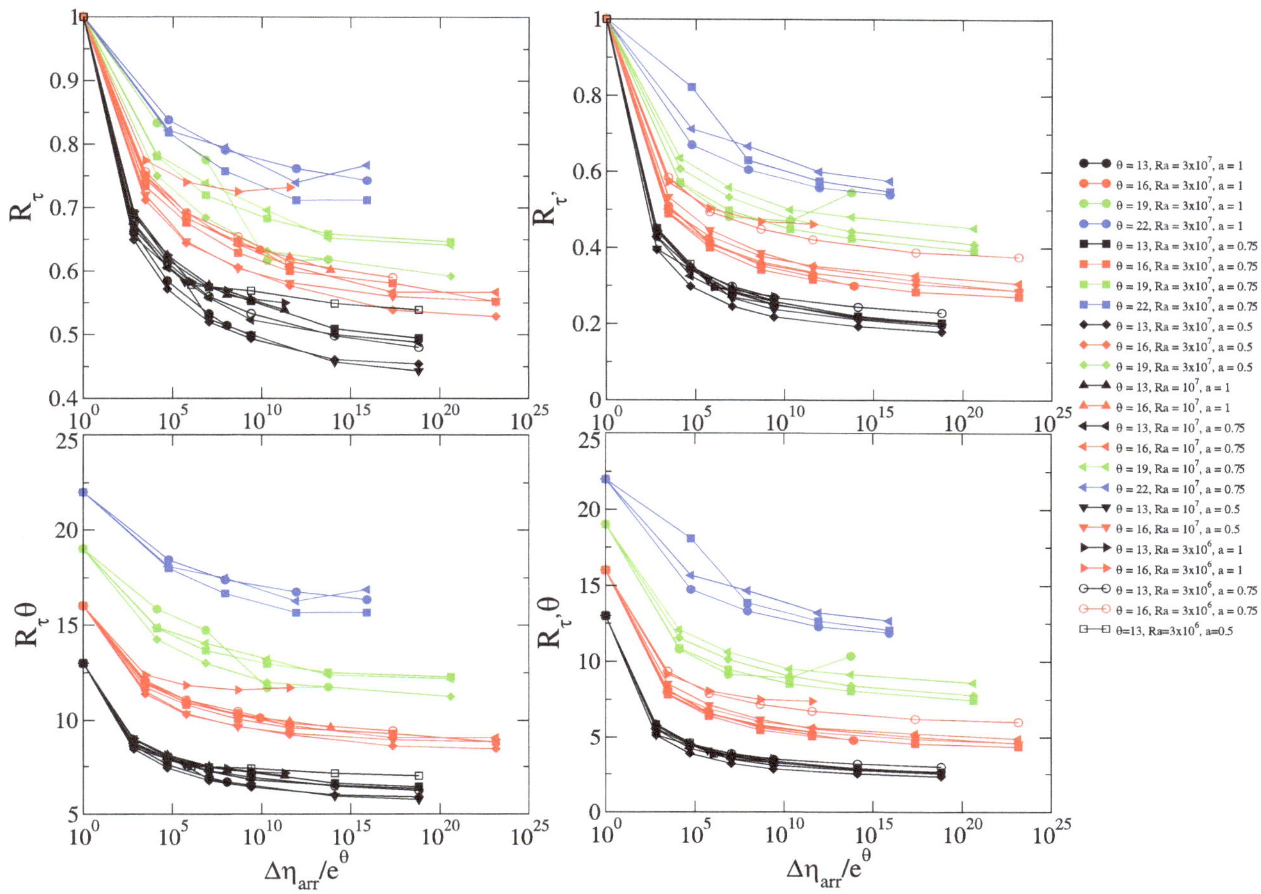

Figure 31 Top figures: ratio of yield stress for Arrhenius viscosity to that for exponential viscosity R_τ and $R_{\tau'}$ as a function of Arrhenius viscosity contrast normalized to exponential viscosity contrast $\Delta\eta_{Arr}/\exp(\theta)$. $\Delta\eta_{Arr}/\exp(\theta) \geq 1$, and it can go up to many orders of magnitude. Bottom figures: ratio of yield stress multiplied by θ. For cases with the same θ but various Ra and a, these ratios do not differ much, meaning that θ is the controlling factor for the difference in yield stress predicted by Arrhenius viscosity and exponential viscosity. Asymptotically towards high $\Delta\eta_{Arr}/\exp(\theta)$, R_τ and R'_τ are approximately proportional to θ.

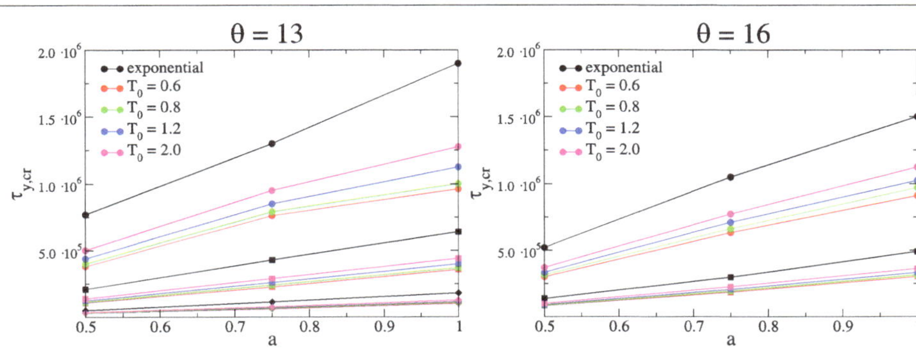

Figure 32 Critical yield stress $\tau_{y,cr}$ as a function of aspect ratio. Circles represent Ra = 3×10^7, squares Ra = 10^7, diamonds Ra = 3×10^6.

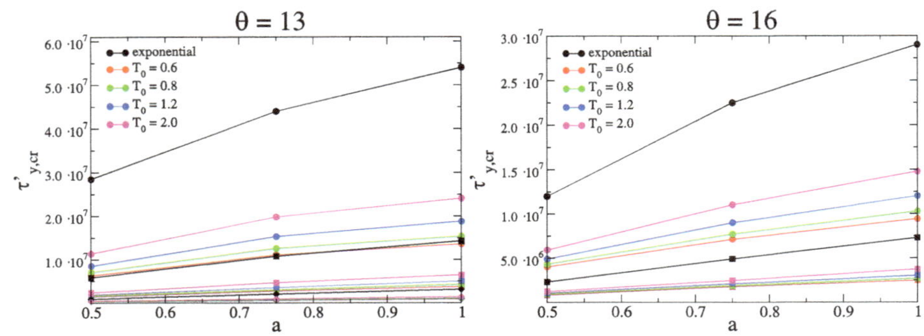

Figure 33 Critical yield stress gradient $\tau'_{y,cr}$ as a function of aspect ratio. Legends are the same as in Figure 32.

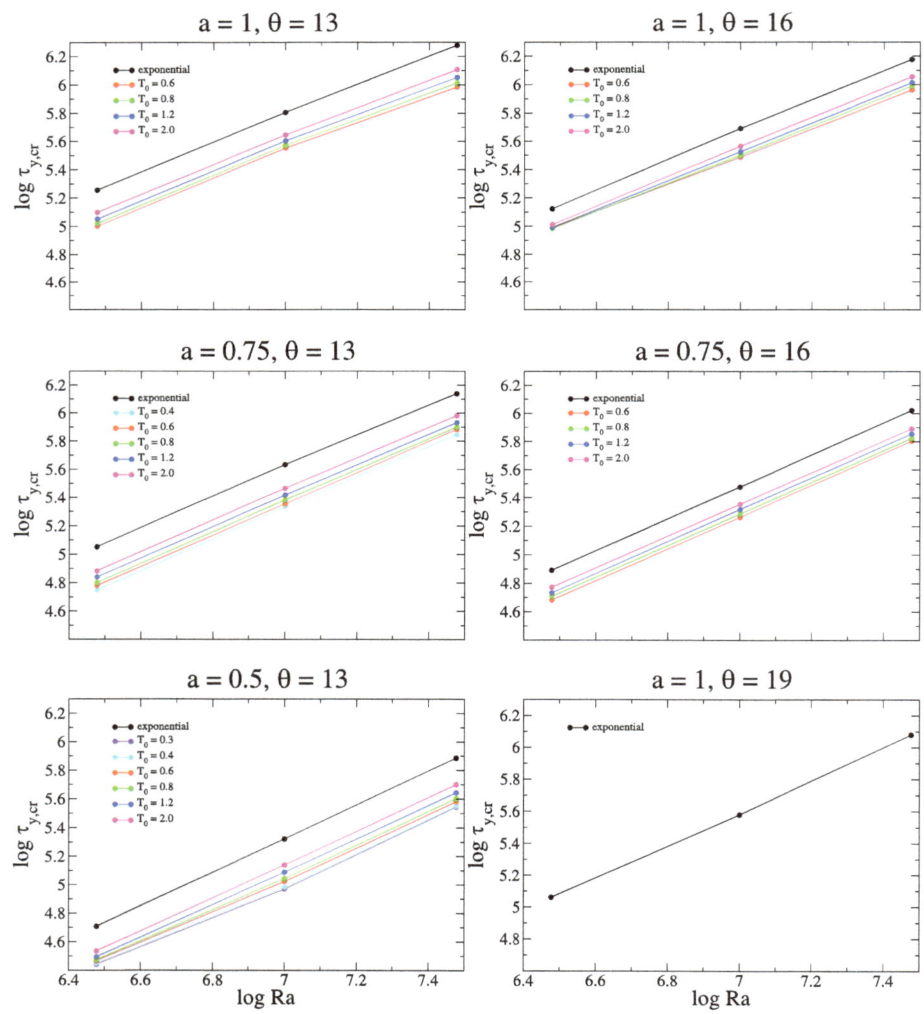

Figure 34 Critical yield stress $\tau_{y,cr}$ as a function of Rayleigh number.

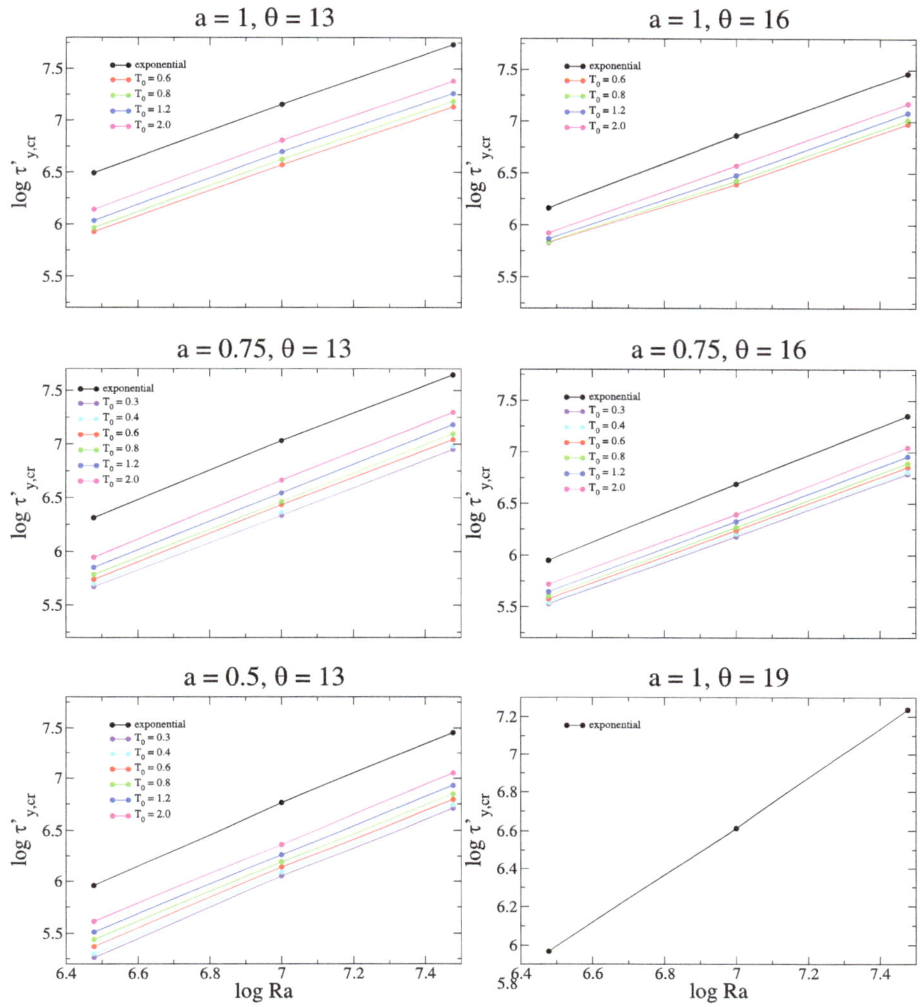

Figure 35 Critical yield stress gradient $\tau'_{y,cr}$ as a function of Rayleigh number.

and in Solomatov (2004a,b):

$$\mu \sim 50\alpha \left(\frac{E}{RT_i^2}\right)^{-2} \Delta T^{-1} \delta_0^{-1} l_{\text{hor}}. \quad (52)$$

Using the typical values of various physical parameters (Table 3), we estimate that the yield strength for the Earth is 5 MPa which is of the same order of magnitude as 3 MPa obtained by Solomatov (2004a,b). To see how variations in various parameters may affect these estimates, it is useful to present the estimates in a different form. Our estimate (Equation 49) can be written as:

$$\tau_{y,cr} \sim 5 \left(\frac{100\text{ km}}{\delta_0}\right)^{0.41} \left(\frac{l_{\text{hor}}}{100\text{ km}}\right)^{1.78} \left(\frac{500\text{ km}}{d}\right)^{0.37} \text{MPa}, \quad (53)$$

and the estimate from Solomatov (2004a,b) (Equations 50) is as follows:

$$\tau_{y,cr} \sim 3 \left(\frac{l_{\text{hor}}}{100\text{ km}}\right) \text{MPa}, \quad (54)$$

from Equation 50.

Our estimates of the critical friction coefficient μ is 8×10^{-3}, which is a factor of 3 larger than 3×10^{-3} obtained in Solomatov (2004a,b). Our estimate can be written as

$$\mu \sim 0.008 \left(\frac{100\text{ km}}{\delta_0}\right)^{1.55} \left(\frac{l_{\text{hor}}}{100\text{ km}}\right)^{1.87} \left(\frac{500\text{ km}}{d}\right)^{0.32}, \quad (55)$$

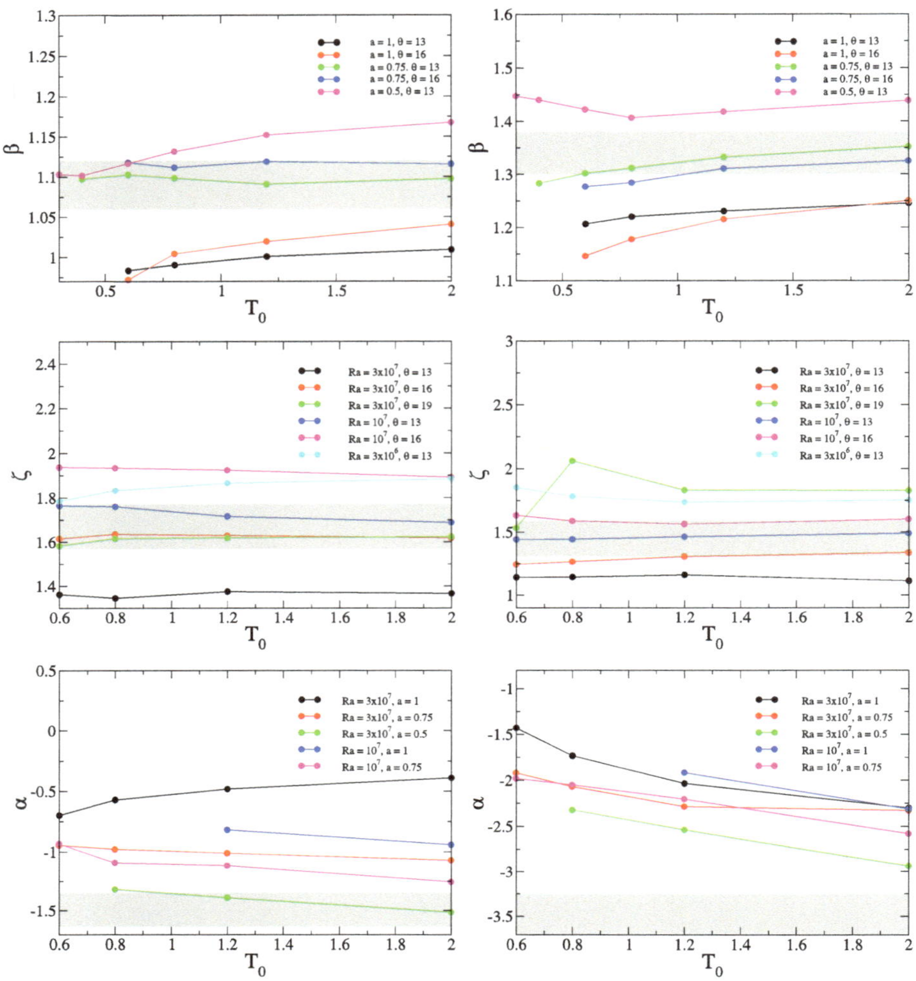

Figure 36 Scaling exponents in $\mathrm{Ra}^{\beta} a^{\zeta} \theta^{\alpha}$ for τ_y (left three figures) and $\tau'_{y,cr}$ (right three figures) with varying T_0 in Arrhenius viscosity. The grey stripes represent the value obtained from exponential viscosity calculations, with the width determined by error bars.

Table 3 Parameters used to estimate $\tau_{y,cr}$ and $\tau'_{y,cr}$ for Earth as in Solomatov (2004a,b)

α	3×10^{-5}
κ	$10^{-6}\ \mathrm{m^2\ s^{-1}}$
δ_{lid}	100 km
k	$3\ \mathrm{W\ m^{-1}\ K^{-1}}$
E	430 kJ mol− 1
d	approximately 500 km
g	$10\ \mathrm{m\ s^{-2}}$
ρ	$3{,}300\ \mathrm{kg\ m^{-3}}$
T_0	300 K
T_i	1,700 K
l_{hor}	100 km

and the estimate from Solomatov (2004a,b) (Equation 50) is as follows:

$$\mu \sim 0.003 \left(\frac{l_{\mathrm{hor}}}{100\ \mathrm{km}}\right)\left(\frac{100\ \mathrm{km}}{\delta_0}\right). \tag{56}$$

If we take into account the fact that Frank-Kamenetskii approximation that we used to derive the scaling laws overestimate the critical yield stress and the critical friction coefficient (Figure 31), then both ours and the estimates in Solomatov (2004a,b) should be further reduced by a factor of 2 (Figure 31), depending on the values of the viscosity parameters and the Rayleigh number.

One major difference between our scaling laws and the scaling laws obtained in Solomatov (2004a) is a much stronger dependence of the critical yield stress and critical friction coefficient on the width of the convecting layer l_{hor} - they scale roughly as approximately l_{hor}^2 as opposed to the previous scaling approximately l_{hor}. This means that the critical values of the yield stress and friction coefficient would increase by 2 to 4 orders of magnitude if the width of the convective cells increased by 1 to 2 orders of magnitude (for example, in the past history of the Earth),and thus, at least in principle, could reach the experimentally observed values that are on the order of 1,000 MPa for $\tau_{y,cr}$ and μ ~0.6 to 0.85 (e.g., Byerlee 1978; Goetze and Evans 1979; Kohlstedt et al. 1995; Mei et al. 2010) and values constrained by loading models with *in situ* stress measurements of Hawaiian Islands (Zhong and Watts 2013) which are 0.25 to 0.7 for μ and 100 to 200 MPa for lithospheric stress. This implies that the chances of plate tectonics might be higher than we thought before. Time-dependent calculations and a more realistic formulation of the problem are required to better understand the implications of these results for plate tectonics initiation.

Uncertainties in stress scaling

The scaling laws derived here are applicable to Newtonian rheology; therefore, the activation energy for diffusion creep is used in our calculations. However, it should be noted that dislocation creep is probably the dominant mechanism in the lithosphere (Karato and Wu 1993). For the Earth, wet dislocation creep may be preferable (Solomatov and Moresi 2000) while for other terrestrial planets such as Venus might have dry lithosphere. To apply on a wider range of planets including icy bodies, scaling laws based on non-Newtonian rheology will be required.

In previous scaling theories, the lid slope is often considered to be small because the lid thickness is assumed to be relatively small. Even in the large lid slope end member case in Fowler's theory, δ_{lid} is assumed to be small relative to the thickness of the convecting layer.

However, our simulation indicates that the slope may be significant, so the derivations may need to be modified to take this into account.

Free-slip boundary conditions are often used in solving equations for thermal convection, but this restricts the vertical motion of the surface. Recent studies have used the free-surface boundary conditions, which is closer to natural surface condition as both normal and shear stress on the surface is reduced to zero (Zhong et al. 1996; Schmeling et al. 2008; Kaus et al. 2010; Crameri et al. 2012; Kramer et al. 2012). It maybe computationally expensive to implement this for the time being, but it could be worthwhile to explore its effect on scaling relations for stresses in the future.

Our numerical results show that $\tau_{y,cr}$ of Arrhenius viscosity approaches that of exponential viscosity as the Frank-Kamenetskii parameter θ increases. This enables us to use exponential viscosity law to extrapolate to high Arrhenius viscosity contrast conditions. Besides the Frank-Kamenestskii approximation, the viscosity contrasts can be reduced in other ways, one of which is to set a cutoff viscosity. The stress structure resulting from the cutoff viscosity will have to be examined. We can then compare the accuracy of these approximations and apply them to extrapolate the results to planetary parameters.

Our results generally support previous conclusions that in order for the convective regime on the terrestrial planets in the inner Solar System to change from stagnant lid convection to plate tectonics, the yield stress of the lithosphere should be much smaller (several MPa) than that predicted by laboratory experiments on rock deformation (hundreds of MPa as predicted by Byerlee's law). However, our results suggest a much stronger dependence of the critical yield stress on the horizontal width of the convective cells. This opens a possibility of subduction initiation even for the large, experimentally measured, lithospheric strength provided that a sufficiently long convective cell forms in a time-dependent mantle convection. In the future, it would be important to investigate the role of initiation conditions and statistical fluctuations of convective cells for the initiation of subduction in time-dependent convection.

Competing interests
The authors declare that they have no competing interests.

Authors' contributions
Both authors read and approved the final manuscript.

Authors' information
TW is a graduate student in Washington University in St. Louis and VS is her advisor.

Acknowledgements
The authors thank NSF for partial support and the two anonymous reviewers for their constructive comments to improve the manuscript.

References
Bercovici D, Ricard Y, Schubert G (2001) A two-phase model for compaction and damage 3. Applications to shear localization and plate boundary formation. J Geophys Res-Solid Earth 106(B5):8925–8939. doi:10.1029/2000JB90043

Bercovici D, Ricard Y (2005) Tectonic plate generation and two-phase damage: void growth versus grain size reduction. J Geophys Res-Solid Earth 110(B3):B03041

Bercovici, D, Ricard Y (2012) Mechanisms for the generation of plate tectonics by two-phase grain-damage and pinning. Phys Earth Planet Int 202–203:27–55. doi:10.1016/j.pepi.2012.05.003

Branlund J, Regenauer-Lieb K, Yuen D (2001) Weak zone formation for initiating subduction from thermo-mechanical feedback of low-temperature plasticity. Earth Planet Sci Lett 190(3–4):237–250

Byerlee J (1978) Friction of rocks. Pure Appl Geophys 116(4–5):615–626. doi:10.1007/BF0087652

Crameri F, Schmeling H, Golabek GJ, Duretz T, Orendt R, Buiter SJH, May DA, Kaus BJP, Gerya TV, Tackley PJ (2012) A comparison of numerical surface topography calculations in geodynamic modelling: an evaluation of the

'sticky air' method. Geophys J Int 189(1):38–54. doi:10.1111/j.1365-246X.2012.05388.x
Frank-Kamenetskii DA (1969) Diffusion and heat exchange in chemical kinetics. Plenum Press, New York. p 574
Fowler A (1985) Fast thermoviscous convection. Stud Appl Math 72(3):189–219
Fowler, A (1993) Boundary-layer theory and subduction. J Geophys Res-Solid Earth 98(B12):21997–22005
Fowler A, O'Brien S (2003) Lithospheric failure on venus. Proc Roy Soc Lond Math Phys Sci 459(2039):2663–2704
Goetze C, Evans B (1979) Stress and temperature in the bending lithosphere as constrained by experimental rock mechanics. Geophys J R Astr Soc 59(3):463–478
Gurnis M, Hall C, Lavier L (2004) Evolving force balance during incipient subduction. Geochem Geophys Geosystems (G^3) 5:07001. doi:10.1029/2003GC000681
Hall C, Gurnis M, Sdrolias M, Lavier L, Muller R (2003) Catastrophic initiation of subduction following forced convergence across fracture zones. Earth Planet Sci Lett 212(1–2):15–30
Hirth G, Kohlstedt D (2003) Rheology of the upper mantle and mantle wedge: a view from the experimentalists. Geophys Monogr Ser 138:83–105
Kanamori H (1994) Mechanics of earthquakes. Annu Rev Earth Planet Sci 22:207–237
Kanamori H, Brodsky E (2004) The physics of earthquakes. Rep Progr Phys 67:1429–1496
Karato S, Wu P (1993) Rheology of the upper mantle - a synthesis. Science 260(5109):771–778
Karato S, Zhang S, Wenk H (1995) Superplasticity in Earth's lower mantle - evidence from seismic anisotropy and rock physics. Science 270(5235):458–461
Kaus BJP, Mulhaus H, May DA (2010) Astabiliplumes algorithm for geodynamic numerical simulations with a free surface. Phys Earth Planet Int 181:12–20
Kemp DV, Stevenson DJ (1996) A tensile flexural model for the initiation of subduction. Geophys J Int 125:73–96
King S (2009) On topography and geoid from 2-D stagant lid convection calculations. Geochem Geophys Geosystems (G^3) 10(3):Q03002. doi:10.1029/2008GC002250
Kohlstedt, D, Evans B, Mackwell S (1995) Strength of the lithosphere - constraints imposed by laboratory experiments. J Geophys Res-Solid Earth 100(B9):17587–17602
Korenaga, J (2007) Thermal cracking and the deep hydration of oceanic lithosphere: a key to the generation of plate tectonics? J Geophys Res-Solid Earth 112(B5):05408. doi:10.1029/2006JB004502
Korenaga J (2009) Scaling of stagnant-lid convection with arrhenius rheology and the effects of mantle melting. Geophys J Int 179:54–170. doi:10.1111/j.1365-246X.2009.04272.x
Korenaga, J (2010a) On the likelihood of plate tectonics on super-earths: does size matter? ApJ 725(1):43–46. doi:10.1088/2041-8205/725/1/L43
Korenaga J (2010b) Scaling of plate tectonic convection with pseudoplastic rheology. J Geophys Res-Solid Earth 115:11405. doi:10.1029/2010JB007670
Kramer SC, Wilson CR, Davies DR (2012) An implicit free surface algorithm for geodynamic simulations. Phys Earth Planet Int 194–195:25–37
Landuyt W, Bercovici D, Ricard Y (2008) Plate generation and two-phase damage theory in a model of mantle convection. Geophys J Int 174(3):1065–1080. doi:10.1111/j.1365-246X.2008.03844.x
McKenzie DP (1977) The initiation of trenches: a finite amplitude instability. In: Talwani M, Pitman WCI (eds). Island Arcs, Deep Sea Trenches, and Back-Arc Basins, Maurice Ewing Series, Chap. 1. AGU, Washington, D. C. pp 57–61. doi:10.1029/ME001
Mei S, Suzuki AM, Kohlstedt DL, NA D, Durham WB (2010) Experimental constraints on the strength of the lithospheric mantle. J Geophys Res:B08204
Morris S (1982) The effects of a strongly temperature-dependent viscosity on slow flow past a hot sphere. J Fluid Mech 124:1–26
Morris S, Canright D (1984) A boundary-layer analysis of benard convection in a fluid of strongly temperature-dependent viscosity. Phys Earth Planet Int 36(SI):355–373
Moresi, L, Solomatov V (1995) Numerical investigation of 2D convection with extremely large viscosity variations. Phys Fluids 7(9):2154–2162
Moresi L, Solomatov V (1998) Mantle convection with a brittle lithosphere: thoughts on the global tectonic styles of the earth and venus. Geophys J Int 133(3):669–682
Mueller S, Phillips R (1991) On the initiation of subduction. J Geophys Res 96(B1):651–665
Nikolaeva K, Gerya T, Marques FO (2010) Subduction initiation at passive margins: numerical modeling. J Geophys Res 115:03406. doi:10.1029/2009JB006549
Ogawa M (1990) Perturbation analysis of convective instability of oceanic lithosphere and initiation of subduction zones. J Geophys Res-Solid Earth 95(B1):409–420
O'Neill C, Lenardic A (2007) Geological consequences of super-sized earths. Geophys Res Lett 34(19):19204
O'Neill C, Jellinek AM, Lenardic A (2007) Conditions for the onset of plate tectonics on terrestrial planets and moons. Earth Planet Sci Lett 261(1–2):20–32. doi:10.1016/j.epsl.2007.05.038
Ratcliff J, Tackley P, Schubert G, Zebib A (1997) Transitions in thermal convection with strongly variable viscosity. Phys Earth Planet Int 102(3–4):201–212
Reese C, Solomatov V, Moresi L (1999) Non-Newtonian stagnant lid convection and magmatic resurfacing on Venus. Icarus 139(1):67–80
Richards M, Yang W, Baumgardner J, Bunge H (2001) Role of a low-viscosity zone in stabilizing plate tectonics: implications for comparative terrestrial planetology. Geochem Geophys Geosystems (G^3) 2(2000GC00043):1040
Regenauer-Lieb K, Kohl T (2003) Water solubility and diffusivity in olivine: its role in planetary tectonics. Mineral Mag 67(4):697–715. doi:10.1180/002646103674012
Regenauer-Lieb K, Yuen D (2004) Positive feedback of interacting ductile faults from coupling of equation of state, rheology and thermal-mechanics. Phys Earth Planet Int 142(1–2):113–135. doi:10.1016/j.pepi.2004.01.00
Regenauer-Lieb K, Yuen D, Branlund J (2001) The initiation of subduction: criticality by addition of water?. Science 294(5542):578–580
Regenauer-Lieb K, Hobbs B, Yuen D, Ord A, Zhang Y, Mulhaus H, Morra G (2006) From point defects to plate tectonic faults. Philos Mag 86(21-22):3373–3392. doi:10.1080/1478643050037515. Symposium on Instabilities Across the Scales, Cairns, Australia, Sep 14–17, 2004
Schmeling H, Babeyko AY, Enns A, Faccenna C, Funiciello F, Gerya T, Golabek GJ, Grigull S, Kaus BJP, Morra G, Schmalholz SM, van Hunen J (2008) A benchmark comparison of spontaneous subduction models–towards a free surface. Phys Earth Planet In 171:198–223. doi:10.1016/j.pepi.2008.06.028. Recent Advances in Computational Geodynamics: Theory, Numerics and Applications
Solomatov V (1995) Scaling of temperature- and stress-dependent viscosity convection. Phys Fluids 7(2):266–274
Solomatov, V (2004a) Initiation of subduction by small-scale convection. J Geophys Res-Solid Earth 109(B1):B01412. doi:10.1029/2003JB002628
Solomatov VS (2004b) Correction to "initiation of subduction by small-scale convection". J Geophys Res 109(B05408). doi:1029/2004JB003143
Solomatov, V S (2012) Localized subcritical convective cells in temperature-dependent viscosity fluids. Phys Earth Planet In 200–201:63–71. doi:10.1016/j.pepi.2012.04.005
Solomatov V, Moresi L (1996) Stagnant lid convection on venus. J Geophys Res-Planets 101(E2):4737–4753
Solomatov, V, Moresi L (2000) Scaling of time-dependent stagnant lid convection: application to small-scale convection on earth and other terrestrial planets. J Geophys Res-Solid Earth 105(B9):21795–21817
Stamenkovic V, Breuer D (2014) The tectonic mode of rocky planets: part 1 driving factors, models & parameters. Icarus In press. doi:10.1016/j.icarus.2014.01.042
Stein C, Hansen U (2008) Plate motions and the viscosity structure of the mantle - insights from numerical modelling. Earth Planet Sci Lett 272(1–2):29–40
Stein, C, Hansen U (2013) Arrhenius rheology versus Frank-Kamenetskii rheology - implications for mantle dynamics. G-cubed 14(8):2757–2770. doi:10.1002/ggge.20158
Stein C, Schmalzl J, Hansen U (2004) The effect of rheological parameters on plate behaviour in a self-consistent model of mantle convection. Phys Earth Planet Int 142(3–4):225–255
Stern R (2004) Subduction initiation: spontaneous and induced. Earth Planet Sci Lett 226(3–4):275–292
Tackley P (2000a) Self-consistent generation of tectonic plates in time-dependent, three-dimensional mantle convection simulations 2. Strain rate weakening and asthenosphere. G-cubed 1:2000–000043. doi:10.1029/2000GC0003

Tackley PJ (2000b) Self-consistent generation of tectonic plates in time-dependent, three-dimensional mantle convection simulations. G-cubed 1(8):1021. doi:10.1029/2000GC000036
Toth J, Gurnis M (1998) Dynamics of subduction initiation at preexisting fault zones. J Geophys Res-Solid Earth 103(B8):18053–18067
Trompert R, Hansen U (1998) Mantle convection simulations with rheologies that generate plate-like behaviour. Nature 395(6703):686–689
Turcotte DL (1977) Lithospheric instability. In: Talwani M, Pitman WCI (eds). Island Arcs, Deep Sea Trenches, and Back-Arc Basins, Maurice Ewing Series. AGU, Washington, D. C. pp 63–69. Chap. 1
Ueda K, Gerya T, Sobolev SV (2008) Subduction initiation by thermal-chemical plumes: numerical studies. Phys Earth Planet Int 171(1–4):296–312. doi:10.1016/j.pepi.2008.06.032
Valencia D, O'Connell RJ (2007) Inevitability of plate tectonics on super-earths. ApJ 670:45–48. doi:10.1086/524012
Valencia, D, O'Connell RJ (2009) Convection scaling and subduction on earth and super-earths. Earth Planet Sci Lett 286(3–4):492–502
van Heck HJ, Tackley P (2011) Plate tectonics on super-earths: equally or more likely than on earth. Earth Planet Sci Lett 310:252–261. doi:10.1016/j.epsl.2011.07.029
Zhong S, Watts AB (2013) Lithospheric deformation induced by loading of the hawaiian islands and its implications for mantle rheology. J Geophys Res 118:1–24. doi:10.1002/2013JB010408
Zhong S, Gurnis M, Moresi L (1996) Free-surface formulation of mantle convection - I. basic theory and application to plumes. Geophys J Int 127:708–718

12

Geochemical approaches to the quantification of dispersed volcanic ash in marine sediment

Rachel P. Scudder[1,6*], Richard W. Murray[1], Julie C. Schindlbeck[2], Steffen Kutterolf[2], Folkmar Hauff[2], Michael B. Underwood[3], Samantha Gwizd[4], Rebecca Lauzon[5] and Claire C. McKinley[6]

Abstract

Volcanic ash has long been recognized in marine sediment, and given the prevalence of oceanic and continental arc volcanism around the globe in regard to widespread transport of ash, its presence is nearly ubiquitous. However, the presence/absence of very fine-grained ash material, and identification of its composition in particular, is challenging given its broad classification as an "aluminosilicate" component in sediment. Given this challenge, many studies of ash have focused on discrete layers (that is, layers of ash that are of millimeter-to-centimeter or greater thickness, and their respective glass shards) found in sequences at a variety of locations and timescales and how to link their presence with a number of Earth processes. The ash that has been mixed into the bulk sediment, known as dispersed ash, has been relatively unstudied, yet represents a large fraction of the total ash in a given sequence. The application of a combined geochemical and statistical technique has allowed identification of this dispersed ash as part of the original ash contribution to the sediment. In this paper, we summarize the development of these geochemical/statistical techniques and provide case studies from the quantification of dispersed ash in the Caribbean Sea, equatorial Pacific Ocean, and northwest Pacific Ocean. These geochemical studies (and their sedimentological precursors of smear slides) collectively demonstrate that local and regional arc-related ash can be an important component of sedimentary sequences throughout large regions of the ocean.

Keywords: Dispersed ash, Caribbean Sea, Equatorial Pacific Ocean, Northwest Pacific Ocean, Ash layers, Volcanic eruptions, DSDP, ODP, IODP

Introduction

In the modern era, marine scientists' interest in volcanic ash can be traced back at least to the *HMS Challenger* expedition (1872–1876). Indeed, "What is marine sediment made of?" is a question that has been asked for 100s of years, and while many aspects of the petrology, mineralogy, and geochemistry of marine sediment have been well characterized, it is remarkable how much remains unknown about some key aspects of ocean mud. In particular, the presence/absence and composition of volcanic ash is challenging given its fine grain size, broad classification as an "aluminosilicate", and common presence, especially since both oceanic and continental arc volcanism are often located physically adjacent to oceanic environments and are known to globally generate and transport widespread ash (Fig. 1). Certainly, civilization's natural curiosity about volcanism has driven interest in the marine records of ash for generations of scientists. Additionally, these studies considered how to link their presence with explosive volcanism, climate, arc evolution, biological productivity, and other geological processes, as described below. Although the term "ash" is normally used in volcanology as a grain size definition (32 μm to 2 mm, e.g., Fisher and Schmincke 1984) for explosive volcanic products (juvenile matter and lithic fragments), we here use the term "ash" to indicate a product (mostly volcanic glass) from explosive volcanism

* Correspondence: rscudder@tamu.edu
[1]Department of Earth & Environment, Boston University, Boston, MA 02215, USA
[6]Department of Oceanography, Texas A&M University, College Station, TX 77843, USA
Full list of author information is available at the end of the article

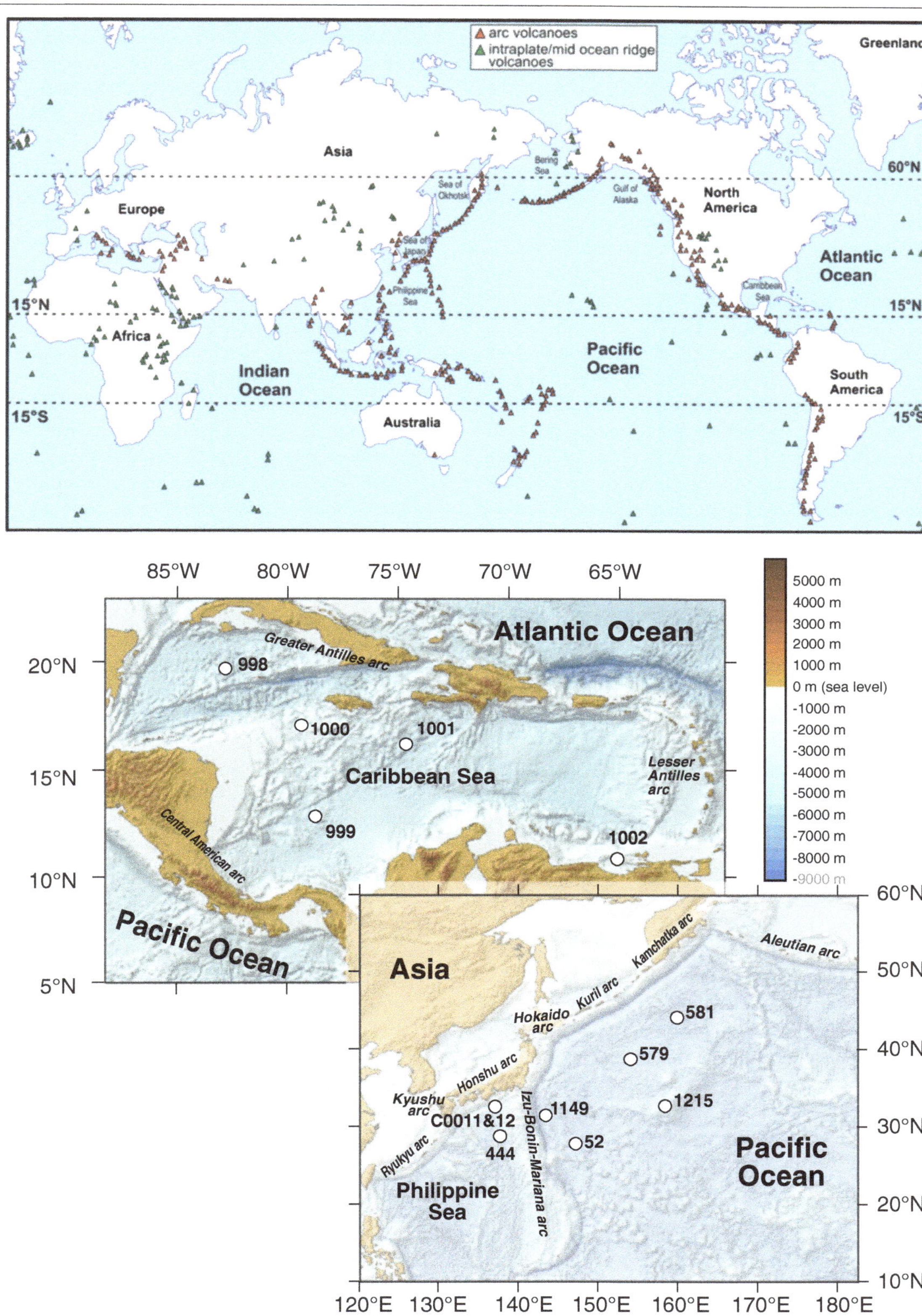

Fig. 1 (*Top*) Global distribution of subaerial volcanoes in different tectonic settings. Map has been modified after http://d-maps.com and volcano positions are taken from the Smithsonian Global Volcanism Program (http://www.volcano.si.edu). (*Bottom*) Locations of sites highlighted in this paper. Data from GeoMapApp (http://www.geomapapp.org); *GMRT* Global Multi-Resolution Topography (Ryan et al. 2009)

no matter which grain size a particle may have. Therefore, in this paper we refer to "discrete ash layers" as a volcanic product that is present as a clearly visible bed in the marine sediments (that is, ash layers of millimeter-to-centimeter or greater thickness, Fig. 2a–f) and their respective glass shards (Fig. 2g–l)) found in sequences at a variety of locations and timescales.

Less widely recognized than the discrete layers of ash is what has been referred to as "dispersed" ash. "Dispersed ash" is defined as volcanic ash material of various grain sizes (including single μm grains) that is mixed throughout the bulk sediment. This ash occurs in addition to the discrete ash layers. The presence of dispersed ash in the marine record has previously been relatively overlooked as it is difficult to identify petrographically due to its commonly extremely fine grain size (Rose et al. 2003, and references therein) and/or alteration to authigenic clay (Sigurdsson et al. 1997; Plank et al. 2000). Such alteration (e.g., Schacht et al. 2008) often results in fine-grained material that is unable to be differentiated from detrital terrigenous clay (non-ash, derived from land) on the basis of visual study alone.

Dispersed ash is the result from bioturbation of preexisting discrete ash layers, the settling of airborne or subaqueous ash through the water column, transport by rivers and currents from terrestrially exposed ash deposits, and other processes (e.g., gravity flows). Because dispersed ash has often been only qualitatively addressed (e.g., via sedimentological smear slides), however it still plays as vital a role as discrete ash layers in many studies on multiple spatial and temporal scales, including (but not limited to) the following:

- Ash plays an important role for geochemical budgets (e.g., "Subduction Factory"), in that studies of geochemical recycling will benefit from an increased understanding of both input (subduction) and output (volcanism) fluxes (e.g., Straub 1997, 2008; Plank and Langmuir 1998; Straub and Schmincke 1998; Bryant et al. 1999, 2003; Straub and Layne 2002, 2003a,b; Hauff et al. 2003; Straub et al. 2004; Stern et al. 2006; Plank et al. 2007; Scudder et al. 2009, 2014; Tamura et al. 2009; Völker et al. 2011, 2014; Freundt et al. 2014). The amount, distribution, and composition of ash in sediment speaks to inputs as well as outputs in the budget, namely, knowing how much ash is entering the Subduction Factory is important for understanding recycling of geological material. Additionally, the ash record in the sediment itself provides an archive of local, regional, and global magmatic evolution.
- The marine repository of explosive volcanic eruptions is a vital record from which the geological history of arc evolution is recorded (e.g., Kennett et al. 1977; Ninkovich et al 1978; Sigurdsson et al. 1980, 2000; Huang 1980; Cambray et al. 1993, 1995; Lee et al. 1995; Bailey 1996; Peters et al. 2000; Straub 2003; Kutterolf et al. 2008a,b,c; Straub et al. 2009, 2010, 2015). An improved understanding of arc evolution has the potential to contribute significantly to knowledge regarding the birth, life, and death of arc magmatism. Furthermore, a full accounting of the ash record (that is, comparison of the dispersed ash record with that of the discrete layers) may provide information about arc evolution and erosional histories of landmasses. Studies of ash distribution are also vitally important to document risk associated with volcanic hazards.
- Reconstructions of eruption intensities and atmospheric wind patterns have been based on grain size characteristics and inferred dispersal patterns of the ash found in discrete layers, both at sea and on land (e.g., Huang et al. 1973, 1975; Shaw et al. 1974; Ledbetter and Sparks 1979; Carey and Sigurdsson 1980, 2000; Carey and Sparks 1986; Reid et al. 1996; Rose et al. 2003; Carey et al. 2010). These finer particles are probably the main constituent of dispersed ash, and therefore documenting the distribution of dispersed ash will constrain inferences of the transport pathways of volcanic material.
- The ability to quantify the amount of dispersed ash will increase knowledge of sedimentary physical properties (e.g., Peacock 1990; Kastner et al. 1991; Underwood and Pickering 1996; Chan and Kastner 2000; Saffer et al. 2008, 2012; Hüpers et al. 2015). Given that ash alters to clay in a hydration reaction, the amount of water bound in the sediment will increase, thus decreasing the shear strength of the mud. Additionally, the fluid budget of subducting sediment will be affected by these and, further, hydration/dehydration reactions. A quantification of the amount of dispersed ash will thus assist interpretation of physical properties of marine sediment and subduction zone modeling.
- Volcanic ash and eolian dust input to the surface ocean contributes nutrients to the surface ocean and are thus important for understanding trace metal cycling, biological productivity, climate change, volcano-climate interactions, and allied biogeochemical studies (e.g., Huang et al. 1974; Shaw et al. 1974; Carey 1997; Frogner et al. 2001; Jones and Gislason 2008; Robock et al. 2009; Duggen et al. 2010; Olgun et al. 2011; Kutterolf et al. 2013; Metzner et al. 2014).

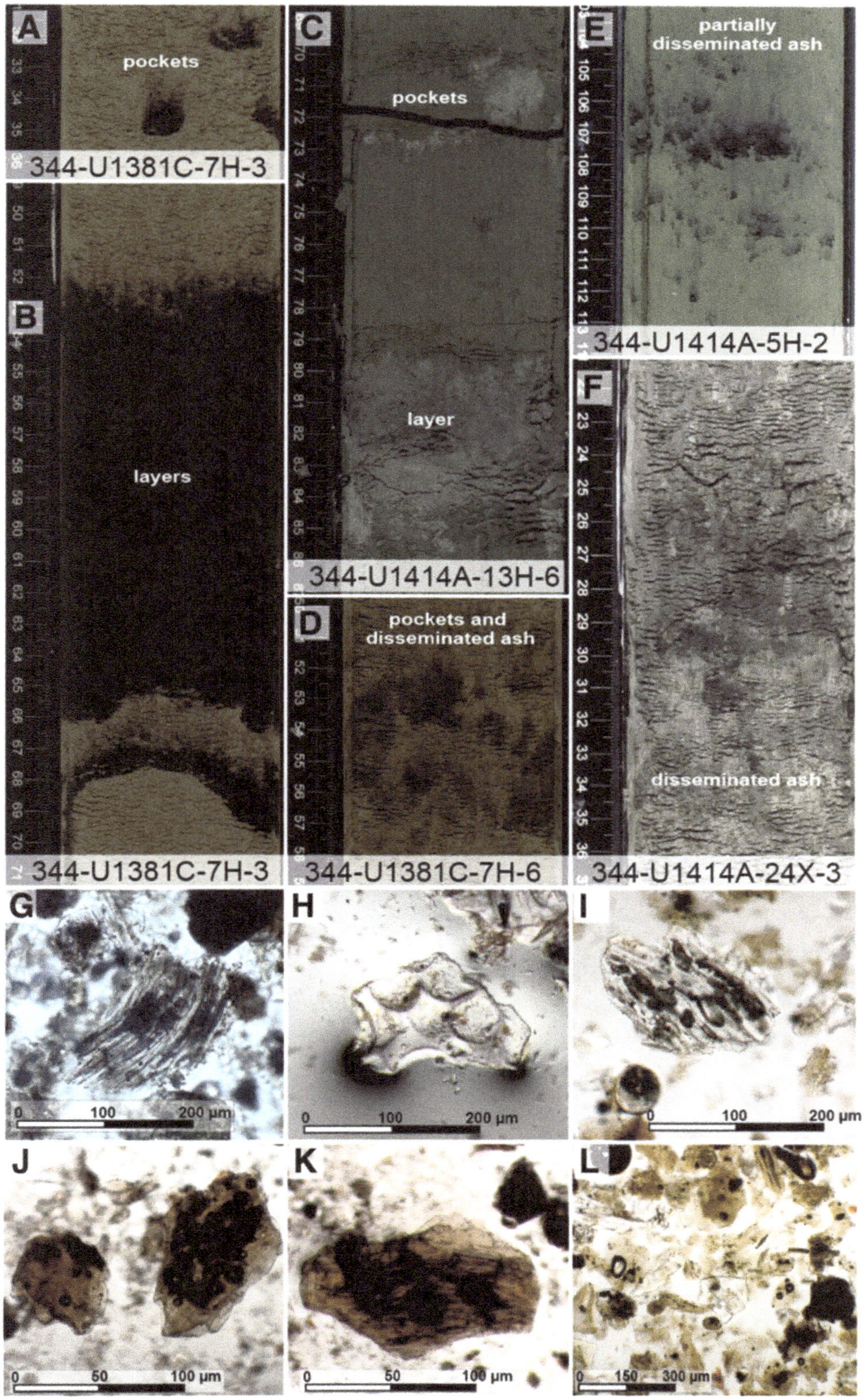

Fig. 2 (See legend on next page.)

(See figure on previous page.)
Fig. 2 Photographs of **a** mafic pockets, **b** discrete mafic ash layers, **c** felsic ash layer and pockets, **d** disseminated mafic ash pockets selected primary mafic ash layers, **e** partially disseminated ash pockets, and **f** disseminated ash. All examples from IODP expedition 344 (CRISP II) offshore Costa Rica. Example smear-slide microphotographs showing glass shard textures of felsic (G-I) and mafic (J-L) marine ash layers. **g** U1381-3H-7, 14–19 cm: felsic ash with tubular transparent to light brown shards. **h** M66-226/78 cm bsf: silicic shard from highly vesicular pumice. **i** M66-222/51 cm bsf: pumice fragment with moderately elongated vesicles. **j** M66-223/305 cm bsf: blocky sideromelane shard with and without vesicles. **k** M66-222/436-441 cm bsf: elongated vesicles in sideromelane shard. **l** M66-222/517–518 cm bsf: overview picture of different glass shards in one sample

Ash is commonly altered on and/or in the seafloor by diagenesis. In addition to the research cited above, for example, in a historically classic study, Gardner et al. (1986) observed many thin, pale green laminae (PGL) in the sediment cored during Deep Sea Drilling Project (DSDP) Leg 90 (Lord Howe Rise), which they identified as bentonites resulting from discrete ash fallout events. They found the temporal distribution of the PGLs to be similar to that of the discrete layers and logically concluded the PGLs were merely altered ash. For dispersed ash, however, the alteration increases the complexity for the application of techniques based on sedimentological and physical separation (e.g., grain size or sequential leachings to arrive at an aluminosilicate "residue"), because a fine-grained altered aluminosilicate ash is difficult to distinguish from, for example, fine-grained aluminosilicate eolian dust. The geochemical techniques summarized in this review, however, will still identify these alteration products as part of the original ash contribution to the sediment. For example, Hein et al. (1978) showed that ash beds in the Bering Sea become more altered with burial and form bentonite layers. They also documented that the chemistry of the bentonite beds reflects the chemistry of the parent ash, and differentiated authigenic smectite, which forms from the *in situ* breakdown of glass in ash layers or glass shards dispersed throughout the sediment, from detrital smectite derived from subaerial breakdown of recycled volcanic debris.

The use of tephra as a stratigraphic and chronologic tool has a long and wide-ranging history (e.g., Cambray et al. 1995; Straub and Schmincke 1998; Lowe 2011) and we do not intend here to repeat those efforts. Rather, in this paper, we (a) present a review of previous geochemical approaches to quantify dispersed ash in the Caribbean Sea and equatorial Pacific Ocean, (b) provide a summary of recent work applying a combined geochemical and multivariate statistical technique to identify dispersed ash in the northwest Pacific Ocean, and (c) describe preliminary results moving toward a regional assessment of inputs to the western Pacific Ocean (Fig. 1). We include some areas of research that are currently unresolved as they are likely to provide further insight into the nature and distribution of this important component of marine sediment. While radiogenic isotopes have proven helpful in studies of dispersed ash and the very fine detrital component (e.g., Ziegler et al. 2008; Scudder et al. 2014), in this paper we focus on major, trace, and rare earth elements since they are more commonly used in the various regions we discuss.

Review

Geochemical approaches: theory and practice

Composition is different from transport

A basic tenet of provenance research is that studies based on chemical, isotopic, or mineralogic composition, or many aspects of sedimentology (e.g., grain size), cannot alone infer the transport pathway (e.g., eolian vs. subaqueous transport) by which a grain of volcanic ash eventually becomes entrained in the sediment. Strictly speaking, transport pathway must be inferred based upon geological or geographic constraints. For example, in the context of eolian transport of eroded continental material, truly detrital aluminosilicate grains must be transported by wind to reach the deepest and most distal north Pacific (Rea and Leinen 1988; Kyte et al. 1993; Rea 1994).

For volcanic ash, however, the situation is more complicated. Subaqueous eruptions can lead to significant volcanic input (e.g., Fiske et al. 2001; Maicher and White 2001; White et al. 2003; Tani et al. 2008). Fiske et al. (2001, their p. 822) in particular note how the finest-grained fraction is missing from caldera deposits in the Izu-Bonin system. This "missing material" undoubtedly ends up as dispersed ash throughout the bulk sediment and due to its grain size has the potential to be advected great distances away from its point of origin. Determining this material's composition in the bulk sediment, a priori will not differentiate between subaerial (eolian) or subaqueous transport.

Subaqueous eruptions can also lead to pumice rafts, which may persist in the ocean for weeks, months, or years (Simkin and Fiske 1983; Risso et al. 2002; Bryan et al. 2004). Even in today's modern era of enhanced ocean and atmospheric observation, such occurrences are surprises (BBC News, 10 Aug. 2012). Yet, although these events may be uncommon on the human timescale, they are most likely significant pathways of ash to the deep sea over longer time frames.

Mixes of mixtures

In the broad context of geologic material, all non-biogenic rocks and sediments on the surface of Earth can be chemically described as falling along a mixing line or mixing curve between families of primordial materials (e.g., basalts) and families of fractionated "upper continental crust". Deviations from such a mixing line or curve are due to local provenance effects, authigenic/diagenetic resetting, weathering, and other processes, but the overall relationship holds true. While such a compositional range may seem relatively easy to differentiate analytically (e.g., felsic vs. mafic), the cases are rare when there are only two aluminosilicate sources and they are sufficiently compositionally distant from each other along the mixing line or curve. Far more common is the challenge of resolving multiple sources that are each at different fractional lengths along a theoretical mixing line or curve. For example, continental crust has been approximated as "granodiorite", "dacite", or other intermediate composition igneous/volcanic rocks (e.g., Taylor and McLennan 1985), and therefore distinguishing eroded continental crust from wind-blown dacitic ash is challenging. Or, still more challenging are scenarios whereby two disparate sources such as a marine basalt and a wind-blown dust of broad continental crust composition are both authigenically altered to, for example, assorted smectites, and thus have the potential to appear as a single source.

The above examples speak to the importance for geochemical studies of dispersed ash to address their attention to refractory elements that are appropriate to unravel mixing problems and that will also be relatively unaffected by diagenetic and authigenic processes. There are no chemical elements that unambiguously record any single geochemical provenance or domain. Within the spectrum of potential elements for use in general characterization as well as the multivariate statistics discussed below, inclusion of certain common major elements such as Al, Ti, and Fe, along with certain common trace elements such as La and Sc, proves to be beneficial. On a case-by-case basis, certain other key elements such as Th, Nb, and Cr can be advantageous as well (e.g., Plank 2005), although interpretations based on only a few elements are likely to be less representative than studies based on suites of many elements. However, common elements involved in reverse weathering reactions (e.g., Mg, Mackenzie and Garrels 1966; Presti and Michalopoulos 2008) should be avoided for such studies of provenance, given their labile behavior.

The contribution from geochemistry teamed with multivariate statistics

A variety of geochemical approaches can help resolve such mixing issues. In certain situations, such as in the Caribbean Sea example below, single element "normative calculations" (e.g., Leinen 1987) are sufficient, while in other cases multiple different analytical and/or computational approaches need to be simultaneously deployed to develop a more holistic understanding of the geochemical variability within datasets. We emphasize here a combined geochemical approach that generates a wide-ranging elemental suite that is examined with multivariate statistics.

The overall approach we summarize in this paper is based on Q-Mode Factor Analysis (QFA) and two forms of multiple linear regression, a Constrained Least Squares (CLS) technique and a Total Inversion (TI) technique (Pisias et al. 2013). The history of these techniques, as well as the specific MATLAB scripts to apply them to marine sediment chemistry (and recommendations for their use), are detailed in Pisias et al. (2013). The interested reader is also pointed toward previous papers by Dymond (1981), Leinen and Pisias (1984), Zhou and Kyte (1992), and Kyte et al. (1993). In our research group, we have successfully used these techniques in marine sediment from the Cariaco Basin, Arctic Ocean, equatorial Pacific, northwest Pacific Ocean, and South Pacific Gyre (Martinez et al. 2007, 2009, 2010; Ziegler and Murray 2007; Ziegler et al. 2007, 2008; Scudder et al. 2009, 2014; Dunlea et al. 2015a, b), demonstrating their utility throughout a variety of oceanic depositional regimes.

QFA is a terrific tool because—among other things—it is an "objective" technique (e.g., Leinen and Pisias 1984) that is minimally affected by inadvertent researcher bias. There are no inputs other than the data itself. It is limited, however, in many cases in that the resultant compositional factor scores may not be geologically reasonable or geologically precise enough across the spectrum of the wide element menus commonly generated in today's world of rapid geochemical analysis. Factor analysis is best used to help identify the number of independent components (also referred to as "end members") and their general composition, but not necessarily the extremely specific composition thereof. Therefore, one of the limitations of using QFA alone is that the user often cannot identify the specific geologic source(s) of the bulk sediment.

The number of end members necessary to best describe the variability of the dataset (that is, the number of factors indicated by the QFA) helps steer the CLS and TI multiple linear regressions. Specifically, in order to identify the exact sources identified by QFA, the CLS or TI linear regressions must be applied. In CLS and TI, the researcher inputs the composition of potential end members and mixes them until the best solution is reached (the differences between CLS and TI are detailed in Pisias et al. (2013)). Thus, we use the QFA to generate an objective "answer" as to the number of sources and their broad composition(s), and then the

CLS and/or TI, based on thoughtful selection of potential sources derived from knowledge of previous work and the local/regional geology, to generate the quantitative mix of specific sources. Libraries of potential end members are typically (a) constructed from the literature of all likely volcanic contributors (that is, individual papers about the geochemistry of particular volcano or region), (b) compiled from international global databases (e.g., EarthChem; http://earthchem.org), and/or (c) developed internally from the study's own data (e.g., using the composition of a discrete ash layer to assess whether it could potentially be a contributor to the dispersed ash component.).

Despite the quantitative attractiveness of applying statistical treatments to a well-measured geochemical dataset, there is no single diagnostic "magic bullet" to fulfill the task of identifying the dispersed ash component within bulk sediment. Instead, a series of interpretative techniques must be used prior to applying the statistical techniques. The use of standard geochemical approaches such as expressing data as concentrations, calculation of fluxes and/or elemental ratios, study of simple coefficient of determination (R^2) matrices, x-y diagrams, and other well-established approaches is extremely important. Such approaches define the boundaries for mass balance, allow comparison with lithologic descriptions and mineralogy, help ground the ensuing statistical treatments, and, in a way that is hard to quantify, ensure that the researcher is intimately familiar with the fundamental underpinnings of each dataset.

Finally, statistical treatments such as those discussed here often lead to non-unique solutions. Contextual oceanographic or geological knowledge plays a major role, however, and can eliminate some of the non-unique solutions. Our group takes a very labor-intensive approach to dealing with this broad subject for both factor analysis and multiple linear regression, strategically built around performing dozens to hundreds of statistical runs per dataset to assess sensitivity of, and variability in, the results (Scudder et al. 2009, 2014; Dunlea et al. 2015a,b). Each dataset is different, and we have found some in which any variation in one element will make a difference; but in another dataset varying a different element will exert a larger lever arm. We take great care to ensure that our results are not dependent upon, or unique to, one or two narrow statistical models nor to one or two specific elements. Doing so would be counter to the very nature of our task at hand—to develop a widely applicable, internally consistent understanding of "What is marine sediment made of?"

Previous work

In this section, we present a review of previously published studies focused on unraveling bulk marine sedimentary compositions. These case studies range from a simple three-component system (Caribbean Sea) in which element normative calculations alone are sufficient to determine the end members, to a very complex system in which the bulk sediment is composed entirely of aluminosilicates with very similar compositions that requires a combination of geochemical and statistical technique in order to identify the individual compositions (Izu-Bonin-Mariana Arc).

Caribbean sea: a simple case

The Caribbean Sea is ringed by volcanic arcs (Fig. 1; Sigurdsson 1999). To the west rises the Central American volcanic arc system, with a spectrum of volcanic edifices ranging in size from basaltic cinder cones to large stratovolcanos and calderas. To the east is the Antilles island arc, from which some of the most famous volcanic eruptions in the world originate (e.g., Lesser Antilles, Carey and Sigurdsson 1978). These combined volcanic systems contribute large amounts of ash to sediment in the Atlantic Ocean and Caribbean Sea from the Miocene on (Carey and Sigurdsson 1980, 2000; Peters et al. 2000; Jordan et al. 2006; Expedition 340 Scientists 2012; Le Friant et al. 2013) Ocean Drilling Program (ODP) Leg 165 cored throughout the Caribbean Sea (Fig. 1) with its most important objectives being related to the Cretaceous-Neogene bolide impact. One surprise that resulted from this research cruise was the startling number of discrete ash layers (Fig. 3, Carey and Sigurdsson 2000; Sigurdsson et al. 2000), their extent in time and space of which was unexpected. In addition to the discrete ash layers, shipboard study observed ash dispersed throughout the background sediment. Bulk chemical analyses of the bulk sediment allowed for the development of a single-element normative calculation scheme to quantify the abundance of this dispersed ash in these carbonate-rich sediments (Peters et al. 2000). Fortuitously, most of the sequence is a relatively simple three-component system (of $CaCO_3$, terrigenous, and ash), which facilitated the study of dispersed ash in parallel to the discrete layers. Petrographically, the bulk sediment appeared to consist of only carbonate and aluminosilicate clay—and the aluminosilicate clay was assumed to be entirely comprised of eroded terrigenous material (Peters et al. 2000).

Bulk chemical analyses, however, demonstrated that the sediment was vastly depleted in Cr beyond that, which could be explained by carbonate dilution. For example, in a sample with an equal (50 and 50 %) mix of $CaCO_3$ and terrigenous clay, one would predict there should be ~55 ppm Cr in the bulk composition (based on typical "average shale" of ~ 110 ppm Cr, using Post-Archean average Australian Shale [PAAS] from Taylor and McLennan (1985) to broadly represent an upper crustal terrigenous source). However, such a sample would instead present <10 ppm Cr, indicating that the brown aluminosilicate

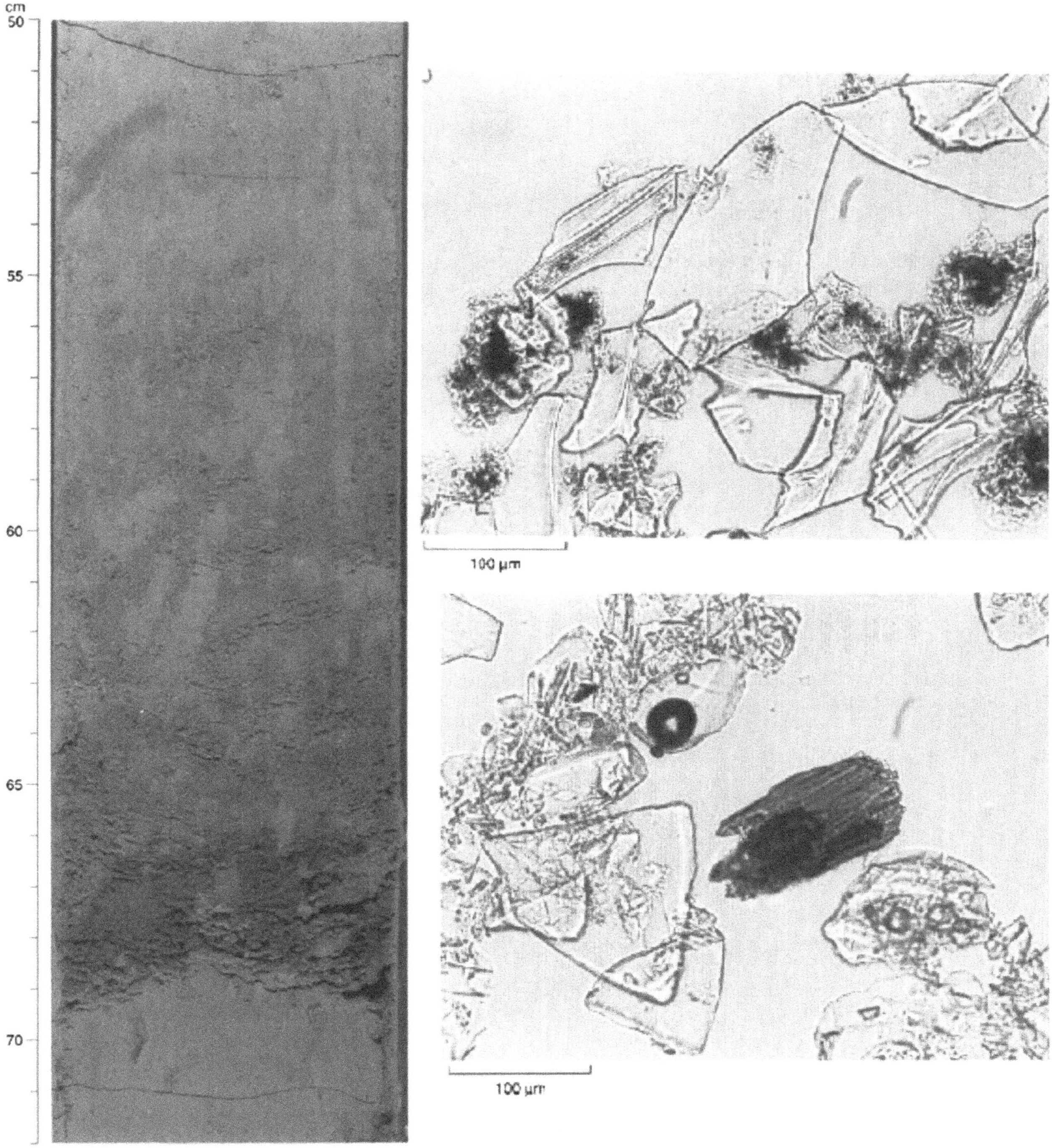

Fig. 3 Closeup photos and photomicrographs of ash layers from ODP Site 998, western Caribbean Sea. *(Left)* Volcanic ash layer in Section 165-998A-10H-4, 50–72 cm. Note the sharp base and transitional top that has been reworked by bioturbation. *(Right, top)* Section 165-998A-19X-3, 55–56 cm. Typical rhyolitic glass shards from a 13.8 Ma volcanic ash fall layer. The width of the field of view is 0.5 mm. *(Right, bottom)* Typical fresh (*clear*) and altered (*cloudy*) glass shards and an amphibole phenocryst in a 15 Ma volcanic ash fall layer in Section 165-998A-21X-3, 51–52 cm. From Sigurdsson, Leckie, Acton, et al. 1997

groundmass was not solely detrital terrigenous material (Peters et al. 2000).

Given the simple three-component system, normative calculations (Leinen 1987) were appropriate to identify the sources to the bulk sediment. Peters et al. (2000) based their calculations on Cr, given that the end member values of Cr in typical shale (110 ppm) and rhyolite (<3 ppm) are strongly divergent. Thus, in a three-component system of $CaCO_3$ (which is quantified precisely by coulometry), terrigenous material (calculated via Cr), and ash, the amount of ash can be calculated by difference, as follows:

$$(\%\,\text{Ash})_{\text{sample}} = 100 - (\%\ \text{CaCO}_3)_{\text{sample}} - (\%\ \text{Terrigenous})_{\text{sample}}$$

where

$$(\%\,\text{Terrigenous})_{\text{sample}} = 100 * (\text{Cr}_{\text{sample}} / \text{Cr}_{\text{PAAS}})$$

Peters et al. (2000) thus documented on a sample-by-sample basis that dispersed ash comprises 15–20 wt.%, with a maximum of 45 wt.%, of the bulk sediment in the western Caribbean. These values were consistent with the sedimentological smear-slide analyses (Sigurdsson et al. 1997) but were more precise.

Peters et al. (2000) also observed that the timing of the accumulation rate of dispersed ash paralleled the sedimentation rate of the discrete layers, although the maxima in dispersed ash preceded the Miocene and Eocene maxima in discrete layers by ~2–4 Ma (Fig. 4). They interpreted the relative timing as recording arc evolution, with the dispersed ash being generated by smaller volcanoes characteristic of the more juvenile arc, and the larger discrete layers representing a mature arc characterized by larger stratovolcanoes and caldera

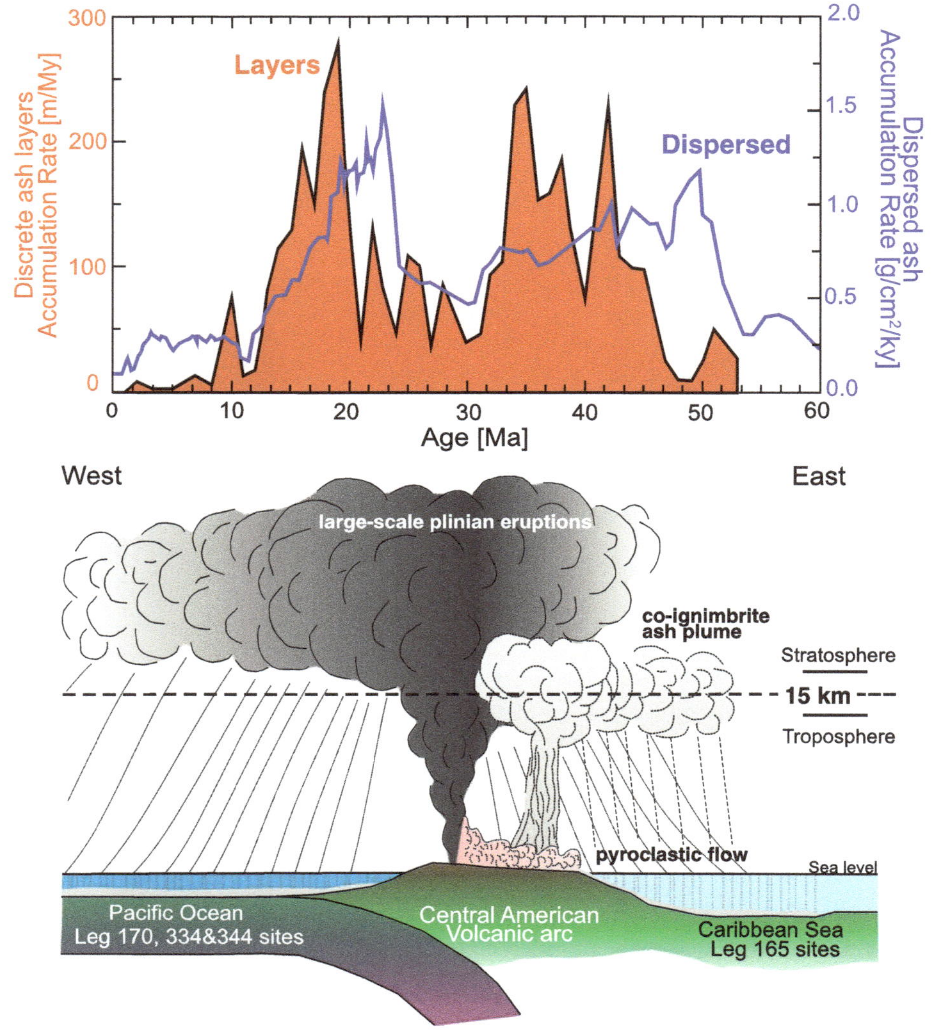

Fig. 4 (*Top*) Dispersed (*blue line*) vs. discrete (*red field*) ash accumulation in the Caribbean Sea. Accumulation of discrete ash layers (Carey and Sigurdsson 2000; Sigurdsson et al. 2000) and dispersed ash (Peters et al. 2000). ODP Site 998, western Caribbean Sea. (*Bottom*) Eruption transport and deposition patterns for recent to Miocene eruptions of the Central American Arc. Large plinian eruption columns (*light to dark gray*) reach deep into the stratosphere and are therefore transported to the west by the prevailing winds. Large ignimbrite-generating eruptions, predominantly in the Miocene, generate co-ignimbrite ash plumes (*pinkish to light gray*) reaching only into the upper troposphere/lower stratosphere and are therefore distributed to the east. Tropopause is shown with a *dashed line* at 15 km altitude. *Gray fields* below the sea level demonstrate submarine ash deposits

systems, as suggested by Carey and Sigurdsson (2000) and Sigurdsson et al. (2000). These are capable of injecting large plumes of Central American co-ignimbrite ash into the lower stratosphere and upper troposphere, that is, the ideal atmospheric height to facilitate west-to-east blowing wind transport (e.g., rather than transport from east-to-west in the higher stratosphere) (Fig. 4). This interpretation is consistent with the previous studies that have shown that Paleogene and Neogene Central American volcanism is predominantly characterized by larger ignimbrite-forming eruptions (Jordan et al. 2006) that may have frequently produced co-ignimbrite ash plumes reaching lower stratospheric heights (~15 to 20 km; e.g., Woods and Wohletz 1991; Bursik 2001), in contrast to the higher stratospheric eruption columns of Plinian eruptions (>>20 km; e.g., Kutterolf et al. 2008a).

Equatorial pacific ocean: a complex case

Our group also has a long-standing interest studying the equatorial Pacific (e.g., Murray and Leinen 1993, 1996; Murray et al. 1993, 1995, 2000; Kryc et al. 2003; Ziegler and Murray 2007; Ziegler et al. 2007, 2008, plus others). The non-biogenic component of these sediments includes terrigenous clay and ash, which are challenging to resolve in sediment with more than 98 % biogenic material (carbonate and biogenic silica). Some of our work specifically targeted how to resolve the eolian, ash, and authigenic components of these sediments and built upon the classic research of the long-standing Michigan group and their progeny (e.g., Hovan et al. 1991; Rea 1994; Rea et al. 1998).

For example, Ziegler et al. (2007) detailed chemical criteria that can be applied to differentiate authigenic, terrigenous, and volcanogenic aluminosilicates from each other. Sequential extractions were performed at ODP Site 1215 (Fig. 1) to remove non-aluminosilicate components. The material remaining upon completion of the extraction procedure was termed the "residual" component, consistent with established protocols. La-Sc-Th ternary diagrams and PAAS-normalized rare earth element (REE) patterns (Fig. 5a) and reference REE patterns (Fig. 5b) were used to aid in determination of the sedimentary components. In the residual component, Ziegler et al. (2007) quantified upper continental and lower crustal components using selected rare earth element (REE), Sc, and Th abundances. In old (>50 Ma) nannofossil ooze, the residual component exhibited a large La contribution. Additionally, the residual component of the nannofossil ooze fell outside the boundaries of a simple two component mixing between upper and lower crust components. REE patterns for these samples exhibit sharp, seawater-like negative Ce anomalies, providing a clear indication that an authigenic phase is present in the residual component (Fig. 5c). The old age of these sediments may contribute to the formation of the authigenic phase, given that such samples have likely experienced significant diagenesis. In this way, the geochemical approach clearly identified samples in which it is virtually impossible to differentiate eolian material from volcanic ash (Fig. 5c, d).

Izu-Bonin-Mariana Arc (ODP Site 1149 and DSDP Site 52)

One of the goals of ODP Leg 185 was to test whether subducted sediments control along-strike geochemical differences in Izu-Bonin-Mariana arc composition (Fig. 1, Plank et al. 2000). In addition to the potential influence of the terrigenous (non-ash) sedimentary component, the amount of ash in the sediment column is critical to constrain, not only for the determination of absolute geochemical fluxes into the "Subduction Factory" but also in order to determine how much of the input is "recycled" from the arc itself. Due to the long-standing interest in the region from a variety of geological perspectives, there is also a wealth of tectonic- and arc-related information to draw upon in the context of both ash layers and dispersed ash.

Preliminary work based on a Nb-based normative calculation showed that 33 ± 9 wt.% of the sediment at ODP Site 1149 is comprised of dispersed ash (Plank et al. 2000). As described in Scudder et al. (2009, 2014), in order to determine the sources to the bulk sediment, we followed the QFA and TI approaches and MATLAB scripts from Pisias et al. (2013). The elemental suite we targeted was composed of refractory elements predominantly associated with aluminosilicate components (Al, Ti, Sc, Cr, Ni, Nb, La, Th) as these are the most likely to identify differences between source components. The QFA results indicate that four factors (end members) explain 97 % of the variability of the bulk sediment (Scudder et al. 2014, Fig. 6). Given that QFA alone is not sufficient to identify the source of these factors, TI modeling was employed. The TI modeling identifies a combination of Chinese Loess (CL), rhyolitic ash from the Honshu arc (HR), mafic ash from the Izu-Bonin Front Arc (IBFA), and a second eolian dust (termed "Eolian 2") that best explain the total chemical composition of the bulk sediment.

Based on the TI results, the dispersed ash mass accumulation rate (MAR, $g/cm^2/ky$) at Site 1149 can be calculated and compared to the MAR of a series of ash layer parameters. At this site, the number of ash layers most closely tracks the total dispersed MAR (that is the sum of HR + IBFA). That the number of layers is most similar to the dispersed MAR suggests that eruption *frequency*, rather than eruption *size*, is the driving mechanism for the dispersed ash record (Scudder et al. 2014, Fig. 7).

The MAR patterns of the ash components are consistent with published eruption records of both the

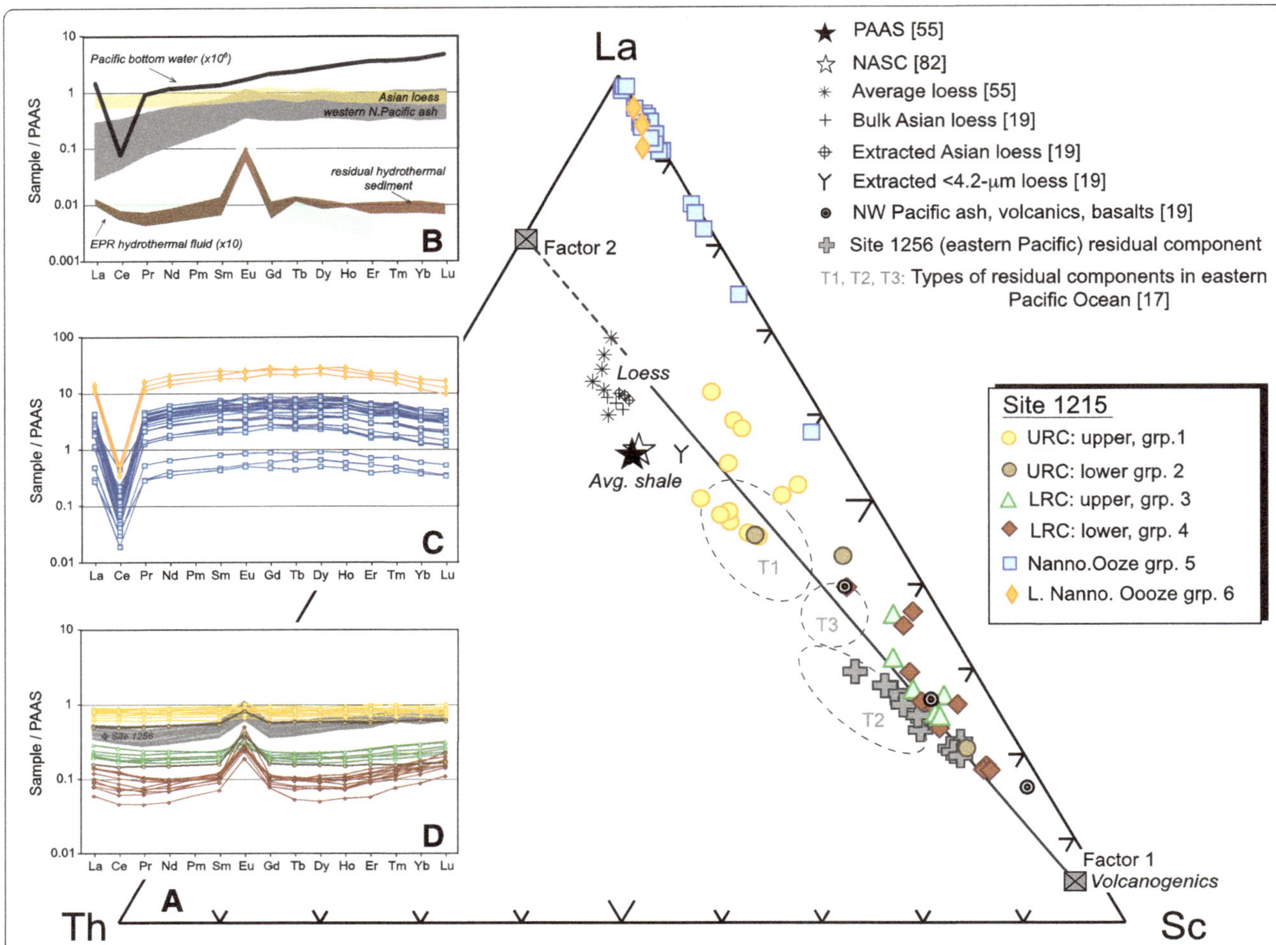

Fig. 5 **a** La-Sc-Th ternary diagram and PAAS-normalized rare earth element (REE) patterns from Site 1215 (central Pacific) and Site 1256 (eastern Pacific) from Ziegler et al (2007). **b** Reference REE patterns. Asian loess (Jahn et al. 2001), western Pacific ash (Bailey 1993), EPR and Pacific bottom water (Piepgras and Jacobsen 1992), hydrothermal residual sediment (Severmann et al. 2004), and hydrothermal fluid (Klinkhammer et al. 1994). **c** Residual component extracted from early-mid Eocene biogenic sediment at Site 1215 as defined in Ziegler et al. (2007). Note the significantly negative Ce anomaly indicating a seawater source. **d** Residual component extracted from the red clay unit from Site 1215 and the silty nannofossil ooze from Site 1256 as defined in Ziegler et al. (2007). See legend for symbol description. Groupings are based on REE patterns. Factor 1 is interpreted as lower crust island arc volcanic arc material and Factor 2 is similar to upper continental crust. T1, T2, and T3 are eolian dust samples classified based on REE patterns, Eu/Sm, and La/Sc ratios from Hyeong et al. (2005). *URC* upper red clay; *LRC* lower red clay

Izu-Bonin and Honshu arcs, and we can interpret these changes in accumulation rate in terms of arc history (Fig. 8). Focusing first on the Honshu Rhyolite, we observe that this component is the dominant dispersed ash, even though the Honshu arc is relatively far away. Given Site 1149's distal location, eruptions from the Honshu arc could have been large and yet resulted in only thin layers. These layers may subsequently be mixed into the bulk sediment to create the dispersed ash component. Alternately, the dispersed ash could simply represent an increase in deposition of ash from the atmosphere. The MAR of HR is consistent with the tectonic history of the Honshu arc, particularly through the younger portions of the record (e.g., Taira 2001). For the IBFA component, the models closely track the temporal changes in the tectonic record of Izu-Bonin (e.g., Sibuet et al. 1987). MAR patterns indicate that a gradual increase in IBFA volcanism occurred from ~18–10 Ma, with a burst of activity from ~4.5–3 Ma followed by a steady increase beginning at ~2 Ma (Scudder et al. 2014, Fig. 8).

We can compare and contrast Site 1149 to DSDP Site 52 in the northern Marianas arc (Fig. 1). Site 52 contains extremely high abundances of volcanic ash not directly linked to a specific eruption style (upwards of 50 % volcanic glass mixed in the brown clay; Fischer et al. 1971). Unfortunately, rotary drilling at Site 52 left only a few ash layers intact (all in the upper 20 mbsf; Fischer et al. 1971). Applying QFA with the same chemical suite as used at Site 1149 indicates that four factors (end

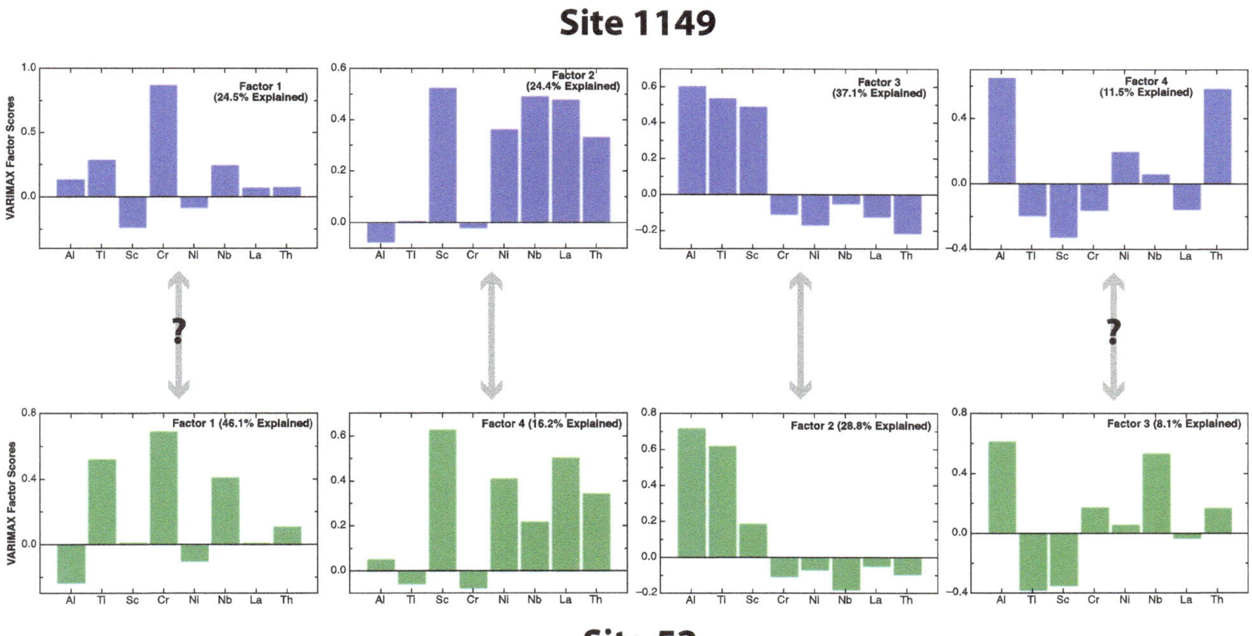

Fig. 6 VARIMAX factor scores from QFA, *(Top)* Site 1149, *(Bottom)* Site 52, from Scudder et al. (2014). Note the similarities between Site 1149 Factor 2 and Site 52 Factor 4 as well as Site 1149 Factor 3 and Site 52 Factor 2

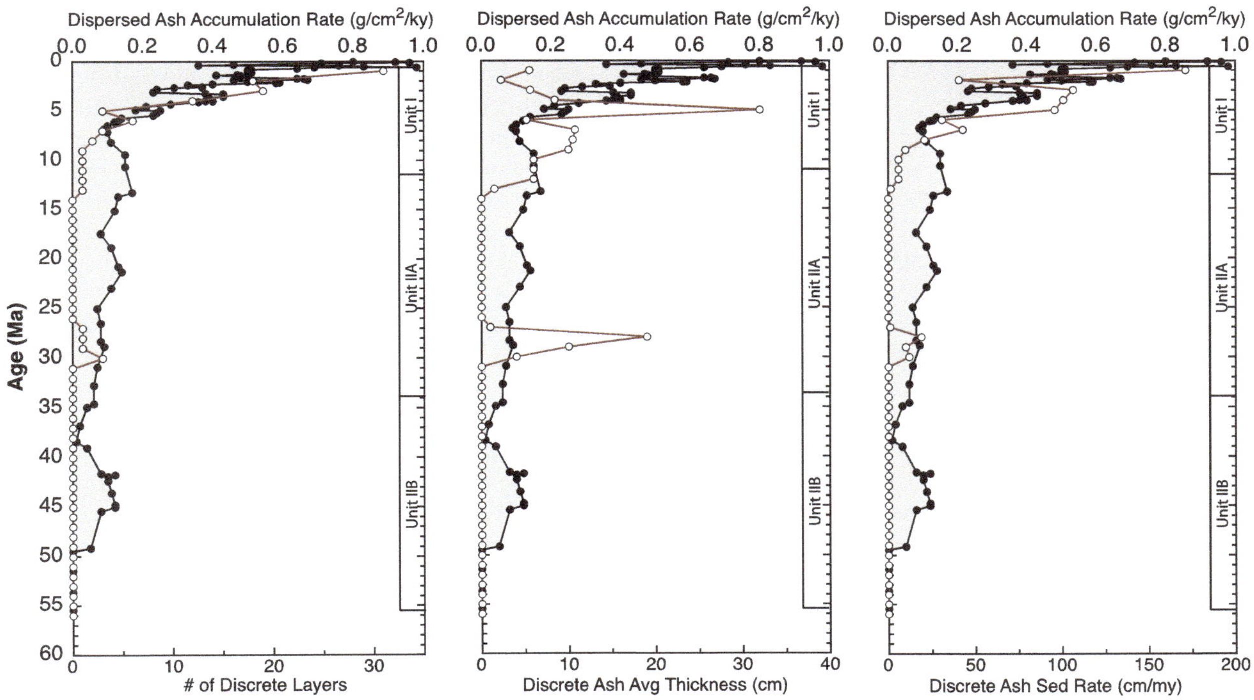

Fig. 7 Site 1149. The accumulation rate of total dispersed ash (*gray shaded* with *filled circles* in g/cm^2/ky) plotted against parameters associated with the sedimentation of discrete ash layers per 1 Ma (*open circles*) from Scudder et al. (2014)

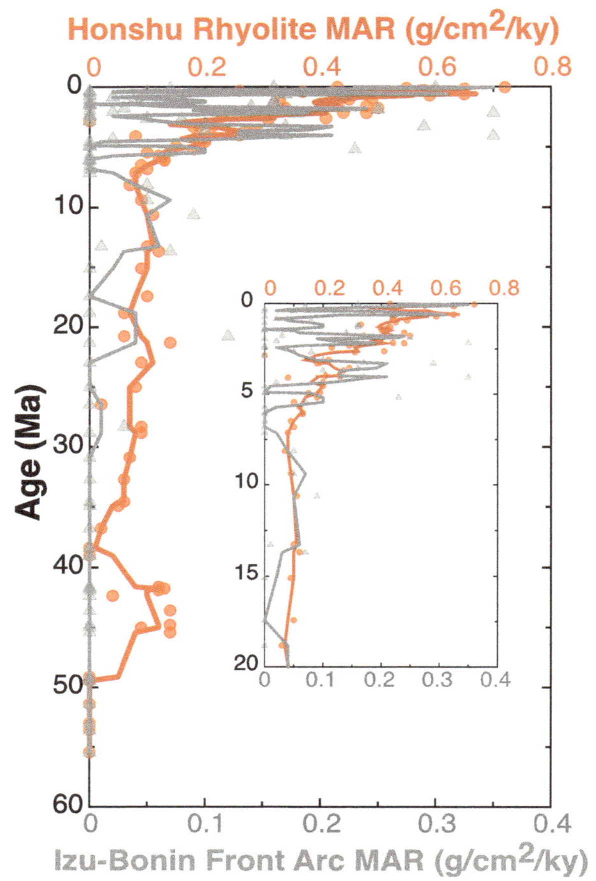

Fig. 8 Individual data and three-point averages of MARs of each dispersed ash end member, Site 1149. *Inset* focuses on the younger portion of each record. Note color-coded different *x*-axis scales and labels. From Scudder et al. (2014)

members) are present (Fig. 6). Two of these factors closely resemble those found at Site 1149, while the other two do not correspond well, if at all (Scudder et al. 2014). Given the limitations of QFA described previously, this interpretation is based on the factor scores (the weight of each element on the discrimination of a single factor) and the broad compositional scores produced by the modeling. TI linear modeling confirms this interpretation, identifying a mix of Chinese Loess, IBFA, dispersed boninite from the Izu-Bonin arc (BNN), and a dispersed felsic ash with the same composition as felsic layers from Site 52 (referred to as Felsic52) as best explaining the bulk sediment chemical composition.

Site 52 presents some key similarities and differences from Site 1149. First, the total amount of dispersed ash (regardless of composition) at both sites is very high, averaging 30 ± 17 wt.% at Site 1149 and 36 ± 18 wt.% at Site 52. Second, whereas at Site 1149 the two ash sources (HR, IBFA) are found both as layers and as dispersed ash, at Site 52 at least three volcanic sources are required to explain the bulk composition of the sediment. The discrete layers appear to be IBFA and felsic ash layers from Site 52, while the dispersed ash is comprised of IBFA, Felsic52, and an additional component, average Izu-Bonin Boninite (BNN). Therefore, at Site 52, the ash layers and the dispersed ash are partially compositionally decoupled from each other (Scudder et al. 2014).

Ongoing research

In this section, we present new data and ongoing observations from a variety of settings in the Northwest Pacific. While some of this research may still be developing, we include it here in order to show the strength of the combined geochemical/statistical technique in moving toward a regional perspective of inputs to the Western Pacific.

Nankai Trough, IODP Sites C0011 and C0012

IODP Sites C0011 and C0012 (Fig. 1), drilled as part of the Integrated Ocean Drilling Program (IODP) NanTroSEIZE transect, are located in the Shikoku Basin ~ 100 km southeast of the Kii Peninsula and ~160 km west of the Izu-Bonin arc on the Kashinosaki Knoll, a prominent bathymetric high. Located near the crest of the Kashinosaki Knoll, Site C0012 represents a condensed sediment section in comparison to Site C0011, which is located on the northwestern flank. The main goal of NanTroSEIZE is to drill across the up-dip limit of the seismogenic and tsunamigenic zone over the Nankai Trough subduction boundary, along which mega-thrust earthquakes are known to occur (Tobin and Kinoshita 2006). Shipboard sedimentological smear slide analyses at both Sites C0011 and C0012 estimate that on average dispersed ash constitutes ~ 25–30 wt.% through Units I and II, and decreases to ~7–15 wt.% in the hemipelagic facies of Unit III (Saito et al. 2010; Henry et al. 2012).

Statistical analyses of newly acquired data from Sites C0011 and C0012 were performed using the same element suite as that applied to Site 1149. At Site C0011, QFA indicates that three factors explain 99 % of the variability of the bulk sediment (Fig. 9). While we cannot determine the exact composition of the end members from QFA, one of the modeled sources appears to be intermediate (roughly equal contributions of most of the elements in the model, that is, of Sc, Cr, Ni, Nb, La, and Th) in composition while the other two are broadly felsic (contributions driven by Al, Ti, and Sc, with lesser covarying Nb, La, and Th). CLS multiple linear regression analysis suggest that the three sources at Site C0011 are consistent with one being PAAS, and the other two being ashes having compositions similar to a

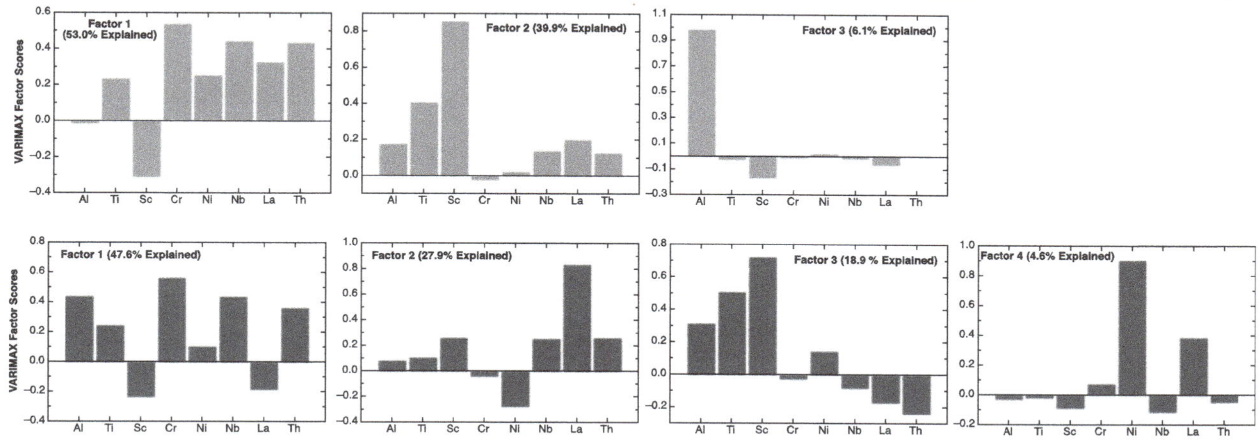

Fig. 9 VARIMAX factor scores from QFA. (*Top*) Site C0011. (*Bottom*) Site C0012

representative Rhyo-Dacite and Rhyolite from the Honshu Arc (HR) (Fig. 10). These latter two sources may in fact be dispersed ash but could also be ash that has since been altered and yet retained its refractory chemical signature. Regardless, these two types of dispersed ash and/or clay minerals with comparable geochemical signatures comprise 36 ± 8 wt.% of the bulk sediment. Evidence for multiple sources of volcanic material, and/or the chemically equivalent clay-alteration products, is consistent with studies of tuffaceous and volcaniclastic sandstones from this site, which calls for volcanic material from both the Izu-Bonin and Honshu Arcs (e.g., Pickering et al. 2013; Schindlbeck et al. 2013; Kutterolf et al. 2014).

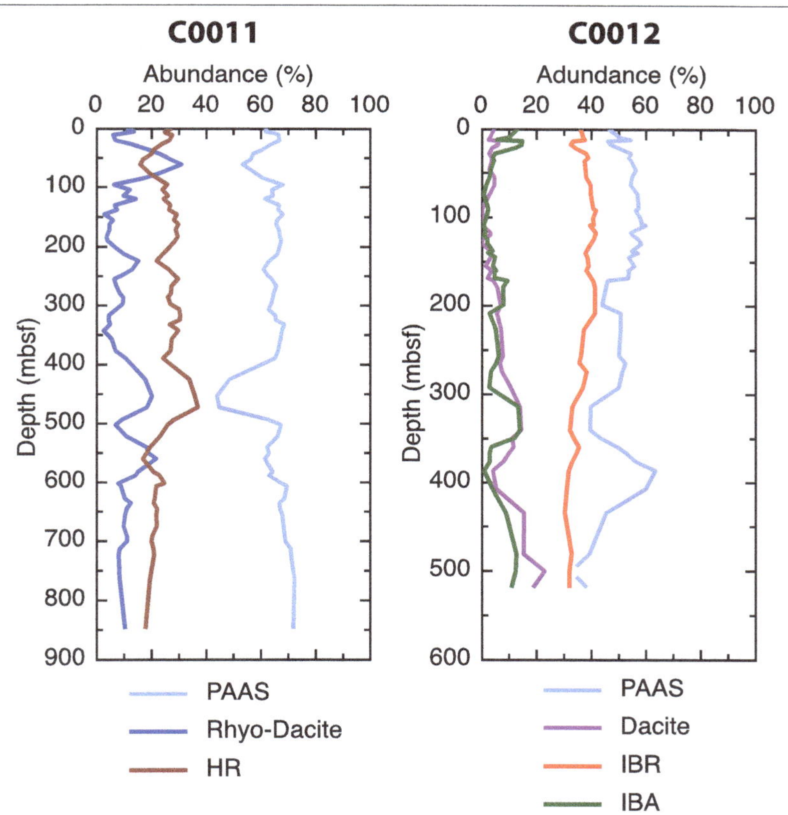

Fig. 10 Abundance (weight %) of sources identified by constrained least squares multiple linear regressions (CLS). (*Left*) Site C0011. (*Right*) Site C0012

Applying the same methods to Site C0012, QFA indicates that four factors (end members) explain 99 % of the variability in the bulk sediment (Fig. 9). The modeled compositions of the bulk sediment at Site C0012 identify two intermediate and two felsic end members for the geochemical proxies. CLS indicates that these end members are best explained by mixing PAAS and three ashes: the dacitic ash layers from Site C0012, rhyolite from the Izu-Bonin Arc (IBR), and andesite from the Izu-Bonin Arc (IBA). Collectively, the ash compositions (and/or their chemically equivalent clay-mineral alteration products) contribute approximately one-half (49 ± 10 wt.%) of the bulk sediment (Fig. 10). Here, it is important to recall that our geochemical approach cannot resolve the pathway by which the ashes have reached these sites, as some of these ashes may be eroded from terrestrial deposits in the Japanese archipelago.

We compare the mass accumulation rate of the dispersed ash (or its clay alteration) component to a number of common ash layer parameters, including the number of ash layers per 0.2 Ma, the thickest ash layer per 0.2 Ma, and the total thickness of ash layers per 0.2 Ma. We binned the data by 0.2 Ma units, in order to generate a discrete ash layer dataset of approximately the same temporal resolution as that of our dispersed ash record (Fig. 11).

When compared to various ash layer parameters, we find that the MAR of the total dispersed ash component at Site C0011 correlates best with "Thickest Layer" in Unit I. This suggests that through this interval, eruption volume, rather than the frequency of explosive eruptions, drives the dispersed ash accumulation (Fig. 11). In contrast, in Unit I of Site C0012, the number of discrete ash layers rather than "thickness" is best correlated to the dispersed ash MAR. At both sites, the number and thickness of ash layers vastly decreases below the Unit I/II boundary while the dispersed ash component remains high as a geochemical proxy. This is reflective of time periods when the volcanism in SW Japan was reduced or eliminated (Mahony et al. 2011), when the sites were influenced by siliciclastic turbidites (Units II, IV, and V), or when dispersal paths from the main detrital sources were further away from eruptive fronts (Unit III). Changes in any of these factors may have resulted in either fewer and/or thinner layers that would be more susceptible to being homogenized (by bioturbation or otherwise). That the geochemical signature equivalent to dispersed ash is high at times when so few ash layers were deposited clearly demonstrates that clay minerals derived from volcanic ash and volcanic rock (e.g., smectite group) are a much greater contributor to the sedimentary sequence than is recorded only by the ash layers.

Shikoku Basin, DSDP Site 444

DSDP Site 444 was drilled during Leg 58 to test hypotheses about the origin of marginal basins (Fig. 1, deVries Klein and Kobayashi 1980; deVries Klein et al. 1980). In particular, the Shikoku Basin portion of Leg 58 was drilled in order to answer questions regarding the basin formation by evaluating various spreading models, the ages of the oceanic crust, sediment evolution, and the paleoceanographic origin of the region (deVries and Kobayashi 1980). Site 444 was selected for our research on dispersed ash because of its abundant volcanic ash layers (for example, in comparison to Sites 442 and 443).

Site 444 contains basaltic, andesitic, and rhyolitic ash (deVries and Kobayashi 1980; Furuta and Arai 1980; and our own studies). Most tephra in other northwestern Pacific DSDP holes are rhyolitic and andesitic, which makes the basaltic nature of the Site 444 tephra geologically and tectonically unusual (White et al. 1980). In the Pleistocene and Pliocene, the ash layers are non-alkali rhyolites. As felsic ash is believed to mainly be a product of explosive volcanic eruptions, these layers could have originated from relatively large-scale eruptions in an island arc setting, most likely located westward of Site 444 and transported in the prevailing westerlies. This direction of transport is evident as grain size increases toward the west and decreases toward the east (Furuta and Arai 1980).

In the Miocene, the tephra is of non-alkali tholeiitic basaltic as well as alkali basaltic composition. These basaltic layers imply that another volcanic source was involved in the Miocene. This source could have been closer to the drilling sites than those sources of rhyolitic and andesitic tephra (possibly the Kinan Seamounts and the Shichito-Iwo Jima volcanic arc), but it could also be the distal evidence for large mafic eruptions from a more oceanic arc source. The knowledge that the occurrence of such large mafic plinian eruptions with a wide dispersion is not limited to arc settings alone and can also be found at ocean island settings (e.g., the Galápagos archipelago, Schindlbeck et al. 2015). Such observations are especially important for future studies of dispersed mafic ash in marine sediments. We note, however, that at least some of these ashes may also be subaqueous in origin. For example, certain volcanic arcs near the Shikoku basin (e.g., Izu-Bonin) contain subaqueous volcanoes (Hochstaedter et al. 2001).

Ash layers were carefully identified throughout the 310 meters of sediment at Site 444 using core photos and the shipboard lithologic description (deVries and Kobayashi 1980). Site 444 was cored using the Rotary Core Barrel (RCB) system, as it was occupied during the early-mid stages of the DSDP prior to the advent of the advanced piston corer. Because of the RCB drilling, the

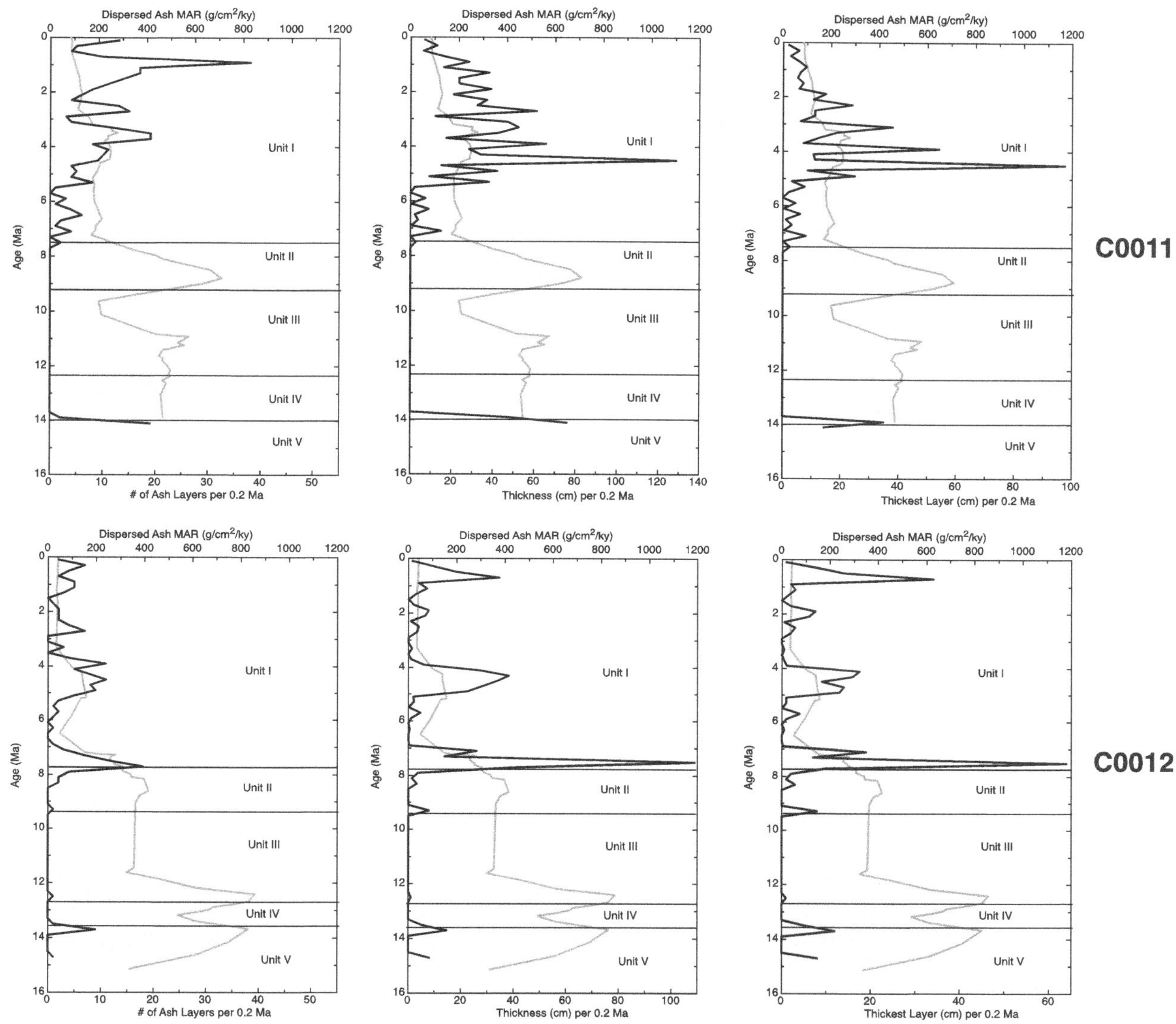

Fig. 11 The accumulation rate of total dispersed ash (*gray shaded line*, $g/cm^2/ky$, the same in each panel by site) plotted against parameters associated with the sedimentation of discrete ash layers per 0.2 Ma (*black line*, parameters are different in each panel as shown in the *bottom* x-axis labels). A moving 0.2 Ma window was chosen to bin each discrete ash layer parameter to approximate the resolution of the modeled dispersed ash record. (*Top*) Site C0011. (*Bottom*) Site C0012

recovered core is highly disturbed. Therefore, we classified the ash deposits as either an "ash layer" or an "ash pocket". Anything that is ash in distinct form but that is not in the shape of a full horizontal layer (e.g., touching both sides of the core liner) is termed an "ash pocket". The pockets are often, though not always, circular in form. Ash layers and ash pockets correlate with each other with depth. Thus, we interpret that the pockets are likely to have originated from ash layers mixed by RCB drilling.

A total of 128 ash layers and 76 ash pockets were identified. Of the 28 ash layers chemically analyzed in this study, four (4) of them are categorized as ash pockets. This is important to note when assessing their compositions as any seemingly unusual compositions can be evaluated for mixing with non-ash sources (e.g., clays). The distribution of ash layers and pockets were parameterized by (1) thickness with depth, (2) number of layers per 10 meters, (3) thickest layer (cm) over successive 10 meter depth increments, and (4) total thickness of ash (cm) in that same 10 meter window (Fig. 12). These parameters collectively should yield a baseline understanding of the distribution of ash layers, and we acknowledge that any one of these parameters is not necessarily preferable to the others. Additionally, we note that there is a relatively thick 31-centimeter-thick ash

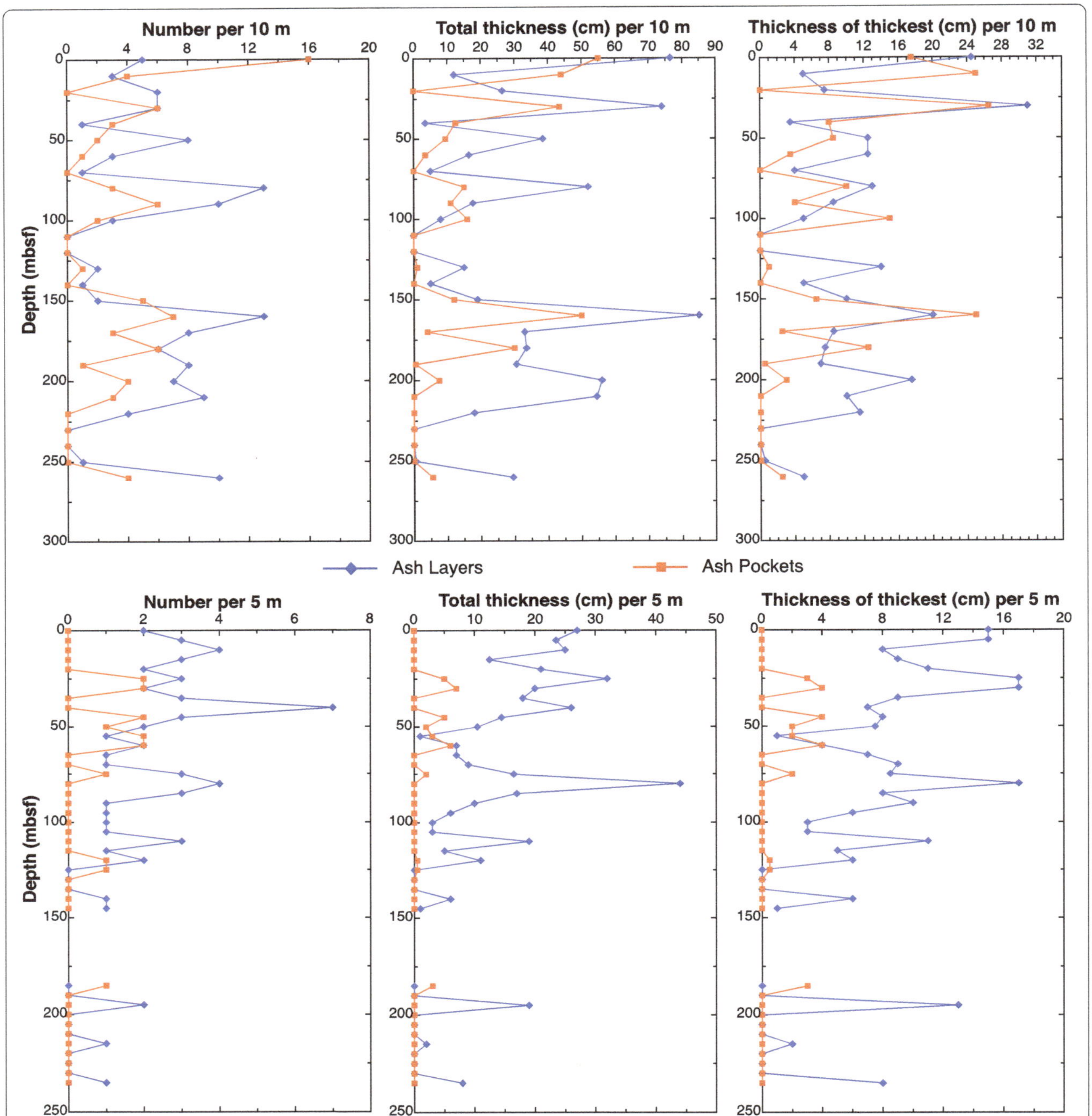

Fig. 12 Ash layer parameters, number of ash layers, total ash thickness (cm), thickness of thickest layer (cm). (*Top*) Site 444. (*Bottom*) Site 579/581

layer at a depth of 30 mbsf (~0.66 Ma, Pleistocene), which may overly influence some of these parameters over that interval.

Ash layers were analyzed for major element glass shard compositions by electron microprobe (average of 15 single shard measurements), and additionally, the bulk ash layers were analyzed by ICP-ES and ICP-MS for major, trace, and REE compositions following the methods of Scudder et al. (2014). Even though the bulk ash data may be compromised by the inclusion of the terrigenous aluminosilicate material (e.g., adjacent pelagic clay mixed into the ash layer or ash pocket; Scudder et al. 2014), we can still infer other key aspects of the distribution of the ash layer/pocket chemistry.

The distribution of key major elements used to differentiate compositional variation, such as SiO_2 or TiO_2, (Fig. 13), suggests the presence of multiple ash populations ranging from mafic to intermediate to felsic in composition. A transition in ash layer composition is observed at ~160–170 mbsf in which the SiO_2 concentration of the bulk ash layers decreases with depth, while the TiO_2 concentration of the bulk ash layers increases with depth. At ~160–170 mbsf, the ash layers are enriched in elements such as Sc and V, and these elemental changes are indicative of a shift to ash of intermediate to mafic composition at this depth (Fig. 13). Ternary diagrams (Fig. 14) indicate at least two populations of ash spanning a mixing line between relatively more felsic (dacites and rhyolite sources) and more mafic (basalt sources) end members.

Other existing data supports the hypothesis of multiple ash populations. Some of the Site 444 cores possess visually black and chemically basaltic ash, which is a unique aspect for the Shikoku Basin. Refractive indices of volcanic glass and associated minerals have been documented as an effective means for correlating widespread tephra layers (Machida and Arai 1983). Based on the refractive indices of glass shards from ash layers at Site 444, Furuta and Arai (1980) (their Figure 12) show a

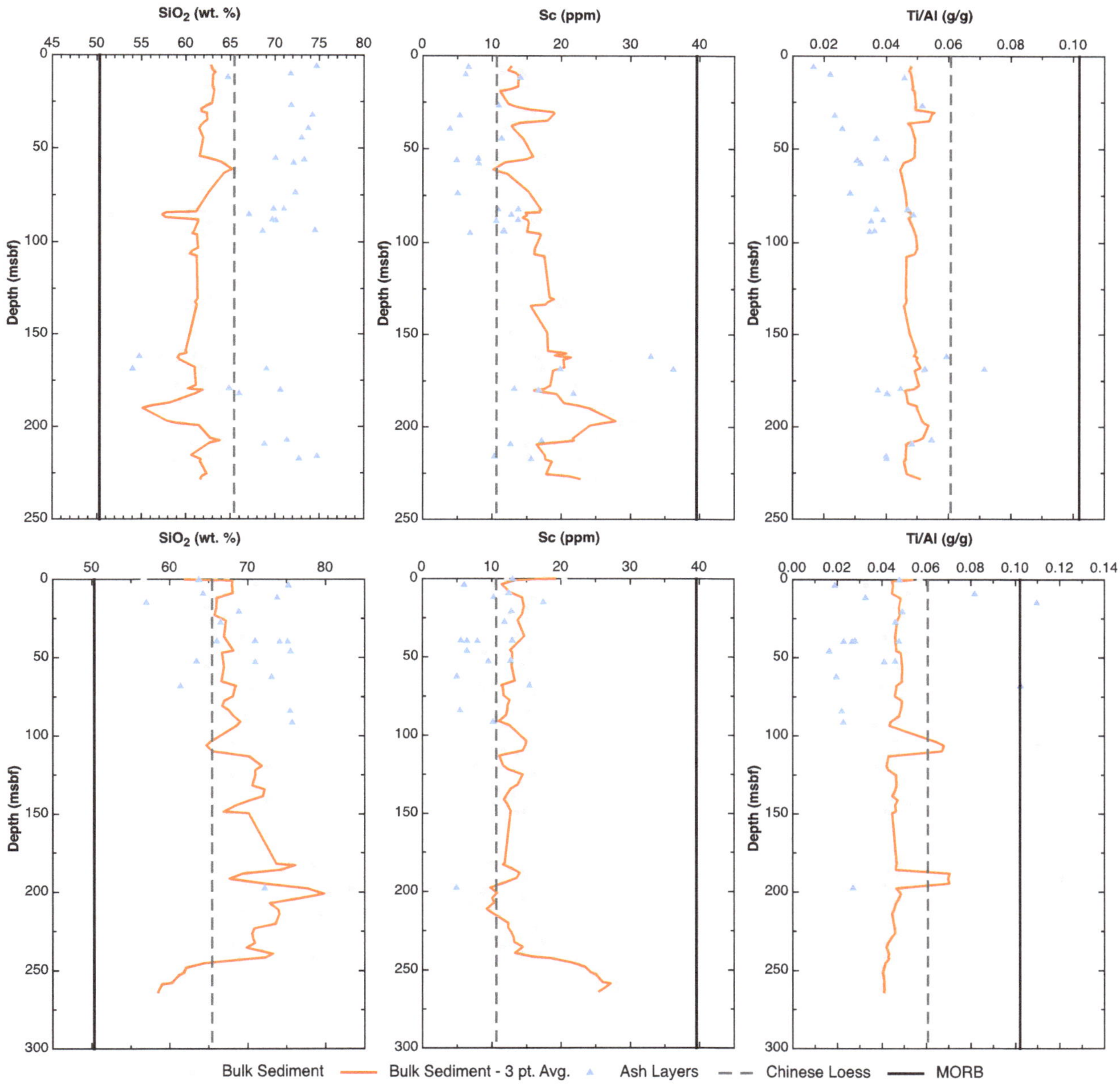

Fig. 13 Representative compositions and elemental ratios plotted with depth. Chinese Loess (Jahn et al. 2001), MORB (Arevalo and McDonough 2010). (*Top*) Site 444. (*Bottom*) Site 579/581

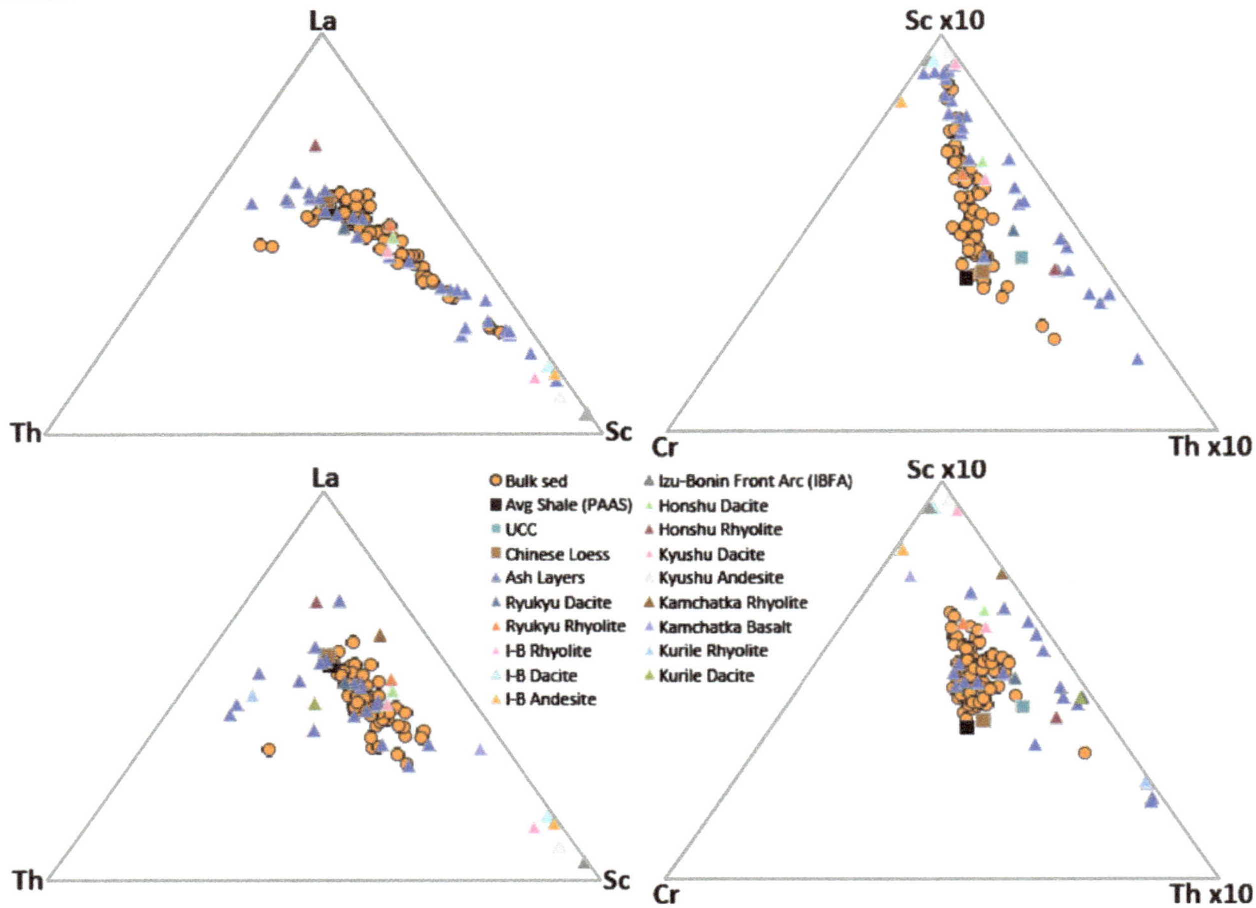

Fig. 14 Ternary diagrams showing potential end members and bulk sediment chemistry at Site 444 (*top*) and Site 579/581 (*bottom*)

linear relationship between refractive index and FeO content. They find that the refractivity of the transition-metal oxides (e.g., FeO, TiO_2, and MnO) at this site is higher than that of SiO_2 and other oxides, thus the refractive index seems to be driven primarily by the transition-metal oxides. The predominate transition-metal oxides in volcanic glass shards are iron-based (i.e., FeO or Fe_2O_3), thus at Site 444 the refractive index of volcanic glass shards can be determined by the content of the iron oxides and indicate that basaltic ashes are present. At Site 444, the ash layers are divided into several groups based on petrographic characteristics, chemical composition, and age. As discussed above, Furuta and Arai (1980) identify rhyolitic and basaltic ash layers at Site 444. Multiple ash populations are also apparent in our microprobe and bulk ash layer chemistry.

Downcore profiles of the bulk sediment do not necessarily provide clear distinction of a single source or two-component mixing (Fig. 13). Additionally, ternary diagrams show that the bulk sediment does not necessarily lie along a mixing line (Fig. 14) with an upper continental component (e.g., Chinese Loess or PAAS) influencing the sediment in addition to the ash components discussed above. Thus, there is likely at least one additional end member beyond the multiple ash populations that influences the bulk sediment composition. This will be explored further in our future studies.

Based on the above discussion of compositional trends, QFA was performed on the data at Site 444 using a modified element suite compared to that used for Site 1149 and Site 52 (Table 1). When performing factor analysis, there is a balance that must be maintained between the number of elements and the number of samples in the dataset. Reimann et al. (2002) outline a number of parameters by which to define the appropriate number of elements. We maintain a very conservative use of these rules, and as such during our study of Site 444, with 71 samples in the dataset, we restricted our element menu to seven (7) elements. Of the refractory elements chosen as representative of the aluminosilicate end members (Al, Ti, Sc, Cr, Ni, Nb, La, Th), Ni overall shows the least variability between potential sources and thus was removed from the element menu for our statistical analysis of Site 444.

Table 1 Bulk sediment elemental composition (ppm) used for QFA analysis, Site 444

Leg	Site	Hole	Core	Type	Section	Top depth (cm)	Bottom depth (cm)	Depth (msbf)	Al (ppm)	Ti (ppm)	Sc (ppm)	Cr (ppm)	Nb (ppm)	La (ppm)	Th (ppm)
58	444	*	2	R	1	24	26	6.24	79,996	3853	11.27	70.41	7.94	29.26	11.58
58	444	*	2	R	2	57	59	8.07	80,559	3832	14.09	68.73	10.92	36.56	10.75
58	444	*	2	R	3	53.5	55.5	9.54	83,039	3853	11.42	77.58	8.36	28.96	12.48
58	444	*	2	R	4	25	27	10.75	84,474	4190	14.67	88.30	12.79	31.16	11.36
58	444	*	3	R	1	124	126	16.74	81,860	3930	15.15	73.85	11.52	32.08	10.63
58	444	*	3	R	2	107.5	109.5	18.08	84,837	4085	11.31	78.40	7.94	30.28	23.13
58	444	*	3	R	3	86	88	19.36	79,428	3889	11.58	63.46	6.84	28.87	13.82
58	444	*	4	R	1	106	108	26.06	82,894	4157	10.43	80.53	6.65	32.58	26.08
58	444	*	4	R	2	108	110	27.58	80,804	3970	14.96	64.32	11.20	29.80	10.11
58	444	*	4	R	3	108	110	29.08	81,114	3965	14.79	66.16	10.00	33.88	9.93
58	444	*	4	R	4	109	111	30.59	78,989	3969	15.99	68.26	11.75	29.27	10.49
58	444	*	4	R	5	46	48	31.46	79,119	5323	25.88	15.44	3.20	10.26	2.67
58	444	*	5	R	1	62	64	35.12	82,536	3834	15.05	66.76	10.52	32.18	11.10
58	444	*	5	R	2	40.5	42.5	36.41	85,835	4087	13.03	81.69	8.47	32.84	11.82
58	444	*	5	R	3	53	55	38.03	76,601	3582	13.45	60.93	11.02	35.68	11.12
58	444	*	5	R	4	64	66	39.64	85,534	4233	11.88	77.77	7.62	32.26	15.33
58	444	*	6	R	1	76	78	44.76	76,374	3713	13.91	67.92	10.15	29.64	10.31
58	444	*	7	R	1	82.5	84.5	54.33	83,010	4126	17.62	68.63	11.54	32.11	10.78
58	444	*	7	R	2	82.5	84.5	55.83	85,068	4160	16.22	66.89	10.94	31.47	10.62
58	444	*	7	R	3	87	89	57.37	78,574	3564	10.59	50.80	5.41	28.56	9.71
58	444	*	7	R	6	86	88	61.31	77,641	3315	8.65	41.39	3.72	24.15	7.96
58	444	*	8	R	1	49	51	63.49	64,239	2914	11.20	21.17	3.33	11.96	4.33
58	444	*	9	R	1	73.5	75.5	73.24	83,958	3897	16.66	44.20	8.50	33.57	10.12
58	444	A	1	R	1	82.5	84.5	82.83	86,300	4026	17.98	65.34	9.35	29.67	10.09
58	444	*	10	R	1	100.5	102.5	83.01	81,435	3842	16.72	72.61	8.44	34.45	8.59
58	444	*	10	R	2	58	60	84.08	85,360	3903	16.11	69.87	11.43	32.10	11.45
58	444	A	1	R	2	86	88	84.36	82,646	4180	17.91	54.83	8.60	26.22	8.16
58	444	A	1	R	3	60	62	85.6	65,077	3067	10.31	47.18	4.61	27.01	7.58
58	444	A	1	R	4	53	55	87.03	84,916	4014	16.79	65.35	10.50	31.42	10.87
58	444	A	1	R	5	77	79	88.27	86,317	4087	16.05	63.91	10.71	30.28	10.35
58	444	A	1	R	6	82	84	89.78	81,209	3897	12.62	57.46	4.99	29.31	8.33
58	444	A	2	R	2	134	136	94.34	80,643	3913	17.20	61.23	8.65	28.27	9.64
58	444	A	2	R	3	96	98	95.46	80,589	4103	15.22	55.74	5.50	28.58	8.52

Table 1 Bulk sediment elemental composition (ppm) used for QFA analysis, Site 444 *(Continued)*

58	444	A	2	R	4	18	20	96.18	78,731	3810	18.30	49.13	8.16	27.02	8.03
58	444	A	3	R	2	74	76	103.24	82,638	4071	17.58	46.40	9.18	25.53	7.54
58	444	A	3	R	3	73	75	104.73	82,775	4320	12.68	55.50	3.79	26.30	7.31
58	444	A	3	R	4	79	81	106.29	85,566	4046	18.19	61.54	8.92	31.96	9.70
58	444	A	3	R	5	53	55	107.53	83,808	3974	17.42	55.99	8.49	30.56	8.95
58	444	A	6	R	1	34	36	129.84	72,532	3238	17.06	57.16	9.60	28.59	9.97
58	444	A	6	R	2	55	57	130.6	84,911	3983	20.53	44.81	7.98	43.70	8.98
58	444	A	6	R	3	94	96	132.49	84,151	4042	19.21	56.38	9.16	29.40	9.67
58	444	A	6	R	4	52	54	133.57	86,082	3811	14.22	65.93	7.13	30.96	11.47
58	444	A	6	R	5	48	50	134.24	87,133	3982	19.72	38.44	7.93	37.25	8.78
58	444	A	8	R	1	74	76	149.24	82,530	3904	12.79	62.99	4.02	27.01	9.03
58	444	A	9	R	1	95.5	97.5	158.96	82,663	4093	21.47	39.19	5.58	18.86	5.19
58	444	A	9	R	2	89	91	160.39	77,146	4000	19.96	47.48	4.25	25.95	6.92
58	444	A	9	R	3	48	50	161.48	76,262	3576	20.69	37.71	5.49	19.33	5.85
58	444	A	9	R	4	65.5	67.5	162.59	74,302	3564	16.57	47.67	4.26	21.70	6.72
58	444	A	9	R	5	37	39	163.8	76,727	4246	26.94	26.36	4.44	16.22	4.59
58	444	A	10	R	1	45	47	167.95	83,868	3826	17.41	49.19	8.77	39.44	9.45
58	444	A	10	R	2	54.5	56.5	169.55	77,551	4036	17.00	28.06	2.93	16.86	4.13
58	444	A	11	R	1	62.5	64.5	177.63	80,760	4019	22.00	43.95	6.23	22.21	5.90
58	444	A	11	R	2	102	104	179.52	77,568	3853	16.32	31.54	3.37	19.58	4.90
58	444	A	11	R	3	30.5	32.5	180.31	73,192	3451	14.22	30.30	2.33	20.26	5.12
58	444	A	11	R	4	68	70	182.18	73,600	3011	17.74	32.50	4.40	18.30	5.17
58	444	A	12	R	1	53.5	55.5	187.04	79,144	4070	26.12	41.96	6.10	19.34	5.91
58	444	A	12	R	2	60	62	188.6	65,137	3165	17.19	39.43	5.67	18.21	5.61
58	444	A	12	R	3	41	43	189.91	76,284	3793	24.56	47.45	6.42	20.07	6.47
58	444	A	13	R	1	95	97	196.95	81,168	4176	30.31	26.61	3.26	12.44	3.16
58	444	A	13	R	2	61	63	198.11	77,704	4165	28.66	39.02	5.47	18.67	5.39
58	444	A	13	R	3	36	38	199.36	76,949	3981	18.65	32.70	8.20	26.95	6.21
58	444	A	14	R	1	83	85	206.33	70,259	3887	25.01	14.61	2.19	9.13	2.06
58	444	A	14	R	2	62	64	207.62	77,583	3748	21.44	30.50	3.88	13.31	3.88
58	444	A	14	R	3	29	31	208.79	73,567	3401	19.02	34.00	4.49	15.12	4.56
58	444	A	14	R	4	22	24	209.64	79,436	3744	15.03	57.19	5.12	22.21	8.21
58	444	A	15	R	1	53	55	215.53	79,550	3659	15.30	31.16	2.72	11.85	4.16

Table 1 Bulk sediment elemental composition (ppm) used for QFA analysis, Site 444 *(Continued)*

58	444	A	15	R	2	125	127	217.75	82,603	3788	22.63	43.75	4.36	17.21	5.07
58	444	A	15	R	3	83	85	218.83	74,140	3484	15.16	37.63	2.89	16.34	4.93
58	444	A	16	R	1	106	108	225.56	79,839	3528	18.10	40.54	8.06	22.48	8.21
58	444	A	16	R	2	79	81	226.79	75,025	3649	20.13	45.88	7.72	20.26	7.33

The results of QFA at Site 444 indicate that three factors (end members) explain 98 % of the variability of the data. This is consistent with the geochemical trends we observe in the bulk data, and the three factors explain 53, 33, and 11 % of the variability of the data respectively (Fig. 15, left). Based on the preliminary modeled compositions of these factors, we interpret that there may be one intermediate crustal source, and at least one dacitic source exhibiting higher Ti/Al ratios, although we cannot differentiate further based on QFA alone. If we were to increase the number of factors, we find that four factors can explain 99 % of the variability of the data, with the four factors explaining 55, 34, 9, and 1 % of the variability of the data, respectively (Fig. 15, right). While this fourth factor is small, we find that it predominantly splits the third factor from the previous QFA and identifies an end member that is dominated by variation in Nb.

In terms of the relative downhole distributions of square factor loadings (Fig. 16), Factor 1 dominates over

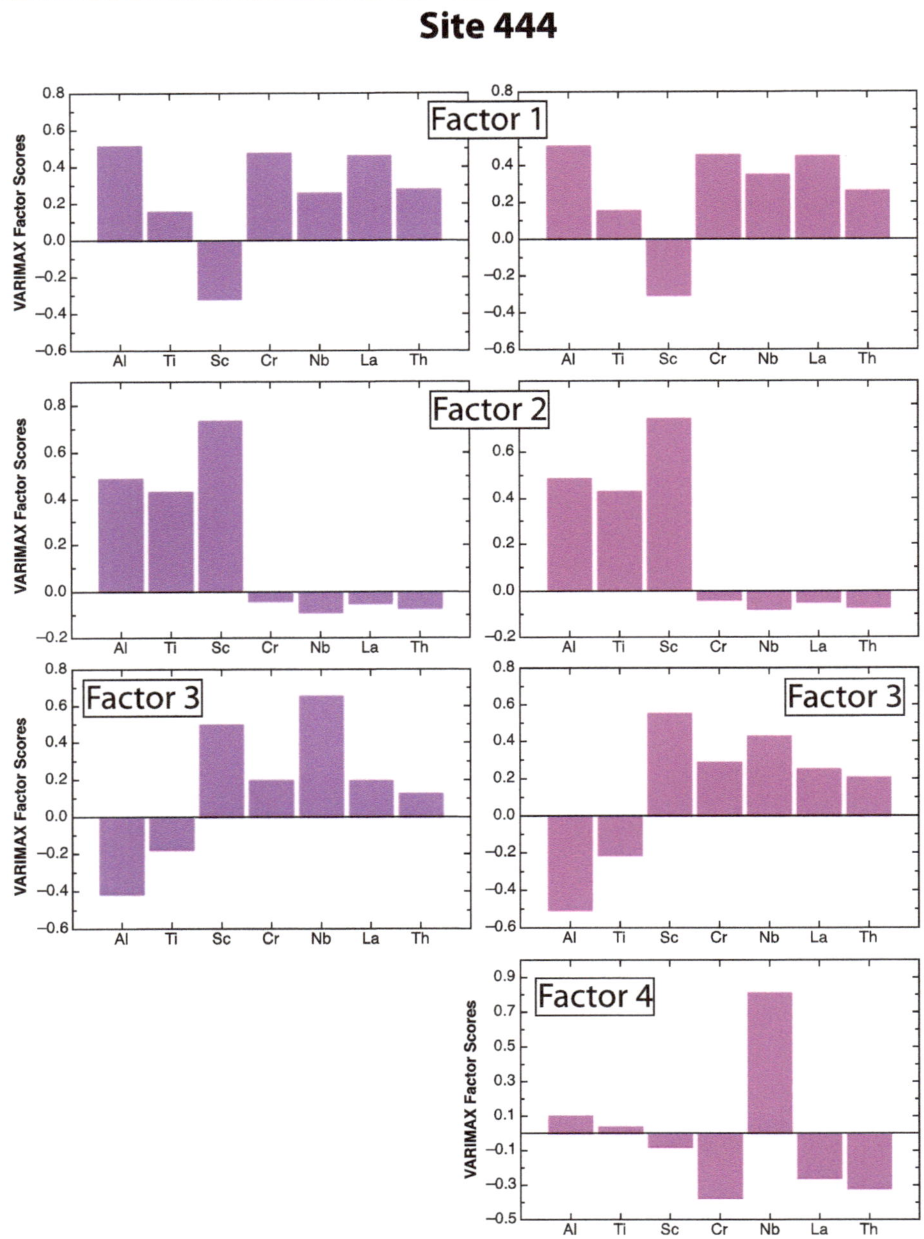

Fig. 15 Site 444, VARIMAX factor scores from QFA. (*Left*) Three-factor model. (*Right*) Four-factor model. As described in the text, Factor 1 is most likely an intermediate crustal source, Factors 2 and/or 3 are most likely dacitic source(s) and Factor 4 is a Nb-rich end member

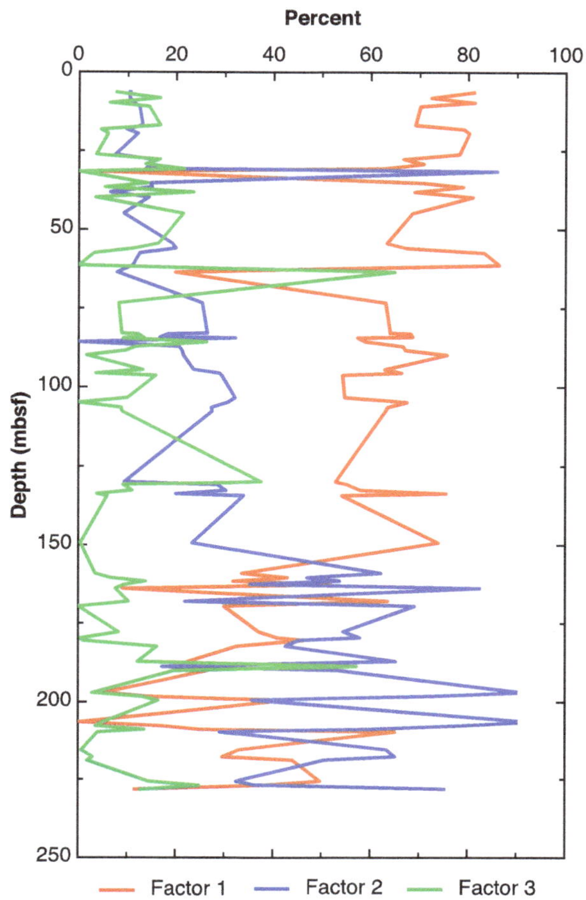

Fig. 16 Depth profiles of square factor loadings from QFA showing the relative contribution of each factor, Site 444

the upper ~150 mbsf and exhibits a broad decrease in importance with depth. Factor 2 is most important in the sequence below ~150 mbsf. Factors 2 and 3 have roughly equal relative contributions over the upper ~50 mbsf, below which Factor 2 exhibits a broad increase while Factor 3 generally decreases. Future statistical analysis by CLS will allow us to determine whether three or four factors are ultimately the appropriate number of end members and will allow us to further determine the sources to the sediment.

Kamchatka and the Kuriles, DSDP Site 579 and Site 581

DSDP Sites 579 and 581 (Fig. 1) were occupied in June 1982 as part of DSDP Leg 86 and were cored as part of a transect across the Kuroshio Current system designed to investigate the paleoceanography of the northwest Pacific. Although spaced relatively far apart, Sites 579 and 581 comprise an excellent composite section of the region (Plank and Langmuir 1998). Site 579 marks the southern margin of the transition between the sediments of the calcareous-siliceous subtropical gyre and the more siliceous subarctic gyre. It was selected for Leg 86 with the goal of compiling a paleoceanographic record of this transition zone for the late Neogene and Quaternary and comparing this record of migrations of the subarctic front with that of Asian eolian inputs. Site 581 is the northernmost of the transect sites drilled during Leg 86 and was designed to provide a high-resolution record of the late Neogene.

As with the other sites in this region, a number of ash layer parameters were examined (Fig. 12). There are no clear trends evident in the number of ash layers through the depth of the site. There are two peaks in the number of ash layers per 5 m, with seven ash layers from 40–45 mbsf and four layers from 80–85 mbsf. This peak at 80 mbsf also corresponds to a peak in total ash thickness (44 cm thick), meaning that not only are more layers present at this depth but also that those layers are thicker. This suggests a period of increased large-scale volcanic eruptions during the Pliocene. However, this is difficult to verify as late Pleistocene glaciations are likely to have obscured the terrestrial record of eruptive history in this area (Volynets et al. 1999). Ash pockets show good correlation with the ash layers in frequency, and we interpret that they originated from disturbance in the core due to the weather during core recovery. The cores had a high occurrence of "flow-in" zones due to heavy seas that may have affected coring and disturbed recovery of some ash layers (Heath et al. 1985).

Interpretation of the bulk ash layers and bulk sediment downhole is not straightforward (Fig. 13). In ternary diagrams, the bulk sediment commonly falls along a mixing line that can be explained by the bulk ash and a relatively more basaltic component, supporting the presence of dispersed ash in the sediment (Fig. 14). Much of the bulk sediment is similar in composition to Chinese Loess, which, while not necessary to explain the composition of the bulk sediment, can be inferred to be almost surely present due to the location of the sites.

Applying QFA to the combined data from these sites, we find that either five or six factors best explain the variability of the data (Table 2, Fig. 17). Comparison of the two sets of factor analysis indicates that Factors 1 through 4 correlate with each other whether there are five or, alternatively, six factors. If there are six factors, Factor 5 from the smaller factor model is split between Factors 5 and 6. Factor 6 explains 1.7 % of the variability of the data in this model, and thus it is interpreted as "real".

Focusing first on the five-factor model, the calculated compositions of the end members identifies Factor 4 (explaining 10.1 % of the variability) as a potential terrigenous source. This is unusual, as in all the other NW pacific sites discussed in this paper the crustal source is the first and/or second end member explaining the

Table 2 Bulk sediment elemental composition (ppm) used for QFA analysis, Sites 579 and 581

Leg	Site	Hole	Core	Type	Section	Top depth (cm)	Bottom depth (cm)	Depth (msbf)	Al (ppm)	Ti (ppm)	Sc (ppm)	Cr (ppm)	Nb (ppm)	La (ppm)	Th (ppm)
86	581	*	1	R	1	28	30	0.28	74,371	4244	20.49	46.23	4.02	16.27	5.36
86	581	*	1	R	1	36	38	0.36	64,845	3268	17.72	39.56	5.38	13.30	4.99
86	579	*	1	H	1	93	95	0.93	70,339	3518	12.17	72.45	6.68	25.01	9.40
86	579	*	1	H	3	75	77	3.75	66,319	2181	10.22	29.39	7.37	8.47	5.93
86	579	*	2	H	1	54	56	8.94	68,056	3532	11.81	58.75	8.02	14.70	6.78
86	579	*	2	H	3	73	75	12.13	73,188	3562	16.02	57.21	8.70	21.20	8.41
86	579	A	1	H	3	94	96	17.94	75,385	3367	15.23	44.72	7.02	15.57	6.65
86	579	A	1	H	7	6	8	23.06	68,144	3364	12.58	54.34	7.94	17.55	6.68
86	579	A	2	H	3	13	15	26.63	69,347	3438	15.07	63.72	9.20	21.34	8.62
86	579	A	3	H	3	73	75	36.73	66,624	2746	13.44	27.85	6.15	10.90	4.88
86	579	A	4	H	3	25	27	45.75	72,574	3398	15.46	59.67	9.37	15.94	7.88
86	579	A	4	H	4	4	6	47.04	43,351	2216	8.89	46.87	8.44	11.55	5.78
86	579	A	5	H	3	107	109	56.07	67,296	3207	14.39	55.31	8.70	21.52	8.26
86	579	A	6	H	3	76	78	65.26	70,026	3410	15.26	57.05	8.74	15.57	7.98
86	579	A	6	H	5	73	75	68.23	53,622	2724	10.20	49.76	6.55	19.78	6.26
86	579	A	6	H	6	10	12	69.1	62,451	2446	8.53	30.52	2.54	17.37	5.54
86	579	A	7	H	3	72	74	74.72	72,871	3585	15.95	59.25	10.44	23.68	9.41
86	579	A	7	H	5	72	74	77.72	68,002	3323	10.60	66.32	4.14	23.43	8.82
86	579	A	8	H	1	52	54	81.02	70,043	3426	11.11	59.48	6.03	23.35	8.69
86	579	A	8	H	3	30	32	83.8	73,144	3565	15.05	61.25	9.79	24.24	9.13
86	579	A	8	H	5	72	74	87.22	59,725	2829	10.54	52.81	6.70	27.02	8.18
86	579	A	9	H	1	110	112	91.1	65,280	3076	10.55	56.03	4.62	21.69	7.78
86	579	A	9	H	3	61	63	93.61	62,214	2294	11.87	38.56	7.77	21.26	8.69
86	579	A	10	H	3	131	133	103.81	64,853	2977	15.04	54.89	7.33	20.29	7.53
86	579	A	10	H	5	72	74	106.22	68,686	7575	17.89	51.60	6.58	15.94	6.28
86	579	A	11	H	1	94	96	109.94	67,054	3167	11.67	57.35	5.18	20.79	8.01
86	579	A	11	H	3	114	116	113.14	60,192	2612	13.67	40.83	6.95	14.15	6.71
86	579	A	12	H	1	83	85	119.33	54,488	2061	7.81	23.08	4.69	19.02	6.70
86	579	A	12	H	3	49	51	121.99	55,246	2474	13.26	41.86	6.21	14.32	6.26
86	579	A	12	H	5	76	78	125.26	63,097	2906	16.01	44.42	5.55	17.65	5.69
86	579	A	13	H	3	74	76	131.74	53,852	2582	13.99	43.75	5.69	12.81	5.30
86	579	A	13	H	5	27	29	134.27	57,088	2580	10.98	50.74	5.24	22.10	6.50
86	579	A	14	H	1	107	109	138.57	54,699	2471	13.11	47.12	6.56	18.01	7.14

Table 2 Bulk sediment elemental composition (ppm) used for QFA analysis, Sites 579 and 581 *(Continued)*

86	579	A	14	H	3	66	68	141.16	55,940	2423	12.06	44.14	6.49	12.61	5.58
86	579	A	14	H	5	70	72	144.2	73,974	3875	10.00	50.23	6.63	19.07	6.45
86	579	A	15	H	2	2	4	148.52	69,270	2925	14.56	55.19	7.95	22.95	8.46
86	579	A	15	H	CC	16	18	149.53	71,892	3146	13.49	62.63	9.86	18.70	9.01
86	581	*	2	R	1	62	64	182.12	42,056	1999	9.89	37.37	2.78	24.82	5.66
86	581	*	2	R	2	2	4	183.02	34,466	1662	12.08	27.39	4.28	18.16	5.07
86	581	*	2	R	3	137	139	185.87	50,664	2181	12.71	28.88	5.89	16.86	5.71
86	581	*	2	R	5	73	75	188.23	61,058	2893	14.22	38.94	5.38	25.91	6.30
86	581	*	3	R	1	58	60	191.58	56,561	6826	14.90	39.01	4.80	14.97	4.67
86	581	*	3	R	3	62	64	194.62	61,976	2550	11.56	33.12	5.22	12.46	4.74
86	581	*	3	R	5	74	76	197.74	27,691	1359	9.00	28.40	4.15	13.98	4.77
86	581	*	4	R	1	57	59	201.07	33,889	1641	8.65	24.95	2.31	15.01	4.71
86	581	*	4	R	3	76	78	204.26	46,571	2231	14.57	47.58	6.51	16.62	7.06
86	581	*	4	R	5	67	69	207.17	30,626	1412	6.83	26.45	1.97	8.75	4.53
86	581	*	5	R	1	126	128	211.26	60,537	2704	9.60	42.27	3.69	12.58	5.50
86	581	*	5	R	3	92	94	213.92	39,316	1781	11.23	32.04	5.22	7.86	4.84
86	581	*	6	R	1	86	88	220.36	51,086	2214	9.28	46.00	2.83	18.85	6.78
86	581	*	6	R	3	67	69	223.17	64,892	3061	16.55	54.76	8.36	18.97	8.96
86	581	*	6	R	5	74	76	226.24	58,477	2715	11.20	50.82	4.27	23.34	9.07
86	581	*	7	R	1	52	54	229.52	54,775	2401	10.65	50.44	3.28	21.27	8.15
86	581	*	7	R	3	36	38	232.36	70,340	2946	17.45	54.41	9.17	16.29	10.10
86	581	*	7	R	5	37	39	235.37	59,378	2529	11.56	50.07	3.26	23.01	8.75
86	581	*	8	R	1	77	79	239.27	63,987	2643	14.31	51.64	5.33	22.36	10.21
86	581	*	8	R	3	11	13	241.61	44,113	2001	14.00	38.89	6.23	17.51	7.56
86	581	*	8	R	3	115	117	242.65	64,594	2703	19.14	47.94	8.11	22.13	10.42
86	581	*	8	R	5	77	79	245.27	83,065	3497	22.35	61.47	10.04	33.19	14.07
86	581	*	9	R	1	16	18	248.16	83,298	3381	22.88	57.87	9.33	31.95	13.20
86	581	*	9	R	3	59	61	251.59	82,323	3316	25.21	57.11	10.80	20.48	13.59
86	581	*	9	R	4	46	48	252.96	85,210	3411	24.70	51.25	9.51	31.48	14.47
86	581	*	10	R	1	42	44	257.92	89,185	3707	25.16	60.99	11.65	34.16	16.84
86	581	*	10	R	1	124	126	258.74	86,271	3549	28.05	55.41	11.20	39.01	17.72
86	581	*	10	R	3	92	94	261.42	88,986	3546	28.26	55.69	11.63	17.12	11.30

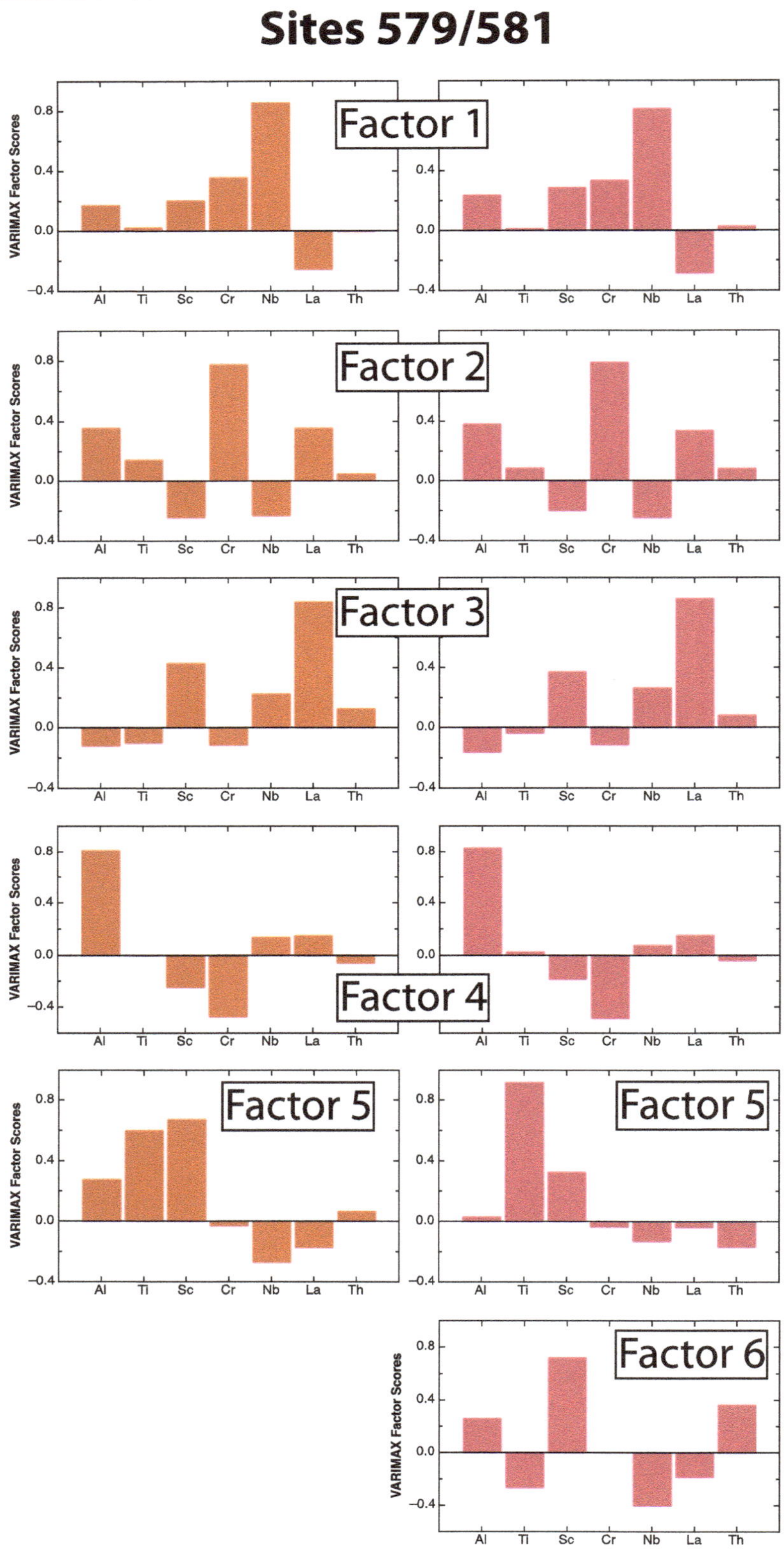

Fig. 17 Sites 579 and 581, VARIMAX factor scores from QFA

majority of the variability in the dataset. This speaks to the importance of dispersed ash at Site 579/581, since the dispersed ash component(s) appear to be controlling most of the variability. This will be further explored in future work. The six-factor model identifies a few factors that are dominated by a single element (Al in Factor 4, Ti in Factor 5) indicating that the division of factors is controlling variability as opposed to their specific concentrations.

As with Site 444, square factor loadings for these Kamchatka/Kurile sites can allow us to assess the relative contribution of the individual factors on a sample-by-sample basis (Fig. 18). In order to assess the relative contributions for this work, we will focus here on the five-factor model. We observe that for most of the sedimentary record, Factor 1 dominates the record, although from ~65 to ~93 mbsf, Factor 2 is the dominate component. Additionally, from ~150 to ~240 mbsf the variance in the record is dominated by Factor 3 (~150 to ~200 mbsf) or Factor 2 (~200 to ~240 mbsf). Factors 4 and 5 are remarkably similar in importance until ~200 mbsf below which Factor 4 increases in relative importance while Factor 5 remains fairly stable. Future statistical analysis by CLS will allow us to determine whether five or six factors are ultimately the appropriate number of end members and will allow us to further determine the sources to the sediment.

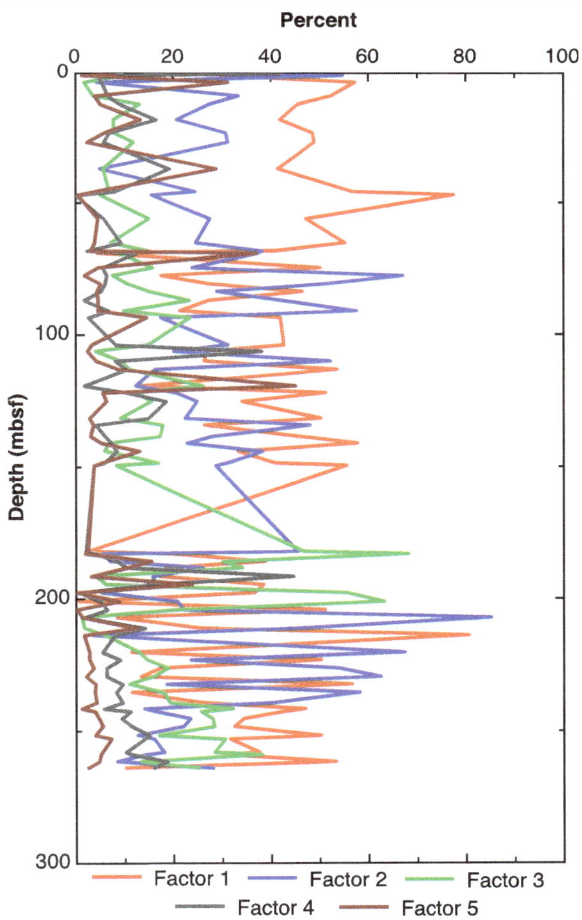

Fig. 18 Depth profiles of square factor loadings from QFA showing the relative contribution of each factor, Sites 579 and 581

Conclusions

Although the presence of dispersed ash in marine sediment has been long known by marine sedimentologists, the overall quantification of its abundance and its composition has recently received renewed attention. The dispersed ash, either altered or unaltered, is extremely difficult to differentiate from detrital/terrigenous/authigenic clay as they are all, fundamentally speaking, "aluminosilicates". The combined geochemical (major, trace, REEs) and statistical (Pisias et al. 2013) approach provides the chemical context for the determination of provenance and the distinctive resolution of the different aluminosilicate components based on their individual geochemical signature(s).

The case studies we present here show that this approach can be useful in multiple different arc systems. We have also shown that by adding the dispersed ash component to the study of discrete ash layers, many additional geological and oceanographic interpretations can be uncovered. Most obviously, we have shown how geochemical modeling of the bulk sediment composition can indicate the presence of an additional ash source that is not recorded in the discrete layer record. Furthermore, in cores where there are no remaining discrete layers (either due to natural bioturbation or man-made coring disturbance), we show that important ash contributions may still be recognized and studied for their composition and accumulation. We have also shown how the accumulation of dispersed ash will either best correlate with the *thickest* ash layer, or alternatively, the *frequency* of the discrete ash layers, which lead to dramatically different interpretations regarding the volcanic mechanisms of a given arc. Finally, when considering these studies throughout the northwest Pacific Ocean, or in marginal basins such as the Japan Sea, our approach may be used to develop a mass balance inventory of the total mass of volcanic ash that is more accurate than estimates based solely on the combined thicknesses of the discrete layers.

There are many other future research directions that are a natural extension of the work described above. There are a number of regions for which the characterization of the abundance and extent of dispersed ash would be of great scientific and societal importance (e.g., Tonga, Costa Rica, and Indonesia), and studying the dispersed ash component in these regions will most certainly lead to an improved understanding of explosive volcanism in those systems. Moreover, most marine volcanoes are beneath the sea surface, either in mid-ocean ridge settings or in convergent margins (e.g., Fiske et al. 2001; White et al. 2003). The vast majority of these eruptions go unobserved and their products are often unrecovered and thus are understudied. Often the only evidence of subaqueous eruptions are widespread pumice rafts. Little is known, however, about the fate of these rafts, yet they too are likely to contribute dispersed ash to the sediment.

Volcanic ash may have significant nutrient fertilization potential as well. Particulate matter from large volcanic eruptions may release enough iron and other nutrients to the surface ocean to stimulate primary production over short time scales (Watson 1997; Olgun et al. 2011). Thus, it would be important to develop an understanding of the effects of iron and other elements from volcanic ash on the biological productivity of the open ocean and, second, to consider the biological effects of volcanic ash-driven elemental fertilization in the open ocean through Earth's history. The use of the dispersed ash record can be applied to achieve these goals.

Additionally, ash, with its major component being volcanic glass, is a metastable phase with high potential biological availability to the subseafloor ecosystem, may provide both electron donors and electron acceptors important to microbial energetics. Characterizing the chemistry of the subseafloor sediment will provide vital information about the nutrients available to subseafloor microbial life as well as the pathways by which these nutrients are utilized.

Abbreviations

BNN: Izu-Bonin Boninite; CL: Chinese Loess; CLS: Constrained Least Squares; CRISP II: Costa Rica Seismogenesis Project Stage II; DSDP: Deep Sea Drilling Project; HR: Honshu Rhyolite; IBA: Izu-Bonin Andesite; IBFA: Izu-Bonin Front Arc; IBR: Izu-Bonin Rhyolite; ICP-ES: Inductively Coupled Plasma-Emission Spectrometer; ICP-MS: Inductively Coupled Plasma-Mass Spectrometer; IODP: Integrated Ocean Drilling Program; LRC: lower red clay MAR, mass accumulation rate; MORB: Mid-Ocean Ridge Basalt; ODP: Ocean Drilling Program; PAAS: Post-Archean average Australian Shale; PGL: pale green laminae; QFA: Q-Mode Factor Analysis; RCB: Rotary Core Barrel; REE: rare earth element; TI: Total Inversion; URC: upper red clay.

Competing interests

The authors' declare that they have no competing interests.

Authors' contributions

RPS is the lead author of the paper, performed the ICP-ES and ICP-MS analyses, and led the research and multiple conversations/interactions with the research team. RWM funded the research (see Acknowledgements), performed the oversight of the project, advised then-Ph.D. student RPS and then-undergraduate students SG, RL, and CCM, facilitated the discussion with co-authors and community, and contributed the text. JCS and SK performed microprobe analyses of volcanic glass from ash layers, discussed the research, contributed revisions of the text, compiled the reference list, created Figs. 1, 2, and 4, and provided the interpretation of the data. SK funded the research (see Acknowledgements) and advised Ph.D. student JCS regarding ash chemistry. FH reviewed and revised the manuscript. MBU discussed the research, contributed the revisions of the text, and provided the interpretation of the data. SG, RL, and CCM assisted in the ICP-ES and ICP-MS analyses, were involved in the preliminary statistics and interpretations of Sites 52, 444, and 579/581, and reviewed and commented on the manuscript. All authors read and approved the final manuscript.

Acknowledgements

This research used samples and/or data provided by the IODP. RPS and RWM thank the US Science Support Program (USSSP) of IODP, NSF OCE-0136855, and NSF OCE-0958002 for financial support, and T. Ireland, A. G. Dunlea, N. Murphy, and J. W. Sparks for laboratory assistance. JCS and SK thank the German Research Foundation for financial support (KU-2685/1-1, 2-1&2) and M. Thöner, K. Strehlow for laboratory assistance. Portions of this material are based upon work supported while RWM was serving at the National Science Foundation. Thanks to those who reviewed and commented on the manuscript.

Author details

[1]Department of Earth & Environment, Boston University, Boston, MA 02215, USA. [2]GEOMAR Helmholtz Centre for Ocean Research Kiel, 24148 Kiel, Germany. [3]Department of Earth and Environmental Science, New Mexico Institute of Mining and Technology, Socorro, NM 87801, USA. [4]Department of Geosciences, Princeton University, Princeton, NJ 08544, USA. [5]Division of Earth and Ocean Sciences, Nicholas School of the Environment, Duke University, Durham, NC 27708, USA. [6]Department of Oceanography, Texas A&M University, College Station, TX 77843, USA.

References

Arevalo R, McDonough WF. Chemical variations and regional diversity observed in MORB. Chem Geol. 2010;271(1):70–85.

Bailey JC. Geochemical history of sediments in the northwestern Pacific Ocean. Geochem J. 1993;27:71–90.

Bailey JC. Role of subducted sediments in the genesis of Kurile-Kamchatka island arc basalts: Sr isotopic and elemental evidence. Geochem J. 1996; 30(5):289–321.

Bryan SE, Cook A, Evans JP, Colls PW, Wells MG, Lawrence MG, et al. Pumice rafting and faunal dispersion during 2001-2002 in the Southwest Pacific: record of a dacitic submarine explosive eruption from Tonga. Earth Planet Sci Lett. 2004;227(1-2):135–54.

Bryant CJ, Arculus RJ, Eggins SM. Laser ablation-inductively coupled plasmamass spectrometry and tephras: a new approach to understanding arc-magma genesis. Geology. 1999;27(12):1119–22.

Bryant CJ, Arculus RJ, Eggins SM. The geochemical evolution of the Izu–Bonin arc system: a perspective from tephras recovered by deep-sea drilling. Geochem Geophys Geosyst. 2003;4(11):1094. doi:10.1029/2002GC000427.

Bursik M. Effect of wind on the rise height of volcanic plumes. Geophys Rese Lett. 2001;28(18):3621–4.

Cambray H, Cadet JP, Pouclet A. Ash layers in deep-sea sediments as tracers of arc volcanic activity: Japan and Central America as case studies. Island Arc. 1993;2:72–86.

Cambray H, Pubellier M, Jolivet L, Pouclet A. Volcanic Activity Recorded in Deep-Sea Sediments and the Geodynamic Evolution of Western Pacific Island Arcs, in Active Margins and Marginal Basins of the Western Pacific (eds B. Taylor and J. Natland), American Geophysical Union, Washington, DC; 1995. doi:10.1029/GM088p0097.

Carey SN. Influence of convective sedimentation on the formation of widespread tephra fall layers in the deep sea. Geology. 1997;25(9):839–42.

Carey SN, Sigurdsson H. Deep-sea evidence for distribution of tephra from the mixed magma eruption of the Soufriére on St Vincent 1902: ash turbidites and airfall. Geology. 1978;6:271–4.

Carey SN, Sigurdsson H. The Roseau ash: Deep-sea tephra deposits from a major eruption on Dominica, Lesser Antilles arc. J Volcanol Geoth Res. 1980;7:67–86.

Carey SN, Sigurdsson H. Grain size of Miocene volcanic ash layers from Sites 998 999 and 1000: Implications for source areas and dispersal. Proc Ocean Drill Program Sci Results. 2000;165:101–10.

Carey SN, Sparks RSJ. Quantitative models of fallout and dispersal of tephra from volcanic eruption columns. Bull Volcanol. 1986;48:109–25.

Carey RJ, Houghton BF, Thordarson T. Tephra dispersal and eruption dynamics of wet and dry phases of the 1875 eruption of Askja Volcano Iceland. Bull. 2010;72(3):259–78.

Chan LH, Kastner M. Lithium isotopic compositions of pore fluids and sediments in the Costa Rica subduction zone: implications for fluid processes and sediment contribution to the arc volcanoes. Earth Planet Sci Lett. 2000;183(1):275–90.

deVries Klein G, Kobayashi K. Initial Reports of the Deep Sea Drilling Project v. 58. Washington DC: US Government Printing Office; 1980. p. 1022.

deVries Klein G, McConville RL, Harris JM, Steffensen CK. Petrology and Diagenesis of Sandstones Deep Sea Drilling Project Site 445 Daito Ridge. Init Rep DSDP. 1980;58:609–16. doi:10.2973/dsdpproc581121980.

Duggen S, Olgun N, Croot P, Hoffmann LJ, Dietze H, Delmelle P, et al. The role of airborne volcanic ash for the surface ocean biogeo- chemical iron-cycle: A review. Biogeosci. 2010;7:827–44. doi:10.5194/bg-7-827-2010.

Dunlea AG, Murray RW, Sauvage J., Pockalny RA, Spivack AJ, Harris RN, et al. Cobalt-based age models of pelagic clay in the South Pacific Gyre. Geochem Geophys Geosys. 2015a;16. doi:10.1002/2015GC005892.

Dunlea AG, Murray RW, Sauvage J, Pockalny RA, Spivack AJ, Harris RN, et al. Dust, volcanic ash, and the evolution of the South Pacific Gyre through the Cenozoic. Paleocean. 2015b; 30. doi:10.1002/2015PA00282.

Dymond J. Geochemistry of Nazca plate surface sediments: an evaluation of hydrothermal biogenic detrital and hydrogenous sources. Geol Soc Am Memoirs. 1981;154:133–74.

Expedition 340 Scientists Lesser Antilles volcanism and landslides: Implications for hazard assessment and long-term magmatic evolution of the arc. Prelim Rep Integrated Ocean Drill Program 340. 2012. doi:10.2204/iodpppr3402012.

Fischer AG, Heezen BC, Boyce RE, Bukry D, Douglas RG, Garrison RE, et al. Site 52. Initial Rep Deep Sea Drill Proj 6:247–290. 1971. doi:10.2973/dsdpproc61101971.

Fisher RV, Schmincke H-U. Pyroclastic rocks. New York: Springer-Verlag; 1984. p. 472.

Fiske RS, Naka J, Iizasa K, Yuasa M, Klaus A. Submarine silicic caldera at the front of the Izu-Bonin arc Japan: voluminous seafloor eruptions of rhyolite pumice. Geol Soc Amer Bull. 2001;113:813–24.

Freundt A, Grevemeyer I, Rabbel W, Hansteen TH, Hensen C, Wehrmann H, et al. Volatile (H_2O, CO_2, Cl, S) budget of the Central American subduction zone. Int J Earth Sci. 2014;103(7):2101–27. doi:10.1007/s00531-014-1001-1.

Frogner P, Gíslason SR, Óskarsson N. Fertilizing potential of volcanic ash in ocean surface water. Geology. 2001;29:487–90.

Furuta T, Arai F. Petrographic and geochemical properties of tephras in Deep Sea Drilling Project cores from the north Philippine Sea. Init Rep DSDP. 1980;58: 617–27. doi:10.2973/dsdpproc581131980.

Gardner JV, Nelson CS, Baker PA. Distribution and character of pale-green laminae in sediment from Lord Howe Rise: A probable late Neogene and Quaternary tephrostratigraphic record. In Blakeslee JH (ed) Initial Reports of the DSDP 90, Wahsington, 1986. p. 1145–1158.

Hauff F, Hoernle K, Schmidt A. Sr-Nd-Pb composition of Mesozoic Pacific oceanic crust (Site 1149 and 801 ODP Leg 185): Implications for alteration of ocean crust and the input into the Izu-Bonin-Mariana subduction system. Geochem Geophys Geosyst. 2003;4(8):8913. doi:10.1029/2002GC000421.

Heath GR, Kovar RB, Lopez C, Campi GL. Elemental composition of Cenozoic pelagic clays from deep sea drilling project sites 576 and 578 Western North Pacific. Init Rep DSDP. 1985;86:605–48. doi:10.2973/dsdpproc861271985.

Hein JR, Scholl DW, Barron JA, Jones MG, Miller J. Diagenesis of late Cenozoic diatomaceous deposits and formation of the bottom simulating reflector in the southern Bering Sea. Sedimentol. 1978;25(2):155–81.

Henry P, Kanamatsu T, Moe K, Expedition 333 Scientists. Proceedings IODP 333, Tokyo Integrated Ocean Drilling Program Management, International Inc Washington DC; 2012. doi:10.2204/iodpproc3332012.

Hochstaedter A, Gill J, Peters R, Broughton P, Holden P, Taylor B. Across-arc geochemical trends in the Izu-Bonin arc: Contributions from the subducting slab Geochem Geophys Geosys. 2001;2(7). doi:10.1029/2000GC000105.

Hovan SA, Rea DK, Pisias NG. Late Pleistocene continental climate and oceanic variability recorded in northwest Pacific sediments. Paleoceanography. 1991;6(3):349–70.

Huang TC. A volcanic sedimentation model: Implications of processes and responses of deep-sea ashes. Mar Geol. 1980;38:103–22.

Huang TC, Watkins ND, Shaw DM, Kennett JP. Atmospherically transported volcanic dust in South Pacific deep sea sedimentary cores at distances over 3000 km from the eruptive source. Earth Planet Sci Lett. 1973;20(1):119–24.

Huang TC, Fillon RH, Watkins ND, Shaw DM. Volcanism and silicious microfaunal diversity in the southwest Pacific during the Pleistocene period. Deep-Sea Res. 1974;21:377–84.

Huang TC, Watkins ND, Shaw DM. Atmospherically transported volcanic glass in deep-sea sediments: volcanism in sub-antarctic latitudes of the south Pacific during late Pliocene and pleistocene time. Bull Geol Soc Am. 1975;86(9):1305–15.

Hüpers A, Ikari MJ, Dugan B, Underwood MB, Kopf AJ. Origin of a zone of anomalously high porosity in the subduction inputs to Nankai Trough. Mar Geol. 2015;361:47–162.

Hyeong K, Park S, Yoo CM, Kim K. Mineralogical and geochemical compositions of the eolian dust from the northeast equatorial Pacific and their implications on paleolocation of the Intertropical Convergence Zone. Paleoceanography. 2005;20(0883-8305). doi:10.1029/2004PA001053.

Jahn BM, Gallet S, Han JM. Geochemistry of the Xining Xifeng and Jixian sections Loess Plateau of China: eolian dust provenance and paleosol evolution during the last 140 ka. Chem Geol. 2001;178(1–4):71–94.

Jones MT, Gíslason SR. Rapid releases of metal salts and nutrients following the deposition of volcanic ash into aqueous environments. Geochim Cosmochim Acta. 2008;72:3661–80. doi:10.1016/jgca200805030.

Jordan BR, Sigudsson H, Carey SN, Rogers R, Ehrenborg J. Geochemical correlation of Caribbean Sea tephra layers with ignimbrites in Central AmericaGeological Society of America Special Papers. 2006; 402:175–208. doi:10.1130/2006.2402(08).

Kastner M, Elderfield H, Martin JB. Fluids in convergent margins: what do we know about their composition origin role in diagenesis and importance for oceanic chemical fluxes? Philos Trans R Soc Lond A. 1991;335(1638):243–59.

Kennett JP, McBirney AR, Thunell RC. Episodes of Cenozoic volcanism in the circum-Pacific region. J Volc Geo Res. 1977;2:145–63.

Klinkhammer GP, Elderfield H, Edmond JM, Mitra A. Geochemical implications of rare earth element patterns in hydrothermal fluids from mid-ocean ridges. Geochim Cosmochim Acta. 1994;58(23):5105–13.

Kryc KA, Murray RW, Murray DW. Al-to-oxide and Ti-to-organic linkages in biogenic sediment: relationships to paleo-export production and bulk Al/Ti. Earth Planet Sci Lett. 2003;211(1):125–41.

Kutterolf S, Freundt A, Peréz W, Mörz T, Schacht U, Wehrmann H, et al. The Pacific offshore record of Plinian arc volcanism in Central America part 1: Along-arc correlations. Geochem Geophys Geosyst. 2008a;9(2). doi:10.1029/2007GC001631.

Kutterolf S, Freundt A, Peréz W. The Pacific offshore record of Plinian arc volcanism in Central America part 2: Tephra volumes and erupted masses. Geochem Geophys Geosys. 2008b;9(2). doi:10.1029/2007GC001791.

Kutterolf S, Freundt A, Schacht U, Bürk D, Harders R, Mörz T, et al. The Pacific offshore record of Plinian arc volcanism in Central America part 3: Application to forearc geology. Geochem Geophys Geosys. 2008c;9(2). doi:10.1029/2007GC001826.

Kutterolf S, Jegen M, Mitrovica JX, Kwasnitschka T, Freundt A, Huybers P. A detection of Milankovitch frequencies in global volcanic activity. Geology. 2013;41(2):227–30. doi:10.1130/G334191.

Kutterolf S, Schindlbeck JC, Scudder RP, Murray RW, Pickering KT, Freundt A, et al. Large volume submarine ignimbrites in the Shikoku Basin: an example for explosive volcanism in the Western Pacific during the Late Miocene. Geochem Geophys Geosys. 2014;15(5):1837–51.

Kyte FT, Leinen M, Heath GR, Zhou L. Cenozoic sedimentation history of the central North Pacific: inferences from the elemental geochemistry of core LL44-GPC3. Geochim Cosmochim Acta. 1993;57:1719–49.

Le Friant A, Ishizuka O, Stroncik NA and the Expedition 340 Scientists (2013) Proceedings of Integrated Ocean Drilling Program vol 340, Tokyo. doi:10. 2204/iodp.proc.340.104.2013

Ledbetter MT, Sparks RSJ. Duration of large-magnitude explosive eruptions deduced from graded bedding in deep-sea ash layers. Geology. 1979;7:240–4.

Lee J, Stern RJ, Bloomer SH. Forty million years of magmatic evolution in the Mariana arc: the tephra glass record. J Geophys Res. 1995;100(B9):17671–87.

Leinen M. The origin of paleochemical signatures in North Pacific pelagic clays: partitioning experiments. Geochim Cosmochim Acta. 1987;51(2):305–19.

Leinen M, Pisias N. An objective technique for determining end-member compositions and for partitioning sediments according to their sources. Geochim Cosmochim Acta. 1984;48(1):47–62.

Lowe DJ. Tephrochronology and its application: a review. Quat Geochronol. 2011; 6:107–53. doi:10.1016/j.quageo.2010.08.003.

Machida H, Arai F. Extensive ash falls in and around the Sea of Japan from large late Quaternary eruptions. J Volc Geotherm Res. 1983;18(1):151–64.

Mackenzie FT, Garrels RM. Chemical mass balance between rivers and oceans. Am J Sci. 1966;264(7):507–25.

Mahony SH, Wallace LM, Miyoshi M, Villamor P, Sparks RSJ, Hasenaka T. Volcano-tectonic interactions during rapid plate-boundary evolution in the Kyushu region SW Japan. Geol Soc Am Bull. 2011;123:2201–23.

Maicher D, White JDL. The formation of deep-sea Limu o Pele. Bull Volcanol. 2001;63:482–96.

Martinez NC, Murray RW, Thunell RC, Peterson LC, Muller-Karger F, Astor Y, et al. Modern climate forcing of terrigenous deposition in the tropics (Cariaco Basin Venezuela). Earth Planet Sci Lett. 2007;264(3):438–51.

Martinez NC, Murray RW, Dickens GR, Kölling M. Discrimination of sources of terrigenous sediment deposited in the central Arctic Ocean through the Cenozoic. Paleoceanography. 2009;24(1). doi:10.1029/2007PA001567.
Martinez NC, Murray RW, Thunell RC, Peterson LC, Muller-Karger F, Lorenzoni L, et al. Local and regional geochemical signatures of surface sediments from the Cariaco Basin and Orinoco Delta Venezuela. Geology. 2010;38(2):159–62.
Metzner D, Kutterolf S, Toohey M, Timmreck C, Niemeier U, Freundt A, et al. Radiative forcing and climate impact resulting from SO2 injections based on a 200,000 year record of Plinian eruptions along the Central American Volcanic Arc. Int J Earth Sci. 2014;103(7):2063–79. doi:10.1007/s00531-012-0814-z.
Murray RW, Leinen M. Chemical transport to the seafloor of the equatorial Pacific Ocean across a latitudinal transect at 135 W: tracking sedimentary major trace and rare earth element fluxes at the Equator and the Intertropical Convergence Zone. Geochim Cosmochim Acta. 1993;57(17):4141–63.
Murray RW, Leinen M. Scavenged excess aluminum and its relationship to bulk titanium in biogenic sediment from the central equatorial Pacific Ocean. Geochim Cosmochim Acta. 1996;60(20):3869–78.
Murray RW, Leinen M, Isern A. Biogenic flux of Al to sediment in the central equatorial Pacific Ocean: evidence for increased productivity during glacial periods. Paleoceanography. 1993;8(5):651–70.
Murray RW, Leinen M, Murray DW, Mix AC, Knowlton CW. Terrigenous Fe input and biogenic sedimentation in the glacial and interglacial equatorial Pacific Ocean. Global Biogeochem Cycles. 1995;9(4):667–84.
Murray RW, Knowlton C, Leinen M, Mix AC, Polsky CH. Export production and carbonate dissolution in the central equatorial Pacific Ocean over the past 1 Myr. Paleoceanography. 2000;15(6):570–92.
Ninkovich D, Sparks RSJ, Ledbetter MT. The exceptional magnitude and intensity of the Toba eruption Sumatra: an example of the use of deep-sea tephra layers as a geological tool. Bull Volcanol. 1978;41:286–98.
Olgun N, Duggen S, Croot PL, Delmelle P, Dietze H, Schacht U, et al. Surface ocean iron fertilization: the role of subduction zone and hotspot volcanic ash and fluxes into the Pacific Ocean Global Biogeochemical Cycles. 2011;25. doi:10.1029/2009GB003761
Peacock SA. Fluid processes in subduction zones. Science. 1990;248(4953):329–37.
Peters JL, Murray RW, Sparks J, Coleman DS. Terrigenous matter and dispersed ash in sediment from the CaribbeanSea: results from Leg 165. Proc Ocean Drill Program Sci Results. 2000;165:115–24. doi:10.2973/odpprocsr1650032000.
Pickering KT, Underwood MB, Saito S, Naruse H, Kutterolf S, Scudder RP, et al. Depositional architecture provenance and tectonic/eustatic modulation of Miocene submarine fans in the Shikoku Basin: results from Nankai Seismogenic Zone experiment. Geochem Geophys Geosys. 2013;14:6. doi:10.1002/ggge.20107.
Piepgras DJ, Jacobsen SB. The behavior of rare earth elements in seawater: precise determination of variations in the North Pacific water column. Geochim Cosmochim Acta. 1992;56(5):1851–62.
Pisias NG, Murray RW, Scudder RP. Multivariate statistical analysis and partitioning of sedimentary geochemical data sets: General principles and specific MATLAB scripts. Geochem Geophys Geosys. 2013;14(10):4015–20. doi:10.1002/ggge20247.
Plank T. Constraints from Thorium/Lanthanum on sediment recycling at subduction zones and the evolution of continents. J Petrol. 2005;46:921–44.
Plank T, Langmuir CH. The chemical composition of subducting sediment and its consequences for the crust and mantle. Chem Geol. 1998;145(3–4):325–94.
Plank T, Ludden JN, Escutia C, et al. Proc Ocean Drill Program Init Results. 2000; 185:1–190. doi:10.2973/odpprocir1852000.
Plank T, Kelley KA, Murray RW, Stern LQ. Chemical composition of sediments subducting at the Izu-Bonin trench. Geochem Geophys Geosyst. 2007;8: Q04I16. doi:10.1029/2006GC001444.
Presti M, Michalopoulos P. Estimating the contribution of the authigenic mineral component to the long-term reactive silica accumulation on the western shelf of the Mississippi River Delta. Cont Shelf Res. 2008;28(6):823–38.
Rea DK. The palcoclimatic record provided by eolian deposition in the deep sea: the geologic history of wind. Rev Geophys. 1994;32:159–95.
Rea DK, Leinen M. Asian aridity and the zonal westerlies: Late Pleistocene and Holocene record of eolian deposition in the northwest Pacific Ocean. Palaeogeog Palaeoclimatol Palaeoecol. 1988;66(1-2):1–8.
Rea DK, Snoeckx H, Joseph LH. Late Cenozoic eolian deposition in the North Pacific: Asian drying Tibetan uplift and cooling of the northern hemisphere. Paleoceanography. 1998;13(3):215–24.
Reid P, Carey SN, Ross DR. Late quaternary sedimentation in the Lesser Antilles island arc. Geol Soc Am Bull. 1996;108:78–100.
Reimann C, Filzmoser P, Garrett RG. Factor analysis applied to regional geochemical data: problems and possibilities. App Geochem. 2002;7(3):185–206.
Risso C, Scasso RA, Aparicio A. Presence of large pumice blocks on Tierra del Fuego and South Shetland Islands shorelines from 1962 South Sandwich Islands eruption. Mar Geol. 2002;186:413–22.
Robock A, Ammann CM, Oman L, Shindell D, Levis S, Stenchikov G. Did the Toba volcanic eruption of ~ 74ka BP produce widespread glaciation? J Geophys Res: Atmospheres (1984–2012). 2009;114(D10). doi:10.1029/2008JD011652.
Rose WI, Riley CM, Dartevelle S. Sizes and shapes of 10-Ma distal fall pyroclasts in the Ogallala Group Nebraska. J Geol. 2003;111(1):115–24.
Ryan WBF, Carbotte SM, Coplan JO, O'Hara S, Melkonian A, Arko R, et al. Global Multi-Resolution Topography synthesis. Geochem Geophys Geosys. 2009; 10(3). doi:10.1029/2008GC002332.
Saffer DM, Underwood MB, McKiernan AW. Evaluation of factors controlling smectite transformation and fluid production in subduction zones: application to the Nankai Trough. Island Arc. 2008;17(2):208–30.
Saffer DM, Lockner DA, McKiernan A. Effects of smectite to illite transformation on the frictional strength and sliding stability of intact marine mudstones. Geophys Res Lett. 2012;39:L11304. doi:10.1029/2012GL051761.
Saito S, Underwood MB, Kubo Y. Expedition 322 Scientists. Proceedings IODP 322, Tokyo Integrated Ocean Drilling Program Management, International Inc Washington DC. 2010. doi:10.2204/iodpproc3221042010.
Schacht U, Wallmann K, Kutterolf S, Schmidt M. Volcanogenic sediment–seawater interactions and the geochemistry of pore waters. Chem Geol. 2008; 249(3–4):321.
Schindlbeck JC, Kutterolf S, Freundt A, Scudder RP, Pickering KT, Murray RW. Emplacement processes of submarine volcaniclastic deposits (IODP Site C0011 Nankai Trough). Mar Geol. 2013;343:115–24. doi:10.1016/j.margeo.2013.06.017.
Schindlbeck JC, Kutterolf S, Freundt A, Straub SM, Wang K-L, Jegen M, et al. The Miocene Galápagos ash layer record of IODP Legs 334&344: Ocean-island explosive volcanism during plume-ridge interaction. Geology. 2015. doi:10.1130/G366451.
Scudder RP, Murray RW, Plank T. Dispersed ash in deeply buried sediment from the northwest Pacific Ocean: an example from the Izu-Bonin arc (ODP Site 1149). Earth Planet Sci Lett. 2009;284(3–4):639–48.
Scudder RP, Murray RW, Schindlbeck JC, Kutterolf S, Hauff F, McKinley CC. Regional-scale input of dispersed and discrete volcanic ash to the Izu-Bonin and Mariana subduction zones. Geochem Geophys Geosys. 2014;15(11):4369–79.
Severmann S, Johnson CM, Beard BL, German CR, Edmonds HN, Chiba H, et al. The effect of plume processes on the Fe isotope composition of hydrothermally derived Fe in the deep ocean as inferred from the Rainbow vent site Mid-Atlantic Ridge 36 14' N. Earth Planet Sci Lett. 2004;225(1):63–76.
Shaw DM, Watkins ND, Huang TC. Atmospherically transported volcanic glass in deep-sea sediments: theoretical considerations. J Geophysi Res. 1974;79(21):3087–94.
Sibuet JC, Letouzey J, Barbier F, Charvet J, Foucher JP, Hilde TW, et al. Back arc extension in the Okinawa Trough. J Geophys Res: Solid Earth (1978–2012). 1987;92(B13):14041–63.
Sigurdsson H. Volcanic episodes and rates of volcanism. In: Sigurdsson H, Houghten HB, McNutt SR, Ryme H, Stix J, editors. Encyclopedia of volcanoes. New York: Academic Press; 1999. p. 271–9.
Sigurdsson H, Sparks RSJ, Carey SN, Huang TC. Volcanogenic sedimentation in the Lesser Antilles Arc. J Geol. 1980;88:523.
Sigurdsson H, Leckie RM, Acton GD, et al. Proceedings of the Ocean Drilling Program: Initial reports Volume 165. College Station Texas: Ocean Drilling Program; 1997. p. 865.
Sigurdsson H, Kelley RM, Carey S, Bralower T, King J. History of circum-Caribbean explosive volcanism: $^{40}Ar/^{39}Ar$ dating of tephra layers. Proc ODP Sci Results. 2000;165:299–314.
Simkin T, Fiske RS. Krakatau 1883: the volcanic eruption and its effects. Washington DC: Smithsonian Institution Press; 1983. p. 464.
Stern RJ, Kohut EJ, Bloomer SH, Leybourne M, Fouch M, Vervoort J. Subduction Factory processes beneath the Guguan Cross-chain Mariana Arc: no role for sediments are serpentinites important? Contrib Mineral Petrol. 2006;151:202–21.
Straub SM. Multiple sources of Quaternary tephra layers in the Mariana Trough. J Volcanol Geotherm Res. 1997;76(3–4):251–76.
Straub SM. The evolution of the Izu Bonin–Mariana volcanic arcs (NW Pacific) in terms of major element chemistry. Geochem Geophys Geosyst. 2003;4. doi:10.1029/2002GC000357.

Straub SM. Timescales and causes of secular change in the Izu Bonin–Mariana volcanic arcs. Geochim Cosmochim Acta. 2008;72 (12). A905 Goldschmidt Conference Vancouver Canada.

Straub SM, Layne GD. The systematics of boron isotopes in Izu arc front volcanic rocks. Earth Planet Sci Lett. 2002;198(1–2):25–39.

Straub SM, Layne GD. The systematics of chlorine fluorine and water in Izu arc front volcanic rocks: Implications for volatile recycling in subduction zones. Geochim Cosmochim Acta. 2003b;67(21):4179–203.

Straub SM, Layne GD. Decoupling of fluids and fluid-mobile elements during shallow subduction: evidence from halogen-rich andesite melt inclusions from the Izu arc volcanic front. Geochem Geophys Geosyst. 2003a;4. doi:10.1029/2002GC000349.

Straub SM, Schmincke HU. Evaluating the tephra input into Pacific Ocean sediments: distribution in space and time. Geol Rundsch. 1998;87(3):461–76.

Straub SM, Layne GD, Schmidt A, Langmuir CH. Volcanic glasses at the Izu arc volcanic front: new perspectives on fluid and sediment melt recycling in subduction zones. Geochem Geophys Geosyst. 2004;5:Q01007. doi:10.1029/2002GC000408.

Straub SM, Goldstein SL, Class C, Schmidt A. Mid-ocean-ridge basalt of Indian type in the northwest Pacific Ocean basin. Nat Geosci. 2009;2(4):286–9.

Straub SM, Goldstein SL, Class C, Schmidt A, Gomez-Tuena A. Slab and mantle controls on the Sr-Nd-Pb-Hf isotope evolution of the post 42Ma Izu-Bonin-volcanic Arc. J Petrol. 2010;51(5):993–1026.

Straub, SM, Woodhead, JD, Arculus, RJ. Temporal Evolution of the Mariana Arc: Mantle Wedge and Subducted Slab Controls Revealed with a Tephra Perspective. J Petrol. 2015. doi:10.1093/petrology/egv005.

Taira A. Tectonic evolution of the Japanese island arc system. Annu Rev Earth Pl Sc. 2001;29(1):109–34.

Tamura Y, Gill J, Tollstrup D, Kawabata H, Shukuno H, Chang Q, et al. Silicic magmas in the Izu-Bonin oceanic arc and implications for crustal evolution. J Petrol. 2009;50(4):685–723.

Tani K, Fiske RS, Tamura Y, Kido Y, Naka J, Shukuno H, et al. Sumisu volcano Izu-Bonin arc Japan: site of a silicic caldera-forming eruption from a small open-ocean island. Bull Volcanol. 2008;70:547–62.

Taylor SR, McLennan SM. The continental crust: its composition and evolution. Malden: Blackwell; 1985.

Tobin HJ, Kinoshita M. NanTroSEIZE: the IODP Nankai Trough seismogenic zone experiment. Sci Drill. 2006;2(2):23–7.

Underwood MB, Pickering KT. Clay-mineral provenance sediment dispersal patterns and mudrock diagenesis in the Nankai accretionary prism southwest Japan. Clays Clay Miner. 1996;44(3):339–56.

Völker D, Kutterolf S, Wehrmann H. Comparative mass balance of volcanic edifices at the Southern Volcanic Zone of the Andes between 33°S and 46°S. J Volcanol Geotherm Res. 2011;205:114–29. doi:10.1016/jjvolgeores201103011.

Völker D, Wehrmann H, Kutterolf S, Iyer K, Geersen J, Rabbel W, et al. Constraining input and output fluxes of the southern Central Chile Subduction Zone: water chlorine sulfur. Int J Earth Sci. 2014;103(7):2129–53. doi:10.1007/s00531-014-1002-0.

Volynets ON, Ponomareva VV, Braitseva OA, Melekestsev IV, Chen CH. Holocene eruptive history of Ksudach volcanic massif South Kamchatka: evolution of a large magmatic chamber. J Volc Geotherm Res. 1999;91(1):23–42.

Watson AJ. Volcanic Fe CO2 ocean productivity and climate. Nature. 1997;385: 587–8. doi:10.1038/385587b0.

White SM, Chamley H, Curtis DM, deVries Klein G, Mizuno A. Sediment synthesis: Deep Sea Drilling Project Leg 58 Philippine Sea. Init Rep DSDP. 1980;58:963–1014. doi:10.2973/dsdpproc581451980.

White JDL, Smellie JL, Clague DA. Introduction: A Deductive Outline and Topical Overview of Subaqueous Explosive Volcanism, in Explosive Subaqueous Volcanism (eds J.D.L. White, J.L. Smellie and D.A. Clague), American Geophysical Union, Washington, D.C; 2003. doi:10.1029/140GM01.

Woods AW, Wohletz KH. Dimensions and dynamics of co-ignimbrite eruption clouds. Nature. 1991;350:225–7.

Zhou L, Kyte FT. Sedimentation history of the South Pacific pelagic clay province over the last 85 million years inferred from the geochemistry of Deep Sea Drilling Project Hole 596. Paleoceanography. 1992;7(4):441–65.

Ziegler CL, Murray RW. Geochemical evolution of the central Pacific Ocean over the past 56 Myr. Paleoceanography. 2007;22(2). doi:10.1029/2006PA001321.

Ziegler CL, Murray RW, Hovan SA, Rea DK. Resolving eolian volcanogenic and authigenic components in pelagic sediment from the Pacific Ocean. Earth Planet Sci Lett. 2007;254(3):416–32.

Ziegler CL, Murray RW, Plank T, Hemming SR. Sources of Fe to the Equatorial Pacific Ocean from the Holocene to Miocene. Earth Planet Sci Lett. 2008;270:258–70.

13

The competition between Lorentz and Coriolis forces in planetary dynamos

Krista M. Soderlund[1*], Andrey Sheyko[2], Eric M. King[3] and Jonathan M. Aurnou[4]

Abstract

Fluid motions within planetary cores generate magnetic fields through dynamo action. These core processes are driven by thermo-compositional convection subject to the competing influences of rotation, which tends to organize the flow into axial columns, and the Lorentz force, which tends to inhibit the relative movement of the magnetic field and the fluid. It is often argued that these forces are predominant and approximately equal in planetary cores; we test this hypothesis using a suite of numerical geodynamo models to calculate the Lorentz to Coriolis force ratio directly. Our results show that this ratio can be estimated by $\Lambda_d^* \simeq \Lambda_i Rm^{-1/2}$ (Λ_i is the traditionally defined Elsasser number for imposed magnetic fields and Rm is the system-scale ratio of magnetic induction to magnetic diffusion). Best estimates of core flow speeds and magnetic field strengths predict the geodynamo to be in magnetostrophic balance where the Lorentz and Coriolis forces are comparable. The Lorentz force may also be significant, i.e., within an order of magnitude of the Coriolis force, in the Jovian interior. In contrast, the Lorentz force is likely to be relatively weak in the cores of Saturn, Uranus, Neptune, Ganymede, and Mercury.

Keywords: Planetary dynamos; Core convection; Magnetic fields; Rotation; Magnetostrophic force balance

Background

Magnetic fields are common throughout the solar system and provide a unique perspective on the internal dynamics of planetary interiors. Planetary magnetic fields are driven by the conversion of kinetic energy into magnetic energy; this process is called dynamo action. Kinetic energy is derived from thermo-compositional convection of an electrically conducting fluid although mechanical mechanisms, such as libration and precession, may also drive core flow in smaller bodies (e.g., Le Bars et al. 2015). The geodynamo is the most studied planetary magnetic field, yet the mechanisms that control its strength, morphology, and secular variation are still not well understood. Towards explaining these observations, dimensionless parameters are often used to characterize the force balances present in the core and their link to processes that govern convective dynamics and dynamo action. Two forces that are particularly influential on core processes are the Coriolis force, which tends to organize the core fluid motions, and the Lorentz force which back-reacts on the fluid motions to equilibrate magnetic field growth.

In order to better understand the manifestation of Coriolis and Lorentz forces on core flow, let us consider the Boussinesq momentum equation. In this commonly used approximation, the material properties are assumed to be constant and the density is treated as constant except in the buoyancy force of the momentum equation (e.g., Tritton 1998):

$$\frac{D\mathbf{u}}{Dt} + 2\mathbf{\Omega} \times \mathbf{u} = -\nabla\Pi - \alpha T'\mathbf{g} + \nu\nabla^2\mathbf{u} + \frac{1}{\rho_o}\mathbf{J} \times \mathbf{B}. \quad (1)$$

Here, $\mathbf{u}$ is the velocity vector, $\mathbf{\Omega}$ is the rotation rate of the reference frame, Π is the non-hydrostatic pressure normalized by the background density ρ_o, $\mathbf{J}$ is the electric current density, $\mathbf{B}$ is the magnetic induction, and $\mathbf{g}$ is the acceleration of gravity. Terms from left to right are inertial acceleration, Coriolis acceleration, pressure gradient, buoyancy, viscous diffusion, and Lorentz force. The inertial term consists of the temporal evolution of the velocity field as well as non-linear advection. The pressure gradient term absorbs the hydrostatic mean gravitational component as well as the centrifugal force. The buoyancy force arises from density differences with respect to the density

*Correspondence: krista@ig.utexas.edu
[1] Institute for Geophysics, John A. & Katherine G. Jackson School of Geosciences, The University of Texas at Austin, J. J. Pickle Research Campus, Building 196 (ROC), 10100 Burnet Road (R2200), Austin, Texas, 78758-4445, USA
Full list of author information is available at the end of the article

of the motionless background state (assumed to be constant density ρ_o). These perturbations, $\rho'/\rho_o = -\alpha T'$, are produced by temperature variations, T', from the background state temperature profile; α is thermal expansivity. Viscous diffusion depends on the kinematic viscosity of the fluid, ν. The Lorentz force is due to interactions between the magnetic field and current density.

Planetary cores are rapidly rotating and the Coriolis force is expected to be large compared to the viscous, inertial, and buoyancy forces at global scales (the former two forces are scale dependent quantities). If we neglect these forces and magnetic fields in the Boussinesq momentum equation (Eq. 1), the system is in geostrophic balance:

$$2\boldsymbol{\Omega} \times \mathbf{u} = -\nabla\Pi. \tag{2}$$

This balance requires horizontal fluid motions to follow lines of constant pressure. Further insight can be obtained by taking the curl of Eq. 2 and using the continuity equation ($\nabla \cdot \mathbf{u} = 0$) to obtain $\boldsymbol{\Omega} \cdot \nabla\mathbf{u} = 0$. If we assume $\boldsymbol{\Omega} = \Omega\hat{\mathbf{z}}$ where Ω is constant, this equation simplifies to

$$\frac{\partial \mathbf{u}}{\partial z} = 0. \tag{3}$$

This is the Taylor-Proudman theorem and states that the fluid motion will be invariant along the direction of the rotation axis. However, in order for convection to occur, the Taylor-Proudman constraint cannot strictly hold as there must be slight deviations from two-dimensionality, at least in the boundary layers, to allow overturning motions in the fluid layer (e.g., Zhang 1992; Olson et al. 1999; Grooms et al. 2010). Further beyond the onset of convection, convective flows break down into an anisotropic rotating turbulence (e.g., Sprague et al. 2006; Julien et al. 2012; Stellmach et al. 2014; Cheng et al. 2015; Ribeiro et al. 2015).

Linear asymptotic analyses of quasigeostrophic convection predict that the azimuthal wavenumber of these columns relative to the thickness of the fluid shell varies as $\ell_U/D \propto E^{1/3}$, where the Ekman number, $E = \nu/(2\Omega D^2)$, represents the ratio of viscous to Coriolis forces (e.g., Roberts 1968; Julien et al. 1998; Jones et al. 2000; Dormy et al. 2004). Thus, in rapidly rotating systems such as planetary cores (where $E \lesssim 10^{-12}$), it is often predicted that quasigeostrophic convection occurs as tall, thin columns (e.g., Kageyama et al. 2008; cf. Cheng et al. 2015).

In planetary dynamos, however, magnetic fields are also thought to play an important dynamical role on the convection and zonal flows. When strong imposed magnetic fields and rapid rotation are both present, the dominant force balance is magnetostrophic—a balance between the Coriolis, pressure gradient, and Lorentz terms:

$$2\boldsymbol{\Omega} \times \mathbf{u} = -\nabla\Pi + \frac{1}{\rho_o}\mathbf{J} \times \mathbf{B}. \tag{4}$$

In contrast to Eq. 3, the curl of the magnetostrophic balance Eq. 4 yields

$$\frac{\partial \mathbf{u}}{\partial z} = -\frac{1}{2\,\rho_o\,\Omega}\nabla \times (\mathbf{J} \times \mathbf{B}). \tag{5}$$

This equation shows that magnetic fields can relax the Taylor-Proudman constraint, allowing global-scale motions that differ fundamentally from the small-scale axial columns typical of non-magnetic, rapidly rotating convection (e.g., Cardin and Olson 1995; Olson and Glatzmaier 1996; Zhang and Schubert 2000; Roberts and King 2013).

Magnetic fields are typically considered to be important dynamically when the ratio of Lorentz to Coriolis forces is of order unity. This ratio is represented by the Elsasser number, Λ, which can be approximated by assuming a characteristic current density, magnetic induction, density, rotation rate, and velocity scale:

$$\Lambda = \frac{\text{Lorentz}}{\text{Coriolis}} = \frac{\left|\frac{1}{\rho_o}\mathbf{J} \times \mathbf{B}\right|}{|2\boldsymbol{\Omega} \times \mathbf{u}|} \approx \frac{JB}{2\rho_o\Omega U}. \tag{6}$$

In systems where the magnetic field is not strongly time variant (i.e., $\partial\mathbf{B}/\partial t = -\nabla \times \mathbf{E} \sim 0$), the current density can be approximated via Ohm's Law, $J \sim \sigma UB$ where $\sigma = (\mu_o\eta)^{-1}$ is electrical conductivity and η is magnetic diffusivity, such that

$$\Lambda_i = \frac{B^2}{2\rho_o\mu_o\eta\Omega}. \tag{7}$$

This formulation is thus appropriate for steady *imposed* magnetic fields (e.g., Olson and Glatzmaier 1996; King and Aurnou 2015). In dynamos, which tend to exhibit significant temporal variability, it is more appropriate to approximate the current density using Ampere's law under the MHD approximation, $J \sim B/\mu_o\ell_B$ where ℓ_B is the characteristic length scale of magnetic field variations (e.g., Cardin et al. 2002; Soderlund et al. 2012). In this *dynamic* case, Eq. 6 becomes

$$\Lambda_d = \frac{B^2}{2\rho_o\mu_o\Omega U\ell_B} = \frac{\Lambda_i}{Rm}\frac{D}{\ell_B}, \tag{8}$$

indicating that the Lorentz to Coriolis force ratio is length scale dependent. In the latter equality, $Rm = UD/\eta$ is the global ratio of magnetic induction to magnetic diffusion. This parameter must exceed approximately 10 in order for dynamo action to occur (e.g., Roberts 2007). Thus, on global scales, where $\ell_B \simeq D$, the dynamic Elsasser number will always be less than Λ_i by at least an order of magnitude for active dynamos. However, the influence of magnetic fields will become increasingly important at smaller scales since $\Lambda_d \propto \ell_B^{-1}$ in (Eq. 8).

The quantity ℓ_B/D cannot be measured for planetary cores, however. It is, therefore, important to provide a scaling estimate for Λ_d in terms of quantities that can be estimated from observations, such as Λ_i and Rm. The length scale ℓ_B can be approximated using the magnetic induction equation:

$$\frac{\partial \mathbf{B}}{\partial t} = \nabla \times (\mathbf{u} \times \mathbf{B}) + \eta \nabla^2 \mathbf{B} \tag{9}$$

Assuming a balance between induction and diffusion yields $UB/D \sim \eta B/\ell_B^2$. Rearranging, the dimensionless length scale is proportional to the inverse square root of the magnetic Reynolds number:

$$\ell_B/D \sim Rm^{-1/2} \tag{10}$$

(e.g., Galloway 1978; King and Roberts 2013). Using this relationship in Eq. 8, we estimate the dynamic Lorentz to Coriolis force ratio to be

$$\Lambda_d^* \sim \Lambda_i Rm^{-1/2}. \tag{11}$$

This scaling estimate will be referred to as the *modified dynamic* Elsasser number.

In this paper, we will investigate the relative magnitudes of Lorentz and Coriolis forces to better understand the dynamics of planetary cores. Towards this end, we calculate directly the Lorentz to Coriolis force ratio in a suite of geodynamo models described in the "Methods" section, test Eq. 11 to predict this ratio in the "Results" section, and extrapolate the results to Earth and other planetary cores in the "Discussion" section.

Methods

We analyze 34 dynamo models presented in Soderlund et al. (2012, 2014) and five dynamo models presented in Sheyko (2014); these datasets will be referred to as SKA and S14, respectively. These simulations of Boussinesq convection and dynamo action are carried out in thick, rotating spherical shells with thickness D, rotation rate Ω, and no-slip boundaries. Gravity is assumed to vary linearly with radius; g_o is acceleration at the top of the fluid shell. The outer boundary at radius r_o is assumed electrically insulating and the solid inner core at radius r_i has the same electrical conductivity as the convecting fluid region.

The SKA dataset considers three rotation rates (Ekman numbers in the range $10^{-3} \geq E \geq 10^{-5}$), two magnetic diffusivities (magnetic Prandtl numbers of $Pm = 2, 5$), and thermal forcing near onset (Ra_c) to more than 1000 times critical ($1.9 < Ra/Ra_c < 1125$); the shell geometry is fixed to $\chi = r_i/r_o = 0.4$ with isothermal boundaries. This survey is complemented by the S14 dataset focusing on low Ekman, low magnetic Prandtl number simulations: $3 \times 10^{-7} \lesssim E \lesssim 10^{-6}$, $0.05 \leq Pm \leq 1$, and $10^8 \lesssim Ra \lesssim 10^{10}$ with $\chi = 0.35$, an isothermal inner shell boundary, and fixed heat flux on the outer shell boundary. For both datasets, the thermal (κ) and momentum (ν) diffusivities are assumed equal, such that the Prandtl number, $Pr = \nu/\kappa$, is unity. Here, $E = \nu/\left(2\Omega D^2\right)$ is the ratio of viscous to Coriolis forces; $Pm = \nu/\eta$ is the ratio of viscous to magnetic diffusivities, and $Ra = \alpha g_o \Delta T D^3/(\nu\kappa)$ is the ratio of buoyancy due to superadiabatic temperature contrast ΔT to diffusion. The combination of these datasets, given in Table 1, results in one of the broadest surveys of supercriticality, Ekman, and magnetic Prandtl numbers in the literature.

The SKA simulations are carried out on numerical grids that range from 37 fluid shell radial levels and 42 spherical harmonic degrees (near onset at $E = 10^{-3}$) to 73 radial levels and 213 spherical harmonic degrees (highest Ra for $E = 10^{-5}$). Hyperdiffusion is used for the three most extreme cases ($Ra \geq 2 \times 10^8$ for $E \leq 10^{-4}$); additional details are given in Soderlund et al. (2012). The S14 simulations are carried out on numerical grids with 512–528 radial levels and 256 spherical harmonic degrees. All simulations are carried out using pseudo-spectral methods with no azimuthal symmetries assumed (Wicht 2002; Christensen and Wicht, 2007; Sheyko 2014). Once the initial transient has subsided, global diagnostic quantities (Λ_i, Rm, ℓ_B, Λ_d, and Λ_d^*) are time-averaged to minimize statistical errors.

In our models, the imposed Elsasser number is calculated from the dimensionless magnetic energy density, $\mathcal{E}_M$, within the fluid shell of volume V:

$$\Lambda_i = 2PmE\mathcal{E}_M \text{ where } \mathcal{E}_M = \frac{1}{2\,Pm\,E\,V}\int_V \mathbf{B}\cdot\mathbf{B}dV, \tag{12}$$

while the magnetic Reynolds number is calculated from the dimensionless kinetic energy density, $\mathcal{E}_K$:

$$Rm = \sqrt{2\mathcal{E}_K}Pm \text{ where } \mathcal{E}_K = \frac{1}{2V}\int_V \mathbf{u}\cdot\mathbf{u}dV. \tag{13}$$

These quantities are non-dimensionalized using distance, magnetic induction, velocity, and energy density scales of $D = r_o - r_i$, $\sqrt{2\rho\mu_o\eta\Omega}$, ν/D, and $\rho\nu^2/D^2$, respectively. We also characterize ℓ_B as a quarter wavelength of the typical magnetic field wavenumber (Soderlund et al. 2012):

$$\ell_B/D \sim \frac{(\pi/2)}{\sqrt{\overline{l_B}^2 + \overline{m_B}^2}}, \text{ where } \overline{l_B} = \frac{\sum l(\mathbf{B}_l\cdot\mathbf{B}_l)}{2\,Pm\,E\,\mathcal{E}_M} \text{ and } \overline{m_B} = \frac{\sum m(\mathbf{B}_m\cdot\mathbf{B}_m)}{2\,Pm\,E\,\mathcal{E}_M}. \tag{14}$$

Here, $\mathbf{B}_l$ and $\mathbf{B}_m$ are magnetic induction at spherical harmonic degree l and order m.

Two approaches are used to calculate the Lorentz to Coriolis force ratio in situ. The first integrates the Lorentz

Table 1 Input (E, Pm, Ra, $Ro_c = \sqrt{RaE^2/Pr}$) and output (Λ_i, Λ_d, Λ_d^*, Λ^I, Λ^P) parameters for our dataset. $Pr = 1$ is fixed for all simulations. Additional simulation information is given in Soderlund et al. (2012, 2014) for cases with $E \geq 10^{-5}$ and Sheyko (2014) for cases with $E \lesssim 10^{-6}$

E	Pm	Ra	Ro_c	Λ_i	Λ_d	Λ_d^*	Λ^I	Λ^P
10^{-3}	5	6.50×10^4	0.26	4.34	0.44	0.47	0.39	0.26
10^{-3}	5	9.70×10^4	0.31	3.91	0.30	0.35	0.43	0.32
10^{-3}	5	1.12×10^5	0.33	3.25	0.20	0.25	0.40	0.29
10^{-3}	5	1.32×10^5	0.36	0.61	0.03	0.04	0.05	0.01
10^{-3}	5	5.00×10^5	0.71	4.14	0.15	0.17	0.29	0.16
10^{-3}	5	1.96×10^6	1.40	23.0	0.51	0.64	0.79	0.46
10^{-3}	5	1.96×10^7	4.43	248	2.62	4.00	4.20	3.12
10^{-4}	2	1.42×10^6	0.12	1.30	0.14	0.12	0.18	0.12
10^{-4}	2	2.12×10^6	0.15	2.10	0.17	0.16	0.21	0.14
10^{-4}	2	2.83×10^6	0.17	2.42	0.17	0.16	0.24	0.16
10^{-4}	2	3.54×10^6	0.19	2.37	0.15	0.14	0.27	0.19
10^{-4}	2	3.68×10^6	0.19	2.36	0.15	0.13	0.21	0.17
10^{-4}	2	3.75×10^6	0.19	2.27	0.14	0.12	0.21	0.15
10^{-4}	2	3.82×10^6	0.20	0.14	0.01	0.007	0.01	0.006
10^{-4}	2	3.96×10^6	0.20	0.18	0.01	0.009	0.02	0.009
10^{-4}	2	4.11×10^6	0.20	0.22	0.01	0.01	0.02	0.009
10^{-4}	2	4.24×10^6	0.21	0.24	0.02	0.01	0.02	0.006
10^{-4}	2	6.00×10^6	0.24	0.68	0.04	0.03	0.06	0.03
10^{-4}	2	8.50×10^6	0.29	1.53	0.07	0.06	0.11	0.06
10^{-4}	2	1.42×10^7	0.38	3.44	0.13	0.11	0.21	0.11
10^{-4}	2	2.10×10^7	0.46	5.48	0.18	0.16	0.28	0.17
10^{-4}	2	4.24×10^7	0.65	11.4	0.29	0.26	0.44	0.27
10^{-4}	2	8.00×10^7	0.89	20.7	0.40	0.39	0.59	0.37
10^{-4}	2	1.42×10^8	1.19	36.3	0.58	0.58	0.81	0.53
10^{-4}	2	4.24×10^8	2.06	106	0.84	1.28	1.18	0.77
10^{-4}	2	8.48×10^8	2.91	197	1.11	2.10	1.63	1.07
10^{-5}	2	3.10×10^7	0.06	4.84	0.30	0.30	0.39	0.33
10^{-5}	2	5.89×10^7	0.08	7.07	0.31	0.31	0.45	0.37
10^{-5}	2	8.20×10^7	0.09	8.28	0.32	0.30	0.45	0.34
10^{-5}	2	8.50×10^7	0.09	8.54	0.32	0.31	0.47	0.39
10^{-5}	2	9.50×10^7	0.10	9.02	0.32	0.30	0.45	0.35
10^{-5}	2	1.05×10^8	0.10	9.63	0.33	0.31	0.47	0.34
10^{-5}	2	1.50×10^8	0.12	12.4	0.36	0.35	0.52	0.37
10^{-5}	2	2.00×10^8	0.14	14.8	0.38	0.37	0.56	0.36
1.2×10^{-6}	1	1.86×10^8	0.02	5.73	0.21	0.28	0.39	0.30
1.2×10^{-6}	0.2	1.86×10^8	0.04	0.07	0.005	0.007	0.01	0.009
1.2×10^{-6}	0.2	9.28×10^8	0.09	0.61	0.03	0.03	0.08	0.04
1.2×10^{-6}	0.2	5.57×10^9	0.02	9.92	0.24	0.25	0.47	0.30
3.0×10^{-7}	0.05	2.23×10^{10}	0.04	3.61	0.13	0.17	0.32	0.24

($\mathbf{F_L}$) and Coriolis ($\mathbf{F_C}$) forces over the spherical shell volume following Soderlund et al. (2012):

$$F_L^I = \int_V \left(\left(\mathbf{F_L} \cdot \hat{\mathbf{r}}\right)^2 + \left(\mathbf{F_L} \cdot \hat{\theta}\right)^2 + \left(\mathbf{F_L} \cdot \hat{\phi}\right)^2 \right)^{1/2} dV \quad (15)$$

$$F_C^I = \int_V \left(\left(\mathbf{F_C} \cdot \hat{\mathbf{r}}\right)^2 + \left(\mathbf{F_C} \cdot \hat{\theta}\right)^2 + \left(\mathbf{F_C} \cdot \hat{\phi}\right)^2 \right)^{1/2} dV \quad (16)$$

where the volume-*integrated* force ratio is $\Lambda^I = F_L^I/F_C^I$. Alternatively, the root mean square (RMS) Lorentz and Coriolis forces are calculated at each grid point to determine the force ratio locally (cf. Dharmaraj and Stanley 2012; Dharmaraj et al. 2014). A representative value is determined by taking the most probable ratio in the histogram; this *probability* approach yields Λ^P (Fig. 1). Histogram bins are uniformly spaced to contain equal ranges of $\log_{10}\left(F_L^{RMS}/F_C^{RMS}\right)$ values. The probability of each bin is defined as the volume of the fluid shell that is occupied by the respective range of force ratio values relative to the total volume of the fluid shell. For both approaches, velocity and magnetic fields from three random snapshots in time are used to calculate the force ratios for the SKA dataset, with Λ^I and Λ^P being the average values. Single snapshots are used for the S14 dataset.

Results

A Lorentz to Coriolis force ratio histogram is evaluated for a representative dipole-dominated dynamo case at a snapshot in time in Fig. 1. The histogram indicates that the

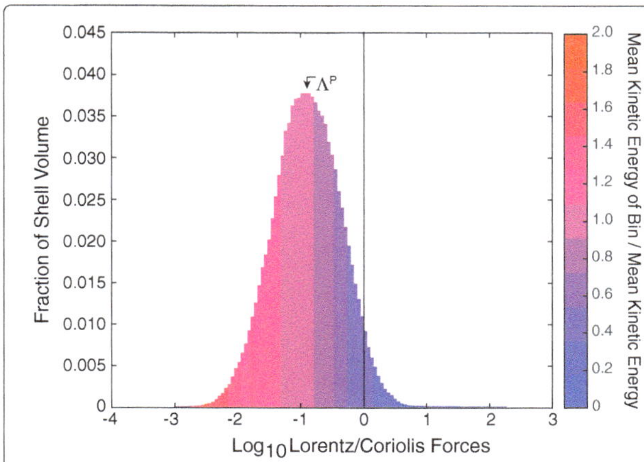

Fig. 1 Histogram of Lorentz to Coriolis force ratios on a point-by-point basis. Histogram showing the distribution of Lorentz to Coriolis forces at each point in our simulation with $E = 10^{-4}$, $Ra = 1.9Ra_c$, $Pr = 1$, and $Pm = 2$. *Color* denotes the mean kinetic energy per force ratio bin relative to the mean kinetic energy of the system. The probability Elsasser number corresponds to the most probable force ratio bin, $\Lambda^P = 0.12$. In contrast, the imposed Elsasser number is $\Lambda_i = 1.31$, the integrated Elsasser number is $\Lambda^I = 0.18$, the dynamic Elsasser number is $\Lambda_d = 0.14$, and the modified dynamic Elsasser number is $\Lambda_d^* = 0.12$

Coriolis force is dominant for the majority of grid points, most frequently by an order of magnitude. Moreover, the points with Lorentz to Coriolis force ratios greater than unity tend to have relatively low kinetic energies, denoted by coloration of the bins. Thus, the Lorentz force is expected to have a secondary influence on the convective (non-zonal) dynamics (Soderlund et al. 2012). This does not mean, however, that the magnetic field cannot have an important, local-scale dynamical impact (cf. Sreenivasan and Jones 2011).

Figure 2 shows the ratio of Lorentz to Coriolis forces in our simulations as characterized by the *integrated* Λ^I, *probability* Λ^P, *imposed* Λ_i, *dynamic* Λ_d, and *modified dynamic* Λ_d^* Elsasser numbers (see Table 2 for a summary of definitions). For comparison, ratios of these parameters are plotted in Fig. 3. The Λ_i Elsasser number, which is traditionally used to test for magnetostrophic balance, substantially overestimates the Lorentz to Coriolis force ratio in all of our dynamo models (i.e., typically by an order of magnitude; see Fig. 3a). In contrast, we find that all of the other Lorentz to Coriolis force ratio definitions (Λ^I, Λ^P, Λ_d, Λ_d^*) agree to within a factor of five; a factor of 1.4 is the mean deviation between definitions. Comparing the in situ calculations, Λ^I exceeds Λ^P in all cases investigated. This result implies that, independent of the Ekman and magnetic Prandtl numbers and in agreement with the literature (e.g., Aubert et al. 2008, Sreenivasan et al. 2014), the Lorentz force is concentrated in discrete regions that

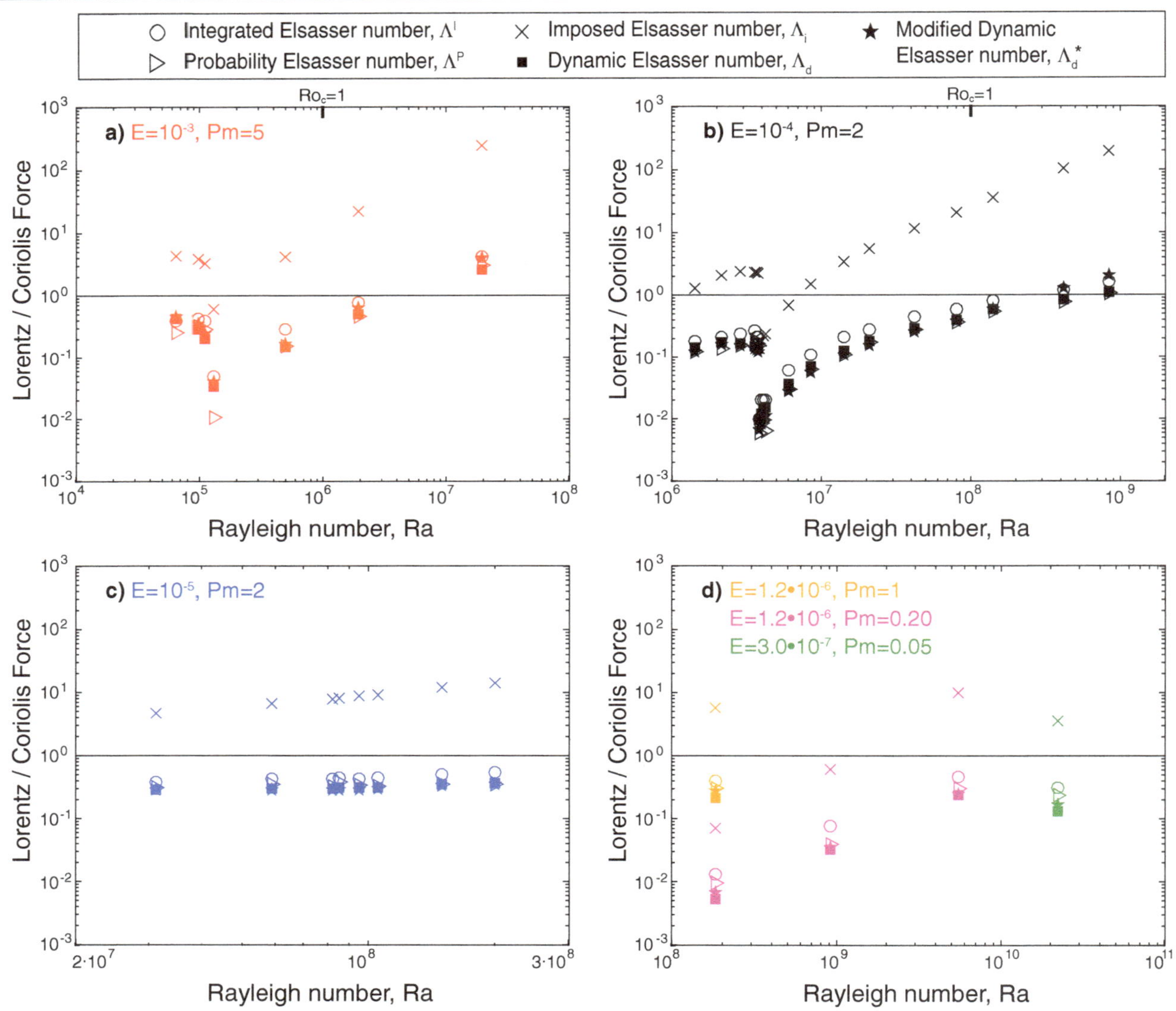

Fig. 2 a–d Direct calculations and parameterizations of the Lorentz to Coriolis force ratio in our simulations. Lorentz to Coriolis force ratios as a function of Rayleigh number for five Ekman numbers and five magnetic Prandtl numbers. The modified dynamic Elsasser number includes the length scale pre-factor: $\Lambda_d^* = 0.77\Lambda_i Rm^{-1/2}$. The Rayleigh numbers corresponding to $Ro_c = \sqrt{RaE^2/Pr} = 1$ are denoted on the top axes

Table 2 Elsasser number definitions: (1) Imposed Λ_i, (2) Dynamic Λ_d, (3) Modified dynamic Λ_d^*, (4) Integrated Λ^I, and (5) Probability Λ^P

Λ_i	Λ_d	Λ_d^*	Λ^I	Λ^P
$\frac{B^2}{2\rho_o\mu_o\eta\Omega}$	$\frac{\Lambda_i}{Rm}\frac{D}{\ell_B}$	$0.77\frac{\Lambda_i}{Rm^{1/2}}$	$\frac{\int_V \lvert\mathbf{F_L}\rvert dV}{\int_V \lvert\mathbf{F_C}\rvert dV}$	$\frac{\lvert\mathbf{F_L}\rvert}{\lvert\mathbf{F_C}\rvert}$, max probability

are more influenced by magnetic fields than the rest of the fluid volume. Similarly, the dynamic Elsasser number is also found to be in better agreement with Λ^P than Λ^I (Fig. 3b, c); the mean (maximum) percent differences between Λ_d and Λ^P are 28 % (68 %) for $E = 10^{-3}$, 18 % (58 %) for $E = 10^{-4}$, 9 % (21 %) for $E = 10^{-5}$, and 49 % (80 %) for $E \lesssim 10^{-5}$.

The Lorentz to Coriolis force ratios in Fig. 2 have distinct minima for cases with $E \geq 10^{-4}$. This sharp drop as the Rayleigh number is increased from onset occurs when the dipole-dominance of the dynamo breaks down. The Lorentz force increases more rapidly than the Coriolis force as Ra is further increased (see Fig. 4a of Soderlund et al. 2012), which explains the subsequent increase in Elsasser numbers.

Before discussing the modified Elsasser number, the assumption that the dimensionless length scale representative of magnetic field gradients in our simulations is approximately equal to the theoretical prediction (Eq. 10) must be assessed. This comparison is made in Fig. 4. Agreement within a factor of two—and typically much less, 16 % difference on average—between the ℓ_B/D values calculated directly from our simulations and the scaling estimate occurs when a pre-factor of 1.3 is included in Eq. 10:

$$\ell_B/D = 1.3Rm^{-1/2}. \quad (17)$$

This pre-factor is determined by minimizing the least squares residual between the calculated and predicted values across our dataset. Roberts and King (2013) also investigated the length scale of magnetic field variations in geodynamo models and found a similar dependence on $Rm^{-1/2}$ with a pre-factor of 3 due to their slightly modified definition of ℓ_B/D. More importantly, they also show that the magnetic length scale is weakly dependent on the magnetic field strength Λ_i, which may explain some of the spread in our model-prediction deviations.

The modified dynamic Elsasser number is given by solid, starred markers in Fig. 2; here, Λ_d^* has been modified compared to Eq. 11 to include the pre-factor derived from the length scale calculations (Eq. 17):

$$\Lambda_d^* = 0.77\Lambda_i Rm^{-1/2}. \quad (18)$$

For all Ekman numbers, the dynamic Λ_d and modified dynamic Λ_d^* Elsasser numbers differ by less than a factor of two (Fig. 5): the mean (maximum) percent difference is 23 % (53 %) for $E = 10^{-3}$, 21 % (89 %) for $E = 10^{-4}$, 4 % (6 %) for $E = 10^{-5}$, and 19 % (30 %) for

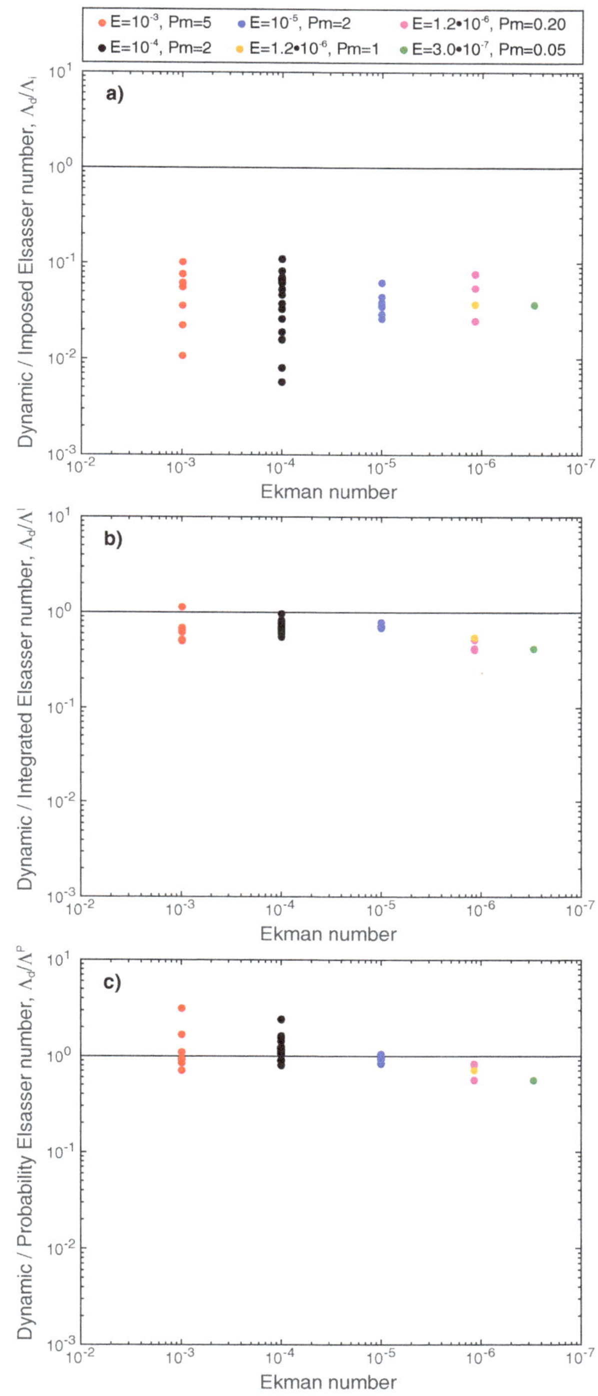

Fig. 3 Comparison of Elsasser number definitions in our simulations. Ratio of the dynamic Elsasser number Λ_d to the **a** imposed Elsasser number Λ_i, **b** integrated Elsasser number Λ^I, and **c** probability Elsasser number Λ^P as a function of Ekman number

$E \lesssim 10^{-6}$. The differences between Λ_d^* and the probability Elsasser number Λ^P tend to be comparable: 28 % (73 %) for $E = 10^{-3}$, 16 % (49 %) for $E = 10^{-4}$, 13 % (28 %) for $E = 10^{-5}$, and 25 % (43 %) for $E \lesssim 10^{-6}$. Thus, the modified dynamic Elsasser number Λ_d^* accurately captures the ratio of Lorentz to Coriolis forces in our suite of simulations.

In order to apply the modified dynamic Elsasser number to planetary cores, its applicability to core conditions must first be evaluated. Our datasets cover a broad range of diagnostic physical parameters: magnetic field strengths of $10^{-1} \lesssim \Lambda_i \lesssim 10^2$, magnetic induction to diffusion ratios of $10^2 \lesssim Rm \lesssim 10^4$, and Lorentz to Coriolis force ratios of $10^{-2} \lesssim \Lambda_d \lesssim 1$. However, these ranges span only a fraction of the planetary parameter estimates, which vary over $10^{-5} \lesssim \Lambda_i \lesssim 10^{10}$, $10^2 \lesssim Rm \lesssim 10^5$, and $10^{-6} \lesssim \Lambda_d^* \lesssim 10^7$ as detailed in the "Discussion" section. The most fundamental deviation is for $\Lambda_d > 1$, which implies a different dominant force balance. While our dataset includes cases with $\Lambda_d > 1$, these simulations are strongly driven with inertia exceeding both Lorentz and Coriolis forces (see Fig. 4 of Soderlund et al. 2012), which is not expected for most planets (cf. Soderlund et al. 2013).

Figure 5 shows the ratio of the dynamic to modified dynamic Elsasser numbers as a function of the convective Rossby number, $Ro_c = \left(RaE^2/Pr\right)^{1/2}$, which characterizes the ratio of buoyancy-driven inertial forces to Coriolis forces (e.g., Gilman 1977, Aurnou et al. 2007). The Λ_d^* approximation works well when the Coriolis force dominates; in contrast, the largest discrepancy occurs when $Ro_c > 1$. Geodynamo simulations from Dormy (2014) provide a Λ_d^* test for cases with $\Lambda_d > 1$ and $Ro_c < 1$ at an Ekman number of $E = 1.5 \times 10^{-4}$ and a Prandtl number of unity. Here, large magnetic Prandtl numbers ($Pm \geq 12$) are required to obtain dynamo action since the inertial forces are relatively weak (i.e., quasi-laminar flows). Application of Eq. 18 to his dataset yields $0.85 \leq \Lambda_d/\Lambda_d^* \leq 1.3$, implying that our results are applicable to strong field systems with dominant Lorentz forces.

A discrepancy between input parameters also exists between the simulations and planetary cores. In particular, all numerical dynamo models are necessarily limited to massively overestimated kinematic viscosities, which prohibits simulations with realistic Ekman and magnetic Prandtl numbers: $10^{-19} \lesssim E \lesssim 10^{-12}$ and $10^{-8} \lesssim Pm \lesssim 10^{-6}$ are estimated for planets with active dynamos (e.g., Schubert and Soderlund 2011). In contrast, our dataset considers $10^{-7} \lesssim E \lesssim 10^{-3}$ and $10^{-1} \lesssim Pm \lesssim 10^1$. Importantly, no clear Ekman or magnetic Prandtl number dependence is identified in Fig. 5. We, therefore, hypothesize that the modified dynamic Elsasser number (Eq. 18) can be extrapolated to extreme planetary parameters.

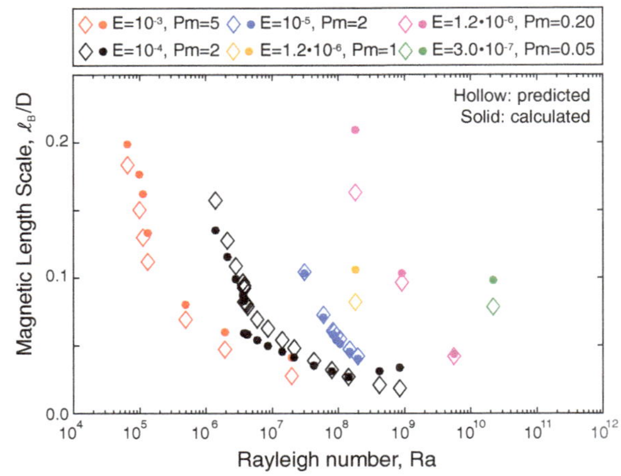

Fig. 4 Calculated and predicted magnetic length scales. Magnetic length scale as a function of the Rayleigh number with Ekman and magnetic Prandtl numbers denoted by *color*. A best-fit between the calculated and predicted values is obtained when a pre-factor is included in (Eq. 10): $\ell_B/D = 1.3Rm^{-1/2}$ (included here)

Discussion

The modified dynamic Elsasser number can be used to estimate the Lorentz to Coriolis force ratio in planetary cores when Λ_i and Rm are known (see Fig. 6 and Table 3). The imposed Elsasser number is traditionally used to quantify magnetic field strengths of planets since the poloidal component (B_p) can be measured by spacecraft and estimates for the other components (density ρ, magnetic diffusivity η, and rotation rate Ω) are relatively

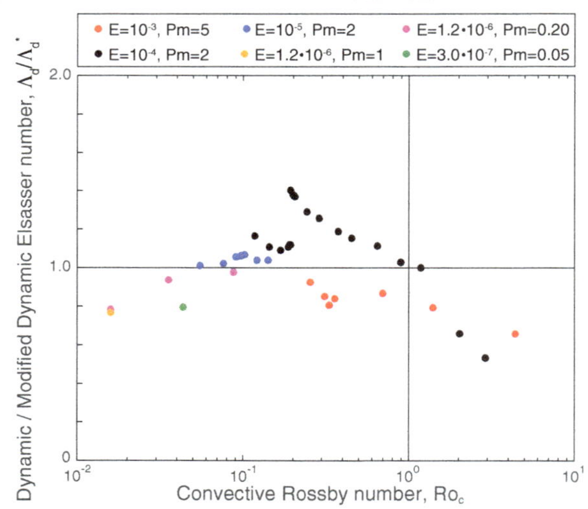

Fig. 5 Comparison of predicted and calculated Elsasser numbers as a function of the relative strengths of buoyancy and Coriolis forces. Ratio of Λ_d/Λ_d^* plotted versus the convective Rossby number, $Ro_c = \sqrt{RaE^2/Pr}$

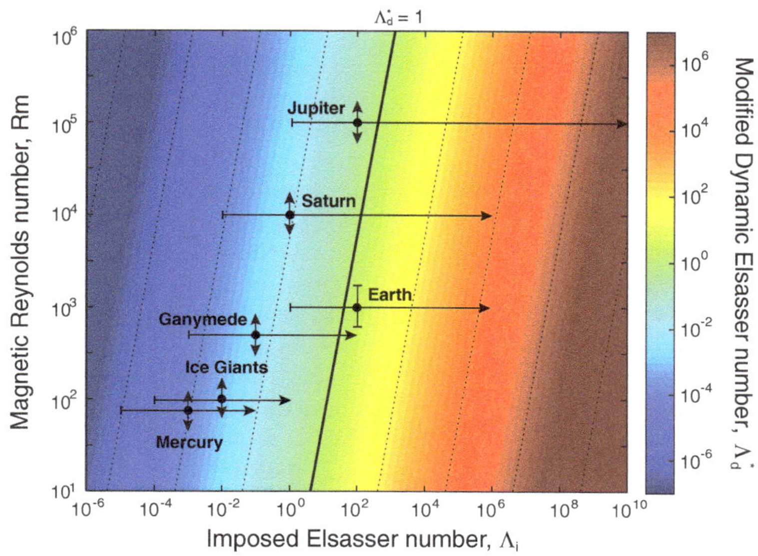

Fig. 6 Lorentz to Coriolis force ratio predictions for planetary cores. Modified dynamic Elsasser number as a function of imposed Elsasser number and magnetic Reynolds number across a range of values that may be relevant to planets in our solar system as well as exoplanets. The approximate ranges of anticipated Λ_i and *Rm* values are given for planets with active dynamos. *Barred line ends* for Λ_i denote the observed magnetic field strengths; *arrowed line ends* denote upper bounds. *Barred line ends* for *Rm* denote constrained bounds; *arrowed line ends* denote a larger range of uncertainties. Black circles denote the most probable Λ_i and *Rm* values

straightforward (see Schubert and Soderlund 2011 for a review); these measurements constitute a lower bound since the toroidal component is not included and small-scale poloidal contributions are below the resolution of detection. Toroidal magnetic fields (B_T) can be amplified by stretching of the poloidal field due to azimuthal differential rotation (zonal flows). This so-called ω-effect can generate toroidal magnetic fields up to $B_T \sim RmB_p$ (Roberts 2007), which sets an Λ_i upper bound. For a best estimate, we assume the total core field strength B to be an order of magnitude larger than the direct measurements (i.e., Λ_i increases by a factor of 100). This assumption is motivated both by geomagnetic data (e.g., Shimizu et al. 1998) and geodynamo modeling results (e.g., Aubert et al. 2009).

Earth: Geodynamo core flow inversions based on secular variation of the magnetic field imply velocities of $U \sim 5 \times 10^{-4}$ m/s (e.g., Holme 2007; Aubert 2013), which corresponds to $Rm \sim 10^3$ assuming $\eta \sim 1$ m^2/s for liquid metals. Ensemble inversions further suggest RMS magnetic field strengths in the cylindrical direction of $B_S \sim 2$ mT (Gillet et al. 2010), while toroidal magnetic field constraints of $1 < B_T/B_S < 100$ have been inferred by Shimizu et al (1998). Thus, $B \sim 10B_S \sim 20$ mT appears to be a reasonable estimate of the core's total field strength such that $\Lambda_i \sim 10^2$. In contrast, direct magnetic field measurements suggest a lower bound of $\Lambda_i \sim 1$ and application of the ω-effect suggests an upper bound of $\Lambda_i \sim 10^6$. We, therefore, predict a Lorentz to Coriolis force ratio of $\Lambda_d^* \sim 1$ for the geodynamo, with a possible range of $10^{-2} \lesssim \Lambda_d^* \lesssim 10^4$. This result suggests that the Lorentz force cannot be neglected dynamically in the core and, more likely, that the core is in magnetostrophic balance.

Table 3 Order of magnitude estimates of the magnetic Reynolds number *Rm*, the imposed Elsasser number Λ_i, and the Lorentz to Coriolis force ratio as estimated by Λ_d^* for planetary cores with active dynamos

	Mercury	Earth	Ganymede	Jupiter	Saturn	Ice Giants
Rm	75	10^3	500	10^5	10^4	10^2
Λ_i	10^{-3}	10^2	10^{-1}	10^2	10^0	10^{-2}
Λ_d^*	10^{-4}	10^0	10^{-3}	10^{-1}	10^{-2}	10^{-3}

Mercury and Ganymede: No observation-based velocity constraints are available currently for Mercury or Ganymede, so dynamo models that capture many of their magnetic field characteristics may serve as a basis for determining the magnetic Reynolds number. For example, Cao et al. (2014) obtain a Mercury-like dynamo when convection is driven by volumetrically distributed buoyancy sources and the core-mantle boundary heat flux peaks at low latitudes; their best fitting case has $Rm = 75$. Similarly, Christensen (2015) find Ganymede-like dynamos driven by iron snow with a stably stratified layer below the outer shell boundary; $Rm \approx 500$ in his optimal simulation. Using the upper and lower bound constraints derived from the ω-effect and observations, respectively, the imposed Elsasser number is estimated to vary between

$\Lambda_i \sim 10^{-5}$ and $\sim 10^{-1}$ for Mercury, which corresponds to $\Lambda_d^* \lesssim 10^{-2}$ and implies that magnetostrophic balance is unlikely in the Hermian core. Conversely, the possible magnetic field strengths for Ganymede range from $\Lambda_i \sim 10^{-3}$ to $\sim 10^2$, yielding $10^{-5} \lesssim \Lambda_d^* \lesssim 1$. However, the best estimate for the imposed Elsasser number is $\Lambda_i \sim 10^{-1}$ such that $\Lambda_d^* \sim 10^{-3}$. As a result, magnetostrophic balance is possible, but unlikely, in Ganymede's core.

Gas Giants: Jovian velocities can be estimated through scaling arguments and potentially secular variation. Jones (2014) suggests flow velocities of $U \gtrsim 10^{-3}$ m/s, leading to values of $Rm \gtrsim 10^5$ given the planet's large size. In this case, magnetic field strengths range from $\Lambda_i \sim 1$ via core field measurements to $\Lambda_i \sim 10^{10}$ when the ω-effect is included, with a best estimate of $\Lambda_i \sim 10^2$. The corresponding Lorentz to Coriolis force ratio then spans the range $10^{-3} \lesssim \Lambda_d^* \lesssim 10^7$, with $\Lambda_d^* \sim 0.1$ being the best estimate. Thus, barring the large uncertainties, the Lorentz force is predicted to be sub-dominant, but not negligible, in Jupiter's core.

Starchenko and Jones (2002) predict similar flow speeds for the Saturnian interior, which corresponds to a lower $Rm \sim 10^4$ value due to the smaller size of the dynamo region. Saturn's magnetic field is also weaker than that of Jupiter; the best estimate imposed Elsasser number is $\Lambda_i \sim 1$ with a feasible range of $10^{-2} \lesssim \Lambda_i \lesssim 10^6$. In the most likely scenario, the Lorentz force is predicted to be two orders of magnitude smaller than the Coriolis force ($\Lambda_d^* \sim 10^{-2}$) with uncertainties extending the possible range to $10^{-4} \lesssim \Lambda_d^* \lesssim 10^4$.

Ice Giants: Uranus and Neptune have relatively weak core magnetic fields with $\Lambda_i \sim 10^{-4}$ based on spacecraft observations, but there are few constraints on the flow speeds. We, therefore, assume $U \sim 10^{-3}$ m/s based on the other planetary estimates such that $Rm \sim 10^2$ to give an upper bound of $\Lambda_i \sim 1$ when the toroidal estimate is included. For these Λ_i estimates, the predicted Lorentz to Coriolis force ratio is always less than unity: $10^{-5} \lesssim \Lambda_d^* \lesssim 10^{-1}$. These results thus imply that the ice giants are not in magnetostrophic balance.

Conclusions

The Lorentz to Coriolis force ratio is an important parameter for the dynamics of planetary cores since dynamos with dominant Coriolis forces at global scales are expected to be driven by fundamentally different archetypes of fluid motions than those with dominant (or co-dominant) Lorentz forces (Roberts and King 2013, Calkins et al. 2015). We hypothesize that a representative global estimate of the Lorentz to Coriolis force ratio can be predicted by $\Lambda_d^* = 0.77\Lambda_i Rm^{-1/2}$. An advantage of this formulation is that it depends on quantities that can be estimated for planetary cores (Table 3). Our results suggest that the Earth's core is likely to be in magnetostrophic balance where the Lorentz and Coriolis forces are comparable. The Lorentz force may also be substantial in Jupiter's core, where it is predicted to be a factor of ten less than the Coriolis force. Magnetic fields become increasingly sub-dominant for the other planets: the Coriolis force is predicted to exceed the Lorentz force by at least two orders of magnitude within the cores of Saturn, Uranus/Neptune, Ganymede, and Mercury.

These conclusions are subject to large uncertainties, however. Core flow speeds are difficult to estimate, while total magnetic field strengths cannot be measured directly. The applicability of Eq. 18 for the Lorentz to Coriolis force ratio may also break down at extreme planetary core conditions that cannot be explored numerically or in the laboratory due to technological limitations. In order to mitigate these uncertainties, we have considered a range of possible core magnetic field strengths (Λ_i) and included state-of-the-art simulations (Sheyko 2014; cf. Nataf and Schraeffer 2015).

Abbreviations

MHD: Magnetohydrodynamics; RMS: Root mean square; SKA: Soderlund et al. (2012, 2014); S14: Sheyko 2014.

Competing interests

The authors declare that they have no competing interests.

Authors' contributions

KMS and JMA designed the study. KMS and AS carried out and analyzed the numerical simulations. All authors interpreted the data. All authors read and approved the final manuscript.

Acknowledgements

The authors thank two anonymous referees for their thoughtful reviews, Hao Cao for helpful suggestions, and Wolfgang Bangerth for enlightening discussions at the 2014 Study of Earth's Deep Interior (SEDI) meeting in Kanagawa, Japan. KMS gratefully acknowledges Japan Geoscience Union the National Science Foundation to attend this symposium as well as research support from the National Science Foundation (grant AST-0909206). JMA acknowledges the support of the National Science Foundation Geophysics Program (grant EAR-1246861). Computational resources supporting this work were provided by the NASA High-End Computing (HEC) Program through the NASA Advanced Supercomputing (NAS) Division at Ames Research Center and by the Swiss National Supercomputing Centre (CSCS) under project ID s225. This is UTIG contribution 2858.

Author details

[1]Institute for Geophysics, John A. & Katherine G. Jackson School of Geosciences, The University of Texas at Austin, J. J. Pickle Research Campus, Building 196 (ROC), 10100 Burnet Road (R2200), Austin, Texas, 78758-4445, USA. [2]Institut für Geophysik, ETH Zürich, NO H 11.2, Sonneggstrasse 5 8092 Zürich, LA59 Switzerland. [3]U.S. Global Development Lab, U.S. Agency for International Development, Washington, DC, 20004, USA. [4]Department of Earth, Planetary, and Space Sciences, University of California, Los Angeles, 595 Charles Young Drive East, Los Angeles, California, 90095-1567, USA.

References

Aubert J, Aurnou JM, Wicht J (2008) The magnetic structure of convection-driven numerical dynamos. Geophys J Int 172:945–956

Aubert J, Labrosse S, Poitou C (2009) Modelling the palaeo-evolution of the geodynamo. Geophys J Int 179:1414–1428

Aubert J (2013) Flow throughout the Earth's core inverted from geomagnetic observations and numerical dynamo models. Geophys J Int 192:537–556

Aurnou JM, Heimpel MH, Wicht J (2007) The effects of vigorous mixing in a convective model of zonal flow on the Ice Giants. Icarus 190:110–126

Calkins MA, Julien K, Tobias SM, Aurnou JM (2015) A multiscale dynamo model driven by quasi-geostrophic convection. J Fluid Mech. doi:10.1017/jfm.2015.464

Cao H, Aurnou JM, Wicht J, Dietrich W, Soderlund KM, Russell CT (2014) A dynamo explanation for Mercury's anomalous magnetic field. Geophys Res Lett 41(12):4127–4134

Cardin P, Olson PL (1995) The influence of toroidal magnetic field on thermal convection in the core. Earth Planet Sci Lett 133:167–181

Cardin P, Brito D, Jault D, Nataf HC, Masson JP (2002) Towards a rapidly rotating liquid sodium dynamo experiment. Magnetohydrodynamics 38:177–189

Cheng JS, Stellmach S, Ribeiro A, Grannon A, King EM, Aurnou JM (2015) Laboratory-numerical models of rapidly rotating convection in planetary cores. Geophys J Int 201:1–17

Christensen U. R, Wicht J (2007) Numerical Dynamo Simulations. In: Schubert G (ed). Treatise on Geophysics, Core Dynamics. Elsevier, Amsterdam Vol. 8. pp 245–282. Chap. 8

Christensen UR (2015) Iron snow dynamo models for Ganymede. Icarus 247:248–259

Dharmaraj G, Stanley S (2012) Effect of inner core conductivity on planetary dynamo models. Phys Earth Planet Int 212–213:1–9

Dharmaraj G, Stanley S, Qu AC (2014) Scaling laws, force balances and dynamo generation mechanisms in numerical dynamo models: influence of boundary conditions. Geophys J Int 199:514–532

Dormy E, Soward AM, Jones CA, Jault D, Cardin P (2004) The onset of thermal convection in rotating spherical shells. J Fluid Mech 501:43–70

Dormy E (2014) Strong field spherical dynamos. arXiv:1412.4090v1

Galloway DJ, Proctor MRE, Weiss NO (1978) Magnetic flux ropes and convection. J Fluid Mech 87:243–261

Gilman PA (1977) Nonlinear dynamics of Boussinesq convection in a deep rotating spherical shell – I. Geophys Astrophys Fluid Dyn 8:93–135

Gillet N, Jault D, Canet E, Fournier A (2010) Fast torsional waves and strong magnetic field within the Earth's core. Nature 465:74–77

Grooms I, Julien K, Weiss JB, Knobloch E (2010) Model of convective Taylor columns in rotating Rayleigh-Bénard convection. Phys Rev Lett 104:224501

Holme R (2007) Large-scale flow in the core. In: Schubert G (ed). Treatise on Geophysics, Core Dynamics. Elsevier, Amsterdam Vol. 8. pp 107–130. Chap. 4

Jones CA, Soward AM, Mussa AI (2000) The onset of thermal convection in a rapidly rotating sphere. J Fluid Mech 405:157–179

Jones CA (2014) A dynamo model of Jupiter's magnetic field. Icarus 241:148–159

Julien K, Knobloch E, Werne J (1998) A new class of equations for rotationally constrained flows. Theoret Comput Fluid Dynamics 11:251–261

Julien K, Rubio AM, Grooms I, Knobloch E (2012) Statistical and physical balances in low Rossby number Rayleigh-Bénard convection. Geophys Astrophys Fluid Dyn 106:392–428

Kageyama A, Miyagoshi T, Sato T (2008) Formation of current coils in geodynamo simulations. Nature 454:1106–1108

King EM, Aurnou JM (2015) Magnetostrophic balance as the optimal state for turbulent magnetoconvection. Proc Natl Acad Sci 112:990–994

Le Bars M, Cebron D, Le Gal P (2015) Flows driven by libration, precession, and tides. Annu Rev Fluid Mech 47:163–193

Nataf H-C, Schaeffer N (2015) Turbulence in the core. In: Olson P, Schubert G (eds). Treatise on Geophysics, 2nd ed., Core Dynamics. Elsevier BV, Amsterdam. pp 161–181. Chap. 6

Olson PL, Glatzmaier GA (1996) Magnetoconvection and thermal coupling of the Earths core and mantle. Philos Trans R Soc London Ser A 354:1413–1424

Olson PL, Christensen UR, Glatzmaier GA (1999) Numerical modeling of the geodynamo: Mechanisms of field generation and equilibration. J Geophys Res 104:10383–10404

Ribeiro A, Fabre G, Guermond JL, Aurnou JM (2015) Canonical models of geophysical and astrophysical flows: Turbulent convection experiments in liquid metals. Metals 5:289–335

Roberts PH (1968) On the thermal instability of a rotating-fluid sphere containing heat sources. Philos Trans R Soc London Ser A 264:93–117

Roberts, PH (2007) Theory of the geodynamo. In: Schubert G (ed). Treatise on Geophysics, Core Dynamics. Elsevier, Amsterdam Vol. 8. pp 67–105. Chap. 3

Roberts PH, King EM (2013) On the genesis of the Earth's magnetism. Rep Prog Phys 76(9):096801

Schubert G, Soderlund KM (2011) Planetary magnetic fields: Observations and models. Phys Earth Planet Int 187:92–108

Sheyko A (2014) Numerical investigations of rotating MHD in a spherical shell. Dissertation, ETH ZURICH

Shimizu H, Koyama T, Utada H (1998) An observational constraint on the strength of the toroidal magnetic field at the CMB by time variation of submarine cable voltages. Geophys Res Lett 25:4023–4026

Soderlund KM, King EM, Aurnou JM (2012) The influence of magnetic fields in planetary dynamo models. Earth Planet Sci Lett 333-334:9–20

Soderlund KM, Heimpel MH, King EM, Aurnou JM (2013) Turbulent models of ice giant internal dynamics: Dynamos, heat transfer, and zonal flows. Icarus 224:97–113

Soderlund KM, King EM, Aurnou JM (2014) Corrigendum to "The influence of magnetic fields in planetary dynamo models". Earth Planet Sci Lett 392:121–123

Sprague M, Julien K, Knobloch E, Werne J (2006) Numerical simulation of an asymptotically reduced system for rotationally constrained convection. J Fluid Mech 551:141–174

Sreenivasan B, Jones CA (2011) Helicity generation and subcritical behavior in rapidly rotating dynamos. J Fluid Mech 688:5–30

Sreenivasan B, Sahoo S, Gaurav D (2014) The role of buoyancy in polarity reversals of the geodynamo. Geophys J Int 199:1698–1708

Starchenko SV, Jones CA (2002) Typical velocities and magnetic field strengths in planetary interiors. Icarus 157:426–435

Stellmach S, Lischper M, Julien K, Vasil GM, Cheng JS, Ribeiro A, King EM, Aurnou JM (2014) Approaching the asymptotic regime of rapidly rotating convection: Boundary layers versus interior dynamics. Phys Rev Lett 113:254501

Tritton DJ (1998) Physical Fluid Dynamics. Oxford University Press, Oxford

Wicht J (2002) Inner-core conductivity in numerical dynamo simulations. Phys Earth Planet Int 132:281–302

Zhang K (1992) Spiraling columnar convection in rapidly rotating spherical fluid shells. J Fluid Mech 236:535–554

Zhang K, Schubert G (2000) Magnetohydrodynamics in rapidly rotating spherical systems. Annu Rev Fluid Mech 32:409–443

14

Impacts of cloud microphysics on trade wind cumulus: which cloud microphysics processes contribute to the diversity in a large eddy simulation?

Yousuke Sato*, Seiya Nishizawa, Hisashi Yashiro, Yoshiaki Miyamoto, Yoshiyuki Kajikawa and Hirofumi Tomita

Abstract

This study investigated the impact of several cloud microphysical schemes on the trade wind cumulus in the large eddy simulation model. To highlight the differences due to the cloud microphysical component, we developed a fully compressible large eddy simulation model, which excluded the implicit scheme and approximations as much as possible. The three microphysical schemes, the one-moment bulk, two-moment bulk, and spectral bin schemes were used for sensitivity experiments in which the other components were fixed. Our new large eddy simulation model using a spectral bin scheme successfully reproduced trade wind cumuli, and reliable model performance was confirmed. Results of the sensitivity experiments indicated that precipitation simulated by the one-moment bulk scheme started earlier, and its total amount was larger than that of the other models. By contrast, precipitation simulated by the two-moment scheme started late, and its total amount was small. These results support those of a previous study. The analyses revealed that the expression of two processes, (1) the generation of cloud particles and (2) the conversion from small droplets to raindrops, were crucial to the results. The fast conversion from cloud to rain and the large amount of newly generated cloud particles at the cloud base led to evaporative cooling and subsequent stabilization in the sub-cloud layer. The latent heat released at higher layers by the condensation of cloud particles resulted in the development of the boundary layer top height.

Keywords: Large eddy simulation; Shallow clouds

Background

The effect of clouds is one of the most uncertain factors in climate projection and numerical weather prediction. Shallow clouds (e.g., stratus, stratocumulus, shallow cumulus) play particularly important roles in the energy budget of the earth through radiation process because they cover a broad area of the earth (e.g., Randall et al. 1984). The 5th Intergovernmental Panel on Climate Change (IPCC) report suggested that the uncertainties with regards to shallow clouds should be reduced for reliable assessments (IPCC AR5, Stocker et al. 2013).

In global scale models (e.g., general circulation model (GCM)) and regional models with coarse grid spacing, shallow clouds are usually expressed by parameterizations (e.g., Tiedtke 1993; Considine et al. 1997; Kain 2004), but these parameterizations have not been able to effectively simulate the shallow cloud cover observed from satellites (e.g., Chepfer et al. 2008; Naud et al. 2010).

To improve the expression of shallow cloud, the results of large eddy simulation (LES) models have been utilized. For example, Bretherton and Park (2009) used the results of an LES model to develop a new moist turbulent parameterization. Suzuki et al. (2004) and Posselt and Lohmann (2008) introduced an autoconversion parameterization that was typically used in an LES model (Khairoutdinov and Kogan 2000) into a global scale model. Many studies using LES models have been conducted to determine the characteristics of shallow cloud and improve large-scale modeling (e.g., Wang and Feingold 2009; Xue et al. 2008, Savic-Jovcic and Stevens 2008, Yamaguchi and Randall 2012). However, the

* Correspondence: yousuke.sato@riken.jp
RIKEN Advanced Institute for Computational Science, 7-1-26 Minatojima-Minami-machi, Chuo-ku, Kobe, Hyogo 650-0047, Japan

results of LES models are diverse, as indicated by several LES model intercomparison studies targeting shallow cloud (e.g., Stevens et al. 2005; Ackerman et al. 2009; van Zanten et al. 2011; Siebesma et al. 2003). It has been suggested that the difference in the microphysical schemes used in LES models is one of the reasons for the diversity in the results (Ackerman et al. 2009; van Zanten et al. 2011). However, it is difficult to investigate the exact effect of the different cloud microphysical schemes (i.e., the effect on the results that comes from changing the cloud microphysical scheme while keeping all other components fixed), because each LES model uses a different scheme, not only in cloud microphysics but also many other components (e.g., governing equation, turbulent scheme, advection scheme, and others). Thus, it is also difficult to determine which microphysical processes contribute most to the diversity of the LES results. Sensitivity experiments, which change only the cloud microphysical scheme, are required to better understand the exact effects of the cloud microphysical scheme.

The kinematic driver (KiD) model developed by Shipway and Hill (2012) enables us to conduct sensitivity simulations by changing only the cloud microphysics. Using KiD, we can consider the exact effects of the differences in cloud microphysical schemes. However, KiD ignores feedbacks to the atmosphere, even though the feedbacks of microphysics can affect the microphysical properties of shallow clouds and the turbulent structure of the boundary layer (e.g., Stevens et al. 1998; Wang et al. 2010). It is necessary to consider the feedback of microphysical processes on the dynamics when determining which process causes the diversity in the results of LES models. We should use the model that excludes approximation and implicit schemes as much as possible, because these features also affect the cloud microphysical properties simulated by the model.

This study developed a model that satisfies these requirements. Using the model, we attempted to reproduce the diversity in the results of the LES model used in van Zanten et al. (2011) and to determine impact of the various cloud microphysical schemes. Three types of cloud microphysical schemes (one-moment bulk, two-moment bulk, and spectral bin schemes) were considered in this study, because these three schemes have been used in previous intercomparison studies.

Of the several types of shallow cloud, we focused on trade wind cumulus because their variability in the results of LES models was larger than that of other types of shallow clouds. For example, the variability of surface precipitation in an intercomparison of trade wind cumulus (van Zanten et al. 2011) was larger than that in an intercomparison of stratocumulus (Ackerman et al. 2009). van Zanten et al. (2011) proposed that one of the reasons for the diversity in the microphysical properties of trade wind cumuli was the different cloud microphysical models. They interpreted that a simple (one-moment bulk) microphysical scheme produced large amounts of precipitation (i.e., Table 3 of van Zanten et al. (2011)) and liquid water simulated by one-moment bulk schemes tended to be distributed in the lower layer. By contrast, the liquid water was located in the higher layer, and the precipitation flux was small in most of the two-moment schemes (i.e., Figure 6a of van Zanten et al. (2011)). In this study, we confirmed the validity of their interpretation through a simulation using our new fully compressible LES model and determined the main processes contributing to the diversity in the results of LES model intercomparison studies.

Methods

Experimental setup, dynamic framework, turbulence model, and external forcing

This section describes the common parts of the model with its experimental setup (i.e., the dynamic framework, turbulence model, and external forcing). The different parts of the microphysical schemes are highlighted in the subsequent section.

The dynamic model used in this study is an LES model that is included in the Scalable Computing for Advanced Library and Environment library (SCALE). Henceforth, we call this LES model SCALE-LES. The details of SCALE-LES are found at http://scale.aics.riken.jp/.

A fully compressible system is adopted for the governing equation of SCALE-LES. The prognostic variables are the three-dimensional momentum (ρu, ρv, ρw), total density (ρ), mass-weighted potential temperature ($\rho\theta$), and mass concentration of tracers (ρq_s), where q_s includes specific humidity, ratio of mass, and number concentration ratio of hydrometeors to total mass. Explicit time integration is used in all directions. Furthermore, the fourth-order central difference scheme is adopted for spatial discretization to avoid the numerically implicit diffusion that would be induced by odd-ordered difference schemes. The three-step Runge–Kutta scheme is adopted. To retain stability of the model, fourth-order superviscosity/diffusion is applied for all prognostic variables. Using SCALE-LES, developed as described above, we exclude the effects of approximation in the governing equation system and implicit diffusion as far as possible, enabling a consideration of the effects of the target component (i.e., cloud microphysics in this study).

To guarantee monotonicity, the flux-corrected transport (FCT) scheme (Zalesak 1979) is applied for all prognostic variables except for density. The effects of sub-grid scale turbulence are calculated using a Smagorinsky-type scheme based on Brown et al. (1994) and Scotti et al. (1993).

The experimental setup is almost the same as that of the previous Global Energy and Water Exchange project (GEWEX) Cloud System Study (GCSS) intercomparison of Rain in Cumulus over the Ocean (RICO) (van Zanten et al. 2011). The calculation domain covers 12.8×12.8 km^2 horizontally with a double periodic boundary condition and 4.0 km vertically. The horizontal and vertical grid intervals are 100 and 40 m, respectively. Rayleigh damping is applied for three-dimensional momentum over the upper atmosphere of $z > 3.5$ km. The strength of the numerical diffusion is set as 1.25×10^5 m^4 s^{-1}, which is determined by the sensitivity experiment (the results of the sensitivity experiment are shown in Appendix 1 of this paper). The simulations are conducted for 24 h with time steps (Δt) of 0.05, 0.15, and 0.5 s for dynamics, cloud microphysics, and other physics. The Δt for cloud microphysics is determined as the largest time step to avoid artificial noise, and the ratio of Δt for dynamics to Δt for other physics is set to 10 based on the sensitivity experiments (see Appendix 2 of this paper for details of the sensitivity experiment).

The external forcing of the radiation, the surface flux, and the large-scale horizontal advection are applied in the same way as in van Zanten et al. (2011). The forcing of the large-scale subsidence is applied for all prognostic variables including density, except for u and v. In nature, large-scale subsidence generates a divergence of total density, which makes the air mass flow out from the domain. However, it cannot flow out in the compressible model using the configuration of van Zanten et al. (2011) due to the periodicity in the lateral boundary condition. Although this problem can be ignored in the anelastic and Boussinesq systems due to the fixed density, it is necessary to consider this problem in the compressible system. Although no description is available for this problem in van Zanten et al. (2011), the density of each layer should be reduced according to the divergence. The density reduction rate and the equation system with large-scale forcing are given below.

Large-scale subsidence (w_{LS}) was given by van Zanten et al. (2011) as:

$$w_{LS} = \begin{cases} -\dfrac{0.005}{2260}z & (z < 2260\,\mathrm{m}) \\ -0.005 & (z \geq 2260\,\mathrm{m}) \end{cases}$$

where z is the height. Instead of this formulation, we give the subsidence formulated directly to vertical momentum as:

$$\rho w_{LS} = \begin{cases} -\dfrac{0.005}{2260}z & (z < 2260\,\mathrm{m}) \\ -0.005 & (z \geq 2260\,\mathrm{m}) \end{cases}$$

Consequently, the density reduction rate (D) is given as:

$$D \equiv \frac{\partial(\rho w_{LS})}{\partial z} = \begin{cases} -\dfrac{0.005}{2260} & (z < 2260\mathrm{m}) \\ 0 & (z \geq 2260\mathrm{m}) \end{cases}$$

The continuous equation modified with the subsidence term is given by

$$\frac{\partial \rho}{\partial t} + \frac{\partial(\rho u)}{\partial x} + \frac{\partial(\rho v)}{\partial y} + \frac{\partial(\rho(w + w_{LS}))}{\partial z} = D \qquad (1)$$

The density reduction derived from the divergence by the large-scale subsidence is added in the right-hand side (rhs) of Eq. (1). Note that this equation is identical to the equation without large-scale subsidence. The Lagrangian conservation equation of the scalar quantities (ϕ) is given as:

$$\rho\frac{\partial \phi}{\partial t} + \rho u\frac{\partial \phi}{\partial x} + \rho v\frac{\partial \phi}{\partial y} + \rho(w + w_{LS})\frac{\partial \phi}{\partial z} = 0 \qquad (2)$$

The prognostic variable of SCALE-LES is a mass-weighted value, and the equation of mass-weighted values is derived from Eqs. (1) and (2) as:

$$\frac{\partial(\rho\phi)}{\partial t} + \frac{\partial(\rho u\phi)}{\partial x} + \frac{\partial(\rho v\phi)}{\partial y} + \frac{\partial(\rho(w + w_{LS})\phi)}{\partial z} = D\phi \qquad (3)$$

As shown in the rhs of Eq. (3), the scalar quantities (ϕ) flow out from each layer of the system by subsidence. The equation for the vertical momentum is modified, as is that of the scalar quantities. The vertical flux $\rho w_{LS}\phi$ at the top boundary can be determined so that such additional convergence of the flux is canceled with $D\phi$ at the top layer.

Three microphysical schemes

To reproduce the diversity in the results of the RICO study, three types of cloud microphysical scheme are used for this study: the one-moment bulk microphysical scheme (Tomita 2008), the two-moment bulk scheme (Seiki and Nakajima 2014), and the spectral bin microphysical scheme (Suzuki et al. 2010). The one-moment bulk scheme and the two-moment bulk scheme are based on Berry (1968) and Seifert and Beheng (2001), respectively. Both of the original bulk schemes were used in the RICO study.

The essential difference among the three schemes is their treatment of the size distribution of the number of hydrometeor particles. The one-moment bulk scheme expresses this value by the Marshall–Palmer distribution, with the assumption of a constant total number of particles. By this assumption, only the mass concentration is needed to determine the size distribution. Although the two-moment bulk scheme conceptually treats the size distribution almost the same way, it differs from the one-moment bulk scheme in the assumption about the type of size distribution function and in the process of its determination. The two-moment bulk scheme assumes the generalized gamma distribution as the size distribution function, and the size distribution itself is determined not only by the mass concentration but also by the number concentration. In this sense, the two-moment bulk scheme is more sophisticated than the one-moment bulk scheme. The spectral bin scheme is an intrinsically sophisticated method compared with the others. It explicitly predicts the size distribution. To compensate for the detailed expression of size distribution, the spectral bin scheme requires about four- to fivefold larger number of prognostic variables compared with the other two schemes.

Although the three microphysical schemes treat both warm and ice phase cloud, the ice phase was not calculated because the temperature at the model top is greater than 273.15 K. In this case, the categorization of hydrometeors for each scheme is as follows. The one-moment and two-moment bulk schemes address two types of hydrometeors: cloud droplet and raindrop for warm rain process. The spectral bin scheme addresses only a type of water drop, which covers the particle size of both cloud droplets and raindrops. For the spectral bin scheme, the radius of newly generated cloud particles by nucleation is set to 3 μm (Suzuki et al. 2006). The size distribution of hydrometeors is discretized to 33 bins; its configuration has been established by several previous studies (e.g., Khain and Sednev 1996; Iguchi et al. 2008). The center of mass of ith bin (m_i) is set by using the i-1-th bin (m_{i-1}) as $m_i = 1.874\ m_{i-1}$. m_1 is the mass of cloud particles whose radius is 3 μm.

All three schemes consider the generation of cloud droplets, condensation, evaporation, and sedimentation of cloud hydrometeors. Although the two-moment bulk scheme also considers the breakup of cloud droplets, this difference is minor based on sensitivity experiments examining the breakup process (figure not shown). Sedimentation is calculated by the first-order upwind scheme for all three schemes.

Since the generation of new cloud droplets is one of the critical processes in this experiment, the difference in this process among the schemes should be noted. The mass of newly generated cloud droplets is calculated by saturation adjustment in the one-moment bulk scheme, which was also used in some one-moment bulk schemes in the RICO study. By contrast, it is calculated by the nucleation schemes in the two-moment bulk and spectral bin schemes. The number concentration of the cloud droplets ($N_{c,nucl}$) generated by the nucleation process is calculated as follows (e.g., Pruppacher and Klett 1997):

$$N_{c,nucl} = N_0 {S_w}^k, \quad (4)$$

where S_w is supersaturation over water. The constants N_0 and k are set as $N_0 = 100 \times 10^6$ m^{-3} and $k = 0.462$. This scheme was also used in several models used in the RICO study. The growth of cloud droplets into raindrops is another key issue for this experiment, as well as the underlying creation process. In the bulk scheme, this is expressed as autoconversion and accretion. To investigate the strength of the impact of the autoconversion and accretion processes, we conducted the same simulation with autoconversion and accretion ratios that were twice as large (0.067-fold smaller) as the two-moment (one-moment) scheme. The autoconversion rate (P_{auto}) in the one-moment bulk scheme is calculated as in Tomita (2008), which was based on Berry (1968):

$$P_{auto} = \frac{1}{\rho}\left[16.7 \times (\rho q_c)^2 \left(5 + \frac{3.6 \times 10^{-5} N_{c,T08}}{D_d \rho q_c}\right)\right] \\ \left[\mathrm{kg\ kg^{-1} s^{-1}}\right], \quad (5)$$

where $D_d = 0.1456–5.964 \times 10^{-2}$ log ($N_{c,T08}$/2000), q_c is the cloud water mixing ratio, and $N_{c,T08}$ is the number concentration of cloud droplets. In Tomita (2008), $N_{c,T08}$ was set as 50 cm^{-3}, but in this study, $N_{c,T08}$ was set as 70 cm^{-3} based on the experimental setup in the RICO study (van Zanten et al. 2011). Another autoconversion scheme from Khairoutdinov and Kogan (2000) was implemented into the one-moment bulk scheme because it had performed better than the scheme of Berry (1968) in a GCM (Suzuki et al. 2004). The scheme was also adopted in some models used in the RICO study. P_{auto} in the Khairoutdinov and Kogan (2000) scheme is calculated as:

$$P_{auto} = 1350 \times q_c^{2.47} \times N_{c,T08}^{-1.79} \quad \left[\mathrm{kgkg^{-1} s^{-1}}\right] \quad (6)$$

Using the schemes shown above, we attempted to reproduce the diversity of the LES results in the RICO study.

Results and discussion

Basic performance of SCALE-LES

The validity of SCALE-LES was confirmed through comparison with a previous study (van Zanten et al. 2011). The results of the spectral bin scheme were regarded as a reference solution of SCALE-LES, because it is the most sophisticated scheme of the three. First, the results of the reference solution were compared with the previous study. As shown in Figs. 1 and 2, the results of our model (SCALE-LES) are within the range between the maximum and minimum of the intercomparison study in terms of temporal evolution and vertical profile

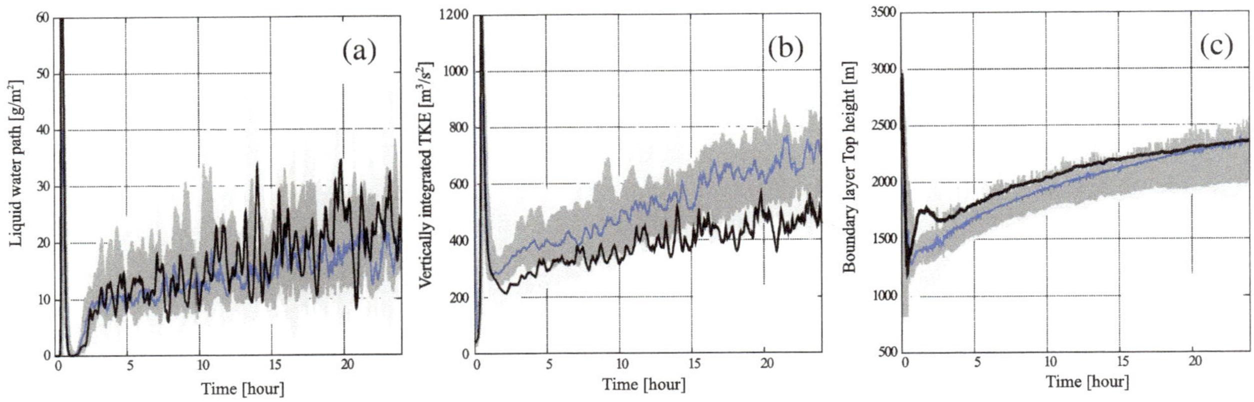

Fig. 1 Comparison of the time evolution between SCALE-LES and a previous intercomparison study. Time evolution of the **a** liquid water path, **b** vertically integrated turbulence kinetic energy, and **c** boundary layer top height averaged over the entire calculation domain simulated by (*black line*) SCALE-LES, with the spectral bin scheme. The *blue line*, *dark gray shading*, and *light gray shading* indicate the median, range between the first and third quartiles, and range between the maximum and minimum values, respectively, of the previous intercomparison study (van Zanten et al. 2011)

for several quantities. This indicates that our model could reproduce the shallow cumulus simulated by the LES models used in the previous study.

As well as the physical performance, the computational performance and the scalability of SCALE were investigated. The elapsed time for a time step (Δt) and the performance efficiency are shown in Table 1. The elapsed time and performance of SCALE-LES do not change when it is used with a large number of Message Passing Interface (MPI) processes (e.g., over 10,000 MPI processes). This indicates that SCALE-LES has excellent scalability and a reasonable performance in massive parallel computing.

Impacts of cloud microphysical scheme on simulation results

Second, we show the results of the same simulation (RICO) by using three microphysical schemes for investigating the impacts of the cloud microphysical scheme. Figure 3 indicates the differences among the three schemes. The precipitation flux simulated by the two-moment bulk scheme is small, and its peak value is distributed in the upper layer. By contrast, the precipitation flux by the one-moment bulk scheme is large and the peak value locates in the lower layer. The precipitation by the spectral bin scheme is between the other two (Fig. 3a). The trend is consistent with the previous study (Figure 6a of van Zanten et al. 2011).

The impacts of the different cloud microphysical schemes on the vertical distribution of the precipitation flux, the liquid water mixing ratio (q_l), and cloud fraction are clearly shown in Fig. 3a–c. The q_l in the one-moment bulk scheme is distributed in the lower layer (the peak value is represented at $z \sim 900$ m). Table 2 shows the surface precipitation averaged during the last 4 h for the three schemes. The precipitation amount in the one-moment bulk scheme is the largest, and the precipitation begins earliest among the three. The surface precipitation over 0.1 W m^{-2} starts at 1.8 h in the one-moment bulk scheme, at 10.05 h in the two-moment bulk scheme, and at 2.63 h in the spectral bin scheme. The liquid water simulated in the two-moment bulk scheme is located in the upper layer (the peak value is located at $z \sim 2400$ m), and only trace amounts of precipitation reach the surface (Table 2). The cloud fraction of the one-moment scheme is located in the lower layer and is small in the upper layer. On the other hand, the positive cloud fraction in the two-moment bulk scheme reaches a higher layer. The spectral bin scheme simulates an intermediate value between the other two, with the same trend as the RICO study. These results are consistent with the result of the previous study (van Zanten et al. 2011).

The large amount of precipitation in the one-moment bulk scheme carries a large amount of liquid water to the lower layer. As shown in Fig. 3d, the total water mixing ratio (q_t) of the lower layer (i.e., $z <$ 1500 m) in the one-moment bulk scheme is larger than that in the others. Despite the difference in the vertical distribution of liquid water, the liquid water path (LWP) shows the same value (Fig. 3e). By contrast, the cloud coverage of the one-moment scheme is smaller than in the other scheme (Fig. 3f). This implies that the amount of cloud that extends horizontally at the top of boundary layer is small in the one-moment scheme, which is because of the liquid water removed from the cloud layer earlier by precipitation. This is also found in Fig. 3c, where the peak of the cloud fraction does not appear around the top of the boundary layer (i.e., $z \sim 2000$ m).

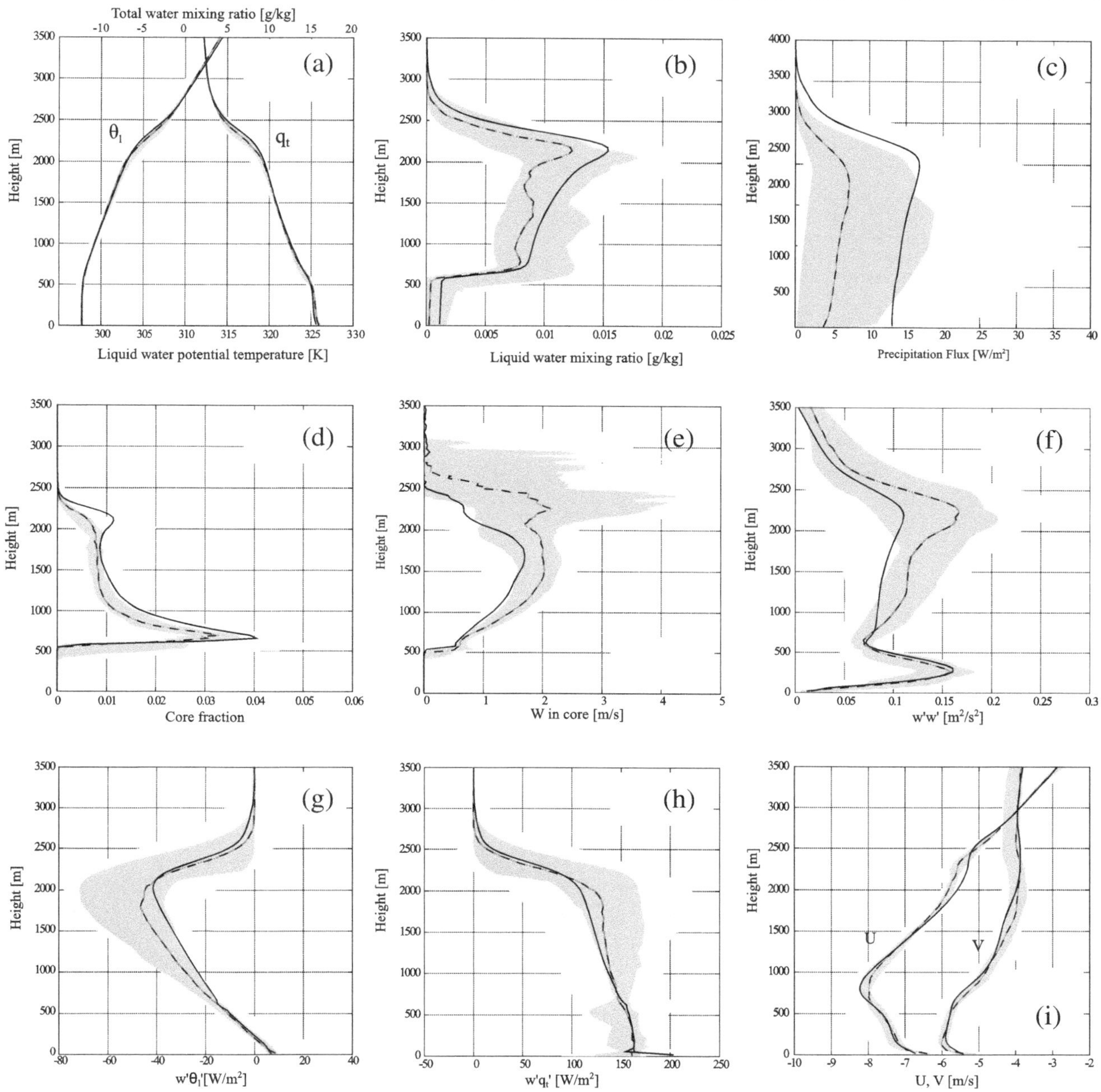

Fig. 2 Comparison of vertical profiles between SCALE-LES and a previous intercomparison study. Horizontally averaged profile of the **a** liquid water potential temperature (θ_l) and total water mixing ratio (q_t), **b** liquid water mixing ratio (q_l), **c** precipitation flux, **d** cloud fraction, **e** vertical velocity in cloud core, **f** variance of resolved w', **g** $w'\theta_l'$, **h** $w'q_t'$, and **i** horizontal wind velocity, averaged during the last 4 h. The *solid line* indicates the results of SCALE-LES. The *dashed line*, *heavy gray shading*, and *light gray shading* indicate the median, range between the first and third quartiles, and range between the maximum and minimum values, respectively, of the previous intercomparison study (van Zanten et al. 2011)

The impacts of cloud microphysics appear not only on the precipitation and vertical distribution of liquid water, but also the turbulent properties such as the turbulence kinetic energy (TKE), the variance in resolved vertical velocity (w'), and boundary layer top height. The boundary layer top height in the two-moment bulk scheme is the highest among the three schemes, whereas the one-moment bulk scheme simulates the lowest boundary layer height. This trend in boundary layer height continues to the end of the simulation time, as shown in Fig. 4a, and the difference among the schemes gradually becomes large. The variance in w' (Fig. 4c) and TKE (Fig. 4d) in the two-moment bulk scheme attributes a larger value to the upper layer. By contrast, those in the one-moment scheme are smaller in the upper layer. Vertically integrated TKE tends to be large and small in the

Table 1 Computational performance of SCALE-LES

Number of core	16	2048	8192	32,768	131,072	663,552[a]
(Number of MPI process)	(2)	(256)	(1024)	(4096)	(16,384)	(82,944)
Elapsed time (s step^{-1})	2.528	2.172	2.443	2.017	1.995	2.113
Performance efficiency (%)	5.5	6.3	5.6	6.8	6.9	6.5

Elapsed time (s step^{-1}) and performance efficiency (%) of SCALE-LES measured using the two-moment bulk microphysical scheme. The performance was measured through an experiment in which the number of grids for each MPI process was the same (weak scaling test) on the K computer

[a]This test used all nodes of the K computer

two-moment and the one-moment bulk schemes, respectively (Fig. 4b).

From the results shown above, we concluded that the turbulent properties of the boundary layer as well as the cloud microphysical properties of cumulus are significantly affected by the different microphysical schemes when other components are unchanged. These results support the proposal of van Zanten et al. (2011) that the use of different cloud microphysical schemes is one of the main reasons for the diversity among LES models. We have confidence in this conclusion because direct effects other than the cloud microphysical schemes were excluded in our experiments.

Reasons for the differences among the results of the three schemes

The impacts of the cloud microphysical schemes are clearly indicated by the differences in the boundary layer top height, vertical distribution of liquid water, and the precipitation flux, as shown in the previous section. The reasons for these differences will be discussed in this section.

Since the impacts of each cloud microphysical scheme originate from the expression of the liquid water in each scheme, an examination of the tendency of q_l and potential temperature (θ) in the cloud microphysical process

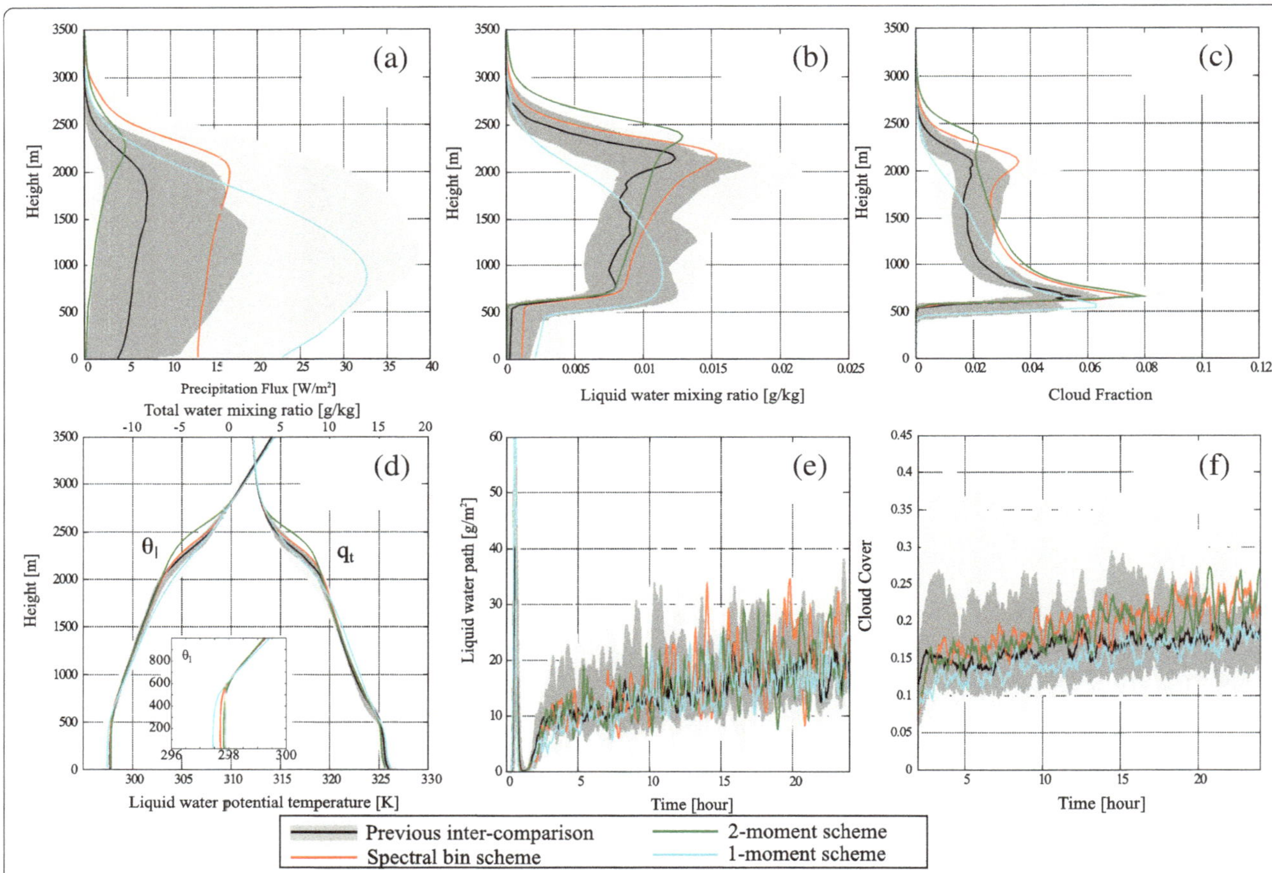

Fig. 3 Comparison of the three cloud microphysical schemes. Horizontally averaged profile of the **a** precipitation flux, **b** liquid water mixing ratio, **c** cloud fraction, **d** total water mixing ratio and liquid water potential temperature, and time evolution of **e** the liquid water path and **f** cloud cover. The *red*, *green*, and *sky-blue lines* show results of the spectral bin scheme, the two-moment bulk scheme, and the one-moment bulk scheme, respectively, and the *black line*, *dark gray shading*, and *light gray shading* indicate the median, range between the first and third quartiles, and range between the maximum and minimum values, respectively, of the previous intercomparison study (van Zanten et al. 2011). The *small figure* in **d** indicates the extension of the profile of liquid water potential temperature below 1000 m

Table 2 Comparison of surface precipitation flux. The surface precipitation flux averaged over the whole calculation domain during the last 4 h of calculation

Scheme	One-moment bulk	Two-moment bulk	Spectral bin
Precipitation flux ($W\ m^{-2}$)	22.89	0.174	13.17

is helpful for understanding the differences among the results. We first investigate the difference during $t = 3$–4 h of the calculation, because the effects of feedbacks of cloud physics to dynamics is not large and it is easy to interpret the difference. The tendencies at each height averaged during $t = 3$ h to $t = 4$ h are shown in Fig. 5a, b. The results from $t = 0$ h to $t = 3$ h were removed to ignore the effects of spin-up. The effect of sedimentation was omitted from the tendencies. The generation of liquid water at the cloud base in the one-moment bulk scheme is more active than that in the others (Fig. 5a). This is attributed to the difference in the mechanism for generating cloud particles. In the one-moment bulk scheme, newly generated cloud particles are calculated by saturation adjustment, whereas in the other schemes, they are calculated based on Eq. (4). Because the saturation adjustment does not permit supersaturation, it can generate larger amounts of liquid water at the cloud base than that can be generated by the scheme based on Eq. (4), which allows for supersaturation. This large amount of liquid water generation results in a large heat release at the cloud base (Fig. 5b) and subsequently a strong vertical velocity (Fig. 5c).

In addition to the large amount of liquid water generated in the one-moment bulk scheme, it is clear from the particle size distribution that the conversion from cloud to rain is fast. Size distributions at the lower part of cloud (i.e., $z = 1000$ m) are shown in Fig. 5d. The generation of drizzle and raindrops (i.e., particles over 40 μm in radius) in the lower part of the cloud is active in the one-moment bulk scheme, whereas small cloud

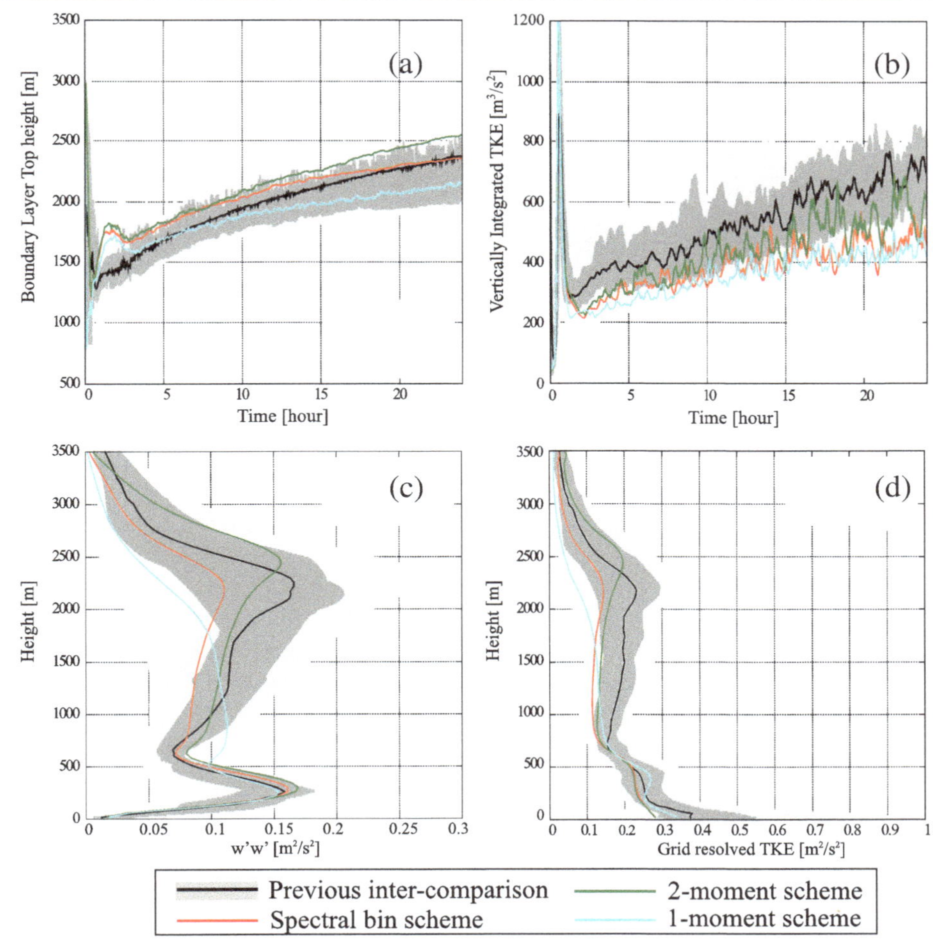

Fig. 4 Comparison of turbulence properties between the three schemes. Time evolution of the **a** boundary layer top height and **b** vertically integrated turbulence kinetic energy (grid resolved + sub-grid scale) and horizontally averaged profile of **c** the variance of resolved w' and **d** grid-resolved turbulence kinetic energy averaged during the last 4 h of calculation

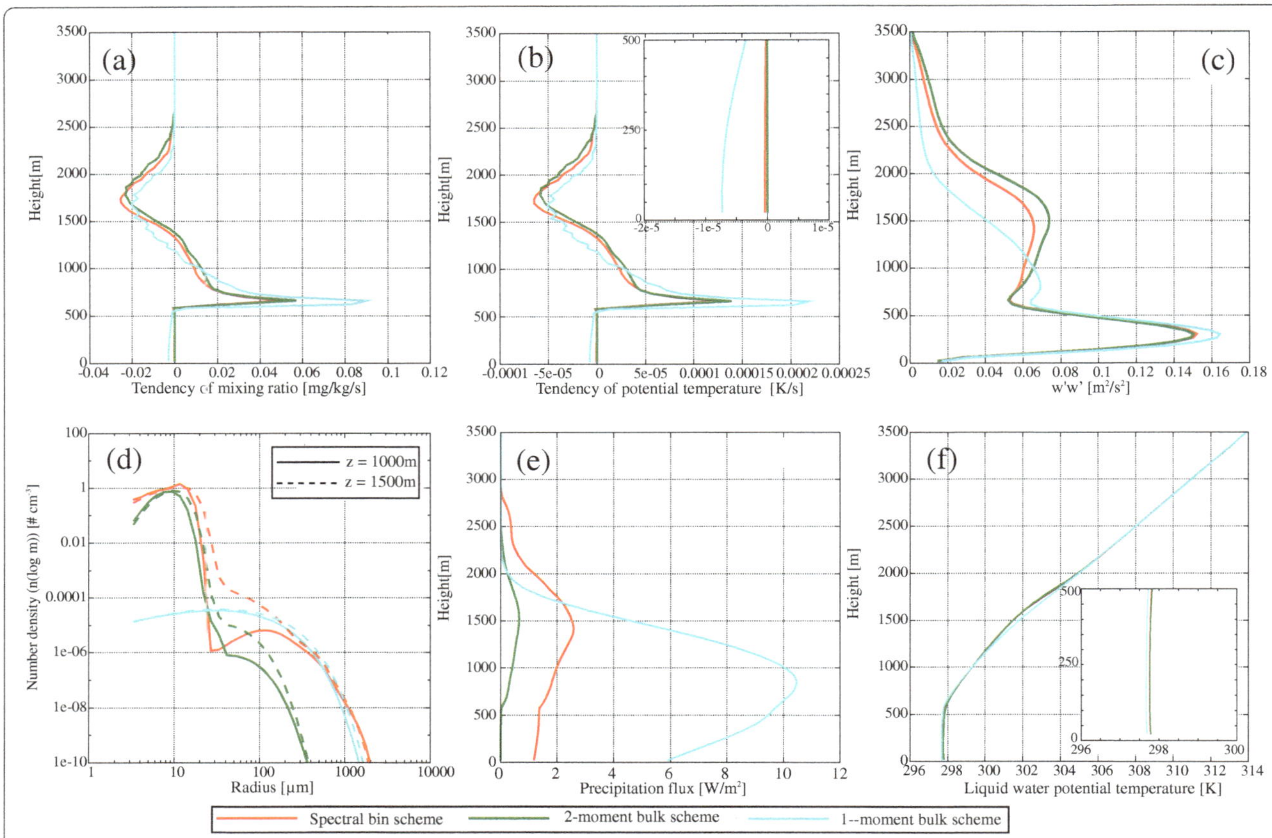

Fig. 5 Profiles and size distribution function averaged during $t = 3$~4 h. Horizontally averaged profile of the **a** tendency of the liquid water, **b** tendency of potential temperature, **c** variance of w', **e** precipitation flux, and **f** liquid water potential temperature averaged during $t = 3$ to 4 h. **d** Number density distribution (n (log m)) averaged over the whole cloudy grid at (*solid*) $z = 1000$ m and (*dashed*) $z = 1500$ m, where n and m are the number concentration and mass of liquid water, respectively. The *red*, *green*, and *sky-blue lines* show results of the spectral bin, two-moment bulk, and one-moment bulk schemes, respectively. The extended figures in **b** and **f** show the tendencies of potential temperature and liquid water potential temperature below 500 m, respectively

particles are dominant in the two-moment bulk schemes. This indicates that the conversion from cloud to rain in the one-moment bulk scheme is faster than that in the others, and larger numbers of raindrops are generated in the lower part of the cloud by the one-moment bulk scheme.

The generation of large raindrops leads to the fast terminal velocity of the hydrometeors and a large precipitation flux in the one-moment bulk scheme (Fig. 5e), which, in turn, leads to the large precipitation flux at the surface and the fast onset of surface precipitation. Figure 5a, b also show that the peak negative tendency of q_l and θ near the cloud top ($z \sim 1700$ m), which corresponds to the evaporation of cloud droplets at the top of the boundary layer, is located in a lower layer in the one-moment scheme than in the others. This indicates that the large precipitation volume and large cloud size in the one-moment bulk scheme restrain the cloud particles from reaching the upper layer. Therefore, the boundary layer top height of the one-moment bulk scheme becomes lower (Fig. 4a).

The large amount of raindrops in the one-moment scheme creates feedback for the thermodynamic structure below the cloud. The water loading due to raindrops leads to active evaporative cooling below the cloud (shown in the negative tendency of q_l and θ below the cloud as shown in Fig. 5a, b and the lower potential temperature below the cloud shown in Fig. 5f). This evaporative cooling stabilizes the boundary layer and suppresses the heat transfer from the ground to the upper part of the boundary layer, which is shown in the fact that the positive tendency of θ in the one-moment bulk scheme does not reach $z > 1200$ m but reaches $z \sim 1400$ m in the other scheme. This supports Stevens et al. (1998), who indicated that precipitation suppresses cloud growth and entrainment. This suppression can limit the development of the boundary layer and results in a more stable boundary layer.

The difference between the two-moment bulk and the spectral bin schemes in the tendency of q_l and θ are relatively small. However, the difference in size distribution function between the two schemes is clearly apparent (Fig. 5d). A larger amount of raindrops ($r > 100$ μm) were present in the spectral bin scheme than in the case in the two-moment bulk scheme. It is indicated that the conversion from cloud droplets to raindrops is slow in the two-moment bulk scheme. Because the amount of large raindrops is small in the two-moment bulk scheme (shown as green lines in Fig. 5d), liquid water is carried to the upper layer more easily than in the other schemes. The liquid water evaporates at the top of the boundary layer. This is shown by the negative tendency of q_l, and by θ locating in a higher layer in the two-moment bulk scheme than in the others (Fig. 5a, b).

The presence of larger particles in the spectral bin scheme subsequently increases the rate of collisions with other particles, which leads to more rapid growth of particles and earlier precipitation in the spectral bin scheme than in the two-moment bulk scheme. This provides feedback due to the large amount of precipitation, which is the same feedback that occurs in the one-moment bulk scheme.

With the feedback, the differences among the three schemes increase with the integration time. The difference in the boundary layer top height in the three schemes becomes large as the integration time increases (Fig. 4a), and the difference in the liquid water potential temperature below clouds during the last 4 h (Fig. 3d) is larger than at $t = 3–4$ h (Fig. 5f). The differences in the tendencies of θ and q_l shown above also become clear (figure not shown).

In summary, the one-moment bulk scheme creates a larger amount of precipitation, because the saturation adjustment was adopted in the one-moment bulk scheme, and raindrops are subsequently produced by the fast conversion from cloud to raindrops. This results in earlier onset and larger amounts of surface precipitation. The evaporative cooling by raindrops, which occurs actively below the cloud, stabilized the boundary layer in the one-moment bulk scheme.

The two-moment scheme creates raindrops more slowly, resulting in a smaller amount of precipitation compared with the other schemes. The smaller amount of precipitation results in an active latent heat release in the higher layer (shown in Fig. 5b) and a high boundary layer. The spectral bin scheme shows an intermediate rate of conversion and creates an intermediate amount of precipitation, with values between those of the two-moment and one-moment bulk schemes.

The large number of cloud particles newly generated at the cloud bottom by saturation adjustment, the difference in the speed of conversion from cloud to rain, and the difference in the timing of the surface precipitation all originate from the variety of microphysical schemes used in this study. The differences in each scheme would not always appear when the results of other one-moment, two-moment, and spectral bin schemes are compared. By contrast, the evaporative cooling and subsequent stabilization of the sub-cloud layer and the suppression of the development of boundary layer height, which appeared in the results of the one-moment scheme used in this study, can be expected if the fast conversion from cloud to rain or the active generation of cloud particles at the cloud base occurs as a result of natural phenomena, regardless of which scheme is used. The active latent heat release and high boundary layer, which appeared in the two-moment scheme used in this study, can also be expected. The results are commonly expected for trade wind cumulus.

Speed of conversion from cloud to rain

The analyses in the previous sections hint that the performance of the bulk microphysical schemes could be modified by changing the nucleation schemes and the conversion speed from cloud to rain. In the bulk scheme, autoconversion and accretion are the main processes involved in the conversion from cloud to rain. To obtain information for the modification of the parameterization of these two processes, a comparison of the autoconversion and accretion rates among the three schemes is helpful. Figure 6 shows the autoconversion and accretion rates

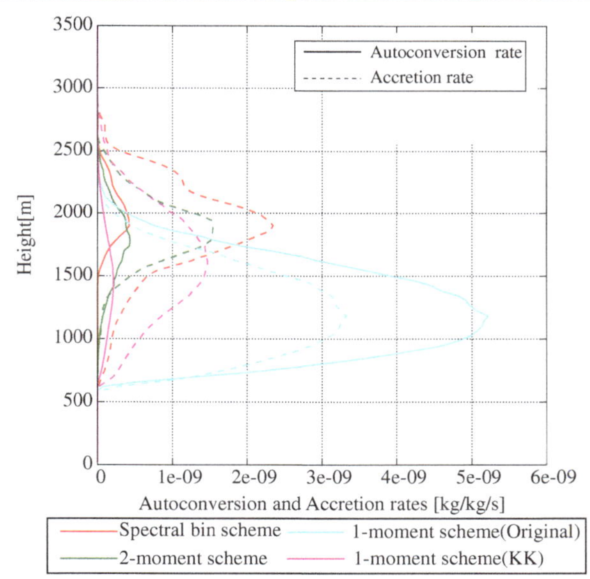

Fig. 6 Autoconversion and accretion rate averaged during $t = 3$~4 h. The horizontally averaged (*solid line*) autoconversion rate and (*dashed line*) accretion rate simulated by the spectral bin scheme (*red*), two-moment scheme (*green*), one-moment scheme (*sky blue*), and one-moment scheme with the Khairoutdinov and Kogan (2000) scheme (*pink*), averaged during $t = 3$~4 h

averaged during $t = 3{\sim}4$ h. The autoconversion rate of the spectral bin scheme is regarded as the rate of increasing mass of raindrops (defined as liquid particles whose radius is larger than 40 μm) generated by the coagulation between cloud particles (defined as liquid particles whose radius is smaller than 40 μm). The accretion rate of the spectral bin scheme is determined as the increasing rate of mass due to the coagulation between cloud particles and raindrops. Figure 6 shows that the autoconversion rate of the one-moment scheme is about 15 times larger than that of the spectral bin scheme. By contrast, the accretion rate of the two-moment scheme is 1.5~2 times smaller than that of the spectral bin scheme. Based on these results, the sensitivity of these two processes in each scheme was investigated and is discussed in the next section.

Sensitivity of the autoconversion to the one-moment bulk scheme

As shown above, the one-moment bulk scheme overestimates the conversion speed from cloud droplet to raindrop. We first investigated the difference between the original autoconversion scheme of Tomita (2008) and that of Khairoutdinov and Kogan (2000) shown in Eq. (6). The results of the one-moment bulk scheme with the Khairoutdinov and Kogan (2000) scheme are more similar to the results of the spectral bin scheme and the previous study than those calculated by Eq. (5) (Fig. 7). This indicates that the conversion speed calculated by Eq. (5) is too fast for shallow clouds because the validity of the one-moment bulk scheme with Eq. (5) was confirmed through experiments with deep convective clouds (Tomita 2008).

In addition to the experiment with the Khairoutdinov and Kogan (2000) scheme, other sensitivity experiments

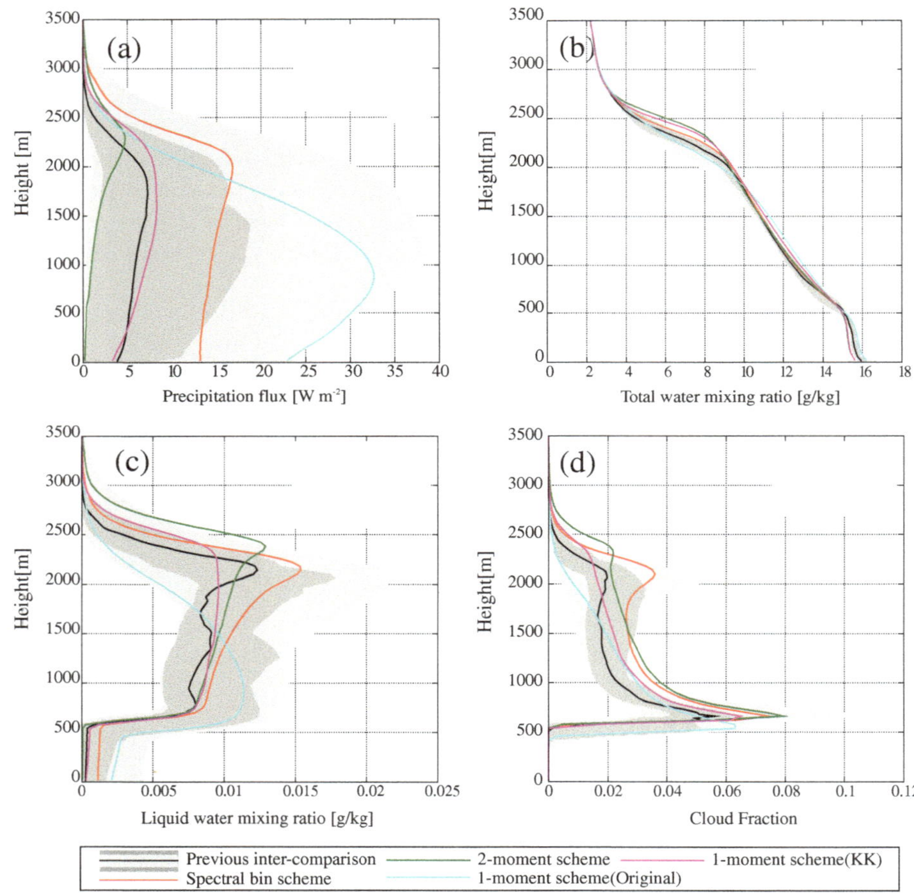

Fig. 7 Cloud microphysical properties simulated by Khairoutdinov and Kogan's auto conversion scheme. Horizontally averaged profile of the **a** precipitation flux, **b** total water mixing ratio (q_t), **c** liquid water mixing ratio, and **d** cloud fraction averaged during the last 4 h. The *red*, *green*, *sky-blue*, and *pink lines* show results by the spectral bin scheme, the two-moment bulk scheme, the one-moment bulk scheme, and one-moment bulk scheme using the Khairoutdinov and Kogan (2000) autoconversion rate, respectively. The *black line*, *dark gray shading*, and *light gray shading* indicate the median, range between the first and third quartiles, and range between the maximum and minimum values, respectively, of the previous intercomparison study (van Zanten et al. 2011)

were conducted by reducing the autoconversion rate. Based on the analyses in Fig. 6, a 0.067-fold (i.e., 1/15) smaller autoconversion rate than the default value was used for the sensitivity experiment. Another sensitivity experiment with an accretion rate that was 0.067-fold smaller than the default value was also conducted. Figure 8 shows the profile of q_l and the precipitation flux, which were calculated using the smaller autoconversion and accretion rate. It can be seen that a small autoconversion rate results in liquid water locating in the upper layer, whereas it does not locate in the upper layer when the accretion rate was reduced to the same extent as the autoconversion rate. Hence, the autoconversion process is more sensitive to the production of raindrops.

Sensitivity of the accretion to the two-moment bulk scheme

Contrasting the one-moment bulk scheme, the two-moment bulk scheme underestimates the conversion speed from cloud to raindrop. We speculate that the faster conversion from cloud droplet to raindrop in the two-moment scheme produced similar results using the spectral bin scheme. Sensitivity experiments were conducted by changing both the autoconversion and accretion rates of the two-moment scheme. Based on the analyses in Fig. 6, a twofold increase in the accretion rate was used, and a twofold increase of the autoconversion rate was also used for the sensitivity experiment. Figure 9 shows the results of these experiments. In the two-moment bulk scheme, both the precipitation rate and q_l increase when the autoconversion rate is doubled. However, doubling the autoconversion rate does not result in a peak value of q_l and a precipitation flux in the lower layer as simulated by the one-moment bulk scheme. By contrast, the precipitation flux simulated when the accretion rate is doubled is considerably larger than both the default and when the autoconversion rate is doubled. This indicates that the accretion process was the major contributor to the creation of liquid water in the lower layer in the two-moment bulk scheme.

In short, the sensitivity of the accretion and the autoconversion processes to the cloud microphysical properties differ between the schemes.

Component level intercomparison

In this study, we investigated the exact effects of the various cloud microphysical schemes on model simulations of trade wind cumulus. If we change components other than the cloud microphysical scheme (e.g., dynamical core, turbulence scheme, advection scheme), the response of the cloud microphysical properties would change as suggested by van Zanten et al. (2011). It is necessary to conduct sensitivity experiments by changing each component while keeping all other components fixed, as we did when targeting the cloud microphysical scheme in this study. We refer to these sensitivity experiments as a "Component level intercomparison".

Grabowski (2014) suggested a piggyback approach to better understand the exact effects of cloud microphysics and the interaction between microphysics and dynamics. This approach can be also applied to the other components. Using a component level intercomparison and the piggyback approach, we can discuss the effects

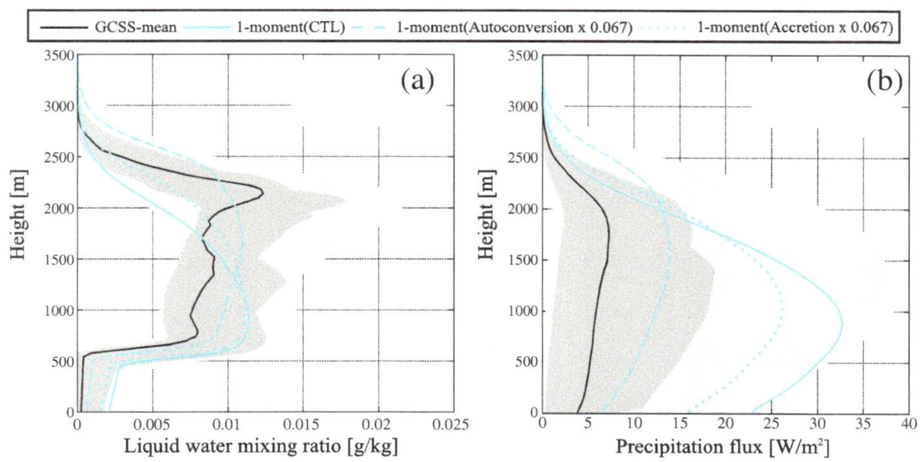

Fig. 8 Cloud microphysical properties simulated in the sensitivity experiment with a varying conversion ratio in the one-moment scheme. Horizontally averaged profile of the **a** liquid water mixing ratio (q_l) and **b** precipitation flux averaged during the last 4 h. The *solid sky-blue*, *dashed sky-blue*, and *dotted sky-blue lines* show the results using the one-moment bulk scheme with the default autoconversion rate, one-moment bulk scheme with an autoconversion rate 0.067-fold (i.e., 1/15) smaller, and one-moment bulk scheme with accretion ratio 0.067-fold smaller, respectively. The *black line*, *dark gray shading*, and *light gray shading* indicate the median, range between the first and third quartiles, and range between the maximum and minimum values, respectively, for a previous intercomparison study (van Zanten et al. 2011)

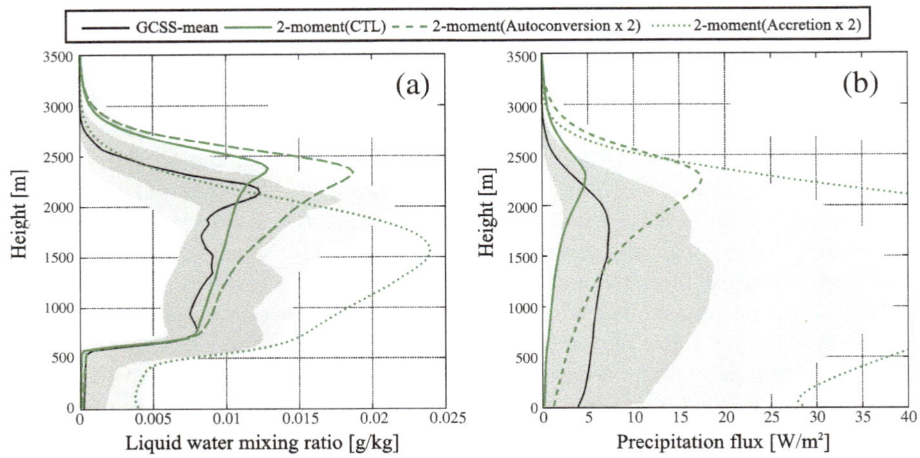

Fig. 9 Cloud microphysical properties simulated in the sensitivity experiment with a changing conversion ratio in the two-moment scheme. The horizontally averaged profile of the **a** liquid water mixing ratio (q_l) and **b** precipitation flux averaged during the last 4 h. The *solid green*, *dashed green*, and *dotted green lines* show the results using the two-moment bulk scheme with the default autoconversion rate, two-moment bulk scheme with an autoconversion rate twice as large as the default, and two-moment bulk scheme with an accretion ratio twice as large as the default. The *black line*, *dark gray shading*, and *light gray shading* indicate the median, range between the first and third quartiles, and the range between the maximum and minimum values, respectively, for a previous intercomparison study (van Zanten et al. 2011)

of each component separately and obtain knowledge that would improve the parameterization of global scale models or regional models with coarse resolution.

Conclusions

In this study, we developed a large eddy simulation model (SCALE-LES), which excludes approximations and implicit schemes as much as possible. The results of a benchmark test indicated that SCALE-LES effectively reproduces the trade wind cumuli simulated in a previous LES intercomparison study (van Zanten et al. 2011). Using SCALE-LES, we investigated the impacts of cloud microphysical schemes on shallow cumulus and investigated which processes were critical for the diversity observed in previous LES intercomparison studies. Three types of cloud microphysical scheme, the one-moment bulk scheme of Tomita (2008), the two-moment bulk scheme of Seiki and Nakajima (2014), and the one-moment spectral bin scheme of Suzuki et al. (2010), were implemented with SCALE-LES for the sensitivity experiments.

The results indicated that the precipitation at the surface increases, in order, from the two-moment bulk scheme to the spectral bin scheme and the one-moment bulk scheme. Surface precipitation begins first in the

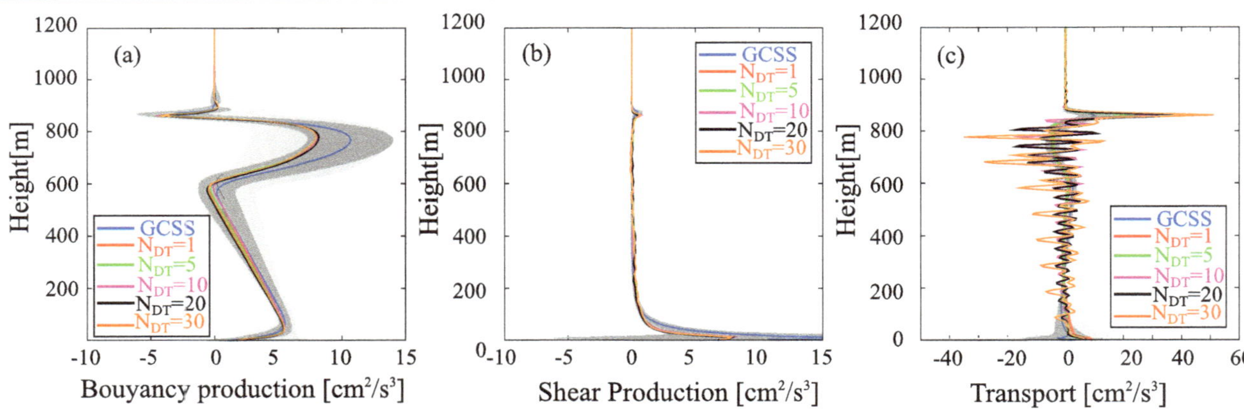

Fig. 10 Cloud microphysical properties in the sensitivity experiment of N_{DT} ($=\Delta t_{microphy}/\Delta t_{dyn}$). Hourly averaged profile of the **a** buoyancy production, **b** shear production, and **c** transport terms averaged over the entire calculation domain during the last hour of each simulation. The *red*, *green*, *pink*, *black* and *orange lines* represent the results of N_{DT} = 1, 5, 10, 20, and 30, respectively. The *blue line*, *dark gray shading*, and *light gray shading* indicate the mean, standard deviation, and the range between the maximum and minimum values, respectively, from a previous intercomparison stucy (Stevens et al. 2005)

Table 3 List of non-dimensional coefficient for sensitivity experiment

Non-dimensional coefficient (γ)	$\gamma = 10^{-3}$	$\gamma = 10^{-5}$	$\gamma = 10^{-7}$
The strength of the numerical filter	1.25×10^5	1.25×10^3	1.25×10^1

The value of the numerical filter ($m^4\ s^{-1}$) for each experiment

one-moment bulk scheme, followed in order by the spectral bin and two-moment bulk schemes. These results support the suggestion of a previous intercomparison study (van Zanten et al. 2011)

Our analyses confirmed that the differences between the schemes were derived mainly from the generation of cloud particles and the speed of conversion from cloud droplets to raindrops. The differences in the two processes originated from the differences in the microphysical schemes used. By contrast, the phenomena generated by this variety, i.e., evaporative cooling and stabilization below the cloud and a low boundary layer, active latent heat release, and a high boundary layer can be expected regardless of the scheme used, if the active conversion from cloud to rain and the active generation of new cloud particles occur in nature.

The sensitivity of the autoconversion and accretion processes to the cloud microphysical properties simulated by the bulk microphysical schemes was also investigated. In the two-moment bulk scheme the accretion process was more sensitive to the cloud microphysical properties, whereas the autoconversion was more sensitive in the one-moment bulk scheme. These results indicate that the tuning method of the microphysical scheme differs from scheme to scheme, and a component level intercomparison is useful to obtain the exact method for each scheme.

Appendix 1

Sensitivity of the ratio of the physical time step to the dynamical time step. In this study, the time step of dynamics (Δt_{dyn}), cloud microphysics ($\Delta t_{microphy}$), and other physics (Δt_{phy}) were set as 0.05, 0.15, and 0.5 s, respectively. The time step of dynamics was determined by the Courant–Friedrichs–Lewy (CFL) condition for the acoustic wave. The time step for physics (except for cloud microphysics) was set as $10 \times \Delta t_{dyn}$ based on the sensitivity experiment (Nishizawa et al., 2015). The $\Delta t_{microphy}$ was determined in sensitivity experiments examining the ratio of $\Delta t_{microphy}$ to Δt_{dyn}. The results of the sensitivity experiments examining the ratio N_{DT} ($=\Delta t_{microphy}/\Delta t_{dyn}$) are shown in this section. For this sensitivity test, the experimental setup of the second Dynamics and Chemistry of Marine Stratocumulus Research Flight 1 (DYCOMS-II RF01) (Stevens et al. 2005) was used. In this study, the experimental setup of the RICO study was used in most cases, but the effects of the acoustic wave appeared more clearly in the DYCOMS-II case (figure not shown). Consequently, the ratio (N_{DT}) was determined using the DYCOMS-II RF01 experimental setup. For this sensitivity experiment, Δt_{dyn} was set as 0.01 s and N_{DT} was swept from 1 to 30 (i.e., $\Delta t_{microphy}$ was set from 0.01 to 0.3 s).

The results of the sensitivity experiment indicated that the effects of N_{DT} were mostly small (figures not shown), except for the TKE budget. Figure 10 shows the profile of the buoyancy production term, shear production term, and transport term of TKE production. The transport term was quite noisy when N_{DT} was large ($N_{DT} > 10$). Noise is also present in the shear production term. This noise was derived from the acoustic wave that was generated at every time step of the microphysical processes. When N_{DT} was

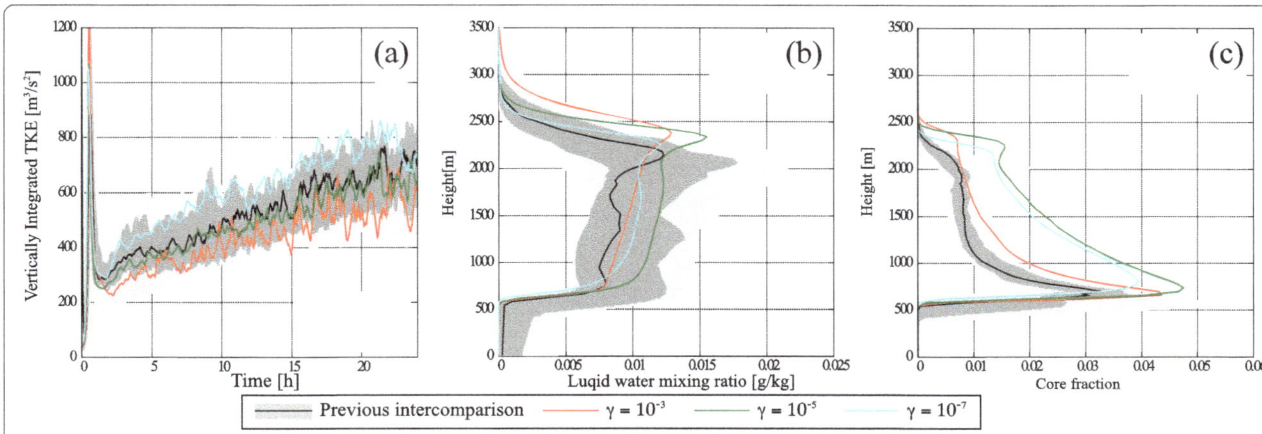

Fig. 11 Results of a sensitivity experiment of the strength of numerical diffusion. Time evolution of the **a** vertically integrated turbulence kinetic energy averaged over the whole calculation domain and the profile of the **b** liquid water mixing ratio and **c** cloud fraction averaged over the whole calculation domain during the last 4 h. The *red*, *green*, and *sky-blue line* shows the results with the coefficient (γ) as 10^{-3}, 10^{-5}, and 10^{-7}, respectively. The *black line*, *thick gray shade*, and *thin gray shade* indicate the median, range between the first and third quartiles, and range between the maximum and minimum values of a previous intercomparison study, respectively (van Zanten et al. 2011)

large, the variation of θ at each step of the microphysical processes was also large. The large variation in θ generates an acoustic wave, which produced the noise. This indicates that N_{DT} should be smaller than 5 to render the effects of the acoustic wave negligibly. From these results and for safety, N_{DT} was set as 3 for the RICO experiment (i.e., $\Delta t_{microphy} = 3 \times \Delta t_{dyn}$).

Appendix 2

Numerical filter. This section describes the sensitivity experiments of the strength of the numerical diffusion. The nth ordered superviscosity/diffusion is defined as:

$$\frac{\partial}{\partial x}\left[\nu\rho\frac{\partial^{n-1}f}{\partial x^{n-1}}\right] \quad f\in\{u,v,w,\theta,q_s\}.$$

$$\frac{\partial}{\partial x}\left[\nu\frac{\partial^{n-1}\rho}{\partial x^{n-1}}\right]$$

The ν is written as:

$$\nu = (-1)^{\frac{n}{2}+1}\gamma\frac{\Delta x^n}{2^n\Delta t}$$

where γ is a non-dimensional coefficient, Δx is the grid spacing, and n is the order of the numerical filter. This study adopted a fourth-order numerical diffusion (i.e., $n = 4$) to dampen the artificial noise. n was set as 4 because of the computational efficiency.

To investigate the sensitivity of the strength of the numerical diffusion, simulations of the RICO study were conducted while changing the strength of the numerical diffusion. Based on the rough estimation of Nishizawa et al. (2015), the non-dimensional coefficient (γ) should be smaller than $O(10^{-3})$–$O(10^0)$ for $n = 4$. Thus, γ was changed from 10^{-3} to 10^{-7}. The strength of the numerical diffusion for each γ is shown in Table 3. The two-moment bulk scheme was used for the sensitivity experiment. As a result of the numerical diffusion, one-dimensional sinusoidal two-grid noise will decay to $1/e$ with a $1/\gamma$ time step.

Figure 11 shows the results of the sensitivity experiment. The vertically integrated TKE increased with a weakening of the numerical filter, even though the TKE of all experiments was located in the range between the maximum and minimum of the previous studies. This is because a strong numerical diffusion can efficiently dampen small-scale turbulence. As well as the TKE, the liquid water mixing ratio of all the sensitivity experiments was also located in the range of previous intercomparison studies regardless of the strength of the numerical diffusion. However, the cloud fraction simulated with a small numerical diffusion (i.e., $\gamma \le 10^{-5}$) was much larger than that of the intercomparison study, and the cloud fraction was completely outside the range of the inter comparison studies. The cloud fraction, core fraction, and variance of w' were also outside the range of the inter comparison studies when γ was smaller than 10^{-5} (figure not shown). The temporal evolution of the cloud fraction indicates that the large cloud fraction, with small numerical diffusion, seems to originate from artificial noise. However, it is difficult to identify the reason for the large cloud fraction being artificial noise. The same experiment must be conducted with a fine grid resolution to divide all of the elements of the wave into a physically meaningful wave and artificial noise. However, computational limitations prevented us from conducting these experiments. Consequently, γ was determined using the results of the intercomparison study (van Zanten et al. 2011) as a reference solution, and γ was set as 10^{-3} (i.e., the strength of the numerical diffusion is 1.25×10^5 m^4 s^{-1}).

Abbreviations

DYCOMS: Dynamics and Chemistry of Marine Stratocumulus; GCSS: GEWEX Cloud System Study; GEWEX: Global Energy and Water Exchange project; KiD: the kinematic driver; LES: large eddy simulation; LWP: liquid water path; MPI: Message Passing Interface; RICO: Rain in Cumulus over the Ocean; SCALE: Scalable Computing for Advanced Library and Environment; TKE: turbulence kinetic energy.

Competing interests

The authors declare that they have no competing interests.

Authors' contributions

YS implemented the cloud microphysical schemes in the SCALE library, designed this study, conducted the numerical simulations, analyzed the results of the simulations, and developed the manuscript. SN, HY, and YM developed the main frames of the SCALE library. YK collaborated with the corresponding author in the creation of the manuscript. HT proposed the development of the SCALE library. All authors read and approved the final manuscript.

Acknowledgement

Part of the results is obtained by the K computer at the RIKEN Advanced Institute for Computational Science. This work was supported by FOCUS Establishing Supercomputing Center of Excellence. SCALE-LES developed by Team-SCALE of the RIKEN Advanced Institute for Computational Sciences. The data from GCSS intercomparison studies used in several figures were downloaded from http://www.knmi.nl/samenw/rico/.

References

Ackerman AS, van Zanten MC, Stevens B, Savic-Jovcic V, Bretherton CS, Chlond A, Golaz J-C, Jiang H, Khairoutdinov M, Krueger SK, Lewellen DC, Lock A, Moeng C-H, Nakamura K, Petters MD, Snider JR, Weinbrecht S, Zulauf M (2009) Large-eddy simulations of a drizzling, stratocumulus-topped marine boundary layer. Mon Weather Rev 137:1083–1110. doi:10.1175/2008MWR2582.1

Berry EX (1968) Modification of the warm rain process. In: Proceeding of First Conference on Weather Modification., pp 81–85

Bretherton CS, Park S (2009) A new moist turbulence parameterization in the Community Atmosphere Model. J Clim 22:3422–3448. doi:10.1175/2008JCLI2556.1

Brown AR, Derbyshire SH, Mason PJ (1994) Large-eddy simulation of stable atmospheric boundary layers with a revised stochastic subgrid model. Q J R Meteorol Soc 120:1485–1512. doi:10.1002/qj.49712052004

Chepfer H, Bony S, Winker D, Chiriaco M, Dufresne JL, Sèze G (2008) Use of CALIPSO lidar observations to evaluate the cloudiness simulated by a climate model. Geophys Res Lett 35:1–6. doi:10.1029/2008GL034207

Considine G, Curry JA, Wielicki B (1997) Modeling cloud fraction and horizontal variability in marine boundary layer clouds. J Geophys Res 102:13517. doi:10.1029/97JD00261

Grabowski WW (2014) Extracting microphysical impacts in large-eddy simulations of shallow convection. J Atmos Sci 71:4493–4499. doi:10.1175/JAS-D-14-0231.1

Iguchi T, Nakajima T, Khain AP, Saito K, Takemura T, Suzuki K (2008) Modeling the influence of aerosols on cloud microphysical properties in the east Asia region using a mesoscale model coupled with a bin-based cloud microphysics scheme. J Geophys Res 113:D14215. doi:10.1029/2007JD009774

Kain, JS (2004) The Kain–Fritsch convective parameterization: An update. J Appl Meteorol. doi:10.1175/1520-0450(2004)043<0170:TKCPAU>2.0.CO;2

Khain AP, Sednev I (1996) Simulation of precipitation formation in the Eastern Mediterranean coastal zone using a spectral microphysics cloud ensemble model. Atmos Res 43:77–110. doi:10.1016/S0169-8095(96)00005-1

Khairoutdinov M, Kogan Y (2000) A new cloud physics parameterization in a large-eddy simulation model of marine stratocumulus. Mon Weather Rev 128:229–243. doi:10.1175/1520-0493(2000)128<0229:ANCPPI>2.0.CO;2

Naud CM, Del Genio AD, Bauer M, Kovari W (2010) Cloud vertical distribution across warm and cold fronts in cloudsat-CALIPSO data and a general circulation model. J Clim 23:3397–3415. doi:10.1175/2010JCLI3282.1

Nishizawa S, Yashiro H, Sato Y, Miyamoto Y, Tomita H (2015) Influence of grid aspect ratio on planetary boundary layer turbulence in large-eddy simulations. Geosci Model Dev Discuss 8:6021–6094. doi:10.5194/gmdd-8-6021-2015

Posselt R, Lohmann U (2008) Introduction of prognostic rain in ECHAM5: design and single column model simulations. Atmos Chem Phys 8:2949–2963. doi:10.5194/acpd-7-14675-2007

Pruppacher, HR, and Klett, JD, 1997, Microphysics of Clouds and Precipitation, 2nd ed., Kluwer Academic Publisher, Dordrecht, The Netherlands, 954pp.

Randall DA, Coakley JA, Lenschow DH, Fairall CW, Kropfli RA (1984) Outlook for research on subtropical marine stratification clouds. Bull Am Meteorol Soc 65:1290–1301. doi:10.1175/1520-0477(1984)065<1290:OFROSM>2.0.CO;2

Savic-Jovcic V, Stevens B (2008) The structure and mesoscale organization of precipitating stratocumulus. J Atmos Sci 65:1587–1605. doi:10.1175/2007JAS2456.1

Scotti A, Meneveau C, Lilly DK (1993) Generalized Smagorinsky model for anisotropic grids. Phys Fluids A Fluid Dyn 5:2306–2308. doi:10.1063/1.858537

Seifert A, Beheng KD (2001) A double-moment parameterization for simulating autoconversion, accretion and self collection. Atmos Res 59–60:265–281. doi:10.1016/S0169-8095(01)00126-0

Seiki T, Nakajima T (2014) Aerosol effects of the condensation process on a convective cloud simulation. J Atmos Sci 71:833–853. doi:10.1175/JAS-D-12-0195.1

Shipway BJ, Hill AA (2012) Diagnosis of systematic differences between multiple parametrizations of warm rain microphysics using a kinematic framework. Q J R Meteorol Soc 138:2196–2211. doi:10.1002/qj.1913

Siebesma A, Bretherton CS, Brown A, Chlond A, Cuxart J, Duynkerke P, Jiang H, Khairoutdinov M, Lewellen D, Moeng C-H, Sanchez E, Stevens B, Stevens DE (2003) A large-eddy simulation intercomparison study of shallow cumulus convection. J Atmos Sci 60:1201–1219. doi:10.1175/1520-0469(2003)060<1201:AALESIS>2.0.CO;B2

Stevens B, Cotton WR, Feingold G, Moeng C-H (1998) Large-eddy simulations of strongly precipitating, shallow, stratocumulus-topped boundary layers. J Atmos Sci 55:3616–3638. doi:10.1175/1520-0469(1998)055<3616:LESOSP>2.0.CO;2

Stevens B, Moeng C-H, Ackerman AS, Bretherton CS, Chlond A, de Roode S, Edwards J, Golaz J-C, Jiang H, Khairoutdinov M, Kirkpatrick MP, Lewellen DC, Lock A, Müller F, Stevens DE, Whelan E, Zhu P (2005) Evaluation of large-eddy simulations via observations of nocturnal marine stratocumulus. Mon Weather Rev 133:1443–1462. doi:10.1175/MWR2930.1

Stocker, TF, Qin D, Plattner G-K, Tignor M, Allen SK, Boschung J, Nauels A, Xia Y, Bex V, and Midgley PM (2013), IPCC, 2013: Climate Change 2013: The Physical Science Basis, Cambridge University Press, Cambridge, United Kingdom and New York, NY, USA.

Suzuki K, Nakajima T, Nakajima TY, Khain A (2006) Correlation pattern between effective radius and optical thickness of water clouds simulated by a spectral bin microphysics cloud model. SOLA 2:116–119. doi:10.2151/sola.2006-030

Suzuki K, Nakajima T, Nakajima TY, Khain AP (2010) A study of microphysical mechanisms for correlation patterns between droplet radius and optical thickness of warm clouds with a spectral bin microphysics cloud model. J Atmos Sci 67:1126–1141. doi:10.1175/2009JAS3283.1

Suzuki K, Nakajima T, Numaguti A, Takemura T, Kawamoto K, Higurashi A (2004) A study of the aerosol effect on a cloud field with simultaneous use of GCM modeling and satellite observation. J Atmos Sci 61:179–194. doi:10.1175/1520-0469(2004)061<0179:ASOTAE>2.0.CO;2

Tiedtke M (1993) Representation of clouds in large-scale models. Mon Weather Rev 121:3040–3061. doi:10.1175/1520-0493(1993)121<3040:ROCILS>2.0.CO;2

Tomita H (2008) New microphysical schemes with five and six categories by diagnostic generation of cloud ice. J Meteorol Soc Japan 86A:121–142. doi:10.2151/jmsj.86A.121

van Zanten MC, Stevens B, Nuijens L, Siebesma AP, Ackerman AS, Burnet F, Cheng A, Couvreux F, Jiang H, Khairoutdinov M, Kogan Y, Lewellen DC, Mechem D, Nakamura K, Noda A, Shipway BJ, Slawinska J, Wang S, Wyszogrodzki A (2011) Controls on precipitation and cloudiness in simulations of trade-wind cumulus as observed during RICO. J Adv Model Earth Syst 3:M06001. doi:10.1029/2011MS000056

Wang H, Feingold G (2009) Modeling mesoscale cellular structures and drizzle in marine stratocumulus. Part II: the microphysics and dynamics of the boundary region between open and closed cells. J Atmos Sci 66:3257–3275. doi:10.1175/2009JAS3120.1

Wang H, Feingold G, Wood R, Kazil J (2010) Modelling microphysical and meteorological controls on precipitation and cloud cellular structures in Southeast Pacific stratocumulus. Atmos Chem Phys 10:6347–6362. doi:10.5194/acp-10-6347-2010

Xue H, Feingold G, Stevens B (2008) Aerosol effects on clouds, precipitation, and the organization of shallow cumulus convection. J Atmos Sci 65:392–406. doi:10.1175/2007JAS2428.1

Yamaguchi T, Randall DA (2012) Cooling of entrained parcels in a large-eddy simulation. J Atmos Sci 69:1118–1136. doi:10.1175/JAS-D-11-080.1

Zalesak ST (1979) Fully multidimensional flux-corrected transport algorithms for fluids. J Comput Phys 31:335–362. doi:10.1016/0021-9991(79)90051-2

15

Diffusion-controlled growth and degree of disequilibrium of garnet porphyroblasts: is diffusion-controlled growth of porphyroblasts common?

Kazuhiro Miyazaki

Abstract

Rate-limiting processes and the degree of disequilibrium during metamorphic mineral growth are key controls on the rate of dehydration and hydration in the Earth's crust. This paper examines diffusion-controlled growth and the degree of disequilibrium of garnet porphyroblasts in the Tsukuba metamorphic rocks of central Japan. The analyzed porphyroblasts have irregular and branching morphologies with clear diffusional haloes, indicating that they grew in a diffusion-controlled regime. Mathematical analysis shows that the dominant wavelength of the interface of a garnet porphyroblast is dependent on the extent of supersaturation ($\Delta\zeta$), which is an index for the degree of disequilibrium. Using the calculated upper and lower limits of the dominant wavelength, the value of $\Delta\zeta$ is estimated to be 0.05×10^{-1}–0.16, which corresponds to a Gibbs free energy (ΔG_r) overstep of 0.9–27 kJ per mole of garnet (12 oxygen atoms) and a temperature overstep (ΔT) of 1.7–50 °C. Using the average value of the dominant wavelength, the following results are obtained: $\Delta\zeta = 0.15 \times 10^{-1}$, $\Delta G_r = 2.7$ kJ per mole of garnet, and $\Delta T = 5$ °C. These values bring into question the importance of diffusion-controlled growth of garnet porphyroblasts, as highly irregular and branching garnet porphyroblasts are rare in most metamorphic belts. After significant overstepping for the nucleation of garnet, the garnet porphyroblasts grow at a high degree of disequilibrium. However, a high degree of disequilibrium under diffusion-controlled growth would be characterized by diffusional instability. The results indicate that garnet porphyroblasts that lack an irregular and branching morphology may grow at a high degree of disequilibrium under interface-controlled growth, provided they are set in a medium where the diffusion and supply of constituent elements are sufficient, such as a sufficient volume of metamorphic fluid.

Keywords: Diffusion-controlled growth; Garnet; Metamorphism; Disequilibrium; Porphyroblast

Background

Rate-limiting processes and the degree of disequilibrium of a metamorphic reaction are crucial factors in the devolatilization and/or hydration of the Earth's crust. The release and consumption of fluids, such as water-rich fluids and melts, have a critical influence on the evolution of mountain building and the transport of energy and materials in the crust. Many studies have assessed the rate-limiting processes for metamorphic reactions (e.g., Ague and Carlson 2013, Walther and Wood 1984). The following three types of rate-limiting processes have been recognized (e.g., Fisher 1978): 1) interface-controlled, 2) diffusion-controlled, and 3) heat-flow-controlled. The formation of garnet is one of the most common devolatilization reactions in the crust (e.g., Ague and Carlson 2013), and diffusion-controlled processes are widely applied to garnet-forming reactions, including the nucleation and growth of the garnet (e.g., Carlson 1989) and Ostwald ripening (Miyazaki 1991, 1996). Although Carlson (1989) and Miyazaki (1991) both assumed a diffusion-controlled process, the degree of disequilibrium differed in their models. In particular, the degree of disequilibrium for the nucleation and growth model is much higher than that for Ostwald ripening.

Correspondence: kazu-miyazaki@aist.go.jp
Geological Survey of Japan, AIST, Higashi 1-1-1, Tsukuba, 305-8567 Ibaraki, Japan

Even if the rate-limiting processes were governed by diffusion, the degree of disequilibrium should directly affect the reaction rate and the patterns of metamorphic textures. The growth of a metamorphic mineral at a low degree of disequilibrium (i.e., a system close to equilibrium, implying a small overstep from the equilibrium temperature and a low degree of supersaturation for the metamorphic reaction) may result in an equigranular metamorphic texture because interfacial energy becomes an important control on crystal morphology and size. On the other hand, growth at a high degree of disequilibrium with a limited number of nuclei may result in a porphyroblastic texture. However, diffusion-controlled growth of a mineral at a high degree of disequilibrium is likely to involve diffusional instability and may lead to highly irregular and branched morphologies. In this paper, I examine the diffusional instability of garnet porphyroblasts (Tsukuba metamorphic rocks, central Japan) that grew in a diffusion-controlled regime.

Methods

Identification of a texture characteristic of diffusion-controlled growth

Diffusion-controlled growth is the result of the sluggish diffusion of the elements required for growth of the product of the reaction and is characterized by the presence of a depleted zone that formed around the growing product (Fig. 1). It is assumed that a uniform medium existed initially before nucleation of the product. After nucleation occurs, the region surrounding the reaction product will supply elements to the growing product. Although a uniform medium continues to exist outside the region immediately surrounding the growing product, the concentration of elements required for growth is reduced in the immediate surroundings by the diffusion of those elements to the product. If the reactants needed for the growth of the product exist in the surrounding region, they should dissolve selectively. Therefore, it is expected that the reactants should be remarkably depleted in the immediate surroundings of the growing product. The resulting depletion zone contains the characteristic texture for diffusion-controlled growth. Therefore, a search for a zone that is depleted in the reactants of a garnet-forming reaction is the first step towards estimating the degree of disequilibrium associated with diffusion-controlled growth of garnet porphyroblasts.

Mathematical analysis of diffusional instability

Diffusional instability should occur when a product grows by a diffusion-controlled process. For example, if the small tip of a growing product occurs in a diffusional medium, the concentration gradient around the tip becomes larger compared with the area around the flat surface of the product (Fig. 2). The growth rate of the tip will be higher than the flat surface because the influx of any reaction-limiting

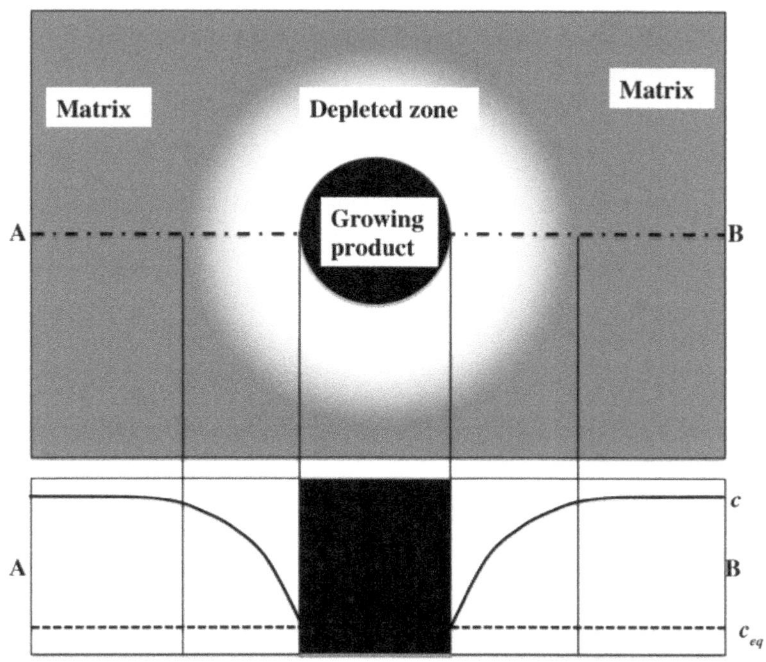

Fig. 1 Schematic diagram of diffusion-controlled growth. A zone depleted in the reaction-limiting component (RLC) should form around the growing product in a diffusion-controlled regime. The terms c and c_{eq} represent the concentrations of the RLC in the diffusional medium and at the flat surface interface of the growing product, respectively

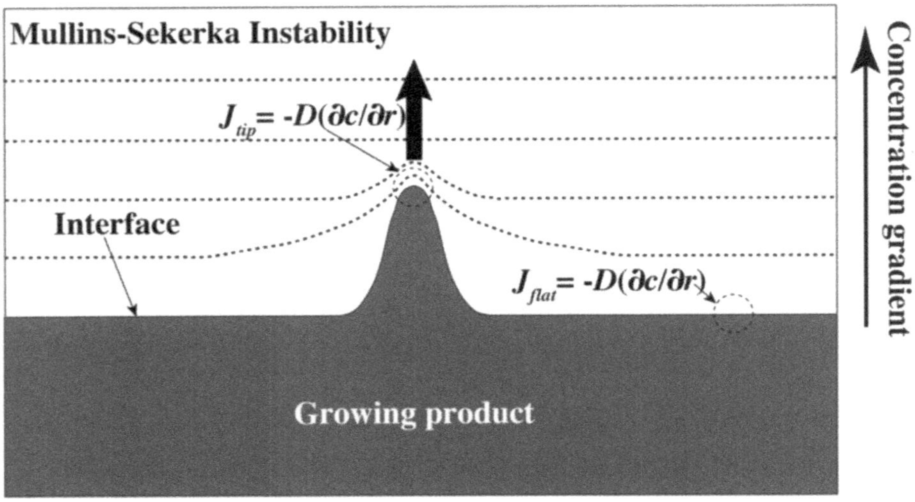

Fig. 2 Schematic diagram of diffusional instability

component (RLC) (Kelly et al. 2013a) is proportional to the concentration gradient of the RLC in the diffusional medium (see Fig. 2). This implies that diffusional instability is always to be expected under diffusion-controlled growth, and that diffusional instability will lead to highly branched textures that emerge from very small perturbations. However, interfacial energy may offset this type of instability.

Assuming the growth of a spherical product (such as a garnet porphyroblast) with radius R_g in a diffusional medium, and a mass balance at the spherical surface, where a growth rate $dV/dt = 4\pi R_g^2\ (dR_g/dt)$ is proportional to the diffusional flux $J = -D\ (\partial c/\partial r)$, the diffusion-controlled growth rate becomes:

$$(C-c_R)\frac{dV}{dt}=4\pi R_g^2 D\left(\frac{\partial c}{\partial r}\right)_{r=R_g},$$
$$\frac{dR_g}{dt}=\frac{D}{(C-c_R)}\left(\frac{\partial c}{\partial r}\right)_{r=R_g}, \tag{1}$$

where D is the diffusion coefficient of the RLC in the medium, C is the concentration of the RLC in the product, and r, c, and c_R are the distance from the center of the product, the concentration of the RLC in the diffusional medium, and the equilibrium concentration of the RLC at the interface of the product with radius R_g, respectively. Assuming the Gibbs–Thomson effect, the concentration c_R is defined as follows:

$$c_R=c_{eq}\left(1+\frac{2\Gamma_D}{R_g}\right), \tag{2}$$

where $\Gamma_D = \gamma\Omega/RT$ is the capillary length, and γ and Ω are the interfacial energy and the molar volume of the product, respectively. The R is the gas constant and T is the temperature. The c_{eq} is the equilibrium concentration of the RLC in the medium at the interface of a product with a flat surface. The c_R is higher than c_{eq} due to the Gibbs–Thomson effect. When concentration c only depends on the radial component, the diffusion equation in polar coordinates becomes $\partial c/\partial t = D\ (\partial^2 c/\partial r^2 + 2/r\ (\partial c/\partial r))$. Assuming steady-state conditions (i.e., $\partial c/\partial t = 0$), the solution of $c = c_m - (c_m - c_R)\ (R_g/r)$ is obtained, where c_m is the concentration of the RLC at a long distance from the product interface. The solution satisfies $c_{r=\infty} = c_m$ and $c_{r=Rg} = c_R$. Using Eqs. 1 and 2, and assuming steady-state diffusion around the spherical product of radius R_g, the growth rate dR_g/dt becomes (e.g., Lifshitz and Slyozov 1961):

$$\frac{dR_g}{dt}=\frac{Dc_{eq}}{(C-c_R)R_g}\left(\Delta\zeta-\frac{2\Gamma_D}{R_g}\right), \tag{3}$$

where $\Delta\zeta$ is the supersaturation. The $\Delta\zeta$ is defined as follows:

$$\Delta\zeta=\frac{c_m-c_{eq}}{c_{eq}}. \tag{4}$$

Mullins and Sekerka (1963) presented a mathematical analysis of diffusional instability, and they introduced a small perturbation δ on a spherical product with radius R_g. The distorted interface of the sphere then becomes:

$$r=R_g+\delta Y_{lm}(\theta,\phi), \tag{5}$$

where r is the distance from the center of the product to the interface, and $Y_{lm}(\theta,\phi)$ is a spherical harmonic function. The l (degree) and m (order) of the spherical harmonic function are integers, where $l > 0$ and $-l \le m \le l$. The wavelength of the perturbation decreases with increasing l. For example, the wavelength of the small perturbation of $Y_{30}\ (\theta,\phi)$ is three (Fig. 3). The following

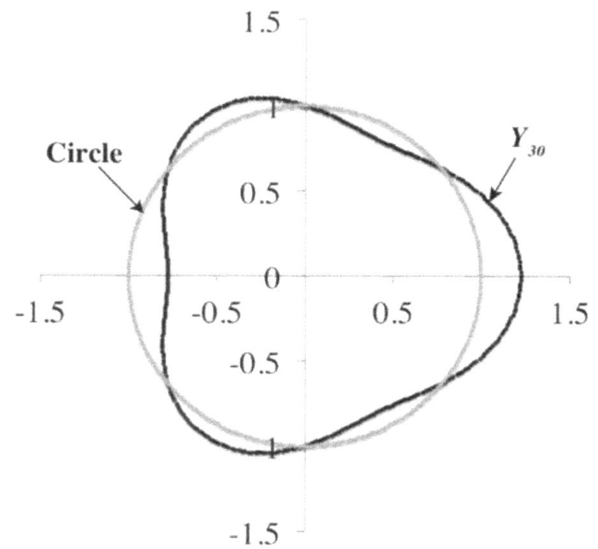

Fig. 3 A sphere perturbed by the harmonic Y_{30}

equation describes the relationship between degree l and wavelength λ_l (Mullins and Sekerka 1963):

$$\lambda_l \simeq \frac{2\pi R_g}{l}. \quad (6)$$

A growth rate $d\delta_l/dt$ for a small perturbation δ_l for the degree l is obtained by assuming steady-state diffusion around the growing product with radius R_g, as follows (Mullins and Sekerka 1963):

$$\frac{d\delta_l}{dt} = \frac{D(l-1)c_{eq}}{(C-c_R)R_g^2}\left\{\Delta\zeta - \Gamma_D\frac{(l+1)(l+2)+2}{R_g}\right\}\delta_l. \quad (7)$$

The first term in the bracket on the right-hand side of Eq. 7 represents an acceleration of the perturbation due to diffusional instability, and the second term shows a reduction in the perturbation due to interfacial energy. Together, these two terms govern the evolution of the small perturbation. Using Eq. 7, the growth rate $d\delta_l/dt$ of the small perturbation can be obtained as a function of the degree l. Using the relationship between l and λ_l (Eq. 6), Fig. 4 implies that the growth rate has a maximum value at l_{max}. The l_{max} is defined as follows (Mullins and Sekerka 1963):

$$l_{max} \equiv \left(\frac{R_g\Delta\zeta}{3\Gamma_D}\right)^{1/2}. \quad (8)$$

Equations 6 and 8 show that the dominant wavelength λ_{max} ($=2\pi R_g/l_{max}$) can be directly extracted from the supersaturation $\Delta\zeta$. In reverse, this means that the observed dominant wavelength of the product interface should be linked to the supersaturation.

The relative growth rate of a small perturbation for a sphere with radius R_g is calculated using Eqs. 3 and 7, as follows:

$$\frac{[(d\delta_l/dt)/\delta_l]}{[(dR_g/dt)/R_g]} = \frac{(l-1)(1-R_l/R_g)}{1-R_c/R_g}, \quad (9)$$

where R_l is the critical radius of the diffusional perturbation of the degree l, and R_c is the critical radius for Ostwald ripening. Under the condition of $R_g > R_c$, the perturbation of the degree l will grow when $R_g > R_l$. In contrast, the perturbation of the degree l will disappear when $R_g < R_l$. Because Eq. 9 becomes zero at $l = 1$, the condition $R_g > R_{l=2}$ will lead to unstable growth, causing an irregular shape to form instead of a sphere. On the

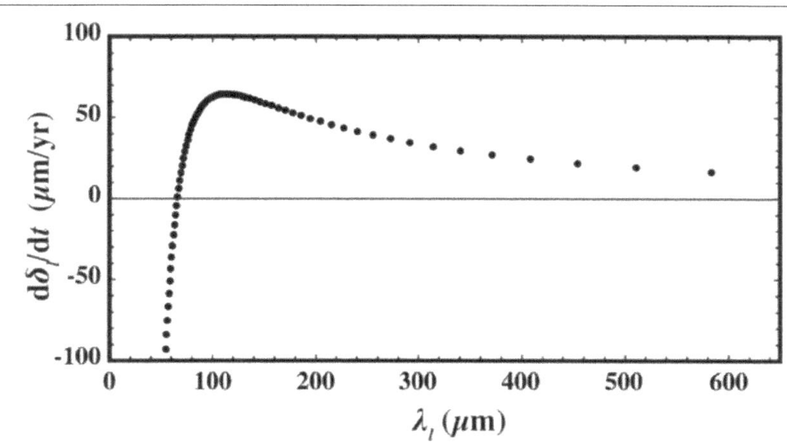

Fig. 4 Relationship between the growth rates of a perturbation on a spherical surface ($d\delta_l/dt$) and wavelength λ_l. The growth rate reaches a maximum value with increasing wavelength. The growth rate and wavelength are calculated using Eqs. 6 and 7 assuming an interfacial energy $\gamma = 1.0$ J/m^2, $C - c_R = 1/\Omega$, molar volume $\Omega = 1.2 \times 10^{-4}$ m^3/mole, a capillary length $\Gamma_D = 0.016$ μm, a diffusion coefficient $D = 10^{-19}$ m^2/s, an equilibrium concentration of the RLC $c_{eq} = 10^{-4}$ mole/m^3, and a radius of the growing product $R_g = 650$ μm

other hand, when $R_g < R_{l=2}$, a compact spherical morphology becomes stable. Assuming a growth rate $d\delta_l/dt = 0$ and $R_g = R_l$ in Eq. 7, R_l becomes:

$$R_l = \{(l+1)(l+2)+2)\}\frac{\Gamma_D}{\Delta\zeta}. \quad (10)$$

Similarly, R_c is given by Eq. 3, as follows:

$$R_c = \frac{2\Gamma_D}{\Delta\zeta}. \quad (11)$$

Combining Eqs. 10 and 11, R_l can be written as follows:

$$R_l = \frac{1}{2}\{(l+1)(l+2)+2\}R_c. \quad (12)$$

Although R_c has a very small value, Eq. 12 implies that R_l becomes very large for a large l.

Measurement of the garnet morphology

Equations 6 and 8 imply that the supersaturation $\Delta\zeta$ for diffusion-controlled growth can be directly obtained from the dominant wavelength λ_{max} of the interface of the growing product. Measurements of the dominant wavelength of garnet produced by diffusion-controlled growth were performed by measuring the radius of local curvature of the garnet interface (method 1), and by measuring the power spectra of the garnet interface (method 2).

In method 1, the local curvature k is characterized by the number of particles n_l belonging to a circle with radius r_s (Koshizuka 2005). Here, the local curvature $k = 1/r_r$ is calculated by counting the pixels included in a circle with radius r_s (Fig. 5), as follows:

$$2\,\theta = \frac{n_i}{n_0}\pi, \quad (13)$$

$$k = \frac{1}{r_r} = \frac{2\cos\theta}{r_s}, \quad (14)$$

where n_i is the number of pixels within circle i with radius r_s. The n_0 represents n_i for a flat surface. The r_s ranges from 3.1 to 21.1 pixel units and is adjusted to obtain the largest value of local curvature at a given position on the garnet interface. The calculated local curvatures are calibrated, as a local curvature of a circle with radius r becomes $1/r$. The calibration length ranges from $r = 2$ to 120 pixel units. The calibrating local curvatures for circles with radii from $r = 2$ to 120 pixel units are basically identical to the inverse of the radii of the target circles (=$1/r$) (Fig. 6). The dominant wavelength of the garnet interface is clearly larger than twice the length of the radius of the local curvature because the wavelength of a tip is always larger than twice the length of the radius of the enveloping circle at the tip. Hence, the lower limit for the dominant wavelength is twice the length of the radius of the local curvature.

Method 2 uses interface positions measured from the garnet barycenter to estimate the power spectra of the interface morphology. Before the measurement of interface positions, inclusions in the garnet are filled. The measured interface positions are represented by the angle θ and the distance r from the barycenter (Fig. 7). A θ value of 0° is defined as downward in the vertical direction from the barycenter, and the value of θ increases counterclockwise. The number of sampling points is 1024 or 2048 between θ = 0° and 360° around the barycenter. The power spectra $f(\theta)$ was estimated using the fast Fourier transform (FFT) of the distance r as a function of θ. The dominant spectra obtained in degrees are converted to lengths by the mean radius from the barycenter of the garnet. The FFT cannot be applied to overhanging parts of the interface; therefore, only

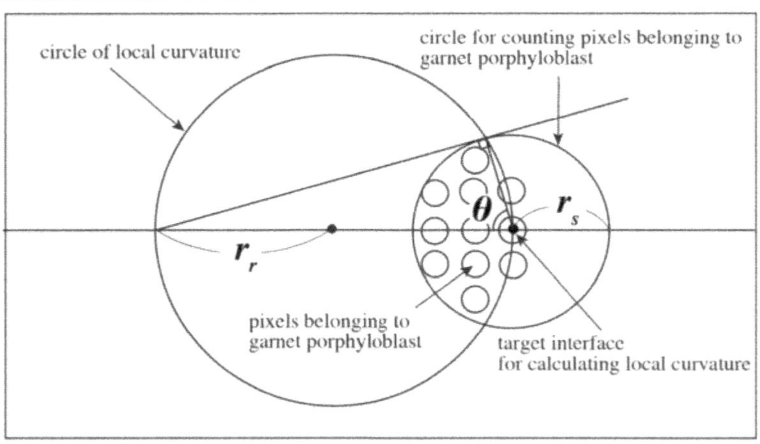

Fig. 5 Geometrical relationships between the radius of curvature and the circle radius where pixels are counted. The r_r is the radius of curvature, and r_s is the radius of the circle in which pixels are counted. Please see text and Eqs.13 and 14 for a detailed explanation

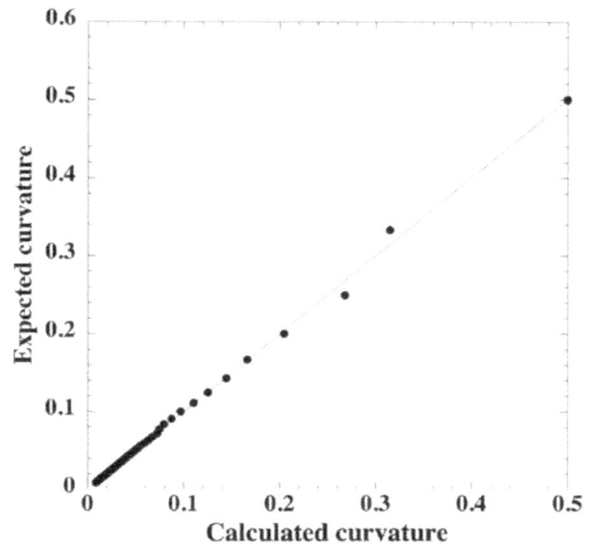

Fig. 6 Results for the calibration of local curvature for circles with radii = 2–120 pixel units. Calculated and expected curvatures represent the calibrated results using Eqs. 13 and 14, and the inverse of the circle radii, respectively

non-overhanging parts were used. This treatment leads to a lack of information on the power spectra in overhanging parts, which have many undulations (Fig. 7). Therefore, the estimates of the power spectra for shorter wavelengths are underestimated in this paper. Here, I used the dominant wavelength obtained from method 2 as the upper limit. Hence, the true value of the dominant wavelength should lie between the values obtained from methods 1 and 2.

Tsukuba metamorphic rock sample

The sample analyzed for this study was collected from the high-grade part of the Late Cretaceous low pressure–high temperature (*LP–HT*) Tsukuba metamorphic rocks that form an eastern extension of the Ryoke metamorphic complex. The locality of the sample (number Ma2-48a) was described by Miyazaki (1999). The protoliths of the Tsukuba metamorphic rocks were mudstones and sandstones of an accretionary complex of Late Jurassic to earliest Cretaceous age (Miyazaki et al. 1996). The Tsukuba metamorphic rocks were intruded by Late Cretaceous to earliest Paleogene granitic plutons. The metamorphic grade increases from southeast to northwest.

Based on the mineral assemblages, the pelitic metamorphic rocks can be assigned to the biotite and sillimanite zones. The mineral assemblage of the biotite zone rocks is biotite + muscovite + quartz + plagioclase ± andalusite, whereas the sillimanite zone rocks contain K-feldspar + sillimanite + biotite + quartz + plagioclase ± cordierite ± garnet. The boundary between the biotite and sillimanite zones is defined by the dehydration reaction of muscovite and quartz, which produces K-feldspar, sillimanite, and water. The sample analyzed for this study belongs to the high-grade part of the sillimanite zone and contains K-feldspar + garnet + cordierite + biotite + plagioclase + quartz + sillimanite. The average pressure–temperature (*P–T*) conditions of samples previously studied from the locality of the present sample are $P = 0.34 \pm 0.1$ GPa and $T = 642 \pm 16$ °C (Miyazaki 1999).

Based on digital maps showing the distribution of minerals, as compiled from X-ray intensity maps of Na, K, Mn, Fe, and Al, the mineral mode of the sample is 57 % quartz, 22 % K-feldspar, 8 % plagioclase, 6–7 % biotite, 4–5 % cordierite, 1–2 % garnet, and <1 % sillimanite (Miyazaki 2001). K-feldspar, biotite, plagioclase, and quartz are distributed randomly as matrix minerals with diameters of 50–60 μm and do not occur as porphyroblasts (Miyazaki 2001). Because of its low abundance in the sample, it is unclear if sillimanite also has a random distribution in the matrix.

Cordierites occur as porphyroblasts (diameter ≤1 mm) and exhibit irregular shapes. The porphyroblasts have a highly branched texture that may have been formed by diffusion-limited aggregation (DLA). The diffusion of Al

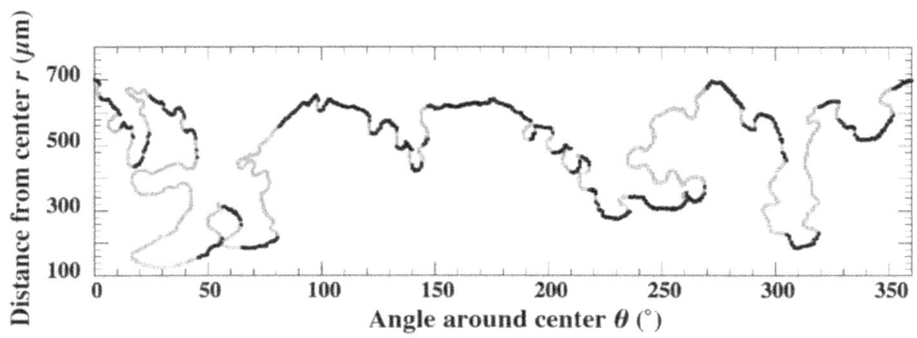

Fig. 7 Locations of the interfaces of a garnet porphyroblast (shown in Fig. 9). The locations are plotted by distance from the barycenter, and the angle around the barycenter. *Black dots* are used for fast Fourier transform (FFT) calculations to obtain power spectra

may be rate-limiting for DLA-like cordierite porphyroblast growth (Miyazaki 2001). Comparisons of the morphologies and fractal dimensions of the cordierites with those of DLA indicate that formation of the cordierite porphyroblasts can be modeled as a DLA process (Miyazaki 2001). A finite-sized growth fluctuation, induced by a random distribution of matrix minerals, is amplified by the diffusional field and results in a DLA-like pattern of porphyroblast growth. Hence, diffusion-controlled growth within a random matrix field is essential for the formation of DLA-like patterns in cordierite porphyroblasts (Miyazaki 2001).

Garnets occur as porphyroblasts (diameters ≤1 mm) and have irregular shapes (Figs. 8 and 9). Biotite is depleted around garnet porphyroblasts (Fig. 8). Quartz, cordierite, K-feldspar, and plagioclase are present in the depletion zone but biotite is rare or has smaller crystal sizes compared with biotite outside the depletion zone. Cordierite in the depletion zone is much smaller in size than the porphyroblastic cordierite (Fig. 9b). The number and size of K-feldspar grains in the depletion zone are also small compared with those outside the depletion zone (Fig. 9a). Sillimanite does not occur in the depletion zone.

The chemical compositions of the constituent minerals were given by Miyazaki (1999). The garnet contains about 7 wt% MnO (73 % almandine, 16 % spessartine, 9 % pyrope, and 2 % grossular). Compositional zoning of MnO, FeO, MgO, and CaO was not detected (Fig. 9), but slightly higher MnO contents occur rarely at the tips on the outer rims of garnet porphyroblasts. Biotites are made up of the components phlogopite (41 %), eastonite (26 %), Ti-biotite (17 %), and muscovite (16 %). The end-member formulae of these components are phlogopite ($KM_3AlSi_3O_{10}[OH]_2$), eastonite ($KM_2Al_2Si_2O_{10}[OH]_2$), Ti-biotite ($KM_{1.5}TiAl_2Si_2O_{10}[OH]_2$), and muscovite ($KAl_3Si_3O_{10}[OH]_2$) (e.g., Ikeda 1990), where M represents the divariant cations of Mg, Fe, and Mn. The cordierites consist of hydrous cordierite (50 %) and Fe-cordierite (50 %). The K-feldspar contains 20 % of the albite component.

Garnet-forming reaction and reaction entropy

To estimate the degree of disequilibrium, the garnet-forming reaction in the sample should be identified, which is crucial for evaluating the extent of reactant depletion in the region surrounding garnet porphyroblasts. In addition, the reaction entropy ΔS_r for the garnet-forming reaction is required to estimate the extent to which the Gibbs free energy ΔG_r was overstepped. Because $dG = (\partial G/\partial P)\ dP + (\partial G/\partial T)\ dT + (\partial G/\partial N)\ dN$, where P is pressure and N is the number of molecules, the Gibbs free energy depends only on $(\partial G/\partial T) = -S$ at constant pressure and bulk composition. Therefore, the reaction entropy for the garnet-forming reaction is required to estimate the overstepping of Gibbs free energy at constant pressure and bulk composition. The ΔG_r is calculated as follows:

$$\Delta G_r = \Delta S_r \Delta T, \tag{15}$$

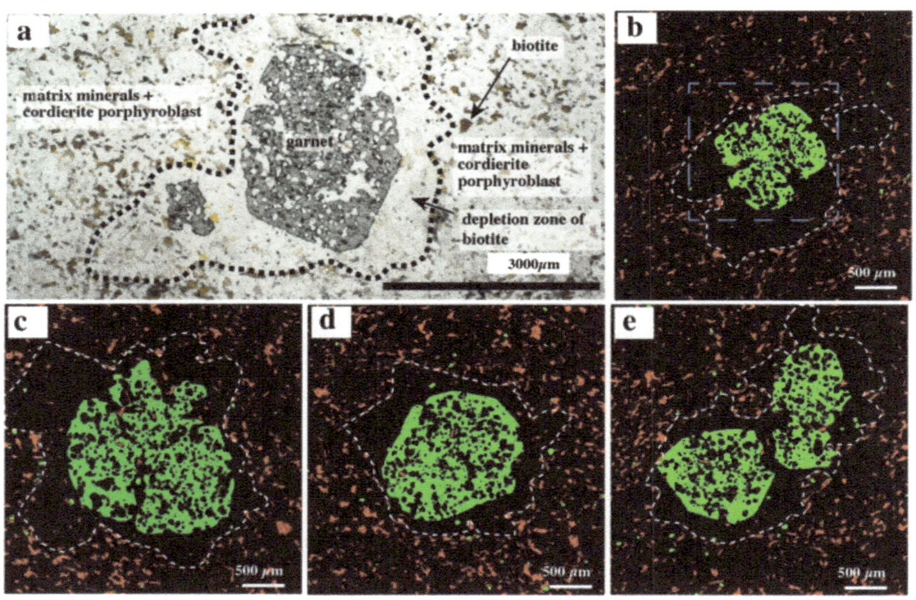

Fig. 8 Irregular and branching morphologies of garnet porphyroblasts and biotite depletion zones. **a** Photomicrograph. **b**–**e** Compositional maps of Fe obtained by electron probe micro-analysis (EPMA). Green and red colors represent garnet and biotite, respectively. *Black broken line* in (**a**) and *white broken lines* in (**b**)–(**e**) denote the area of the biotite depletion zone. *Blue broken line* in (**b**) shows area of Fig. 9a–d

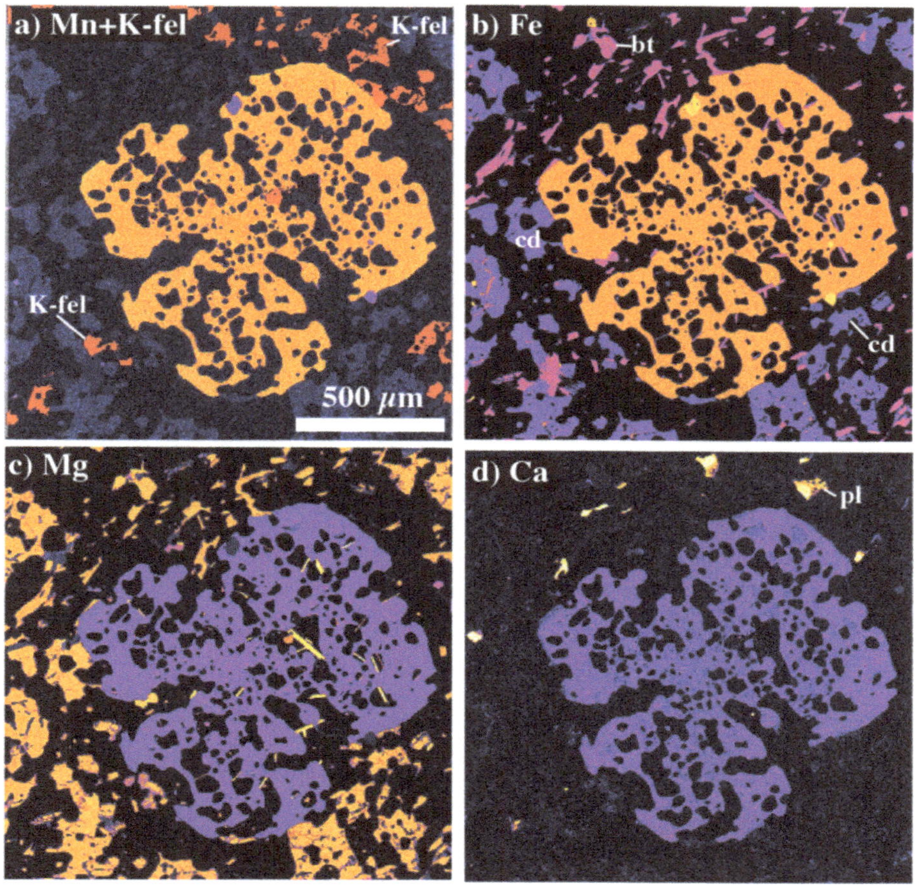

Fig. 9 Compositional maps of a garnet porphyroblast (garnet in Fig. 8b). X-ray mapping images of (**a**) Mn + K-fel: MnKα and distribution of K-feldspar, (**b**) Fe: FeKα, (**c**) Mg: MgKα (**c**), and (**d**) Ca: CaKα. The intensity of each X-ray increases from cooler to warmer colors. The distribution of K-feldspar (K-fel) in (**a**), biotite (bt) and cordierite (cd) in (**b**), and plagioclase (pl) in (**d**) are also shown

where ΔT is the temperature overstep from the equilibrium temperature. Equation 15 shows that a smaller reaction entropy will yield a larger temperature overstep at a constant overstepping of Gibbs free energy.

For simplicity, the garnet-forming reaction for the studied sample is represented by the following univariant reaction in the $K_2O–MgO–FeO–Al_2O_3–SiO_2–H_2O$ system:

$$10\ \text{Sillimanite} + 4\ \text{Biotite} + 15.5\ \text{Quartz} = \text{Garnet} + 4.5\ \text{Cordierite} + 4\ \text{K-feldspar} + 1.75\ \text{water}. \quad (16)$$

The model compositions used for the minerals are listed in Table 1. In this system, garnet, biotite, and cordierite are treated as almandine–pyrope, phlogopite–annite, and hydrous cordierite–Fe-cordierite solid solutions, respectively, with the X_{Fe} of these minerals adjusted to the X_{Fe} observed in the sample. The other minerals are treated as pure phases. The spessartine component in garnet and the albite component in feldspar make minor contributions to the reaction entropy. However, the Ti-biotite, eastonite, and muscovite components in the biotite may contribute significantly to the reaction entropy by increasing the stoichiometric coefficient of water in the garnet-forming reaction. Increasing these components increases the reaction

Table 1 Mineral compositions and entropies at 642 °C (used in this paper)

Mineral	Formula	S (J/mole K)
Garnet (Grt)	$Fe_{2.7}Mg_{0.3}Al_2Si_3O_{12}$	831
Biotite (Bt)	$KFe_{1.8}Mg_{1.2}AlSi_3O_{10}(OH)_2$	913
Sillimanite (Sil)	Al_2SiO_5	279
Quartz (Qtz)	SiO_2	106
K-feldspar (Kfs)	$KAlSi_3O_8$	509
Cordierite (Crd)	$FeMgAl_4Si_5O_{18}{\cdot}(H_2O)_{0.5}$	1192
Water (W)	H_2O	229
Reaction	4Bt + 10Sil + 15.5Qtz = Grt + 4Kfs + 4.5Crd + 1.75 W	542

entropy. To obtain the minimum reaction entropy, I used a phlogopite–annite solid solution for biotite and the maximum contribution from the hydrous cordierite component.

Using the standard-state entropy and heat capacity of the end-member minerals (Holland and Powell 2011) and assuming ideal mixing for garnet, biotite, and cordierite solid solutions, the reaction entropy at 642 °C becomes 542 J/K for the formation of 1 mole of garnet (12 oxygen atoms). The reaction entropy is also a function of temperature at constant pressure. Here, I assumed that the change in reaction entropy between 642 and 742 °C is not significantly different compared with the reaction entropy at 642 °C. For example, the reaction entropy is 541 J/K at 700 °C and 540 J/K at 742 °C.

Results

Diffusional haloes around garnet porphyroblasts

The garnet-forming reaction (Eq. 16) shows sillimanite, biotite, and quartz as the reactants and garnet, cordierite, K-feldspar, and water as the products. As shown in Fig. 8, biotite is scarce in the area surrounding the garnet porphyroblasts. Biotite is the most important reactant that supplies Al, Fe, and Mg to a growing garnet crystal. Therefore, the observed depletion of biotite around garnet porphyroblasts provides direct evidence for diffusion-controlled growth. The depletion zone is recognized as diffusional haloes around garnet porphyroblasts. Sillimanite is also an important reactant that supplies Al to garnet, but it is scarce in the matrix. Because sillimanite is typically found farther away from the garnet, the depletion zone of sillimanite around the garnet is much more difficult to identify than the biotite depletion zone.

Cordierite and K-feldspar are products of the garnet-forming reaction. The cordierite crystals in the depleted zone are much smaller than the porphyroblastic cordierite. The number and size of K-feldspar grains in the depletion zone are also smaller than outside the depletion zone. These observations indicate that growth of cordierite and K-feldspar within the depleted zone was suppressed in comparison with growth of cordierite porphyroblasts and K-feldspar matrix minerals outside the depletion zone.

Dominant wavelength and supersaturation

Histograms of the radii of the local interface curvature of seven garnets within the sample show a peak around 30–70 μm (Fig. 10). This peak corresponds to the highest frequency of radii measured from local curvatures of undulations in the irregular garnets. As mentioned above, the wavelengths of these undulations are larger than twice the radius of the local curvatures. Using the mid-range values of the histogram peaks for the seven garnet crystals, the calculated average radius of the local curvatures is 43 ± 4.5 μm (Table 2). Therefore, the dominant wavelength obtained from method 1 should be greater than 86 μm.

The power spectra of the interface of irregular garnets show a broad maximum around 150–700 μm (Fig. 11). Using the mid-range values of these maxima, the calculated average value is 489 ± 88 μm (Table 2). Therefore, the dominant wavelength obtained from method 2 should be smaller than 489 μm.

Given a dominant wavelength between 86 and 489 μm, interfacial energy $\gamma = 1.0$ J/m^2 (Miyazaki 1991), molar volume $\Omega = 1.2 \times 10^{-4}$ m^3/mole (Miyazaki 1991), $T = 642$ °C, and a capillary length $\Gamma_D = \gamma\Omega/RT = 1.58 \times 10^{-5}$ mm for garnet, Eqs. 6 and 8 yield a supersaturation $\Delta\zeta$ of 0.05×10^{-1}–0.16 for garnet using the observed average garnet radius of 644 μm. By decreasing the supersaturation $\Delta\zeta$, the dominant wavelength becomes larger (Fig. 12). Taking the average value of the lower and upper boundaries for the dominant wavelength (288 μm; see Table 2), a $\Delta\zeta$ value of 0.15×10^{-1} is calculated. The average value of the dominant wavelength is consistent with the crude spacing of undulations along the irregular interface of garnet porphyroblasts (Figs. 8 and 9).

Gibbs free energy and temperature oversteps

The calculated supersaturation can be used to determine the magnitude of Gibbs free energy and temperature oversteps. The sample retains the reactant mineral assemblage of sillimanite + biotite + quartz in areas distant from the garnet porphyroblasts. Conversely, the product mineral assemblage of K-feldspar + cordierite occurs proximal to the garnet porphyroblast. Assuming that reactants far from the garnet porphyroblast are locally in equilibrium with the diffusive medium, the chemical potential difference $\Delta\mu_{\mathrm{dissolution}}$ for dissolution of a reactant is

$$\Delta\mu_{\mathrm{dissolution}} = \mu_{\mathrm{reactant}} - \mu_{\mathrm{IGM}} = 0. \tag{17}$$

Here, μ_{reactant} is the chemical potential of the reactant and μ_{IGM} is the chemical potential of the chemical species dissolved in the intergranular medium. The μ_{IGM} is expressed as follows:

$$\mu_{\mathrm{IGM}} = \sum_i m_i\mu_i + m_{\mathrm{RLC}}\mu_{\mathrm{RLC}}, \tag{18}$$

$$\mu_{\mathrm{RLC}} = \mu^0_{\mathrm{RLC}} + RT\ln X_{\mathrm{RLC},m}, \tag{19}$$

where m_i is the number of moles of chemical species i exclusive of the RLC that enter or leave the intergranular fluid in the reactant-dissolution and product-precipitation reactions, μ_i is the chemical potential of chemical species i exclusive of the RLC, m_{RLC} is the number of moles the RLC, and μ_{RLC} is the chemical potential of the RLC. The summation in Eq. 18 excludes the chemical potential

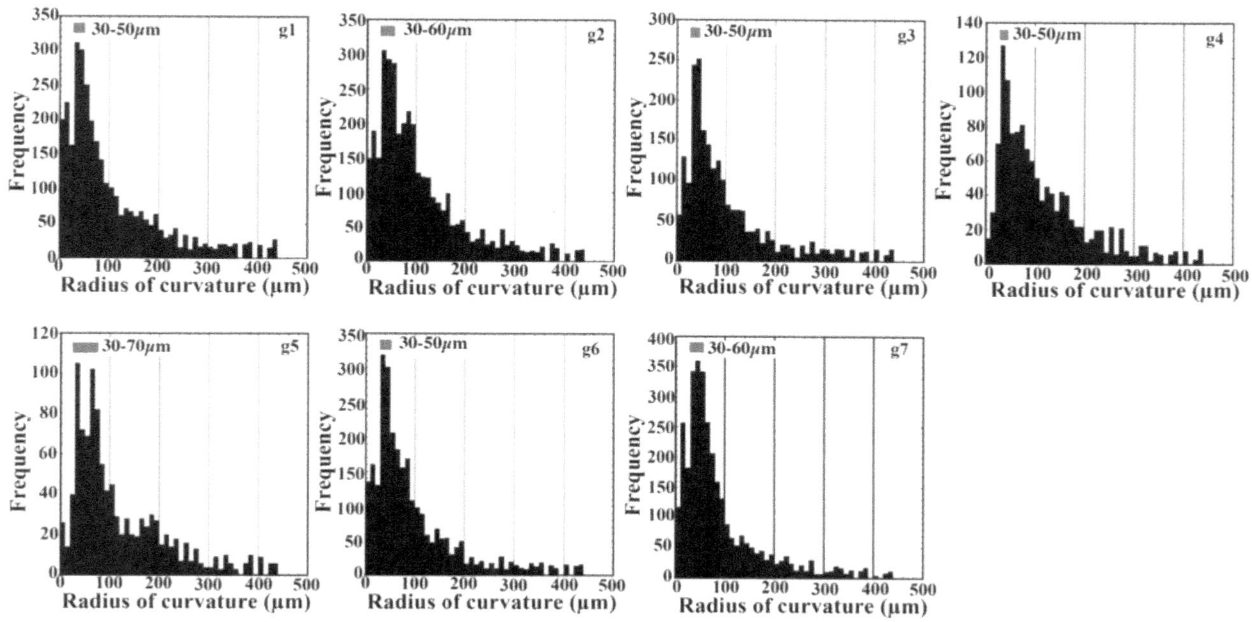

Fig. 10 Histograms of the radius of curvature for irregularly shaped interfaces of garnet porphyroblasts. The *gray-colored horizontal bar* represents the range of the radius peaks for each sample

of the RLC. In Eq. 19, the μ^0_{RLC} is the chemical potential of the pure (unmixed) RLC and $X_{\mathrm{RLC},m}$ is the mole fraction of the RLC in the intergranular medium at a site distant from the garnet porphyroblast. Equation 19 assumes ideal mixing of the RLC in the intergranular medium. In the diffusion-controlled regime, $X_{\mathrm{RLC},m}$ varies between areas proximal to and distant from the garnet porphyroblast. Hence, Eq. 17 becomes

$$\Delta\mu_{\mathrm{dissolution}} = \mu_{\mathrm{reactant}} - \mu_{\mathrm{IGM0}} - m_{\mathrm{RLC}} RT \ln X_{\mathrm{RLC},m} = 0. \tag{20}$$

Assuming the products around the garnet porphyroblast are locally in equilibrium with the diffusive medium, the chemical potential difference $\Delta\mu_{\mathrm{precipitation}}$ for a product surrounding the garnet porphyroblast is

$$\Delta\mu_{\mathrm{precipitation}} = \mu_{\mathrm{product}} - \mu_{\mathrm{IGM0}} - m_{\mathrm{RLC}} RT \ln X_{\mathrm{RLC},grt} = 0\,, \tag{21}$$

where μ_{product} is the chemical potential of the product and $X_{\mathrm{RLC},grt}$ is the mole fraction of the RLC around the garnet porphyroblast. Because garnet growth is controlled by the slowest diffusion of the RLC, the chemical potential differences of the intergranular medium between the areas proximal to and distant from the garnet

Table 2 Radius of curvature and dominant wavelength of garnet porphyroblast interfaces, and the estimated supersaturation (Δζ)

Garnet	R_g: radius (μm)	*a*: mid-range of radius of curvature (μm)	*b*: mid-range of dominant wavelength (μm)	*c*: average of the dominant wavelengths by 2 × a and b (μm)	$\Delta\zeta^{a}$ lower bound calculated with value of *b*	$\Delta\zeta^{*a}$ mid-range value calculated with value of *c*	$\Delta\zeta^{*a}$ upper bound calculated with value of 2 × *a*
g1	780	40	500	270	0.06×10^{-1}	0.02	0.23
g2	904	45	550	298	0.06×10^{-1}	0.02	0.21
g3	536	40	425	233	0.06×10^{-1}	0.02	0.16
g4	541	40	550	295	0.03×10^{-1}	0.01	0.16
g5	479	50	300	175	0.01	0.03	0.09
g6	702	40	550	295	0.04×10^{-1}	0.02	0.20
g7	568	45	550	298	0.04×10^{-1}	0.01	0.13
Average	644	43	489	288[b]	0.05×10^{-1}[c]	0.15×10^{-1}[c]	0.16[c]

[a]Calculated using Eq. 8, with $\Gamma_D = 1.58 \times 10^{-5}$ mm
[b]Calculated with average values of 2 × a and b
[c]Calculated with average values of radius R_g, 2 × a, b, and c

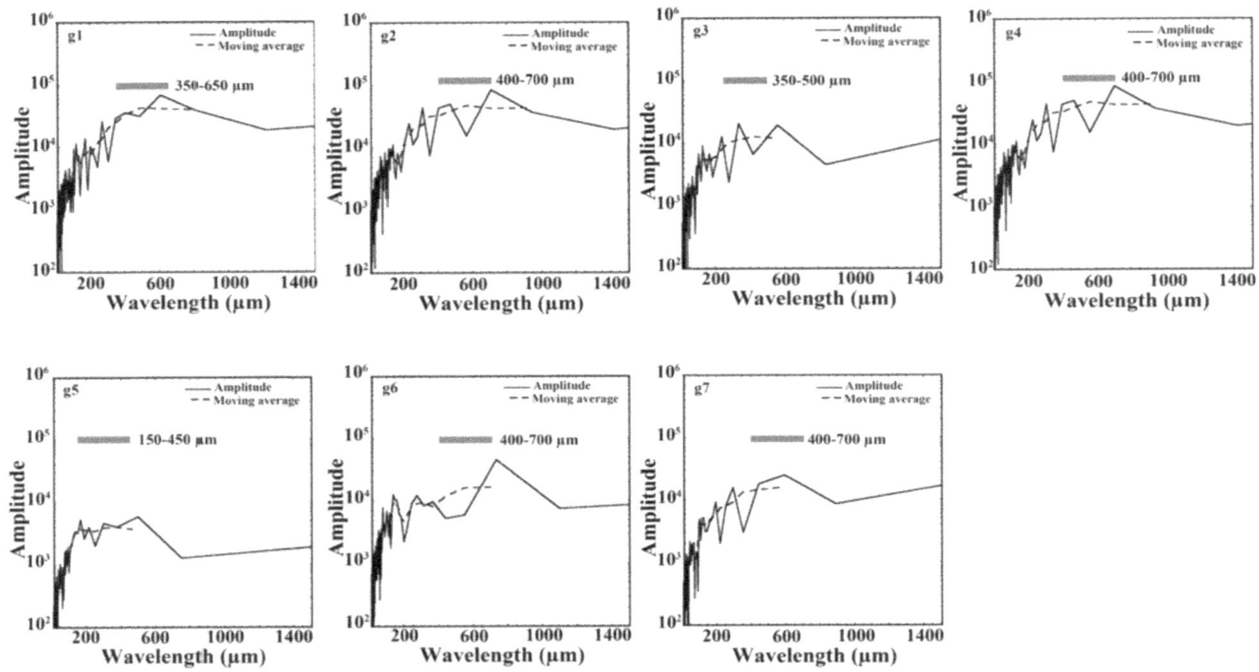

Fig. 11 Power spectra for the wavelengths of irregularly shaped interfaces of garnet porphyroblasts. *Broken lines* show the moving average of the power spectra. The *gray-colored horizontal bar* represents the range of the dominant wavelength for each sample

porphyroblast depend on the chemical potential of the RLC in the medium. This constancy indicates that μ_{IGM0} in Eq. 20 has the same value as in Eq. 21. Using Eqs. 20 and 21, the mole fractions of the most sluggish component of the RLC become

$$X_{RLC,m}=\mathrm{Exp}\left(\frac{\mu_{reactant}-\mu_{IGM0}}{m_{RLC}RT}\right), \quad (22)$$

$$X_{RLC,grt}=\mathrm{Exp}\left(\frac{\mu_{product}-\mu_{IGM0}}{m_{RLC}RT}\right). \quad (23)$$

From Eqs. 4, 22, and 23, the supersaturation $\Delta\zeta$ becomes

$$\Delta\zeta=\frac{c_m-c_{eq}}{c_{eq}}=\frac{aX_{RLC,m}-aX_{RLC,grt}}{aX_{RLC,grt}}, \quad (24)$$

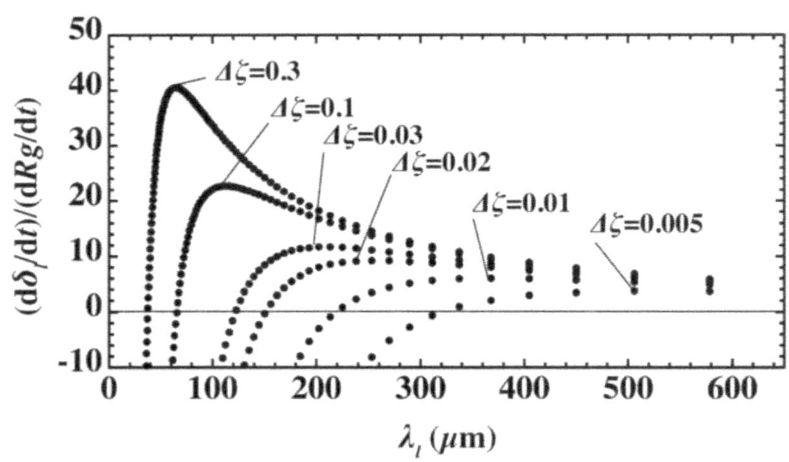

Fig. 12 Relative growth rate of a sphere perturbation relative to the growth rate of a sphere. The relative growth rate $((d\delta_l/dt)/(dR_g/dt))$ is a function of wavelength λ_l. The relative growth rate is calculated using Eqs. 6, 9 , 10, and 11, with $\Gamma_D = 1.58 \times 10^{-5}$ mm, $R_g = 644$ μm, and $\Delta\zeta = 0.05 \times 10^{-1}$–0.3

$$\Delta\zeta = \mathrm{Exp}\left(\frac{\Delta G_r}{m_{\mathrm{RLC}}RT}\right) - 1\ , \qquad (25)$$

where a is a constant for converting from a mole fraction to concentration, and $\Delta G_r = \mu_{\mathrm{reactant}} - \mu_{\mathrm{product}}$ is the Gibbs free energy for the garnet-forming reaction (Eq. 16). The ΔG_r in this study is equivalent to reaction affinity A (e.g., Pattison et al. 2011) and ΔG_r is zero at the equilibrium temperature. I assumed that the RLC is Al, corresponding to m_{RLC} = 24, and that the equilibrium temperature is T = 642 °C. Using Eq. 25, the supersaturation $\Delta\zeta$ of 0.05 × 10^{-1}–0.16 yields a ΔG_r of 0.9–27 kJ per mole of garnet (12 oxygen atoms). The average $\Delta\zeta$ of 0.15 × 10^{-1} yields a ΔG_r of 2.7 kJ per mole of garnet.

Using Eq. 15, ΔG_r, and the reaction entropy ΔS_r of 542 J/K per mole of garnet for the garnet reaction, a temperature overstep ΔT of 1.7–50 °C is calculated using the lower and upper boundaries of the dominant wavelength. When the average dominant wavelength is used, a ΔT of 5 °C is calculated.

Discussion

Estimates of the degree of disequilibrium for the garnet-forming reaction have been reported in previous studies. Pattison and Tinkham (2009) evaluated the temperature and Gibbs free energy oversteps for a garnet-forming reaction in the Nelson aureole (British Columbia) using discrepancies between phase-equilibrium modeling and field observations of the location of the garnet isograd. This approach yielded ΔT = 30 °C and ΔG_r = 4.8 kJ per mole of garnet. The degree of disequilibrium for the Nelson aureole is similar to that calculated for the sample in this study. However, the garnet porphyroblasts in the Nelson aureole are euhedral rather than irregular and branching in the case of the garnets from the Tsukuba metamorphic rocks. Kinetic modeling based on interface-controlled nucleation and growth was used to explain the formation of the garnet in the Nelson aureole (Gaidies et al. 2011). On the other hand, Kelly et al. (2013b) used numerical modeling of diffusion-controlled nucleation and growth to reproduce the crystal sizes and spatial distributions of minerals in 13 porphyroblastic rocks. They calculated ΔT = 5–67 °C and ΔG_r = 0.7–5.8 kJ per mole of garnet. Spear et al. (2014) also evaluated the degree of disequilibrium for the garnet-forming reaction. Using a combination of quartz in garnet barometry (QuiG) and thermodynamic modeling for garnet zone metamorphic rocks in eastern Vermont, they calculated a ΔT of 10 °C, a pressure overstep of 0.6 kbar, and a ΔG_r of 2 kJ per mole of garnet. For the staurolite–kyanite zone in the same metamorphic terrain and for a blueschist sample from Sifnos, Greece, Spear et al. (2014) calculated a ΔT of 50 °C, a pressure overstep of 2–5 kbar, and a ΔG_r of 10–18 kJ per mole of garnet. Although diverse methods have been used to infer the degree of disequilibrium for the garnet-forming reaction, the calculated values suggest that in all cases the garnet nucleated and grew after a significant overstep from equilibrium conditions.

Wilbur and Ague (2006) used Monte Carlo simulations of crystal growth to demonstrate the formation of irregular and branching morphologies in garnet porphyroblasts. Because this simulation assumed random walk of chemical species, the morphology was produced by diffusion-controlled growth. They determined that a minimum Gibbs free energy overstep of about 2 kJ per mole of garnet was needed to produce a branched morphology for silicate minerals (Wilbur and Ague 2006). This value is consistent with the Gibbs free energy overstep calculated in this paper. Spear and Daniel (2001) proposed an amoeba-like growth model for garnet porphyroblasts based on observations of heterogeneous chemical zoning. This model supports the development of an irregular and branching morphology in garnet porphyroblasts. Collectively, the results obtained using these different approaches suggest that the irregular and branching morphology of garnet porphyroblasts is indicative of diffusion-controlled growth.

Ostwald ripening and grain growth are not considered in this paper because the garnets occur as porphyroblasts, meaning that the size of the garnet crystals is much larger than the surrounding matrix minerals. Because grain growth associated with Ostwald ripening takes place due to the reduction of interfacial energy when the system is close to equilibrium, this mechanism should occur at a faster rate for smaller grain sizes. Hence, grain growth due to Ostwald ripening is only important for non-porphyroblastic minerals.

Figure 13 summarizes the relationships among supersaturation, temperature oversteps, and the critical grain diameter for diffusional instability with respect to the diffusion-controlled growth of garnet porphyroblasts in the Tsukuba metamorphic rocks. It is clear that nucleation occurs after a significant overstep from equilibrium conditions for the garnet-forming reaction. Figure 13 shows two different trends in garnet size vs. degree of supersaturation, with path-A and path-B representing nearly constant supersaturation and decreasing supersaturation during garnet growth, respectively. The supersaturation should decrease with consumption of the garnet component in the system. The rate of decrease of the garnet component depends on the relative magnitudes of the garnet precipitation rate and the dissolution rate of reactants in the garnet-forming reaction. Both rates are limited by diffusion of the RLC. For simplicity, it was assumed that the garnet and reactant crystals are the same size.

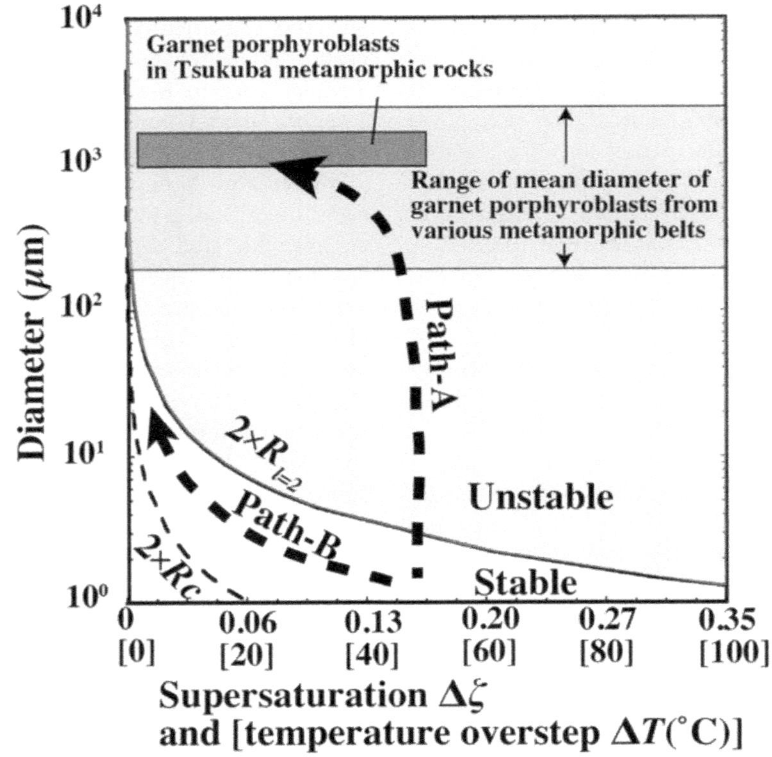

Fig. 13 Relationships among supersaturation, temperature overstep, and critical diameter 2 × $R_{l=2}$ for diffusional instability. The 2 × R_c represents the critical diameter for Ostwald ripening. The $R_{l=2}$ and R_c are calculated using Eqs.10 and 11 , and Γ_D = 1.58 × 10^{-5} mm. The measured diameters of the garnet porphyroblasts, the estimated supersaturation (calculated using Eq. 8 and R_g = 644 μm), and the temperature overstep (calculated using Eqs. 15 and 25, ΔS = 542 J/K per mole of garnet, and assuming 12 oxygen atoms) for the Tsukuba metamorphic rocks are shown (*dark gray box*). Two representative evolution paths (path-A and path-B) are also shown. See the text for detailed explanations of these paths. The range of mean diameters of garnet porphyroblasts from various metamorphic belts is based on the data of Gaidies et al. (2011) and Kelly et al. (2013b) (*gray box*)

If the number of garnet crystals is much fewer than the number of reactant crystals, the evolution of garnet size and degree of supersaturation should follow path-A. For diffusion-controlled growth in this case, a small nucleation rate and large number of reactant crystals favor the growth of a small number of garnet crystals, thus producing unstable porphyroblastic garnet. Large garnet porphyroblasts with irregular and branching morphologies will form. Conversely, if the number of garnet crystals is much larger than the number of reactant crystals, the evolution of garnet size and degree of supersaturation should follow path-B. Due to a high nucleation rate and a small number of reactant crystals, many small garnets should grow rather than porphyroblasts. In addition, path-B promotes a stable garnet morphology, thus preventing the formation of branching and irregular textures.

Kelly et al. (2013b) proposed diffusion-controlled growth for the formation of garnet porphyroblasts whereas Gaidies et al. (2011) proposed interface-controlled growth. Based on the mean radii of garnet porphyroblasts measured in these studies, the mean diameters of garnet porphyroblasts range from 180 to 2400 μm (Fig. 13). Most diameters fall in the range of 1000 to 2000 μm. Therefore, diffusion-controlled growth of garnet porphyroblasts without the development of irregular and branching morphology is difficult to achieve. This finding contradicts the hypothesis that diffusion-controlled growth of garnet porphyroblasts dominates in nature, which is consistent with the rarity of irregular and branching garnet porphyroblasts in metamorphic belts.

Over a scale of 10 cm, temperature heterogeneities produced by the growth of a garnet porphyroblast will be quickly dissipated over a time interval significantly shorter than the long growth period of the porphyroblast. This reasoning indicates that a control on porphyroblast growth by heat flow is unlikely. Therefore, the results of this study suggest that garnet porphyroblasts lacking an irregular and branching morphology (common in many metamorphic belts) may grow at a high degree of disequilibrium under interfacial-controlled growth, provided they are set in a medium in which diffusion is rapid and the supply of elements from metamorphic fluids is sufficient.

Conclusions

Garnet porphyroblasts with an irregular morphology and a diffusional halo of reactants occur in the *LP–HT* Tsukuba metamorphic rocks of central Japan. The features observed suggest diffusion-controlled growth of the porphyroblasts.

Using an analysis of diffusional instability, the degree of disequilibrium can be determined by measuring the dominant wavelength of the interface of the garnet porphyroblasts. The results show that the supersaturation is 0.05×10^{-1}–0.16. From the extent of supersaturation, the calculated overstep of Gibbs free energy from equilibrium is 0.9–27 kJ per mole of garnet (12 oxygen atoms), and the temperature overstep is 1.7–50 °C.

A high degree of disequilibrium for a garnet-forming reaction is expected during growth of a garnet porphyroblast after nucleation. The present results suggest that garnet porphyroblasts that lack irregular and branching morphologies, a typical feature of many metamorphic belts, may grow at a high degree of disequilibrium under interfacial-controlled growth, provided they are set in a medium where diffusion and the supply of elements is sufficient, such as a sufficient volume of metamorphic fluid.

Abbreviations

FFT: fast Fourier transform; QuiG: quartz in garnet barometry; RLC: rate-limiting component.

Competing interests

The author declares that he has no competing interests.

Acknowledgements

The author would like to express his gratitude to Dr. Tetsuo Kawakami and Kazuhiko Ishii who were conveners of the Japan Geoscience Union session on "Deformed Rocks, Metamorphic Rocks and Tectonics" for encouraging me to submit this paper to PEPS. I also thank two anonymous reviewers for their constructive reviews, comments, and suggestions that helped to improve the manuscript.

References

Ague JJ, Carlson WD (2013) Metamorphism as garnet sees it: the kinetics of nucleation and growth, equilibration, and diffusional relaxation. Elements 9:439–445

Carlson WD (1989) The significance of intergranular diffusion to the mechanisms and kinetics of porphyroblast crystallization. Contrib Mineral Petrol 103:181–200

Fisher GW (1978) Rate laws in metamorphism. Geochim Cosmochim Acta 42:1035–1050

Gaidies F, Pattison DRM, de Capitani C (2011) Toward a quantitative model of metamorphic nucleation and growth. Contrib Mineral Petrol 162:974–993

Holland TJB, Powell R (2011) An improved and extended internally consistent thermodynamic dataset for phases of petrological interest, involving a new equation of state for solids. J Metamorph Geol 29:333–383

Ikeda T (1990) Ti endmember composition of biotite. Ti substitution in biotite from Ryoke metamorphic rocks in the Yanai district. J Mineral Petrol Sci 85:357–363

Kelly ED, Carlson WD, Ketchman RA (2013a) Crystallization kinetics during regional metamorphism of porphyroblastic rocks. J Metamorph Geol 31:963–979

Kelly ED, Carlson WD, Ketchman RA (2013b) Magnitudes of departures from equilibrium during regional metamorphism of porphyroblastic rocks. J Metamorph Geol 33:981–1002

Koshizuka S (2005) The MPS method. Maruzen Ltd, Tokyo

Lifshitz IM, Slyozov VV (1961) The kinetics of precipitation from supersaturated solid solutions. J Phys Chem Solids 19:35–50

Miyazaki K (1991) Ostwald ripening of garnet in high P/T metamorphic rocks. Contrib Mineral Petrol 108:118–128

Miyazaki K (1996) A numerical simulation of textural evolution due to Ostwald ripening in metamorphic rocks: a case for small amount of volume of dispersed crystals. Geochim Cosmochim Acta 60:277–290

Miyazaki K (1999) Thermobarometry of the Tsukuba Metamorphic Rocks. Bull Geol Surv Japan 50:515–525

Miyazaki K (2001) Heterogeneous growth of cordierite in low P/T Tsukuba metamorphic rocks from central Japan. J Metamorph Geol 19:155–164

Miyazaki K, Sasada M, Yoshioka T (1996) Geology of the Makabe district, with geological sheet map at 1:50,000. Geological Survey of Japan, Tsukuba

Mullins WW, Sekerka RF (1963) Morphological stability of a particle growing by diffusion or heat flow. J Applied Phys 34:323–329

Pattison DRM, Tinkham DK (2009) Interplay between equilibrium and kinetic in prograde metamorphism of pelites: an example from the Nelson aureole, British Columbia. J Metamorph Geol 27:249–279

Pattison DRM, de Capitani C, Gaides F (2011) Petrological consequences of variations in metamorphic reaction. J Metamorph Geol 29:953–977

Spear FS, Daniel CG (2001) Diffusion control of garnet growth, Harpswell Neck, Maine, USA. J Metamorph Geol 19:179–195

Spear FS, Thomas JB, Hallet BW (2014) Overstepping the garnet isograd: a comparison of QuiG barometry and thermodynamic modeling. Contrib Mineral Petrol 168:1059

Walther JV, Wood BJ (1984) Rate and mechanism in prograde metamorphism. Contrib Mineral Petrol 88:246–259

Wilbur DE, Ague JJ (2006) Chemical disequilibrium during garnet growth: Monte Carlo simulations of natural crystal morphologies. Geology 34:689–692

16

Biogeochemistry and limnology in Antarctic subglacial weathering: molecular evidence of the linkage between subglacial silica input and primary producers in a perennially ice-covered lake

Yoshinori Takano[1*], Hisaya Kojima[2], Eriko Takeda[2], Yusuke Yokoyama[1,3] and Manabu Fukui[2]

Abstract

We report a 6,000 years record of subglacial weathering and biogeochemical processes in two perennially ice-covered glacial lakes at Rundvågshetta, on the Soya Coast of Lützow-Holm Bay, East Antarctica. The two lakes, Lake Maruwan Oike and Lake Maruwan-minami, are located in a channel that drains subglacial water from the base of the East Antarctic ice sheet. Greenish-grayish organic-rich laminations in sediment cores from the lakes indicate continuous primary production affected by the inflow of subglacial meltwater containing relict carbon, nitrogen, sulfur, and other essential nutrients. Biogenic silica, amorphous hydrated silica, and DNA-based molecular signatures of sedimentary facies indicate that diatom assemblages are the dominant primary producers, supported by the input of inorganic silicon (Si) from the subglacial inflow. This study highlights the significance of subglacial water-rock interactions during physical and chemical weathering processes and the importance of such interactions for the supply of bioavailable nutrients.

Keywords: Antarctic ice sheet; Subglacial biogeochemistry; Subglacial limnology; Sedimentary record; Siliceous primary producer

Background

Subglacial Antarctic lakes were first identified by radio-echo sounding in the late 1960s (Robin *et al.* 1970; Oswald and Robin 1973). Since then and especially during the past two decades, researchers have identified numerous subglacial lakes (e.g., Kapitsa *et al.* 1996; Siegert *et al.* 1996; Jouzel *et al.* 1999; Karl *et al.* 1999; Priscu *et al.* 1999; Christner *et al.* 2006) and extensive networks of subglacial meltwater channels in Antarctica (e.g., Anderson *et al.* 2002; Wingham *et al.* 2006). Siegert and coworkers compiled an inventory of 145 subglacial lakes beneath the East Antarctic Ice Sheet and the West Antarctic Ice Sheet (EAIS and WAIS) (Siegert 2000; Siegert *et al.* 2005). The subglacial water, which initially derives from melting caused by geothermal heat (heat flow rates, approximately 50 mW m^{-2}; Siegert *et al.* 2012), is involved in various water-rock interaction processes beneath the ice sheet, and these interactions play important roles in the supply of nutrients, including trace metals, to organisms in Antarctic environments. For example, silicon (Si) is one of the critical elements limiting the growth of Antarctic diatoms (e.g., Nelson and Treguer 1992), likewise nitrogen and other elements (e.g., Hutchins and Bruland 1998). Moreover, glacial input of particulate and dissolved Fe is essential to biological productivity in the Southern Ocean (Raiswell and Canfield 2012). In addition to the nutrient contributed by seasonal snowmelt (Hodson 2006), subglacial meltwater flowing through channels may also influence the productivity and diversity of microbial communities by controlling the concentrations of nutrients and the physicochemical conditions of glacial environments (e.g., Tranter *et al.* 2005; Esposito *et al.* 2006; Mikucki *et al.* 2009; Bentley *et al.* 2011).

At the margins of the EAIS and WAIS, there are a number of locations where former subglacial lakes have

* Correspondence: takano@amstec.go.jp
[1]Department of Biogeochemistry, Japan Agency for Marine-Earth Science and Technology (JAMSTEC), 2-15 Natsushima, Yokosuka, Kanagawa 237-0061, Japan
Full list of author information is available at the end of the article

emerged from beneath the ice after the deglaciation since the Last Glacial Maximum (LGM, ca. 20 ka; Yokoyama and Esat 2011: lake settings, e.g., Hodgson *et al.* 2006; Hodgson *et al.* 2009; Verleyen *et al.* 2011). Glacial lakes affected by the input of subglacial water can be observed at the retreating margins of the Antarctic ice sheet. In the Rundvågshetta area on the Soya Coast of Lützow-Holm Bay, East Antarctica (Figure 1a), freshwater flows from the subglacial drainage channels of the EAIS (Anderson *et al.* 2002, and literature therein). The objective of this study was to examine the interactions between the limnology of the subglacial water input and microbiological responses to the subglacial water discharged into a perennially ice-covered glacial lake at Rundvågshetta (i.e., Lake Maruwan Oike) over the last 6,000 years. By combining geochemical data and the molecular signatures preserved in sediment core samples, this study aimed to clarify the primary factors controlling biological facies patterns in subglacial-water-fed lakes.

Methods

Geological setting of the sampling sites

Since the initial work by Yoshikawa and Toya (1957) and Murayama (1977), studies undertaken in the Lützow-Holm Bay area of East Antarctica have contributed further insight into the limnology of Antarctic lakes and the relative sea-level changes that occurred during the Holocene. Imura *et al.* (2003) described a range of lake types in the Soya Coast region, including freshwater lakes affected by continental glaciers and saline lakes that evaporated after their isolation from seawater during the Holocene glacio-isostatic uplift. Kudoh and Tanabe (2014) reviewed the limnology and ecology of benthic microbial assemblages from saline to glacial lakes in this area. The Holocene marine limit along the Soya Coast is estimated to have been approximately 18 m above the mean sea level (AMSL), based on radiocarbon analyses of *in situ* bivalve fossils from raised beach deposits (*Laternula elliptica*; Miura *et al.* 1998). Cosmogenic-radionuclide-based dating of raised beach exposures showed that Holocene deglaciation occurred locally on the Soya Coast at ca. 10 ka (Yamane *et al.* 2011). Bassett *et al.* (2007) also compared predictive models of glacio-isostatically induced relative sea-level changes at seven other locations around the Antarctic coast, as well as on the Soya Coast.

Rundvågshetta (69°54.5′ S, 39°02′ E; Figure 1a) is a rock headland at the southwest margin of the Rundvåg Glacier (e.g., Sawagaki and Hirakawa 1997; Miura *et al.* 1998). Figure 1b shows the electric conductivity (mS m^{-1}) of the study site, confirming the origin of the lake water inflow. Two lakes on the headland, Lake Maruwan Oike (hereafter, L. Maruwan), with a water level 8 m AMSL (data from the Geographical Survey Institute 1984), and Lake Maruwan-minami (hereafter, L. Maruwan-minami), with a water level >8 m AMSL (the overflow channel of L. Maruwan-minami flows into L. Maruwan; see Figure 2a), provide interesting localities in which to evaluate the influence of subglacial water input on the biogeochemical and sedimentological characteristics of the lakes into which it is discharged (Table 1). It has been reported that the maximum thickness of the surface ice in the Soya Coast region is less than 2 m (Imura *et al.* 2003). Figure 2 shows lakes Maruwan and Maruwan-minami and the surrounding features, including Rundvåg Glacier, a lateral moraine close to the glacier, and the outflows of lakes Maruwan and Maruwan-minami. We also observed a number of erratic boulders derived from glacial transport processes.

Lacustrine sediments were collected using push-type corers (diameter, < 8 cm; see the core images in Figure 3) during the 47th Japanese Antarctica Research Expedition (December 2005). The latitude and longitude of the sampling position in L. Maruwan (hereafter, sample Mw5S; core length, 156 cm) were recorded with GPS as 69.54′27.7″ S and 39.02′46.7″ E, respectively. The water depth and the ice thickness at this site were 20.2 and 1.5 m, respectively (cf. surface sediment description; Watanabe *et al.* 2013). We also collected lake sediments from L. Maruwan-minami (hereafter, sample Ms5S) as a reference site (see Additional file 1). The cores were cut into 3 to 10 cm intervals (to fit into the field refrigeration unit) and stored at 0 to 4°C for later geochemical analysis and at −20°C for later molecular analysis.

Geochemical analysis of sedimentary facies

The major elements (MgO, Al_2O_3, SiO_2, K_2O, CaO, TiO_2, MnO, and Fe_2O_3) were analyzed with X-ray fluorescence (XRF; JEOL JSX-3211; JEOL Ltd., Tokyo Japan), following Takano *et al.* (2012), and calibrated to Geological Survey of Japan (GSJ) standards (JA-1, JA-3, JB-1a, JB-3, JG-2, JG-3, JGb-1, JP-1, JLK-1, JLs-1, JDo-1, JSI-1, JCh-1, JR-3, JMn-1, JSd-2, and JSd-3; Imai *et al.* 1995, 1999). The data presented represent the average values of duplicate analyses. Amorphous hydrated silica (opal-A) identified with X-ray diffraction (XRD; Mac Science Co., Ltd., Yokohama, Japan) was used as a proxy for amorphous biogenic silica derived from diatoms (siliceous primary producers; e.g., Kastner *et al.* 1977; Leng and Barker 2006), with DNA-based molecular signatures as supporting data (Takano *et al.* 2012). Color data were obtained with a digital color meter (Konica Minolta, Tokyo, Japan; SPAD 503) and revised Standard Soil Color Charts (Oyama and Takehara 2005).

The methods used for elemental and bulk isotopic analyses are described by Takano *et al.* (2012). Briefly, the analyses of carbon, nitrogen, and sulfur were performed

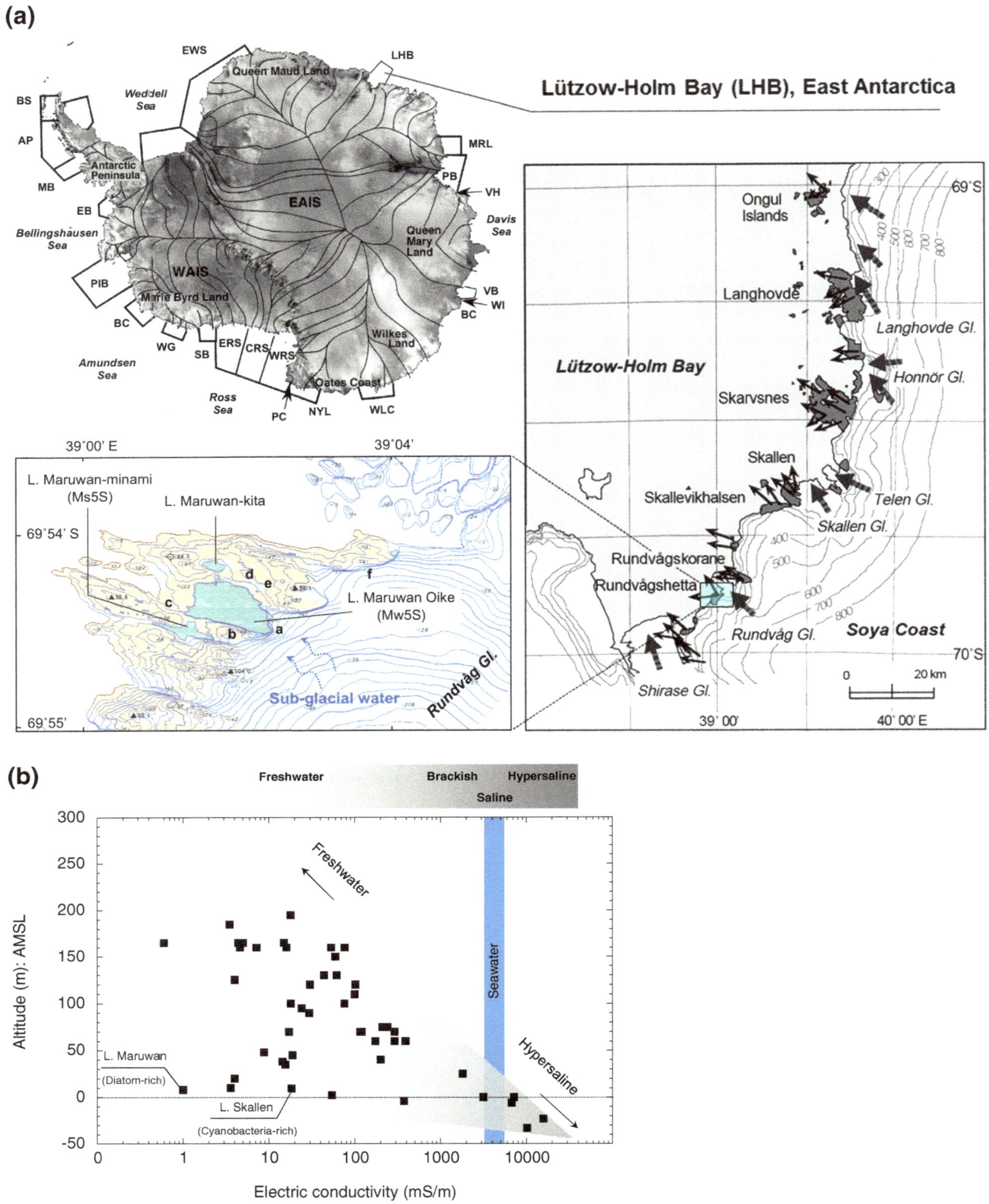

Figure 1 (See legend on next page.)

(See figure on previous page.)
Figure 1 Drainage map of the Antarctic Ice Sheet and electric conductivity (mS m^{-1}) of the study site. (a) (upper left) Drainage map of the Antarctic Ice Sheet showing areas where marine and geological surveys are being conducted to record the extent of the Last Glacial Maximum (LGM). (right) Detail of the Soya coast, Lützow-Holm Bay, East Antarctica, showing the locations of ice-covered (white) and ice-free (shaded) areas at Rundvågshetta, Skallen, Skarvsnes, and Langhovde. Arrows show the flow directions of the present outlet glaciers (after Sawagaki and Hirakawa 1997). (lower left) Topography at Rundvågshetta, showing the locations of lakes Maruwan, Maruwan-minami, and Maruwan-kita. Labels a to f indicate the locations of the images in Figure 2. NWWS, northwestern Weddell Sea; BS, Bransfield Strait; AP, Antarctic Peninsula; MB, Marguerite Bay; EB, Eltanin Bay; PIB, Pine Island Bay; BC, Bakutis Coast; WG, Wrigley Gulf; SB, Sulzberger Bay; WRS, Western Ross Sea; CRS, Central Ross Sea; ERS, Eastern Ross Sea; NVL, Northern Victoria Land; WLC, Wilkes Land Coast; PC, Pennell Coast; BC, Budd Coast; WI, Windmill Islands; PB, Petersen Bank; VB, Vincennes Bay; VH, Vestfold Hills; PB, Prydz Bay; MRL, Mac. Robertson Land; LHB, Lützow-Holm Bay; EWS, eastern Weddell Sea; SWS, southwestern Weddell Sea (modified from Anderson *et al.* 2002). **(b)** Diagram showing the altitudes of lakes (in meters above the mean sea level; AMSL) on the Soya Coast and electrical conductivities of lake waters (in mS/m) obtained by Imura *et al.* (2003), combined with data from L. Maruwan (diatom-rich microflora; this study) and L. Skallen (cyanobacteria-rich microflora; Takano *et al.* 2012). Note that the altitude data for L. Maruwan (8.0 m) and L. Skallen (9.64 ± 0.02 m) were referred from Geographical Survey Institute (1984) and Takano *et al.* (2012), respectively. See the limnological features for the numbers of lakes (Kudoh and Tanabe 2014). Seawater electrical conductivity (e.g., Cox *et al.* 1967; Lee *et al.* 2006), salinity (e.g., Koblinsky *et al.* 2003), and a variety of electrical conductivity measurements in hypersaline lakes (e.g., Williams and Sherwood 1994) are also given.

using a Micro CORDER JM10 (J-Science Lab Co., Ltd, Kyoto, Japan). Carbon and nitrogen isotopic ratios were determined using an elemental analyzer (Costech 4010 or Flash 2000)-isotope ratio mass spectrometer (Thermo Finnigan, Delta Plus, or Thermo Finnigan Delta V Advantage). Carbon and nitrogen isotopic compositions are expressed as per mil (‰) deviations from the standard as:

$$\delta^{13}C = \left[\left({}^{13}C/{}^{12}C\right)_{sample}/\left({}^{13}C/{}^{12}C\right)_{standard}-1\right] \times 1,000(‰ \text{ vs. PDB})$$

$$\delta^{15}N = \left[\left({}^{15}N/{}^{14}N\right)_{sample}/\left({}^{15}N/{}^{14}N\right)_{standard}-1\right] \times 1,000(‰ \text{ vs. Air}),$$

respectively.

The standard deviations for the carbon and nitrogen isotopic compositions obtained using authentic standard reagents (cf. Tayasu *et al.* 2011) were as follows: BG-A ($n = 12$, $\delta^{13}C < \pm 0.26‰$, $\delta^{15}N < \pm 0.25‰$), BG-P ($n = 6$, $\delta^{13}C < \pm 0.05‰$, $\delta^{15}N < \pm 0.24‰$), and BG-T ($n = 9$, $\delta^{13}C < \pm 0.08‰$, $\delta^{15}N < \pm 0.26‰$) in the first validation and BG-A ($n = 10$, $\delta^{13}C < \pm 0.12‰$, $\delta^{15}N < \pm 0.26‰$), BG-P ($n = 6$, $\delta^{13}C < \pm 0.06‰$, $\delta^{15}N < \pm 0.18‰$), and BG-T ($n = 7$, $\delta^{13}C < \pm 0.11‰$, $\delta^{15}N < \pm 0.38‰$) in the second validation. Some of the organic carbon fractions extracted after HCl pretreatment were analyzed to obtain radiocarbon age data, corrected for the $\delta^{13}C$ value, using an accelerator mass spectrometer (AMS) at the University of Tokyo, Japan (Yokoyama *et al.* 2007), or at Beta Analytic Inc., Florida, USA.

DNA-based analyses

DNA was analyzed in 18 sections (0 to 2, 2 to 4, 4 to 6, 8 to 10, 13 to 14, 14 to 16, 22 to 24, 30 to 32, 38 to 40, 49 to 52, 52 to 55, 55 to 58, 58 to 61, 61 to 64, 78 to 80, 95 to 100, 115 to 120, and 135 to 140 cm) of the core sample obtained from L. Maruwan; DNA was extracted from a 0.5 to 1.0 g sample from each section using an UltraClean Soil DNA Isolation Kit (Mobio, Carlsbad, CA, USA). Polymerase chain reaction (PCR) amplification of the 16S rRNA gene, denaturing gradient gel electrophoresis (DGGE), and sequencing of the resulting DGGE bands were performed as described previously (Muyzer *et al.* 1996; Takano *et al.* 2012). The sequences were taxonomically identified with the RDP Classifier from the Ribosomal Database Project-II release 11.

Results

Geochemistry, major elemental compositions, and sedimentary facies

The olive-gray, olive-black, and dark olive-gray laminations in the upper sections of core Mw5S from L. Maruwan are the remains of soft microbial mats. The laminations in the lower sections of the core are largely gray to olive-black (Figure 3a). The distinct laminations represent conditions favorable for preservation (i.e., no bioturbation). Some sediment deformation was observed at depths of 0 to 5, 10, and 50 to 60 cm. The lithological descriptions and the corresponding ^{14}C ages of organic carbon (years BP) are shown in Figure 3b. Geochemical analyses of carbon, nitrogen, sulfur, and major elements were used to identify the sedimentary facies.

Figure 4a shows depth profiles of the concentrations of the major elements and the carbon and nitrogen isotope ratios. Total carbon (TC) and the carbon isotopic composition ($\delta^{13}C$) of bulk organic matter ranged from 0.7 to 3.5 wt% (mean, 2.6 ± 0.6 wt%) and from −17.0‰ to −23.6‰ (vs. PDB), respectively. Total nitrogen (TN) and the nitrogen isotopic composition ($\delta^{15}N$) of the bulk organic matter were <0.48 wt% (mean, 0.3 ± 0.1 wt%) and from +2.2‰ to +6.6‰ (vs. Air), respectively. Total sulfur (TS) ranged from 0.3 to 1.6 wt% (mean, 1.0 ± 0.3 wt%). The profiles show that the increasing trends in carbon and nitrogen with the movement from marine to freshwater facies in L. Skallen and L. Oyako (Takano *et al.* 2012; Matsumoto *et al.* 2014) are not observed in L. Maruwan. The sedimentary facies were not distinguishable on the basis of their nitrogen isotopic

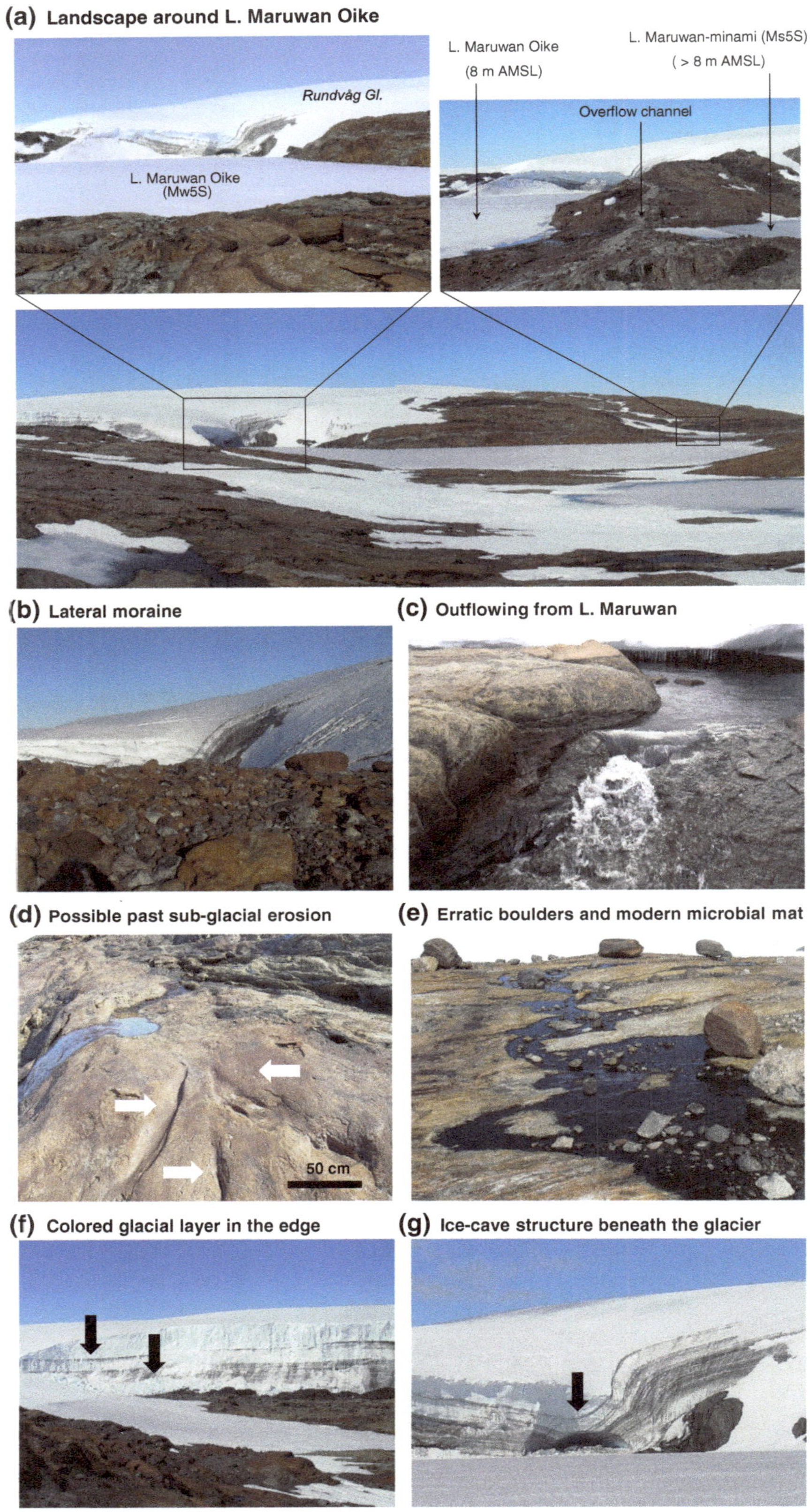

Figure 2 (See legend on next page.)

(See figure on previous page.)
Figure 2 Terrain in the study area, showing lakes Maruwan, Maruwan-minami, and Maruwan-kita and associated geomorphic features. The locations of the images are shown by labels a to g in Figure 1a (lower left panel). **(a)** Overall features of the study area, showing the proximity of the lakes to Rundvåg Glacier, and the overflow points of lakes Maruwan and Maruwan-minami. **(b)** Lateral moraine close to Rundvåg Glacier. **(c)** Outflow of L. Maruwan. **(d)** Possible subglacial erosion features (after Sawagaki and Hirakawa 1997). **(e)** Present-day microbial mat (dark area) and erratic boulders transported and deposited by past glaciers. **(f)** Exposed stratification in glacial ice. **(g)** Detail of (a), showing a cave-like structure beneath the glacier.

compositions because of the relic nitrogen inputs from subglacial water ($\delta^{15}N$ up to +6.6‰ in Mw5S; cf. L. Skallen, in the supporting data of Matsumoto *et al.* 2010).

The trends in the contents of some major elements indicate a marked transition from marine to lacustrine conditions in each of the cores (Figure 4a). The sediments deeper than 50 cm depth in the Mw5S core were substantially enriched in SiO_2 (<83%), indicating high SiO_2 end-member values under marine conditions and low SiO_2 values under freshwater conditions. The mixture of these two end-members can be illustrated on SiO_2-Al_2O_3 and SiO_2-Fe_2O_3 diagrams, in which contrasting mixing trends are observed (Figure 4b,c). The mixing profiles in these diagrams are similar to those of the reference sample obtained from L. Maruwan-minami (Ms5S; Additional file 1), which is located nearby. The Al_2O_3/TiO_2 and Fe_2O_3/TiO_2 ratios (Figure 4d) provide information about the source rocks (Young and Nesbitt 1998) episodic hydrological surge events (at 0 to 5 and 50 to 60 cm) and subglacial discharge variations (e.g., Brown 2002).

X-ray diffraction analyses of the sediments from core Mw5S indicate that opal-A is abundant throughout the section, except at 0 to 5 cm, suggesting that the silicate is of biogenic origin (Figure 5). To confirm this observation, we also obtained DNA-based molecular evidence of biogenic signatures (see Figure 5c). The high end-member SiO_2 concentrations, which also suggest large contributions of biogenic silica to the sediments, were probably derived from abundant diatoms (cf. determination of opal-A in the discussion).

Radiocarbon age determinations of sedimentary organic matter

We determined the AMS radiocarbon (^{14}C) ages of the organic matter in the core samples (Figure 3b and Table 2). The ^{14}C values (years BP) at the top of the core (1,350 ± 40 years in Mw5S at a depth of 0 to 2 cm) indicate significant inputs of relic carbon. Therefore, the sediment age at the top of the core of 1,300 years was subtracted from the raw values to calibrate the radiocarbon dates with the corresponding calendar ages (after Berkman and Forman 1996; Miura *et al.* 2002). Similarly, marine benthic organisms have yielded ^{14}C ages of 1,010 ± 110 to 1,190 ± 90 years BP along the Soya Coast, as reported by Yoshida and Moriwaki (1979). Using this protocol, we estimated the age at which the first sediments were deposited in the bedrock-scoured glacial basin of L. Maruwan (Figures 1 and 4a) to be at least 4,807 to 5,204 cal BP (2σ). Based on the major elemental compositions (Figure 4b), we also obtained an emergence age (i.e., the transition from marine to freshwater) of 3,382 to 3,560 cal BP (2σ); the median age of 3,471 cal BP occurs at a depth of 46 to 49 cm. Here, we note the possibility that the period from 4,000 to 1,350 cal BP represented a hiatus in the sedimentation processes in L. Maruwan. The sedimentary facies in L. Maruwan-minami, in contrast, did not show a similar profile (Additional file 1). Assuming the continuous inflow of subglacial water, the transitions from marine to freshwater conditions *via* temporary brackish conditions are indistinct in the L. Maruwan cores compared with the transitions observed at lakes Skallen and Oyako, also located on the Soya Coast (Takano *et al.* 2012).

Preliminary results for biological facies and molecular signatures from DGGE analysis

A PCR-DGGE analysis was performed to assess the shifts in the composition of residual DNA in the sediments of L. Maruwan (Figure 5c). Depth-related changes were apparent in regions shallower than 6 cm. In the layers deeper than 4 cm, two predominant bands (labeled with arrows in Figure 5c) were consistently observed throughout the range analyzed. The nucleotide sequences of the two bands were identical, corresponding to the chloroplasts of the marine diatom *Chaetoceros*, and were also identical to the diatom sequence detected in the deeper sediment layers of L. Skallen (Takano *et al.* 2012). The other sequenced DGGE bands (labeled with triangles in Figure 5c) represented taxa of nonphototrophic bacteria determined with a 16S rRNA analysis (unpublished data).

Discussion

Primary producers and estimates of biogenic silica

The lake sediments in core Mw5S (L. Maruwan) record a continuous input of relic carbon, nitrogen, and bioavailable silica from subglacial water, indicated by their chemical constituents, including major elements and isotopes. Hodson *et al.* (2010) experimentally verified the chemical weathering and solute production processes in Antarctic glacial meltwater based on aspects of water-rock interactions and hydrochemistry. Stumpf *et al.* (2012) and Tranter *et al.* (2005) also suggested the important influence of subglacial weathering on glacial meltwater chemistry, and the potential

Table 1 Concentrations of major elements (average values of duplicate analyses), TC, TN, and TS in core Mw5S from L. Maruwan

Depth (cm)	Mid-depth	MgO (wt%)	Al_2O_3	SiO_2	K_2O	CaO	TiO_2	MnO	Fe_2O_3	Carbon (wt%)	Nitrogen	Sulfur	C/N	^{14}C age (years BP)	1σ	$\delta^{13}C$ (‰ vs. PDB)	$\delta^{15}N$ (‰ vs. Air)
0 to 2	1.0	3.0	10.5	46.6	3.4	2.8	0.91	0.17	7.5	1.3	0.00	0.3	-	1,350	40	−18.0	+6.6 ($n = 2$)
2 to 4	3.0	2.4	9.1	53.4	2.9	2.8	0.78	0.11	6.5	1.6	0.00		-				
4 to 6	5.0	0.5	4.1	67.6	1.6	2.1	0.35	0.05	3.9	2.5	0.30	1.2	8.5				+4.3 ($n = 2$)
6 to 8	7.0	0.0	3.2	67.2	1.4	1.9	0.30	0.05	3.7	2.8	0.36		7.9	3,950	30	−21.6	+3.1
8 to 10	9.0	0.7	5.9	59.3	2.2	2.5	0.50	0.07	5.0	2.3	0.27	1.6	8.7				
10 to 12	11.0	1.8	6.9	58.9	2.2	2.7	0.50	0.07	4.8	2.0	0.24		8.3	4,200	30	−20.8 ($n = 2$)	+4.9 ($n = 2$)
12 to 13	12.5	1.4	6.2	60.7	1.9	2.6	0.43	0.07	4.3	1.9	0.23	1.2	8.1				
13 to 14	13.5	1.3	5.4	63.7	1.7	2.4	0.39	0.06	4.0	2.0	0.26		7.5				
14 to 16	15.0	1.7	7.0	58.3	1.9	2.9	0.45	0.07	4.8	0.7	0.00	1.1	-			−19.3	+5.9
16 to 18	17.0	0.0	2.0	67.2	1.0	1.8	0.21	0.04	3.0	2.3	0.27		8.7				
18 to 20	19.0	0.0	2.1	74.4	1.1	1.9	0.24	0.04	3.2	2.4	0.30	1.0	8.1	3,920	40	−23.6	+3.2
20 to 22	21.0	0.1	1.9	74.6	1.1	1.8	0.23	0.04	3.3	2.8	0.35		8.0				
22 to 24	23.0	0.3	3.8	65.8	1.5	2.1	0.31	0.05	3.9	3.1	0.37	1.5	8.5				
24 to 26	25.0	0.3	3.6	66.6	1.5	2.1	0.31	0.05	3.6	2.2	0.24		9.2	4,410	30	−20.5	+2.2
26 to 28	27.0	0.2	3.3	70.9	1.4	2.0	0.29	0.05	3.6	3.2	0.38	1.3	8.3				
28 to 30	29.0	0.0	1.4	74.7	0.9	1.6	0.22	0.04	2.6	3.0	0.36		8.2			−21.0	+4.3
30 to 32	31.0	0.0	0.9	67.3	0.7	1.5	0.13	0.05	2.8	2.7	0.35	1.1	7.8				
32 to 34	33.0	0.0	3.0	64.6	1.3	2.0	0.30	0.06	3.7	2.4	0.30		8.0	4,630	30	−21.0	+4.8
34 to 36	35.0	0.0	3.9	65.8	1.6	2.1	0.32	0.06	3.8	2.6	0.36	1.5	7.1			−20.9	+5.1
36 to 38	37.0	0.0	2.8	69.6	1.3	2.0	0.28	0.05	3.8	3.0	0.39		7.6				
38 to 40	39.0	0.1	3.1	71.9	1.3	2.0	0.30	0.05	3.7	2.4	0.32	1.1	7.5				
40 to 43	41.5	0.0	2.6	73.4	1.2	1.9	0.26	0.04	3.3	2.9	0.40		7.2			−20.4	+4.2 ($n = 2$)
43 to 46	44.5	0.0	2.7	60.8	1.2	1.9	0.23	0.04	2.8	2.6	0.36	1.3	7.1				
46 to 49	47.5	0.6	4.4	65.6	1.7	2.1	0.37	0.06	3.6	3.5	0.43		8.0	4,540	40		
49 to 52	50.5	1.0	6.1	62.6	2.0	2.3	0.46	0.06	4.0	2.7	0.31	1.0	8.8			−21.2	+5.8
52 to 55	53.5	1.1	6.8	59.0	2.2	2.5	0.51	0.07	4.2	2.3	0.28		8.1				
55 to 58	56.5	2.6	8.3	55.2	2.6	2.7	0.64	0.08	4.6	1.9	0.22	0.8	8.7				
58 to 61	59.5	1.8	8.1	55.6	2.6	2.7	0.62	0.08	4.5	2.4	0.24		9.9			−21.1	+5.7
61 to 64	62.5	0.2	3.3	72.0	1.3	1.9	0.27	0.05	2.9	2.8	0.36	0.9	7.9				
64 to 67	65.5	0.0	1.3	76.4	0.9	1.6	0.17	0.04	2.8	3.1	0.43		7.3				
67 to 70	68.5	0.0	1.5	76.4	0.9	1.7	0.18	0.04	2.8	3.0	0.40	1.1	7.6			−21.2	+5.4

Table 1 Concentrations of major elements (average values of duplicate analyses), TC, TN, and TS in core Mw5S from L. Maruwan *(Continued)*

70 to 75	72.5	0.0	0.7	79.0	0.7	1.5	0.14	0.03	2.1	3.8	0.50		7.7	5,220	40		
75 to 80	77.5	0.0	0.5	78.5	0.6	1.5	0.12	0.03	2.0	2.8	0.34	0.7	8.4				
80 to 85	82.5	0.0	0.4	80.9	0.6	1.4	0.10	0.03	1.7	3.3	0.48		6.9			−20.6	+4.9
85 to 90	87.5	0.0	0.0	83.2	0.4	1.4	0.07	0.02	1.4	3.1	0.38	0.6	8.2				
90 to 95	92.5	0.0	0.3	82.1	0.5	1.4	0.09	0.02	1.6	2.9	0.35		8.3	5,520	90	−21.2	
95 to 100	97.5	0.0	0.2	80.9	0.6	1.5	0.11	0.03	1.7	3.1	0.37	0.7	8.3			−21.8	+4.5
100 to 105	102.5	0.0	1.0	77.4	0.8	1.6	0.15	0.03	2.2	3.1	0.39		7.8				
105 to 110	107.5	0.0	0.8	78.5	0.8	1.6	0.14	0.03	2.0	2.8	0.35	0.8	7.9	5,400	30	−20.9	
110 to 115	112.5	0.0	0.3	79.4	0.7	1.6	0.12	0.03	2.0	3.2	0.40		8.0				
115 to 120	117.5	0.0	0.4	80.6	0.6	1.5	0.11	0.03	1.7	3.0	0.40	0.8	7.6			−20.9	+4.2
120 to 125	122.5	0.0	0.6	80.7	0.6	1.4	0.11	0.02	1.6	3.3	0.44		7.5				
125 to 130	127.5	0.0	0.3	81.7	0.5	1.4	0.10	0.02	1.4	3.1	0.42	0.7	7.4	5,350	40	−17.0	+2.4
130 to 135	132.5	0.0	1.7	75.3	0.9	1.7	0.18	0.04	2.3	2.6	0.32		8.2				
135 to 140	137.5	0.0	1.7	76.0	0.9	1.7	0.18	0.03	2.3	2.7	0.34	0.9	7.9				
140 to 145	142.5	0.0	1.0	79.1	0.7	1.6	0.13	0.03	1.9	2.8	0.38		7.3				
145 to 150	147.5	0.0	1.2	77.7	0.9	1.8	0.17	0.04	2.4	2.4	0.31	0.8	7.8				
150 to 156	153.0	0.0	0.6	79.8	0.7	1.6	0.12	0.03	1.8	2.7	0.33		8.3	6,010	70	−21.4 ($n = 2$)	+4.5

Radiocarbon (^{14}C age), carbon, and nitrogen isotopic compositions at corresponding depths are also shown.

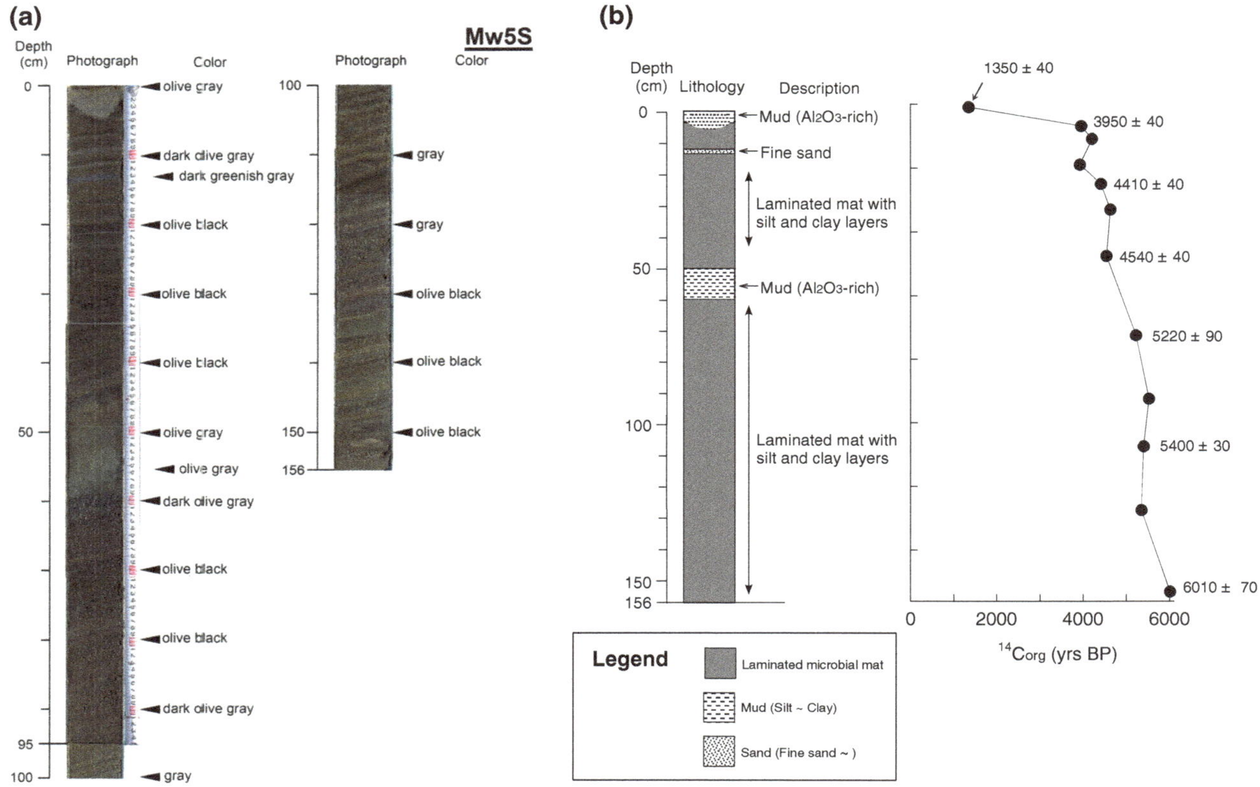

Figure 3 Core images of lacustrine sediments and lithological descriptions. (a) Cross-section of a core sample obtained from L. Maruwan (sample Mw5S) and **(b)** lithological descriptions and uncorrected ^{14}C ages (years BP) of organic carbon. The color data in (a) were obtained with a digital color meter (Konica Minolta, SPAD 503) and revised Standard Soil Color Charts (Oyama and Takehara 2005).

roles of meltwater constituents as sources of energy (e.g., organic carbon) and nutrients (e.g., nitrogen, sulfur, and phosphorus). The concentration of bulk sulfur was also significant in the freshwater sedimentary stage (Figure 4). Using a functional gene marker, Watanabe *et al.* (2013) pointed out the importance of sulfur cycling and microbial interactions in freshwater environments, including in the modern sediments of L. Maruwan. With respect to the input from physical weathering, clay-sized particles of aluminosilicate minerals are generated by subglacial abrasion processes. These solid-phase materials are transported as suspended sediments into the lake, where they liberate dissolved silica, which can be utilized by diatoms (Figure 6).

With respect to the input from chemical weathering, the simple weathering of rock minerals, including hydrolysis processes involving carbonates, silicates, and aluminosilicates, probably control the composition of the inputs (cf. chemical models: Raiswell 1984). These processes include the following:

$$CaCO_3(s) + H^+(aq) \rightarrow Ca^{2+}(aq) + HCO_3^-(aq)$$
$$Mg_2SiO_4(s) + 4H^+(aq) \rightarrow 2\ Mg^{2+}(aq) + H_4SiO_4(aq)$$
$$2NaAlSi_3O_8(s) + 2H^+(aq) + 9H_2O \rightarrow 2Na^+(aq) + 4H_4SiO_4(aq) + Al_2Si_2O_5(OH)_4(s)$$

where $CaCO_3(s)$, $Mg_2SiO_4(s)$, and $NaAlSi_3O_8(s)$ denote calcite, forsterite, and albite, respectively. We observed increasing trends in the concentrations of CaO, MgO, and K_2O in the lacustrine facies after the marine-freshwater transition (Figure 4). The water-rock interactions involving forsterite (Satish-Kumar and Wada 2000) and albite (cf. Shiraishi *et al.* 1987) in the Lützow-Holm complex will eventually produce dissolved silica as H_4SiO_4 (aq) and a clay mineral (kaolinite) as $Al_2Si_2O_5(OH)_4(s)$, respectively. It is particularly relevant that the dissolved silica is biologically available as an essential nutrient and growth element for diatoms.

Because biogenic opal-A is a major component of the L. Maruwan sediments, we were able to estimate the concentrations of biogenic silica in the sediment. We calculated the flux of biogenic silica in core Mw5S based on the end-member SiO_2 and Al_2O_3 concentrations, assuming that: (i) Al is only contained in the detrital component (of glacial inorganic origin) because Al_2O_3

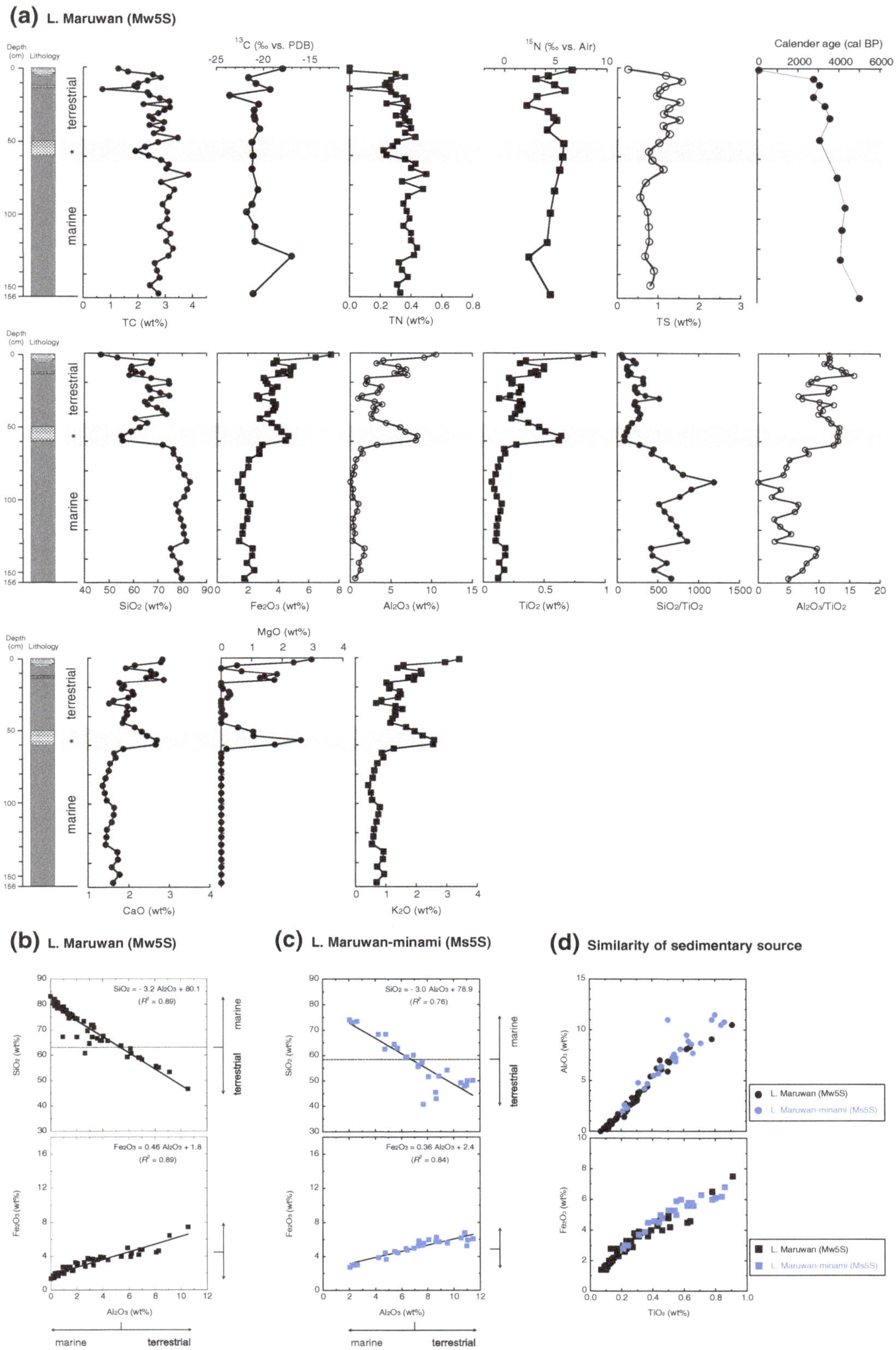

Figure 4 (See legend on next page.)

(See figure on previous page.)

Figure 4 Vertical profiles of some major elements, SiO_2-Al_2O_3 and SiO_2-Fe_2O_3 diagrams, bivariate TiO_2-Al_2O_3 and TiO_2-Fe_2O_3 plots. **(a)** Vertical profiles of lithologies, total carbon (TC), $\delta^{13}C_{org}$ (‰ vs PDB), total nitrogen (TN), $\delta^{15}N$ (‰ vs air), total sulfur (TS), ^{14}C ages, wt% of major elements (SiO_2, Fe_2O_3, Al_2O_3, TiO_2, CaO, MgO, K_2O), and major elemental ratios (SiO_2/TiO_2 and Al_2O_3/TiO_2) in core Mw5S from L. Maruwan, showing the transition between the marine and terrestrial stages *via* possible brackish conditions (*based on microflora differences at a depth of 60 cm, determined with a DGGE analysis). The raw data are provided in Table 1. **(b), (c)** Al_2O_3-SiO_2 and Al_2O_3-Fe_2O_3 diagrams for L. Maruwan (core Mw5S) and Maruwan-minami (core Ms5S) sediments. Possible end-members of marine and terrestrial major elements are noted on the axes. **(d)** The similarity of the source rocks of L. Maruwan (core Mw5S, black symbols: $R^2 = 0.97$ and 0.92) and L. Maruwan-minami (core Ms5S, blue symbols: $R^2 = 0.85$ and 0.89) was confirmed with bivariate TiO_2-Al_2O_3 and TiO_2-Fe_2O_3 plots.

concentrations are constant throughout the core at 10.5% (representing the freshwater end-member) and (ii) SiO_2 is contained in both the biogenic and detrital components (of glacial inorganic origin), with SiO_2 concentrations in the detritus constant at 46.6% (representing the freshwater end-member). Therefore,

$$\text{Total silica}(\%) = \text{Biogenic silica}(\%) + \text{Detrital silica}(\%)$$

where the detrital ratio of the end-member (Figure 4b,c) is as follows:

$$\text{Detrital silica}(\%)/\text{Detrital } Al_2O_3(\%) = 46.6(\%)/10.5(\%)$$

Hence, for L. Maruwan:

$$\text{Biogenic silica}(\%) = \text{Total silica}(\%) - 4.43 \text{ Detrital } Al_2O_3(\%)$$

and in L. Maruwan-minami:

$$\text{Biogenic silica}(\%) = \text{Total silica}(\%) - 4.36 \text{ Detrital } Al_2O_3(\%)$$

Based on the TiO_2-normalized plots of Al_2O_3 and Fe_2O_3, the similarity of the source rocks in the two lakes is high (Figure 4d), suggesting the discharge of homogenous meltwater into both lakes.

Figure 5a shows the depth profiles of the reconstructed biogenic silica concentrations in L. Maruwan. After the marine-terrestrial transition, diatoms were the main primary producers, even in the freshwater/brackish environments. Based on the DGGE and molecular analyses, the major phototrophic producers were identical to those detected in the deeper sediment layers of L. Skallen (i.e., diatoms) (Takano *et al.* 2012). Therefore, the origin of Si in the Antarctic lakes is of particular interest because dissolved silica is essential for the growth of diatoms (i.e., diatoms are the dominant group of phytoplankton with amorphous opal constituents).

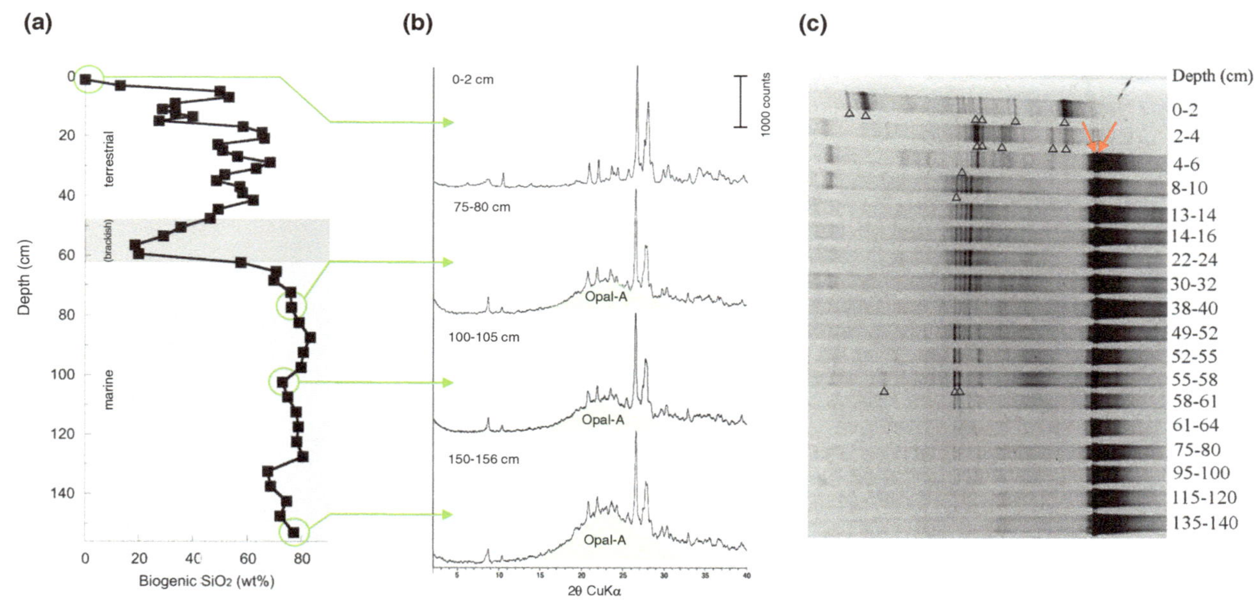

Figure 5 Depth profiles of biogenic silica concentrations and X-ray diffraction analyses of sediments from core Mw5S. **(a)** Vertical profiles of biogenic silica content (wt%), **(b)** opal-A signatures determined with X-ray diffraction analysis, and **(c)** denaturing gradient-gel electrophoresis (DGGE) analysis of L. Maruwan sediments (core Mw5S). Note that the direction of electrophoresis is from right to left.

Table 2 Radiocarbon (^{14}C) age data obtained with AMS for the organic carbon fractions from core Mw5S from L. Maruwan

Depth (cm)	Mid-depth	Conventional radiocarbon age ($\delta^{13}C$ corrected) (years BP)	1σ ±	Calendar age (2σ range) (cal BP)	Relative area under probability function
0 to 2	1.0	1,350	40	24 to 141	0.73
				220 to 262	0.25
6 to 8	7.0	3,950	30	2,737 to 2,798	0.93
				2,819 to 2,844	0.06
10 to 12	11.0	4,200	30	2,950 to 3,160	1.00
18 to 20	19.0	3,920	40	2,706 to 2,813	0.94
				2,816 to 2,844	0.03
				2,618 to 2,633	0.03
24 to 26	25.0	4,410	30	3,256 to 3,392	1.00
32 to 34	33.0	4,630	30	3,475 to 3,637	1.00
46 to 49	47.5	4,540	40	3,382 to 3,560	1.00
70 to 75	72.5	5,220	40	3,790 to 4,053	1.00
90 to 95	92.5	5,520	90	4,056 to 4,565	1.00
105 to 110	107.5	5,400	30	4,042 to 4,274	1.00
125 to 130	127.5	5,350	40	3,950 to 4,218	1.00
150 to 156	153.0	6,010	70	4,807 to 5,204	1.00

Conventional radiocarbon dating based on organic carbon values ($^{14}C_{org}$, years BP) was corrected for the marine reservoir effect (ΔR = 1,300 years for the marine stage sequence) using core-top data. Calendar age data were calculated using the calibration programs Calib Rev 6.0.1 (Stuiver and Reimer 1993; Stuiver *et al.* 1998), INTCAL09 and MARINE09 (e.g., Hughen *et al.* 2004; Reimer *et al.* 2004, 2009). See discussions of the biogeochemical recycling processes for relic carbon in the Antarctic region (e.g., Ingolfsson *et al.* 1998; Berkman *et al.* 1998) and the Soya Coast region (e.g., Miura *et al.* 2002).

Subglacial abrasion and weathering cause continuous input of silica

Sawagaki and Hirakawa (1997) have described the formation of glacial erosional bedforms in the Rundvågshetta area. The erosional bedforms are accompanied by small erosional marks, which support the interpretation that subglacial meltwater has contributed to the erosion at the base of Antarctic glaciers (cf. Figure 6). It is important to note that some subglacial erosional features may have resulted from subglacial streams, as shown in Figure 2d. Anderson *et al.* (2002) have extensively reviewed the features of subglacial meltwater drain channels beneath continental ice sheets. Observations and interpretations of these bedforms have been used to reconstruct the historical development of glacial erosional bedforms (e.g., Shaw 2002), and to understand the significance and implications of subglacial water-rock interactions. Further support for this finding comes from the cosmogenic radionuclide dating of local glacial erratics and bedrock (Yamane *et al.* 2011). Both bedrocks and erratics show similar exposure ages, suggesting the presence of warm basal ice during the last glacial period, which would have included subglacial water channels.

We suggest that the continuous input of subglacial meltwater influenced the chemical compositions of areas marginal to the ice sheet (also, see de Mora *et al.* 1994; Brown 2002). Nelson and Treguer (1992) reported that silicon is an important limiting nutrient for Antarctic diatom productivity. Total biogenic silica (%) and total carbon contents are positively correlated in L. Maruwan (R = 0.75; this study). When inferring the past constraints on the primary production by diatoms in the ice-marginal lakes, we concluded that the subglacial weathering of silicate and aluminosilicate minerals supplied significant levels of minerals, nitrogen, and relic carbon to subglacial meltwaters (e.g., Hutchins and Bruland 1998). Therefore, dissolved inorganic silicates are constantly available for utilization by the diatom communities that are the primary producers in the perennially ice-covered lakes at Rundvågshetta.

Conclusions

[1] This study provides the first published data on the subglacial limnology and Si biogeochemistry of L. Maruwan, Rundvågshetta. The reservoir effects of relic carbon derived from the subglacial water beneath the Rundvåg Glacier were estimated at 1,300 years (cf. core-top age 1,350 years in Mw5S). The nitrogen isotopic composition of the bulk sediment was slightly ^{15}N-enriched relative to the dinitrogen in atmospheric air, even in freshwater

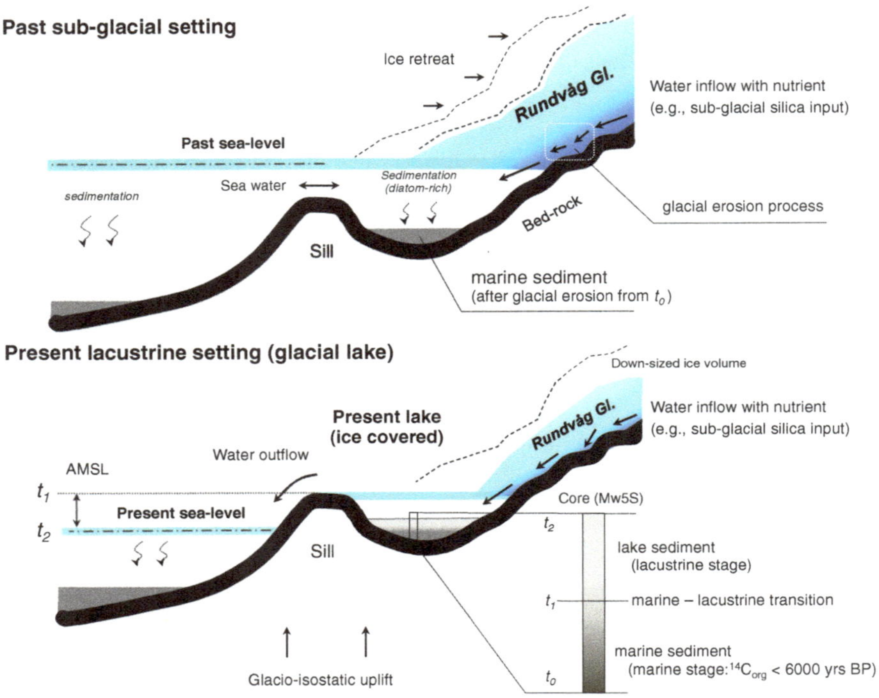

Figure 6 Inflow of subglacial meltwater containing solid-phase materials and dissolved nutrients, including silica, relic carbon, and nitrogen. Schematic models of **(a)** past lake setting after glacial erosion (i.e., the initiation of sedimentation processes during the marine phase) and **(b)** present lake setting (lacustrine stage). The relative sea level is controlled by glacio-isostatic uplift. For the present sea ice conditions during the summer season (December 2005; photograph taken from a helicopter), please see the description of the landscape of Rundvågshetta (Additional file 1).

environments (cf. sedimentary $\delta^{15}N$ in freshwater at L. Skallen). Based on the similarities in the compositions of their major sedimentary elements, lakes Maruwan and Maruwan-minami share the same meltwater source composition.

[2] We suggest that the amount of bioavailable silica flowing out from subglacial drainage channels under the EAIS and into subglacial and proglacial lakes in the area is extremely limited, underscoring the biological significance of subglacial physicochemical weathering and water-rock interactions. Inorganic Si is the main source of biogenic silica (up to 83 wt%) for lacustrine diatom communities, indicating the strong relationship between the subglacial material exported from the glacial system and the primary production in these glacial lakes during the Holocene.

Abbreviations
EAIS: East Antarctic Ice Sheet; WAIS: West Antarctic Ice Sheet; LHB: Lützow-Holm Bay; AMSL: above the mean sea level; opal-A: amorphous hydrated silica; AMS: accelerator mass spectrometry; XRF: X-ray fluorescence; XRD: X-ray diffraction; PCR: polymerase chain reaction; DGGE: denaturing gradient gel electrophoresis.

Competing interests
The authors declare that they have no competing interests.

Authors' contributions
YT compiled the geochemical data and wrote the paper. HK and ET contributed to the molecular analysis. YY supported the radiocarbon analysis. YT and MF performed the field investigations and designed the entire research project, in discussion with the coauthors. All authors read and approved the final manuscript.

Authors' information
YT is a senior research scientist at the Department of Biogeochemistry, Japan Agency for Marine-Earth Science and Technology (JAMSTEC). HK is an assistant professor at the Institute of Low Temperature Science, Hokkaido University. ET was a graduate school student at the Institute of Low Temperature Science, Hokkaido University. YY is a professor at the Atmosphere and Ocean Research Institute (AORI), University of Tokyo. MF is a professor at the Institute of Low-Temperature Science, Hokkaido University.

Acknowledgements
We thank J. Sickman (Univ. California, Riverside), J. Tyler (Univ. Adelaide), and an anonymous reviewer for critical and constructive comments which helped to improve an earlier version of the manuscript. We also thank T. Sato (Hiroshima Univ.) and S. Imura (Natl. Inst. Polar Res., NIPR) for logistical

supports, T. Sawagaki (Hokkaido Univ.) for the onsite field guide regarding past glacial erosion processes, N. Suzuki (Hokkaido Univ.) and N. Ohkouchi (JAMSTEC) for discussions, and members of the 47th Japan Antarctica Research Expedition (Expedition Leader: K. Shiraishi, NIPR) for logistical assistance. The AMS experiments were partially supported by Dr. H. Matsuzaki (Univ. Tokyo). Part of this work has been presented at the AGU Chapman Conference 2010 on the Exploration and Study of Antarctic Subglacial Aquatic Environment at Baltimore, USA. This research was supported in part by the Japan Society for the Promotion of Science (Y.T.) and NEXT program No. GR031 (Y.Y).

Author details

[1]Department of Biogeochemistry, Japan Agency for Marine-Earth Science and Technology (JAMSTEC), 2-15 Natsushima, Yokosuka, Kanagawa 237-0061, Japan. [2]The Institute of Low Temperature Science, Hokkaido University, N19, W8, Kita-ku, Sapporo, Hokkaido 060-0819, Japan. [3]Atmosphere and Ocean Research Institute, University of Tokyo, 5-1-5 Kashiwanoha, Kashiwa, Chiba 277-8564, Japan.

References

Anderson JB, Shipp SS, Lowe AL, Wellner JS, Mosola AB (2002) The Antarctic ice sheet during the Last Glacial Maximum and its subsequent retreat history: a review. Quaternary Sci Rev 21:49–70

Bassett S, Milne G, Bentley M, Huybrechts P (2007) Modelling Antarctic sea-level data to explore the possibility of a dominant Antarctic contribution to meltwater pulse IA. Quaternary Sci Rev 26:2113–2127

Bentley M, Christoffersen P, Hodgson D, Smith AM, Tulaczyk S, Le Brocq A (2011) Subglacial lake sediments and sedimentary processes: potential archives of ice sheet evolution, past environmental change, and the presence of life. Geophys Monogr Ser 192:83–110

Berkman PA, Forman SL (1996) Pre-bomb radiocarbon and the reservoir correction for calcareous marine species in the Southern Ocean. Geophys Res Lett 23:363–366

Berkman PA, Andrews JT, Bjorck S, Colhoun EA, Emslie SD, Goodwin ID, Hall BL, Hart CP, Hirakawa K, Igarashi A, Ingolfsson O, Lopez-Martinez J, Lyons WB, Mabin MCG, Quilty PG, Taviani M, Yoshida Y (1998) Circum-Antarctic coastal environmental shifts during the Late Quaternary reflected by emerged marine deposits. Antarctic Sci 10:345–362

Brown GH (2002) Glacier meltwater hydrochemistry. Appl Geochem 17:855–883

Christner BC, Royston-Bishop G, Foreman CM, Arnold BR, Tranter M, Welch KA, Lyons WB, Tsapin AI, Studinger M, Priscu JC (2006) Limnological conditions in subglacial Lake Vostok, Antarctica. Limnol Oceanogr 51:2485–2501

Cox R, Culkin F, Riley J (1967) The electrical conductivity/chlorinity relationship in natural sea water. Deep Sea Res Oceanogr Abst 14:203–220

de Mora SJ, Whitehead R, Gregory M (1994) The chemical composition of glacial melt water ponds and streams on the McMurdo Ice Shelf, Antarctica. Antarctic Sci 6:17–27

Esposito R, Horn S, McKnight D, Cox M, Grant M, Spaulding S, Doran P, Cozzetto K (2006) Antarctic climate cooling and response of diatoms in glacial meltwater streams. Geophys Res Lett 33:L07406

Geographical Survey Institute (1984) Rundvågshetta: the principal part. Map compilation from air photography, controlled by triangulation points established by JARE, 1971 and 1974

Hodgson DA, Verleyen E, Squier AH, Sabbe K, Keely BJ, Saunders KM, Vyverman W (2006) Interglacial environments of coastal east Antarctica: comparison of MIS 1 (Holocene) and MIS 5e (Last Interglacial) lake-sediment records. Quaternary Sci Rev 25:179–197

Hodgson DA, Roberts SJ, Bentley MJ, Smith JA, Johnson JS, Verleyen E, Vyverman W, Hodson AJ, Leng MJ, Cziferszky A, Fox AJ, Sanderson DCW (2009) Exploring former subglacial Hodgson Lake, Antarctica Paper I: site description, geomorphology and limnology. Quaternary Sci Rev 28:2295–2309

Hodson A (2006) c. Water Resource Res 42, W11406. doi:11410.11029/12005WR004311

Hodson A, Heaton T, Langford H, Newsham K (2010) Chemical weathering and solute export by meltwater in a maritime Antarctic glacier basin. Biogeochem 98:9–27

Hughen KA, Baillie MGL, Bard E, Beck JW, Bertrand CJH, Blackwell PG, Buck CE, Burr GS, Cutler KB, Damon PE, Edwards RL, Fairbanks RG, Friedrich M, Guilderson TP, Kromer B, McCormac G, Manning S, Ramsey CB, Reimer PJ, Reimer RW, Remmele S, Southon JR, Stuiver M, Talamo S, Taylor FW, der Plicht J, Weyhenmeyer CE (2004) Marine04 marine radiocarbon age calibration, 0-26 cal kyr BP. Radiocarbon 46:1059–1086

Hutchins DA, Bruland KW (1998) Iron-limited diatom growth and Si:N uptake ratios in a coastal upwelling regime. Nature 393:561–564

Imai N, Terashima S, Itoh S, Ando A (1995) 1994 compilation values for GSJ reference samples, igneous rock series. Geochem J 29:91–95

Imai N, Terashima S, Itoh S, Ando A (1999) 1998 compilation of analytical data for five GSJ geochemical reference samples. Geostandards Newslett 23:223–250

Imura S, Bando T, Seto K, Ohtani S, Kudoh S, Kanda H (2003) Distribution of aquatic mosses in the Soya Coast region, East Antarctica. Polar Biosci 16:1–10

Ingolfsson O, Hjort C, Berkman PA, Bjorck S, Colhoun E, Goodwin ID, Hall B, Hirakawa K, Melles M, Moller P, Prentice ML (1998) Antarctic glacial history since the Last Glacial Maximum: an overview of the record on land. Antarctic Sci 10:326–344

Jouzel J, Petit J, Souchez R, Barkov N, Lipenkov V, Raynaud D, Stievenard M, Vassiliev N, Verbeke V, Vimeux F (1999) More than 200 meters of lake ice above subglacial Lake Vostok, Antarctica. Science 286:213–2141

Kapitsa A, Ridley J, Robin G, Siegert M, Zotikov I (1996) A large deep freshwater lake beneath the ice of central East Antarctica. Nature 381:684–686

Karl DM, Bird DF, Bjorkman K, Houlihan T, Shackelford R, Tupas L (1999) Microorganisms in the accreted ice of Lake Vostok, Antarctica. Science 286:2144–2147

Kastner M, Keene J, Gieskes J (1977) Diagenesis of siliceous oozes - I. Chemical controls on the rate of opal-A to opal-CT transformation - an experimental study. Geochim Cosmochim Acta 41:1041–1059

Koblinsky C, Hildebrand P, LeVine D, Pellerano F, Chao Y, Wilson W, Yueh S, Lagerloef G (2003) Sea surface salinity from space: science goals and measurement approach. Radio Sci 38:8064, 10.1029/2001RS002584

Kudoh S, Tanabe Y (2014) Limnology and ecology of lakes along the Sôya Coast, East Antarctica. Adv Polar Sci 25:75–91

Lee K, Tong LT, Millero FJ, Sabine CL, Dickson AG, Goyet C, Park GH, Wanninkhof R, Feely RA, Key RM (2006) Global relationships of total alkalinity with salinity and temperature in surface waters of the world's oceans. Geophys Res Lett 33:L19605, 10.1029/2006GL027207

Leng MJ, Barker PA (2006) A review of the oxygen isotope composition of lacustrine diatom silica for palaeoclimate reconstruction. Earth-Sci Rev 75:5–27

Matsumoto G, Tani Y, Seto K, Tazawa T, Yamamuro M, Watanabe T, Nakamura T, Takemura T, Imura S, Kanda H (2010) Holocene paleolimnological changes in Lake Skallen Oike in the Syowa Station area of Antarctica inferred from organic components in a sediment core (Sk4C-02). J Paleolimnol 44:677–693

Matsumoto G, Honda E, Seto K, Tani Y, Watanabe T, Ohtani S, Kashima K, Nakamura T, Imura S (2014) Holocene paleolimnological changes of Lake Oyako-ike in the Soya Kaigan of East Antarctica. Inland Waters 4:105–112

Mikucki JA, Pearson A, Johnston DT, Turchyn AV, Farquhar J, Schrag DP, Anbar AD, Priscu JC, Lee PA (2009) A contemporary microbially maintained subglacial ferrous "Ocean". Science 324:397–400

Miura H, Maemoku H, Igarashi A, Moriwaki K (1998) Late Quaternary raised beach deposits and radiocarbon dates of marine fossils around Lützow-Holm Bay. Special map series of National Institute of Polar Research, pp. 1-46

Miura H, Maemoku H, Moriwaki K (2002) Holocene raised beach stratigraphy and sea-level history at Kizahashi Beach, Skarvsnes, Lutzow-Holm Bay, Antarctica. Royal Soc New Zealand Bull 35:391–396

Murayama H (1977) General characteristics of the Antarctic lakes near Syowa station. Antarctic Record 58:43–62

Muyzer G, Brinkhoff T, Nübel U, Santegoeds C, Schafer H, Wawer C (1996) Denaturing gradient gel electrophoresis (DGGE) in microbial ecology. In: Akkermans ADL, van Elsas JD, de Bruijn FJ (eds) Molecular microbial ecology manual, 3rd edn. Kluwer Academic Publishers, Dordrecht, The Netherlands, pp 1–27, p. 3.4.4

Nelson D, Treguer P (1992) Role of silicon as a limiting nutrient to Antarctic diatoms: evidence from kinetic studies in the Ross Sea ice-edge zone. Mar Ecol Progr Ser 80:255–264

Oswald G, Robin GQ (1973) Lakes beneath the Antarctic ice sheet. Nature 245:251–254

Oyama M, Takehara H (2005) Revised Standard Soil Color Charts: 27th edition (ed. The Ministry of Agriculture, Forestry and Fisheries of Japan)

Priscu JC, Adams EE, Lyons WB, Voytek MA, Mogk DW, Brown RL, McKay CP, Takacs CD, Welch KA, Wolf CF, Kirshtein JD, Avci R (1999) Geomicrobiology of subglacial ice above Lake Vostok, Antarctica. Science 286:2141–2144

Raiswell R (1984) Chemical models of solute acquisition in glacial meltwaters. J Glaciol 30:49–57
Raiswell R, Canfield DE (2012) The iron biogeochemical cycle past and present. Geochem Perspect 1:1–322
Reimer PJ, Baillie MGL, Bard E, Bayliss A, Beck JW, Bertrand CJH, Blackwell PG, Buck CE, Burr GS, Cutler KB, Damon PE, Edwards RL, Fairbanks RG, Friedrich M, Guilderson TP, Hogg AG, Hughen KA, Kromer B, McCormac G, Manning S, Ramsey CB, Reimer RW, Remmele S, Southon JR, Stuiver M, Talamo S, Taylor FW, van der Plicht J, Weyhenmeyer CE (2004) IntCal04 terrestrial radiocarbon age calibration, 0-26 cal kyr BP. Radiocarbon 46:1029–1058
Reimer PJ, Baillie MG, Bard E, Bayliss A, Beck JW, Blackwell PG, Bronk RC, Buck CE, Burr GS, Edwards RL (2009) IntCal09 and Marine09 radiocarbon age calibration curves 0-50,000 years cal BP. Radiocarbon 51:1111–1150
Robin GQ, Swithinbank C, Smith B (1970) Radio echo exploration of the Antarctic ice sheet. Inter Assoc Sci Hydrol Publ 86:97–115
Satish-Kumar M, Wada H (2000) Carbon isotopic equilibrium between calcite and graphite in Skallen Marbles, East Antarctica: Evidence for the preservation of peak metamorphic temperatures. Chem Geol 166:173–182
Sawagaki T, Hirakawa K (1997) Erosion of bedrock by subglacial meltwater. Geografiska Annaler Series a—Phys Geogr 79A, Soya Coast, East Antarctica, pp 223–238
Shaw J (2002) The meltwater hypothesis for subglacial bedforms. Quaternary Inter 90:5–22
Shiraishi K, Hiroi Y, Motoyoshi Y (1987) Plate tectonic development of Late Proterozoic paired metamorphic complexes in eastern Queen Maud Land, East Antarctica. Geophys Monogr Ser 40:309–318
Siegert M (2000) Antarctic subglacial lakes. Earth-Sci Rev 50:29–50
Siegert MJ, Dowdeswell J, Gorman M, McIntyre N (1996) An inventory of Antarctic subglacial lakes. Antarctic Sci 8:281–286
Siegert MJ, Carter S, Tabacco I, Popov S, Blankenship DD (2005) A revised inventory of Antarctic subglacial lakes. Antarctic Sci 17:453–460
Siegert MJ, Royston-Bishop G, Woodward J (2012) Antarctic subglacial lakes. Encyclopedia of lakes and reservoirs. Springer, London, pp 37–39
Stuiver M, Reimer PJ (1993) Extended ^{14}C data base and revised Calib 3.0 ^{14}C age calibration program. Radiocarbon 35:215–230
Stuiver M, Reimer PJ, Braziunas TF (1998) High-precision radiocarbon age calibration for terrestrial and marine samples. Radiocarbon 40:1127–1151
Stumpf A, Elwood Madden M, Soreghan G, Hall B, Keiser L, Marra K (2012) Glacier meltwater stream chemistry in Wright and Taylor Valleys, Antarctica: significant roles of drift, dust and biological processes in chemical weathering in a polar climate. Chem Geol 322:79–90
Takano Y, Tyler JJ, Kojima H, Yokoyama Y, Tanabe Y, Sato T, Ogawa ON, Ohkouchi N, Fukui M (2012) Holocene lake development and glacial-isostatic uplift at Lake Skallen and Lake Oyako, Lützow-Holm Bay, East Antarctica: based on biogeochemical facies and molecular signatures. Appl Geochem 27:2546–2559
Tayasu I, Hirasawa R, Ogawa NO, Ohkouchi N, Yamada K (2011) New organic reference materials for carbon and nitrogen stable isotope ratio measurements provided by Center for Ecological Research, Kyoto University, and Institute of Biogeosciences, Japan Agency for Marine-Earth Science and Technology. Limnol 12:261–266
Tranter M, Skidmore M, Wadham J (2005) Hydrological controls on microbial communities in subglacial environments. Hydrol Processes 19:995–998
Verleyen E, Hodgson DA, Sabbe K, Cremer H, Emslie SD, Gibson J, Hall B, Imura S, Kudoh S, Marshall GJ, McMinn A, Melles M, Newman L, Roberts D, Roberts SJ, Singh SM, Sterken M, Tavernier I, Verkulich S, de Vyver EV, Van Nieuwenhuyze W, Wagner B, Vyverman W (2011) Post-glacial regional climate variability along the East Antarctic coastal margin: evidence from shallow marine and coastal terrestrial records. Earth-Sci Rev 104:199–212
Watanabe T, Kojima H, Takano Y, Fukui M (2013) Diversity of sulfur-cycle prokaryotes in freshwater lake sediments investigated using *aprA* as the functional marker gene. Syst Appl Microbiol 36:436–443
Williams W, Sherwood J (1994) Definition and measurement of salinity in salt lakes. International J Salt Lake Res 3:53–63
Wingham D, Siegert M, Shepherd A, Muir A (2006) Rapid discharge connects Antarctic subglacial lakes. Nature 440:1033–1036
Yamane M, Yokoyama Y, Miura H, Maemoku H, Iwasaki S, Matsuzaki H (2011) The last deglacial history of Lutzow-Holm Bay, East Antarctica. Journal of Quaternary Sci 26:3–6
Yokoyama Y, Esat TM (2011) Global climate and sea level enduring variability and rapid fluctuations over the past 150,000 years. Oceanogr 24:54–69
Yokoyama Y, Miyairi Y, Matsuzaki H, Tsunomori F (2007) Relation between acid dissolution time in the vacuum test tube and time required for graphitization for AMS target preparation. Nuclear Instr Methods Phys Res Sec B 259:330–334
Yoshida Y, Moriwaki K (1979) Some consideration on elevated coastal features and their dates around Syowa Station, Antarctica. Memoirs Nat Inst Polar Res 13:220–226
Yoshikawa T, Toya H (1957) Report on geomorphological results of the Japanese Antarctic Research Expedition, 1956-1957. Antarctic Record 1:1–13
Young GM, Nesbitt HW (1998) Processes controlling the distribution of Ti and Al in weathering profiles, siliciclastic sediments and sedimentary rocks. J Sediment Res 68:448–455

17

Incorporation of Mg^{2+} in surface Ca^{2+} sites of aragonite: an *ab initio* study

Jun Kawano[1,2*], Hiroshi Sakuma[3] and Takaya Nagai[2]

Abstract

First-principles calculations of Mg^{2+}-containing aragonite surfaces are important because Mg^{2+} can affect the growth of calcium carbonate polymorphs. New calculations that incorporate Mg^{2+} substitution for Ca^{2+} in the aragonite {001} and {110} surfaces clarify the stability of Mg^{2+} near the aragonite surface and the structure of the Mg^{2+}-containing aragonite surface. The results suggest that the Mg^{2+} substitution energy for Ca^{2+} at surface sites is lower than that in the bulk structure and that Mg^{2+} can be easily incorporated into the surface sites; however, when Mg^{2+} is substituted for Ca^{2+} in sites deeper than the second Ca^{2+} layer, the substitution energy approaches the value of the bulk structure. Furthermore, Mg^{2+} at the aragonite surface has a significant effect on the surface structure. In particular, CO_3 groups rotate to achieve six-coordinate geometry when Mg^{2+} is substituted for Ca^{2+} in the top layer of the {001} surface or even in the deeper layers of the {110} surface. The rotation may relax the atomic structure around Mg^{2+} and reduces the substitution energy. The structural rearrangements observed in this study of the aragonite surface induced by Mg^{2+} likely change the stability of aragonite and affect the polymorph selection of $CaCO_3$.

Keywords: Aragonite; Impurity; Surface structure; First-principles calculation

Background

The formation of calcium carbonate ($CaCO_3$) polymorphs, calcite, aragonite, and vaterite has been extensively investigated due to their importance in geological and biological environments. To account for the formation of a particular polymorph, the role of impurities has been proposed as the controlling factors in many studies (e.g., Kitano 1962; Davis et al. 2000); however, the mechanism for the incorporation of impurities during crystal growth is poorly understood. In this study, we focus on the incorporation of Mg^{2+} in the aragonite surface and analyze its behavior using first-principles calculations.

Many researchers have previously reported that alkaline-earth cations other than Ca^{2+} affect the growth kinetics of $CaCO_3$ (e.g., De Yoreo and Vekilov 2003; Astilleros et al. 2010; Nielsen et al. 2013). In particular, Mg^{2+} has been considered important to the formation of $CaCO_3$ polymorphs. For example, Kitano (1962) indicated that the addition of Mg^{2+} to a solution promoted the metastable formation of aragonite. Recently, detailed atomic force microscopy (AFM) observations of the growth surface of calcite suggested that Mg^{2+} inhibits the crystal growth of calcite by blocking the propagation of kink sites (Nielsen et al. 2013) or by increasing the mineral solubility (Davis et al. 2000). To analyze this phenomenon on calcite surfaces at the atomic level, atomistic simulations were also conducted using static lattice energy minimization (Titiloye et al. 1998), molecular dynamics (MD) (de Leeuw and Parker 2001), and electronic structure calculations based on density functional theory (DFT) (Sakuma et al. 2014).

However, there are relatively few studies focusing on the aragonite surface. To discuss the mechanism for the formation of $CaCO_3$ polymorphs, not only should the atomic behavior on the calcite surface be understood but also that on the aragonite surface. Moreover, the Mg content in coral fossils comprising aragonite has been used to reconstruct the past climatic record (e.g., Mitsuguchi et al. 1996); however, the location of Mg^{2+} in the coral skeleton is strongly debated (Finch and Allison 2008). Therefore, understanding the mechanism for the incorporation of

* Correspondence: j-kawano@mail.sci.hokudai.ac.jp
[1]Creative Research Institution (CRIS), Hokkaido University, N21 N10, Kita-ku, Sapporo, Hokkaido 001-0021, Japan
[2]Faculty of Science, Hokkaido University, N10 W8, Kita-ku, Sapporo, Hokkaido 060-0810, Japan
Full list of author information is available at the end of the article

Mg^{2+} into the aragonite surface is important not only for the mineral and material sciences but also for the biological and environmental sciences.

Divalent cations smaller than Ca^{2+}, such as Mg^{2+}, do not generally enter the aragonite structure, whereas larger cations, such as Ba^{2+}, cannot be incorporated in the calcite structure. However, the structure near a crystal surface differs from the bulk crystal because of its flexibility. Thus, a crystal surface can incorporate ions that are unstable in the bulk structure and play an important role during the formation and subsequent crystal growth of calcium carbonate polymorphs. We investigated the substitution of Mg^{2+} ions at the Ca^{2+} sites of aragonite surfaces. Mg^{2+} is unstable in ninefold coordination in aragonite and does not readily enter into the bulk aragonite structure; however, Mg^{2+} is expected to be substitutable for Ca^{2+} at sites near the surface. Recently, Ruiz-Hernandez et al. (2012) performed MD calculations regarding Mg^{2+} substitution at the aragonite surface. However, they analyzed only the Mg^{2+} substitution into Ca^{2+} sites at the top surface. To discuss the incorporation of an ion into a specific surface, the ion substitution energy for Ca^{2+} sites should be estimated at the top surface and deeper. Furthermore, the substitution of Mg^{2+} for Ca^{2+} may change the surface structure. This could affect the stability relations among polymorphs, as surface energy differences among polymorphs have been proposed to account for their stability field (Navrotsky 2004; Kawano et al. 2009). Hence, an in-depth analysis of these surface structures and their incorporation of ions is important; however, details regarding the surface structural changes when a Mg^{2+} ion is incorporated at the surface are presently lacking. Therefore, in this study, the stability of Mg^{2+} near the aragonite surface and the structure of Mg^{2+}-containing aragonite surface were investigated using first-principles calculations, and the effect of Mg^{2+} on the formation of polymorphs was examined.

Methods

The optimized geometries and total energies of the surfaces were obtained using DFT with the Vienna *ab initio* simulation package (VASP) code (Kresse and Hafner 1993, 1994; Kresse and Furthmüller 1996a, b; Kresse and Joubert 1999) and the Perdew-Burke-Ernzerhof version of the generalized gradient approximation (GGA-PBE) (Perdew et al. 1996). The energy cutoff of the plane-wave basis set was 900 eV, which was tested for energy convergence. The valence states for Ca, Mg, C, and O are $3s^2 3p^6 4s^2$, $2p^6 3s^2$, $2s^2 2p^2$, and $2s^2 2p^4$, respectively, following previous DFT calculations for $CaCO_3$ (Hossain et al. 2009) and $MgCO_3$ (Hossain et al. 2010).

Prior to calculation of the aragonite surface, the structural parameters of aragonite were simulated. The calculated lattice parameters are $a = 5.022$ Å, $b = 8.042$ Å, and $c = 5.816$ Å, whereas the experimental values are $a = 4.962$ Å, $b = 7.969$ Å, and $c = 5.743$ Å (Balmain et al. 1999). The calculated C-O bond lengths are 1.291 to 1.301 Å, which are comparable to previous experimentally and theoretically obtained values (Balmain et al. 1999; Akiyama et al. 2011).

Aragonite surfaces were simulated as repeated slabs. The unit supercell contained 80 atoms and 4 or 5 Ca and CO_3 layers of the unit (2×1) surface structure with a 15 Å thick vacuum layer. The macroscopic dipole was removed, and the neutrality of the supercell was ensured by the two equivalent surfaces on opposite sides of the slab. To calculate the substitution energy, the total energy of the slab was simulated by relaxing the atoms except those on the bottom layer. In the calculation, the supercell parameters were fixed using the calculated aragonite unit cell parameters, because Mg ions are not supposed to be substituted into the bulk Ca sites but only into surface sites; hence, substitution does not affect the lattice constants. Optimization was performed with a convergence threshold of 5.0×10^{-6} eV/atom for the maximum energy change and 0.05 eV/Å for the maximum force. The atomic structures of the slab were drawn with the VESTA software (Momma and Izumi 2008). $CaCO_3$ crystals generally grow in an aqueous solution with H_2O molecules just above the surface; however, the presence of the vacuum layer above the surface was considered here. The validity of this setting will be discussed in the next section.

Results and discussion

Substitution energy of Mg^{2+} for Ca^{2+} near the aragonite surface

We calculated the surface energies of the aragonite {001}, {010}, and {110} surfaces, which are generally expressed in experimental morphology. On the {001} surface, the CO_3 groups (triangles) are arranged parallel to the surface, whereas these groups are perpendicular to the {010} and {110} surfaces. Recently, Akiyama et al. (2011) determined the stable surface structure of calcite and aragonite via DFT calculations. They reported that the most stable configuration of {001} is a CO_3-terminated surface, wherein the coverage of CO_3 ions is 0.5. In contrast, the most stable configurations for {010} and {110} are Ca-terminated planes, wherein the coverage of Ca^{2+} ions is 0.5. The surface energy of these stable configurations were calculated to be 0.49, 0.57, and 0.49 J/m^2 for the {001}, {010}, and {110} surfaces, respectively (Table 1). In the calculation, the positions of all the slab atoms were relaxed, and the surface energies E_{surf} were obtained by dividing the difference between the total energy of the slab E_{slab} and that of the bulk crystal E_{bulk}, including the same number of atoms, by the total area of the calculated surface A, which includes the top and bottom surfaces:

Table 1 Estimated surface energies (J/m^2) of the aragonite surfaces

Surface	This work DFT Pure	Akiyama et al. DFT Pure	de Leeuw and Parker Empirical potential Pure	de Leeuw and Parker Empirical potential Hydrated
{001} $\theta_{CO_3} = 0.5$	0.49	0.58	0.85	0.90
{010} $\theta_{Ca} = 0.5$	0.57	0.73	0.96	0.24
{110} $\theta_{Ca} = 0.5$	0.49	0.64	0.88	0.56

$$E_{surf} = (E_{slab} - E_{bulk})/A. \quad (1)$$

The results suggest that the {001} and {110} surfaces are slightly more stable than the {010} surface, which is consistent with previous DFT and empirical potential calculations (Akiyama et al. 2011; De Leeuw and Parker 1998). Therefore, we focused on the {001} and {110} aragonite surfaces and analyzed the substitution energy of Mg^{2+} for Ca^{2+}. The structures considered in this study are shown in Figure 1. In the first Ca layer of the {001} surface, there are two types of Ca sites because the coverage of CO_3 ions is 0.5 in the most stable structure of this surface. The site labeled A in Figure 1 (denoted as site A hereafter) is the site above which no CO_3 groups are located, whereas the site labeled B in the same figure (site B) is similar to the site in the bulk with respect to the arrangement of surrounding CO_3^{2-} ions, except for the lack of one CO_3^{2-}. In contrast, there are two Ca layers in the {110} surface with different depths between the CO_3 layers.

De Leeuw and Parker (1998) estimated the energies of pure and hydrated surfaces and found that hydration does not stabilize the carbonate-terminated {001} surface and less so the calcium-terminated {110} surface (Table 1). In contrast, the calcium-terminated {010} surface was significantly stabilized by hydration. The surface energies calculated in this study show almost the same trend. Therefore, the substitution energies obtained for nonhydrated and hydrated {001} and {110} surfaces can be considered to have similar features. It is thus reasonable to analyze nonhydrated {001} and {110} surfaces to discuss the exchange energy in aqueous solutions, whereas Nada (2014) recently reported the importance of water layers on the calcite {104} surface.

In an aqueous solution, Mg^{2+} ions are surrounded by H_2O molecules and exist as hydration shells (e.g., Marcus

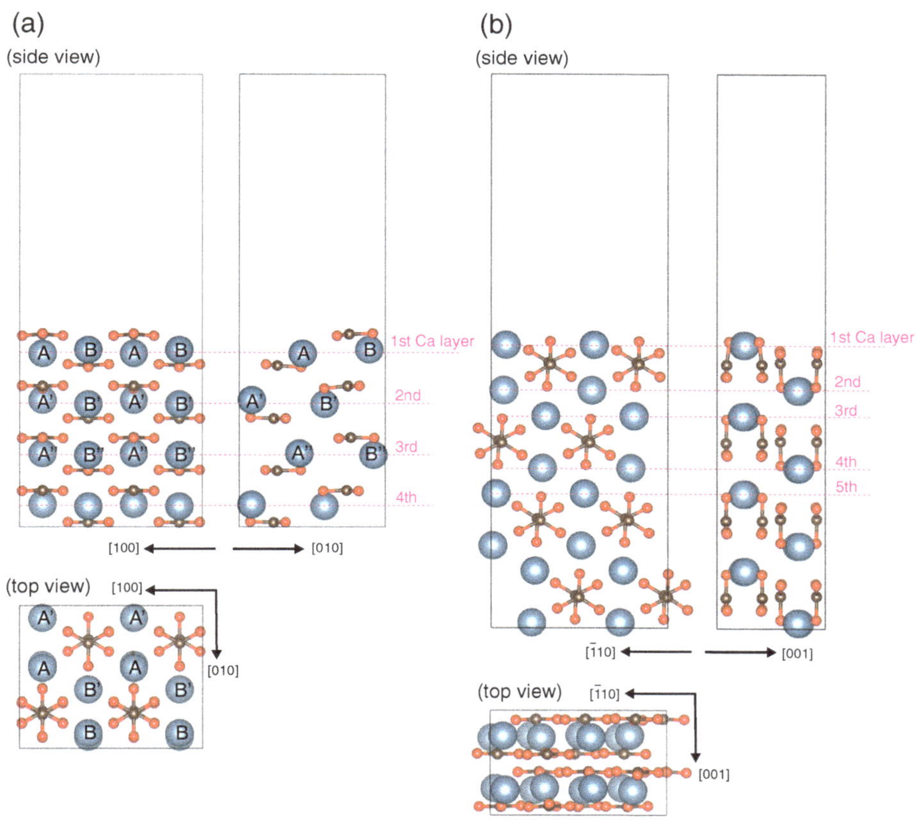

Figure 1 Relaxed atomic arrangement of (a) {001} and (b) {110} surfaces. Red, brown, and blue circles represent O, C, and Ca, respectively. Ca layers are numbered from the top to deeper layers - see text regarding [A] and [B] for the top Ca layer on the {001} surface. In addition, there are two types of sites in the second and third layers of the {001} surface, which are labeled [A′], [B′] and [A″], [B″], respectively.

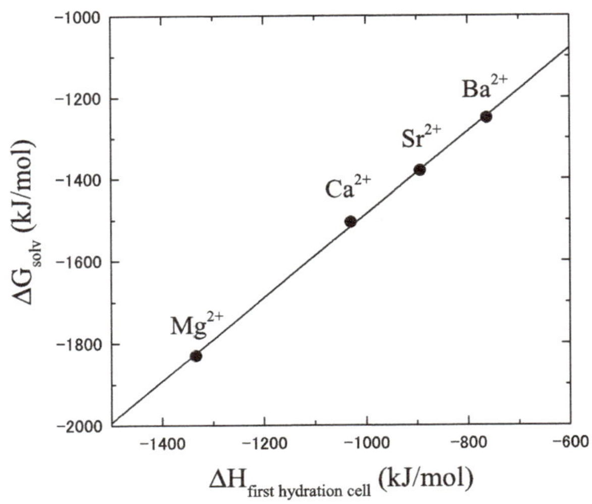

Figure 2 Relation between experimental free energy of solvation and binding enthalpy of the first hydration shell. The binding energies of the hydration shell with six water molecules were calculated.

1987; Rodriguez-Cruz et al. 1999; Bock et al. 2006). Therefore, to substitute into a surface Ca^{2+} site, a Mg atom must be released from the hydration shell. In this study, a six-coordinated complex was assumed for the divalent cations in the hydration shells, and the cohesive energies of the primary hydration shells of Ca^{2+} and Mg^{2+} ions were calculated using the DFT method as references. The calculated and experimental Gibbs free energies for hydration (Marcus 1991) have an excellent 1:1 correspondence, as shown in Figure 2. Following the discussion by Sakuma et al. (2014), such correspondence suggests that the relative binding enthalpy of the first hydration shell with six water molecules can describe the relative free energies of divalent cations in water.

Using these results, the energy ΔE of the Mg^{2+} substitution in surface Ca^{2+} sites of aragonite can be expressed as follows:

$$\Delta E = \left[E_{\mathbf{surface(Mg)}} + E_{\mathbf{hydration\ shell(Ca)}}\right] - \left[E_{\mathbf{surface(Ca)}} + E_{\mathbf{hydration\ shell(Mg)}}\right], \tag{2}$$

where $E_{surface(Ca)}$ and $E_{surface(Mg)}$ represent the total energies of the aragonite slab with and without Mg, and $E_{hydration\ shell(Ca)}$ and $E_{hydration\ shell(Mg)}$ are the energies of the hydration shells of Ca^{2+} and Mg^{2+} with $6H_2O$, respectively. The calculated cohesive energy of the primary hydration shell of Mg^{2+} with $6H_2O$ is approximately −13.8 eV (1,330 kJ/mol) and that for Ca^{2+} is −10.8 eV (1,030 kJ/mol), which indicates that the hydration shell of the smaller cation is more stable than that of the larger cation.

For Mg^{2+} substitution in Ca^{2+} sites of the bulk aragonite structure, the substitution energy obtained from Equation 2 using the energy of the bulk instead of the energy of the slab was calculated as 93.1 kJ/mol. However, the substitution energy for the surface site is much smaller than that for the bulk. Whether Mg ions will actually substitute for Ca depends on the chemical potential difference including entropy, while the entropy of an ion in aqueous solutions is difficult to estimate. Therefore, the negative or positive sign of the substitution energy ΔE does not directly mean that Mg ions enter the crystal or not. However, these energies can provide information regarding the relative stability of Mg in the site around the surface. The results of the present calculations suggest that Mg^{2+} can substitute more easily at surface sites than in the bulk aragonite structure. Figure 3 shows the Mg^{2+} substitution energy ΔE for a Ca^{2+} site as a function of depth for the interface of the {001} and {110} surfaces.

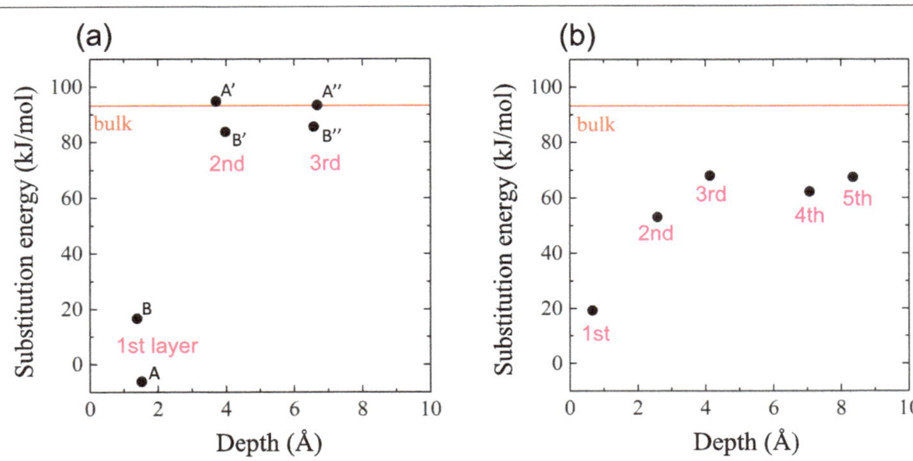

Figure 3 Energies for Mg^{2+} substitution at Ca^{2+} in aragonite surface sites. The substitution energies are shown as a function of the depth from the interface of **(a)** {001} and **(b)** {110} surfaces. Red lines are drawn at 93.1 kJ/mol, which is the energy for Mg^{2+} substitution into bulk aragonite. Layer numbers, [A], [A′], [A″], [B], [B′], and [B″] correspond to the substitution energies for the site shown in Figure 1.

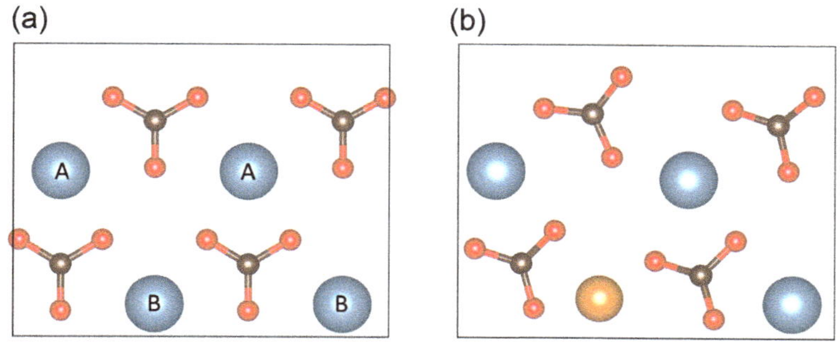

Figure 4 Top view of the relaxed aragonite {001} surface wherein Mg^{2+} substitutes for Ca^{2+}. Comparison of **(a)** surface structure without Mg^{2+} ions and **(b)** with Mg^{2+} substituting for Ca^{2+}. Red, brown, blue, and orange circles represent O, C, Ca, and Mg, respectively. [A] and [B] are labeled as in Figure 1.

We first discuss the {001} surface (Figure 3a). When Mg^{2+} substitutes into site A in the first Ca layer, above which no CO_3 groups are located, the substitution energy is almost zero and Mg^{2+} is easily incorporated into site A, which agrees with the MD calculations (Ruiz-Hernandez et al. 2012). The substitution energy increases when Mg^{2+} is substituted into the B site but is still much lower than that when it enters the Ca site in the bulk aragonite structure. However, for substitution within the deeper layers, the substitution energies increase significantly and reach almost that of the bulk aragonite structure. Thus, Mg^{2+} readily attaches to the first layer of the {001} surface but less so within the deeper layers.

In the {110} surface, the substitution energy of Mg^{2+} into the first Ca layer is almost the same as that for the B site in the {001} surface, and it rapidly increases with substitution in the deeper layers (Figure 3b). However, for this surface, even the energy for the substitution at Ca sites in the fifth layer is smaller than that for the bulk. This suggests that for {110} faces, a slightly higher energy would be required for Mg^{2+} ions to enter the Ca site of the top layer, whereas Mg^{2+} ions would enter the deeper layers more easily than the {001} face.

Structure of aragonite surfaces with Mg^{2+} ions at the Ca^{2+} sites

The structures of aragonite surfaces that contain Mg^{2+} ions were examined next. In the aragonite {001} surface, the CO_3^{2-} ions are arranged parallel to the surface; therefore, the flexibility of the movement and rotation of ions initially seem higher than those on the other surfaces. Figure 4b shows the relaxed structure of the {001} surface containing Mg^{2+} ions in B sites of the first Ca layer, where Mg^{2+} is surrounded by CO_3 ions. Figure 4 shows that the CO_3 groups move from the original positions and rotate 30° in the same direction. Notably, not only the CO_3^{2-} ions above the Mg^{2+} ions, which exist on the surface, but also the CO_3 groups below Mg^{2+}, which make up the second CO_3 layer, rotate in the same manner. The arrangement of CO_3 ions around a Mg atom resembles the structure of calcite. To clarify this, the arrangement of the CO_3 groups around the Mg ions is shown in Figure 5. This figure shows that CO_3 groups around a Mg^{2+} ion move and rotate to assume six-coordinate geometry, which is the arrangement of the MgO_6 octahedron in the calcite structure.

Furthermore, the {110} surface seems to have less flexibility than the {001} surface because the CO_3 groups are arranged perpendicular to the {110} surface. However,

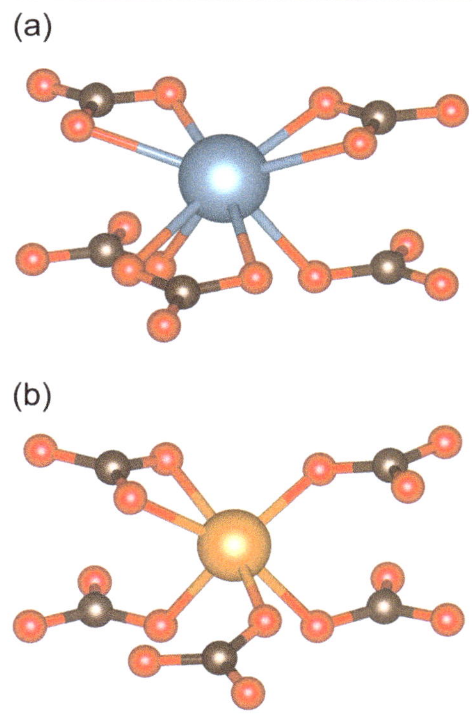

Figure 5 Details of the arrangement of CO_3 groups around the cation on the top {001} surface. (a) Ca and **(b)** Mg in the B site on the top layer of the {001} surface. Mg presumably prefers six-coordinate (octahedral) geometry as in the calcite structure.

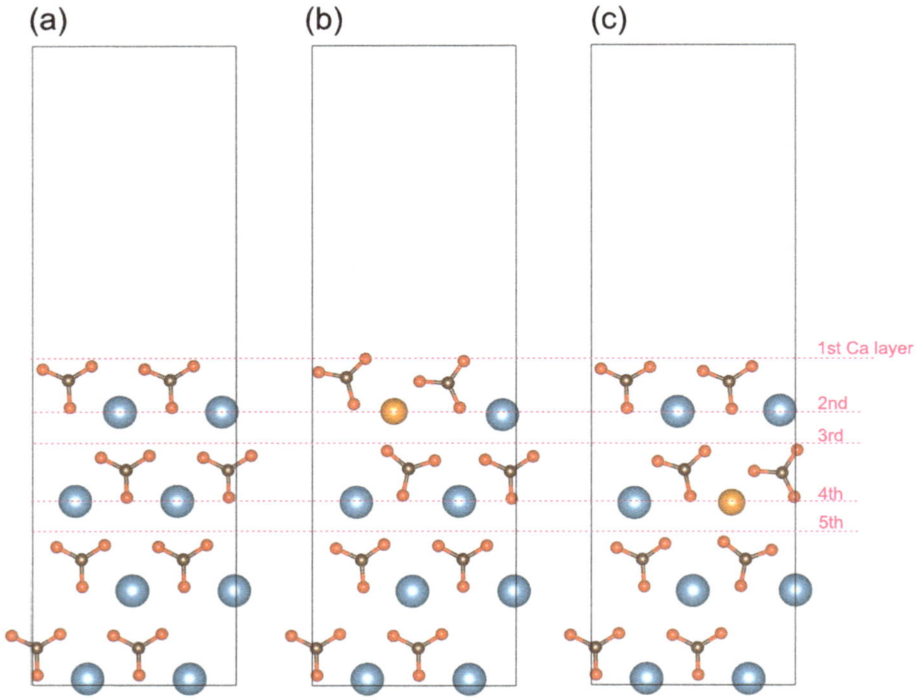

Figure 6 Side views of the relaxed aragonite {110} surface in which Mg^{2+} substitutes for Ca^{2+}. (a) No substitution, **(b)** substitution into the second Ca layer, and **(c)** substitution into the fourth Ca layer. Red, brown, blue, and orange circles represent O, C, Ca, and Mg, respectively.

even in this case, the CO_3 groups move and rotate to achieve six-coordinate geometry, as shown in Figure 6. Moreover, for the {110} surface, when Mg^{2+} substitutes at a Ca site, not only in the first but also in the fifth Ca layer, the CO_3 groups also rotate to achieve six-coordinate geometry but do not form MgO_6 octahedra (Figure 7). This is opposite to the {001} surface, wherein the CO_3 groups do not move to accommodate the Mg substitution into the Ca sites near the {001} surface, except in the top layer. This is probably because atoms can move easily perpendicular to the surface, but less easily when parallel to it because the size of the surface is fixed, even though the layer thickness is not fixed. Therefore, contrary to the first expectation, it is easier for CO_3 groups to rotate in the {110} surface where atoms easily move parallel to the CO_3 groups. In contrast, in the {001} surface, atoms do not readily move in the long direction parallel to the CO_3 groups, because the size of this plane is fixed.

The differences between the {001} and {110} surfaces lead to differences in the substitution energies; when Mg^{2+} ions substitute in the deeper Ca layers, the atomic arrangement near the {110} surface is more relaxed than that near the {001} surface. Hence, the substitution energies for {110} are lower than those for the {001} surface. The energy gained by the rotation of CO_3 is estimated to be around 20 eV, by comparison of the substitution energy for the deep layers of the {110} surface and the value for the substitution in the bulk where CO_3 groups are not rotated.

The results suggest that the CO_3 groups near the surface easily move and rotate relative to their original positions. Moreover, Mg^{2+} ions strongly prefer six-coordinate geometry. Therefore, the presence of Mg^{2+} affects the surface stability of aragonite, and it may further affect the

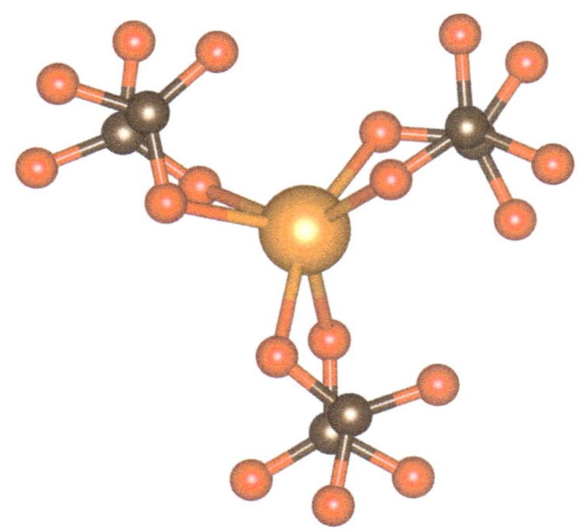

Figure 7 Detailed arrangement of CO_3 groups around Mg on the {110} surface. The arrangement of CO_3 groups around Mg at the Ca site of the fourth layer in the {110} surface is shown, which suggests that Mg achieves six-coordinate geometry but does not form MgO_6 octahedra as in the calcite structure.

structure of the small clusters that appear during the early formation of $CaCO_3$, which has more flexibility than the surface, both of which affect the polymorph selection of $CaCO_3$.

Conclusions

First-principles calculations were performed for Mg^{2+}-containing aragonite surfaces. The results suggest that the substitution energy of Mg^{2+} for Ca^{2+} at the surface is lower than the substitution energy of Mg^{2+} for Ca^{2+} in the bulk structure. However, for the {001} surface, when Mg^{2+} substitutes for Ca^{2+} deeper than the second Ca layer, the substitution energy is almost the same as that for substitution in the bulk aragonite structure. In contrast, for the {110} surface, even when Mg^{2+} ion substitutes into deeper layers, the substitution energy is still lower than the substitution energy in the bulk aragonite structure. Thus, Mg^{2+} ions easily attach onto the {001} surface with lower energy; however, it should be difficult for these ions to move to deeper layers. In contrast, for the {110} surface, a relatively higher energy is required for Mg^{2+} ions to substitute for Ca^{2+} at the top surface sites, whereas they enter more easily to deeper layers than the {001} face. This is probably because the atomic structure of this surface is more relaxed, and the CO_3 groups move and rotate from their original positions even when Mg^{2+} ions are in deeper layers. In contrast, for the {001} surface, the CO_3 groups do not move when Mg^{2+} substitutes for Ca^{2+}, except in the top layer sites where CO_3 groups easily move and rotate to achieve six-coordinate geometry, such as CaO_6 octahedra in the calcite structure. Mg^{2+} generally assumes a preferential six-coordinate geometry, even at the aragonite surface, indicating that it changes the surface stability of aragonite, which may affect the formation of $CaCO_3$ polymorphs.

Abbreviations

AFM: atomic force microscopy; DFT: density functional theory; GGA-PBE: generalized gradient approximation-Perdew-Burke-Ernzerhof; MD: molecular dynamics; VASP: Vienna *ab initio* simulation package.

Competing interests

The authors declare that they have no competing interests.

Authors' contributions

JK proposed the topic, designed the study, conducted the simulations, and interpreted the results. HS contributed to the simulations and data interpretation. TN contributed to the discussions with the corresponding author throughout the study. All authors read and approved the final manuscript.

Acknowledgements

The authors would like to thank Y. Kimura, one of the session conveners of the 2013 JpGU annual meeting, for his recommendation for submission of the present study to PEPS. Thanks are also due to two anonymous reviewers and the editor for their constructive comments. This study was supported by a MEXT grant for the tenure-tracking system and JSPS KAKENHI Grant Number 26870010.

Author details

[1]Creative Research Institution (CRIS), Hokkaido University, N21 N10, Kita-ku, Sapporo, Hokkaido 001-0021, Japan. [2]Faculty of Science, Hokkaido University, N10 W8, Kita-ku, Sapporo, Hokkaido 060-0810, Japan. [3]Environment and Energy Materials Division, National Institute for Materials Science, 1-1 Namiki, Tsukuba, Ibaraki 305-0044, Japan.

References

Akiyama T, Nakamura K, Ito T (2011) Atomic and electronic structures of $CaCO_3$ surfaces. Phys Rev B 84:085248, doi:10.1103/PhysRevB.84.085428

Astilleros JM, Fernández-Díaz L, Putnis A (2010) The role of magnesium in the growth of calcite: an AFM study. Chem Geol 271:52–58

Balmain J, Hannoyer B, Lopez E (1999) Fourier transform infrared spectroscopy (FTIR) and X-ray diffraction analyses of mineral and organic matrix during heating of mother of pearl (nacre) from the shell of the mollusk Pinctada maxima. J Biomed Mater Res 48:749–754

Bock CW, Markham GD, Katz AK, Glusker JP (2006) The arrangement of first- and second-shell water molecules around metal ions: effects of charge and size. Theor Chem Acc 115:100–112

Davis KJ, Dove PM, De Yoreo JJ (2000) The role of Mg^{2+} as an impurity in calcite growth. Science 290:1134–1137, doi:10.1126/science.290.5494.1134

de Leeuw NH, Parker CC (1998) Surface structure and morphology of calcium carbonate polymorphs calcite, aragonite, and vaterite: an atomistic approach. J Phys Chem B 102:2914–2922

de Leeuw NH, Parker CC (2001) Surface-water interactions in the dolomite problem. Phys Chem Chem Phys 3:3217–3221

De Yoreo JJ, Vekilov PG (2003) Principles of crystal nucleation and growth. Rev Mineral Geochem 54(1):57–93

Finch AA, Allison N (2008) Mg structural state in coral aragonite and implications for the paleoenvironmental proxy. Geophys Res Lett 35, L08704, doi:10.1029/2008GL033543

Hossain FM, Murch GE, Belova IV, Turner BD (2009) Electronic, optical and bonding properties of $CaCO_3$ calcite. Solid State Comm 149:1201–1203, doi:10.1016/j.ssc.2009.04.026

Hossain FM, Dlugogorski BZ, Kennedy EM, Belova IV, Murch GE (2010) Electronic, optical and bonding properties of $MgCO_3$. Solid State Comm 150:848–851, doi:10.1016/j.ssc.2010.02.008

Kawano J, Shimobayashi N, Miyake A, Kitamura A (2009) Precipitation diagram of calcium carbonate polymorphs: its construction and significance. J Phys Condens Matter 21:425102, doi:10.1088/0953-8984/21/42/425102

Kitano Y (1962) The behaviour of various inorganic ions in the separation of calcium carbonate from a bicarbonate solution. Bull Chem Soc Japan 35:1973–1980

Kresse G, Furthmüller J (1996a) Efficiency of ab-initio total energy calculations for metals and semiconductors using a plane-wave basis set. Comput Mat Sci 6:15–50

Kresse G, Furthmüller J (1996b) Efficient iterative schemes for ab initio total-energy calculations using a plane-wave basis set. Phys Rev B 54:11169–11186

Kresse G, Hafner J (1993) Ab initio molecular dynamics for liquid metals. Phys Rev B 47:558–561

Kresse G, Hafner J (1994) Ab initio molecular-dynamics simulation of the liquid-metal-amorphous-semiconductor transition in germanium. Phys Rev B 49:14251–14269

Kresse G, Joubert J (1999) From ultrasoft pseudopotentials to the projector augmented-wave method. Phys Rev B 59:1758–1775

Marcus Y (1987) Thermodynamics of ion hydration and its interpretation in terms of a common model. Pure Appl Chem 59:1093–1101

Marcus Y (1991) Thermodynamics of solvation of ions. Part 5.-Gibbs free energy of hydration at 298.15 K. J Chem Soc Faraday Trans 87:2995–2999

Mitsuguchi T, Matsumoto E, Abe O, Uchida T, Isdale PJ (1996) Mg/Ca thermometry in coral skeletons. Science 274:961–963

Momma K, Izumi F (2008) VESTA: a three-dimensional visualization system for electronic and structural analysis. J Appl Crystallogr 41:653–658

Nada H (2014) Difference in the conformation and dynamics of aspartic acid on the flat regions, step edges, and kinks of as calcite surface: a molecular dynamics study. J Phys Chem C 118:14335–14345

Navrotsky A (2004) Energetic clues to pathways to biomineralization: precursors, clusters, and nanoparticles. Proc Natl Acad Sci U S A 101:12096–12101

Nielsen LC, De Yoreo JJ, DePaolo DJ (2013) General model for calcite growth kinetics in the presence of impurity ions. Geochim Cosmochim Acta 115:100–114

Perdew JP, Burke K, Ernzerhof M (1996) Generalized gradient approximation made simple. Phys Rev Lett 77:3865–3868

Rodriguez-Cruz SE, Jockusch RA, Williams ER (1999) Hydration energies and structures of alkaline earth metal ions, $M^{2+}(H_2O)_n$, $n = 5–7$, M = Mg, Ca, Sr, and Ba. J Am Chem Soc 121:8898–8906

Ruiz-Hernandez SE, Grau-Crespo R, Almora-Barrios N, Wolthers M, Ruiz-Salvador AR, Fernandez N, de Leeuw NH (2012) Mg/Ca partitioning between aqueous solution and aragonite mineral: a molecular dynamics study. Chemistry 18:9828–9833, doi:10.1002/chem.201200966

Sakuma H, Andersson MP, Bechgaard K, Stipp SLS (2014) Surface tension alteration on calcite, induced by ion substitution. J Phys Chem C 118:3078–3087, doi:10.1021/jp411151u

Titiloye JO, de Leeuw NH, Parker CC (1998) Atomistic simulation of the differences between calcite and dolomite surfaces. Geochim Cosmochim Acta 62:2637–2641

Heterogeneous interplate coupling along the Nankai Trough, Japan, detected by GPS-acoustic seafloor geodetic observation

Yusuke Yokota[1*], Tadashi Ishikawa[1], Mariko Sato[1], Shun-ichi Watanabe[1], Hiroaki Saito[1], Naoto Ujihara[1], Yoshihiro Matsumoto[1], Shin-ichi Toyama[1], Masayuki Fujita[1], Tetsuichiro Yabuki[1], Masashi Mochizuki[2] and Akira Asada[3]

Abstract

The recurring devastating earthquake that occurs in the Nankai Trough subduction zone between the Philippine Sea plate and the Eurasian plate has the potential to cause an extremely dangerous natural disaster in the foreseeable future. Many previous studies have assumed interplate-coupling ratios for this region along the trench axis using onshore geodetic data in order to understand this recursive event. However, the offshore region that has the potential to drive a devastating tsunami cannot be resolved sufficiently because the observation network is biased to the land area. Therefore, the Hydrographic and Oceanographic Department of Japan constructed a geodetic observation network on the seafloor along the Nankai Trough using a GPS-acoustic combination technique and has used it to observe seafloor crustal movements directly above the Nankai Trough subduction zone. We have set six seafloor sites and cumulated enough data to determine the displacement rate from 2006 to January 2011. Our seafloor geodetic observations at these sites revealed a heterogeneous interplate coupling that has three particular features. The fast displacement rates observed in the easternmost area indicate strong interplate coupling (>75%) around not only the future Tokai earthquake source region but also the Paleo-Zenisu ridge. The slow displacement rates near the trench axis in the Kumano-nada Sea, a shallow part of the 1944 Tonankai earthquake source region, show a lower coupling ratio (50% to 75%). The slow displacement rate observed in the area shallower than the 1946 Nankaido earthquake source region off Cape Muroto-zaki reflects weakening interplate coupling (about 50%) probably due to a subducting seamount. Our observations above the subducting ridge and seamount indicate that the effect of a subducting seamount on an interplate-coupling region depends on various conditions such as the geometry of the seamount and the friction parameters on the plate boundary.

Keywords: Seafloor geodetic observation; Nankai Trough; Interplate coupling; Seamount subduction

Background

Nankai trough subduction zone

In southwestern Japan, the oceanic Philippine Sea plate is subducting beneath the continental Eurasian plate northwestward along the Nankai Trough. In this region, interplate magnitude-8-class earthquakes have been documented with an average recurrence time of about 100 years (e.g., Ando 1975) due to this plate motion. The most recent events are the 1944 Tonankai (M7.9) and the 1946 Nankaido (M8.0) earthquakes. Thus, the next magnitude-8-class earthquake is predicted in the near future.

We analyzed the coseismic slip distributions of the 1944 and 1946 events using triangulation and leveling survey records (e.g., Sagiya and Thacher 1999 and Ito and Hashimoto 2004) and tsunami records (e.g., Tanioka and Satake 2001 and Baba and Cummins 2005). To understand the next disastrous event in advance, several research groups also analyzed the interseismic slip deficit for recent years using onshore global positioning systems (GPS) data (e.g., Nishimura and Hashimoto 2006 and Loveless and Meade 2010). The onshore geodetic

* Correspondence: eisei@jodc.go.jp
[1]Hydrographic and Oceanographic Department, Japan Coast Guard, 2-5-18 Aomi, Koto-ku, Tokyo 135-0064, Japan
Full list of author information is available at the end of the article

data do not have the power to resolve the offshore region as indicated in Yoshioka and Matsuoka (2013) (Figure 1). They pointed out the importance of seafloor geodetic observations for revealing the offshore region along the Nankai Trough.

Seafloor geodetic observation using a GPS-acoustic combination technique

In order to monitor seafloor crustal movements around the offshore plate boundary, the Hydrographic and Oceanographic Department of Japan (JHOD) has been carrying out seafloor geodetic observations using a GPS-acoustic combination technique on the landward side of the major trenches around Japan. Our past observations near the Japan Trench provided great advantage in understanding the 2005 Off-Miyagi Prefecture earthquake (M7.2) (e.g., Matsumoto et al. 2006 and Sato et al. 2011b), the 2011 Tohoku-oki earthquake (M9.0) (e.g., Sato et al. 2011a and Sato et al. 2013b), and others. Important observations have also been made by other research groups in Japan such as Tohoku University (e.g., Kido et al. 2006 and Kido et al. 2011) and Nagoya University (e.g., Tadokoro et al. 2012 and Yasuda et al. 2014).

In the Nankai Trough region, we were operating six seafloor sites along the trench axis at about 50 to 100 km intervals (Figure 1 and Table 1) before the 2011 Tohoku-oki earthquake. In this paper, we report displacement rates on the seafloor surface and discuss interplate coupling along the Nankai Trough during the period before the 2011 Tohoku-oki earthquake and, thus, unaffected by postseismic effects from this enormous event. Our observations directly revealed an

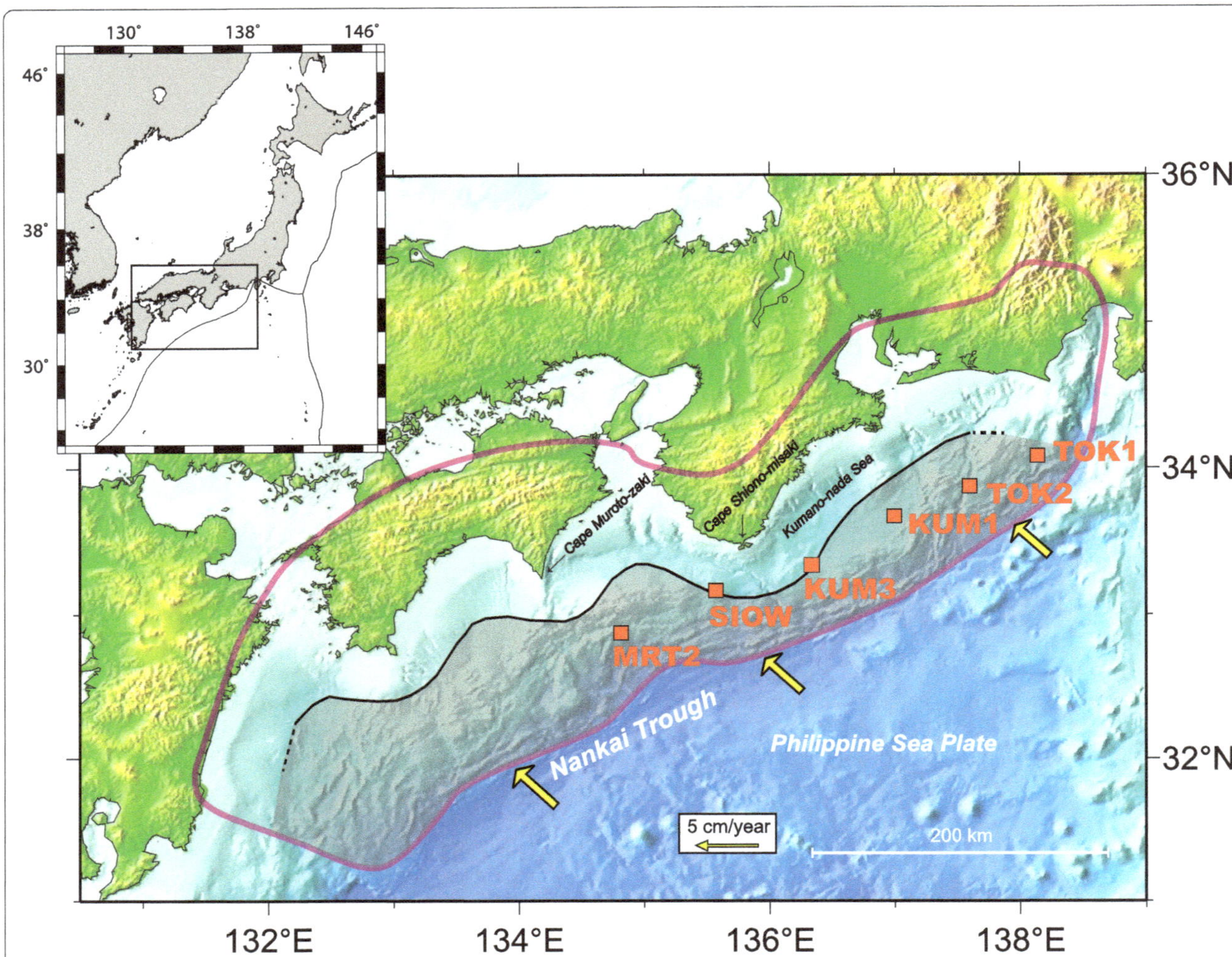

Figure 1 Locations of seafloor geodetic observation network sites along the Nankai Trough. Red squares show the sites set before the 2011 Tohoku-oki earthquake and used in this study. Yellow arrows indicate the convergence rate of the Philippine Sea plate under the Eurasian plate calculated using the NUVEL-1A model (DeMets et al. 1994). The purple region shows the maximum source model of the great Nankai Trough earthquake assumed by the Central Disaster Management Council of the Japanese government (CDMC) (2013). The gray shading near the trench axis indicates an area with small resolution values calculated using the onshore geodetic observations by Yoshioka and Matsuoka (2013).

Table 1 Positions of seafloor sites

Site name	Latitude (degree)	Longitude (degree)	Height (m)
TOK1	34.08231 N	138.13395 E	−2,374
TOK2	33.87698 N	137.59508 E	−1,524
KUM1	33.67026 N	136.99558 E	−1,957
KUM3	33.33287 N	136.34232 E	−1,957
SIOW	33.16056 N	135.57229 E	−1,524
MRT2	32.87264 N	134.81463 E	−1,402

uneven accumulation of interplate coupling along the Nankai Trough.

Methods

Measurement system

The GPS-acoustic combination technique was developed to detect seafloor displacements with an accuracy of a few (2 to 3) centimeters (e.g., Spiess et al. 1998 and Asada and Yabuki 2001). In Japan, our group first succeeded in detecting seafloor movements caused by plate convergence at the Japan Trench (Fujita et al. 2006) and has been deploying a seafloor observation network since 2000.

A schematic picture of our seafloor geodetic observation system is shown in Figure 2. This system consists of a seafloor unit with four acoustic mirror-type transponders and an on-board unit with a GPS antenna/receiver, an undersea acoustic transducer, and a dynamic motion sensor. The onboard acoustic transducers were mounted at the stern of survey vessels using an 8 m-long pole before 2007. We used to carry out surveys using a drifting observation method with the pole system before 2007, as shown in Figure 2a. We improved the observations by

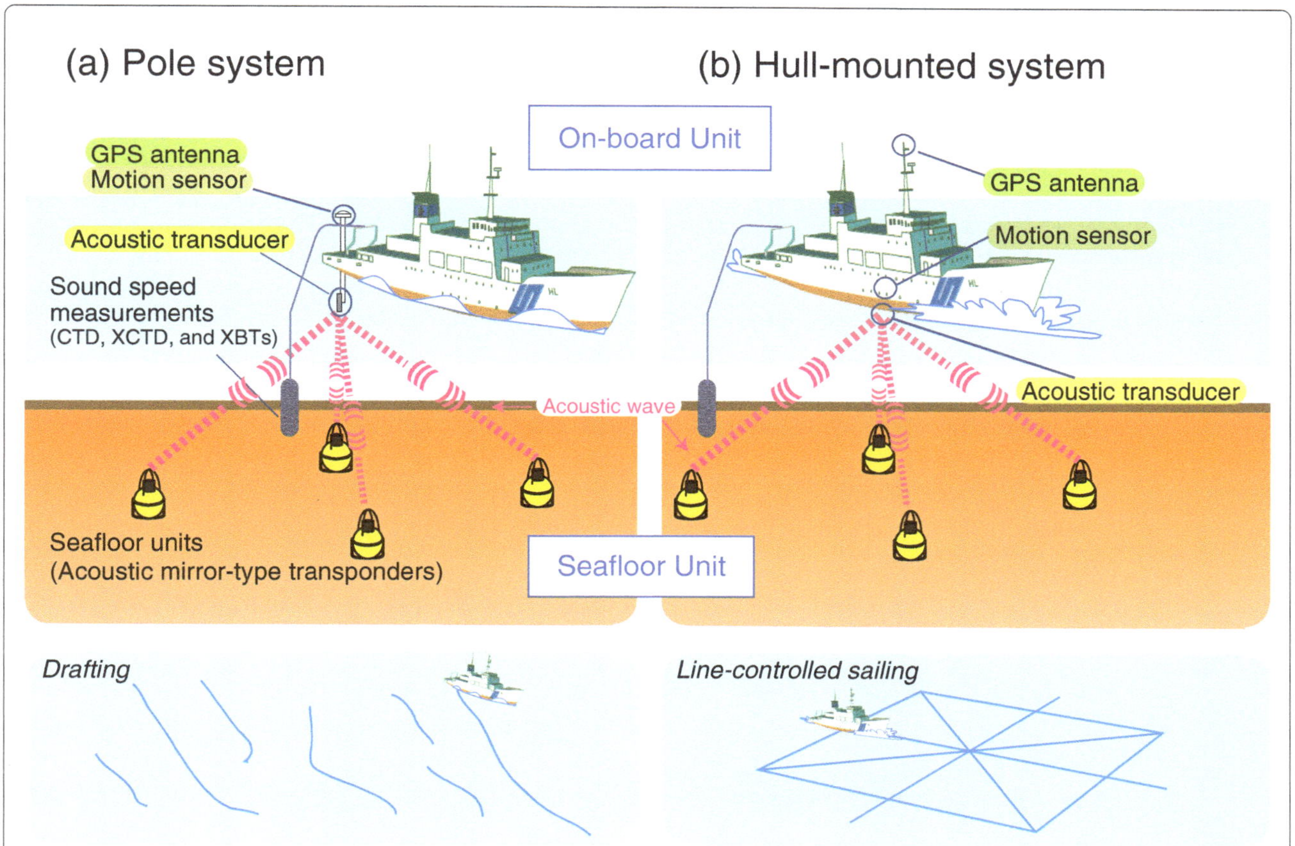

Figure 2 Schematic pictures of the GPS-acoustic seafloor geodetic observation systems. (**a**) Pole system for drifting observation and (**b**) hull-mounted system for line-controlled sailing observation.

switching to a line-controlled sailing observation method with a hull-mounted system after 2008 as shown in Figure 2b. Detail methodologies are introduced in Sato et al. (2013a).

The system measures ranges from the on-board transducer to the seafloor acoustic transponders through roundtrip acoustic travel times. Acoustic velocity profiles in the seawater are necessary to transform travel time into range. These are obtained using temperature and salinity profilers (conductivity temperature depth profiler (CTD), expendable conductivity temperature depth profiler (XCTD), and expendable bathy thermographs (XBTs)) every several hours. Kinematic GPS data are simultaneously gathered to determine the absolute position of the survey vessel. Attitude data on the survey vessel are also acquired on board by a dynamic motion sensor to determine the coordinates of the on-board transducer relative to those of the GPS antenna.

Analytical approach

The data analysis was done in three parts. First, we obtained the range between the on-board transducer and the seafloor transponder using a process that correlates measured acoustic wave data and velocity profiles. Second, the consecutive absolute positions of the GPS antenna on the vessel were determined by kinematic GPS analysis. Third, the position of the seafloor transponder was determined using a linearized inversion method based on a least squares formulation combining the results from prior analyses.

In the early stages of our observations, we performed this inversion analysis constrained by the height of the grouped transponders over all epochs (e.g., Fujita et al. 2006). The present analysis was constrained by the positional relationship of the grouped transponders for all epochs (e.g., Watanabe et al. 2014). The newer method was developed by Matsumoto et al. (2008) (Figure 3). To stabilize the estimates, over 5,000 shots of acoustic wave data per site were required for each observation epoch. We spent approximately 24 h performing an observation. More details on the analytical methodology are presented in Fujita et al. (2006), Matsumoto et al. (2008), Sato et al. (2013a), and Watanabe et al. (2014).

Results and discussion

Observation results

Our seafloor sites along the Nankai Trough are located on the offshore region that is predicted by the Central Disaster Management Council of the Japanese government (CDMC) to undergo the next great Nankai Trough earthquake (CDMC 2013) (available at http://www.bousai.go.jp/jishin/nankai/taisaku/pdf/1_1.pdf, in Japanese) (Figure 1). From the eastern side, sites TOK1, TOK2, KUM1, KUM3, SIOW, and MRT2 were set along the trench axis at 50 to 100 km intervals. TOK1 is located closest to the trench axis.

The observations began with site KUM1 in November 2000, with the other five sites being set between 2001 and 2004. They have been operating steadily since 2006. Here, we show observations before the 2011 Tohoku-oki earthquake. We carried out eight to ten observation campaigns in the period between 2006 and January 2011.

Figure 4 shows the time series of the estimated horizontal coordinates of these six sites. Each circle represents the coordinates of seafloor transponders for each epoch after 2006 relative to their locations in the first campaign. The positions are presented with respect to the stable part of the Eurasian plate, based on the

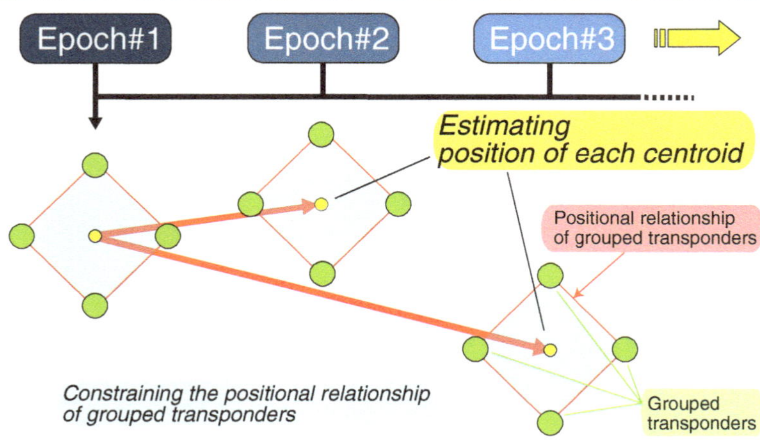

Figure 3 Schematic pictures of the estimation of seafloor positions. The positions were analyzed using the method developed by Matsumoto et al. (2008). In this analysis, we estimated the position (yellow circle) of the centroid of the grouped transponders (green circles) for each epoch, constrained by the positional relationship of the grouped transponders (red diamonds) in all epochs.

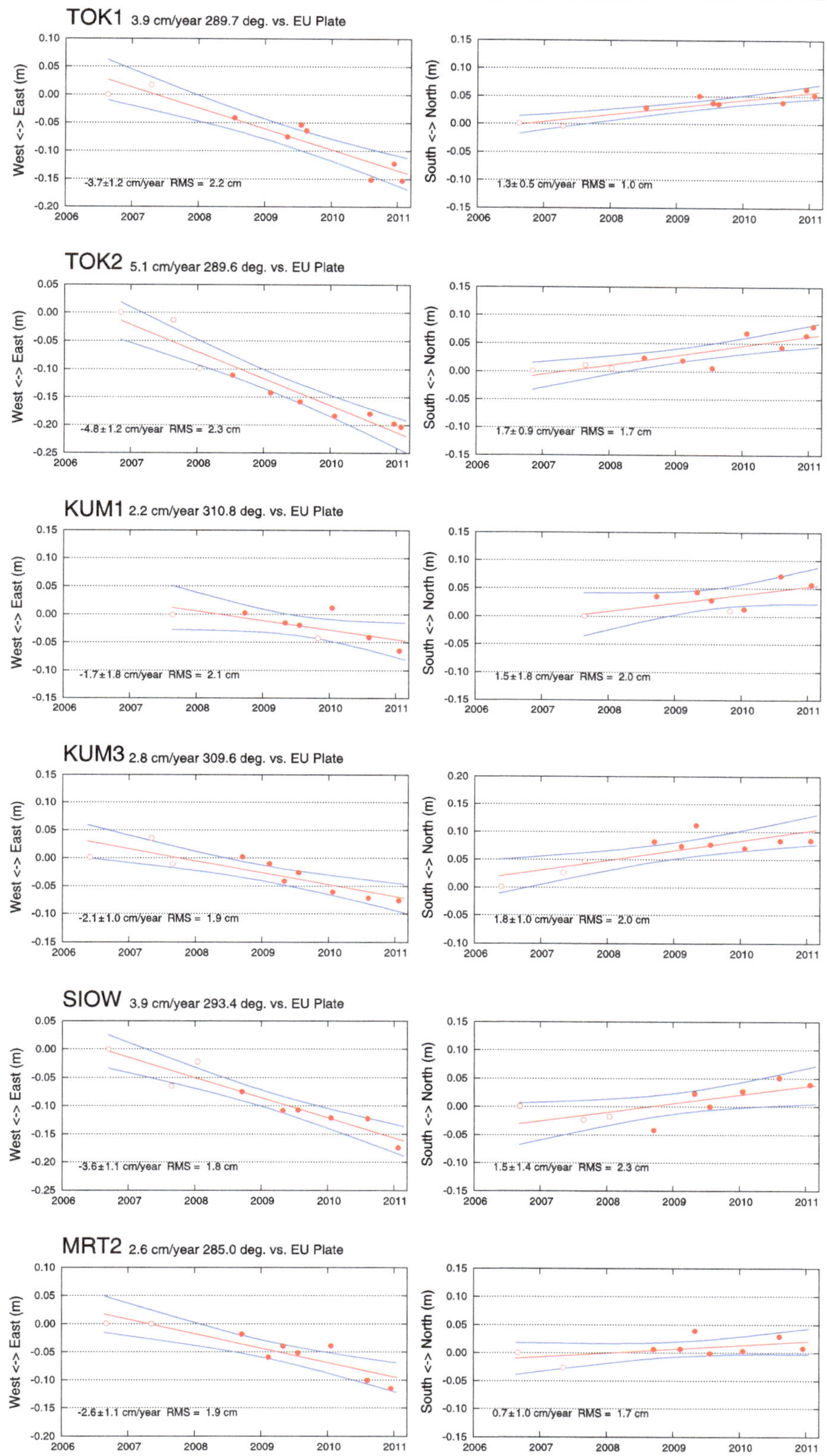

Figure 4 (See legend on next page.)

(See figure on previous page.)
Figure 4 Time series in east-west (left column) and north-south (right column) displacements. These data were obtained at the six seafloor sites TOK1, TOK2, KUM1, KUM3, SIOW, and MRT2 for the period from 2006 to January 2011. The position reference is the Eurasian plate. The open and solid red circles indicate results from drafting observations and results from sailing observations, respectively. The linear trends and the 95% two-sided confidence intervals are shown with red and blue lines, respectively.

NUVEL-1A velocity (DeMets et al. 1994). The data are summarized in Tables 2 and 3. Reference frame in Table 2 is ITRF2005 (Altamimi et al. 2007).

Seafloor displacement rates

Figure 4 also presents linear trends fitted to these time series, which gave us the offshore velocity fields with the 95% two-sided confidence intervals indicated by the blue hyperbolic lines. All the sites moved at stable displacement rates in the northwestward direction. Because no large earthquake occurred in this offshore region after the 2004 off-the-Kii-Peninsula earthquakes (M ~ 7), these movements reflect strain accumulation processes due to the Philippine Sea plate subduction under the upper Eurasian plate.

The estimated horizontal seafloor displacement rates are exhibited by arrows in Figure 5. The ellipses indicate the 2-sigma errors. The slowest and fastest velocities, 2.2 and 5.1 cm/year, were observed at KUM1 and TOK2, respectively, though the error ellipse of KUM1 is larger than that of the other sites because of its shorter observation period.

Figure 5 also presents the onshore velocity fields observed at GEONET stations operated by the Geospatial Information Authority of Japan (GSI) (Sagiya et al. 2000) for comparison. These onshore displacement rates were derived using the averages between March 2006 and February 2011 so as to match to our observation period.

Our seafloor observation results present uneven offshore velocity fields along the Nankai Trough. In particular, the 2 to 3 cm/year displacement rates at KUM1, KUM3, and MRT2 are slower than those of adjacent seafloor sites. In addition, the rates at KUM3 and MRT2 are slower than those of nearby onshore GPS stations.

The displacement rate at KUM1 is smaller than that at TOK2 (95% confidence level (CL)). Similarly, the displacement rates at KUM3 and MRT2 are smaller than those at SIOW (95% CL and 90% CL) and the onshore 'Kushimoto' (99.99% CL) and 'Muroto-4' (99.99% CL) GPS stations as shown in Figure 6. Because the relative velocity between the Eurasian plate and the Philippine Sea plate is approximately 4 to 5 cm/year (e.g., Seno et al. 1993 and DeMets et al. 1994), the plate boundary beneath these three offshore sites does not have high interplate-coupling ratios (<75%). Each observation is discussed in more detail from the eastern side below.

Tokai region

The direction and magnitude of the displacement rate at site TOK1 are approximately consistent with those of nearby onshore stations. In this easternmost plate boundary region beneath TOK1, no major earthquake has occurred in the past 160 years since the last historical earthquake in 1854 (Ando 1975). The CDMC (2013) therefore hypothesized that this will likely be the area of the next large event, the future Tokai earthquake (orange region in Figure 5). Our seafloor observations also confirm a high-coupling ratio (>75%) in this region near the trench axis.

The displacement rate of the next site (TOK2) is larger than not only those of nearby onshore and adjacent offshore sites but also the subducting plate velocity calculated on the basis of the NUVEL-1A model. This is possibly due to the Izu microplate (e.g., Sagiya 1999) and the spray fault along the Nankai Trough (e.g., Park et al. 2002). This observation provides evidence for almost full coupling in this interplate region, though TOK2 was not located in the source region of the future Tokai earthquake. This region also has the potential to produce a large earthquake, similar to the region around TOK1.

Tonankai region

The western side of the future Tokai earthquake source region is the location of the 1944 Tonankai earthquake. Baba and Cummins (2005) estimated the slip distribution (blue contour in Figure 5) using tsunami data. Since the tsunami data used have resolving power in the area near the trench axis, along with the seafloor geodetic data, this offshore slip distribution is reliable and can be compared with the seafloor displacement rates.

The seafloor geodetic sites operated by Nagoya University (Tadokoro et al. 2012) are located on the north side of our seafloor sites in the Kumano-nada Sea (KUM1 and KUM3) (Figure 5). Those seafloor sites moved slower than TOK2, like our result at KUM1, and faster than nearby onshore GPS stations.

Our result at KUM3 differs from Nagoya University's results. The displacement rate at KUM3 is smaller than the nearby 'Kushimoto' GPS station (99.99% CL), and its direction is comparatively clockwise because of a smaller westward component and a larger northward component than at 'Kushimoto' (99.99% CL) as shown in Figure 6. If the coupling ratio of this interplate region beneath KUM3 was the same as in the north deep region beneath Nagoya University's seafloor sites, KUM3 should move faster since it is closer to the plate boundary. Therefore, the coupling ratio is believed to be lower (50% to 75%) there than in the north region beneath the Kumano-nada Sea. This is consistent

Table 2 Estimated relative site positions

Epoch (year)	Eastward (m)	Northward (m)
TOK1		
2006.630[a]	0.000	0.000
2007.285[a]	0.030	−0.015
2008.537	−0.004	−0.001
2009.340	−0.022	0.008
2009.545	0.004	−0.008
2009.627	−0.005	−0.011
2010.595	−0.073	−0.023
2010.948	−0.037	−0.005
2011.068[b]	−0.066	−0.018
TOK2		
2006.847[a]	0.000	0.000
2007.633[a]	0.002	−0.002
2008.030[a]	−0.076	−0.014
2008.526	−0.078	−0.003
2009.104	−0.098	−0.017
2009.548	−0.105	−0.037
2010.063	−0.120	0.017
2010.595	−0.106	−0.017
2010.964	−0.116	−0.001
2011.066	−0.122	0.015
KUM1		
2007.638[a]	0.000	0.000
2008.726	0.025	0.018
2009.337	0.019	0.017
2009.551	0.019	−0.001
2009.827[ab]	0.002	−0.025
2010.041	0.060	−0.025
2010.597	0.019	0.025
2011.055	0.004	0.003
KUM3		
2006.405[a]	0.000	0.000
2007.337[a]	0.053	0.011
2007.655[a]	0.014	0.027
2008.704	0.047	0.046
2009.112	0.043	0.032
2009.334	0.017	0.065
2009.551	0.036	0.028
2010.060	0.011	0.013
2010.600	0.011	0.018
2011.058	−0.037	0.001
SIOW		
2006.690[ab]	0.000	0.000
2007.644[a]	−0.045	−0.039
2008.036[a]	0.005	−0.039
2008.707	−0.034	−0.074
2009.323	−0.055	−0.018
2009.553	−0.049	−0.045
2010.047	−0.053	−0.025
2010.603	−0.043	−0.010
2011.063	−0.085	−0.029
MRT2		
2006.666[a]	0.000	0.000
2007.348[a]	0.013	−0.037
2008.707	0.023	−0.026
2009.110	−0.010	−0.031
2009.332	0.014	−0.002
2009.556	0.007	−0.046
2010.049	0.030	−0.049
2010.603	−0.021	−0.031
2010.956	−0.028	−0.058

Reference frame is ITRF2005 (Altamimi et al. 2007). [a]Epoch with drafting observations. [b]Relatively unreliable due to less data.

with the 1944 Tonankai earthquake source region, which has a peak slip region in the deep side. These features are different from the easternmost offshore region around TOK1 and TOK2.

Nankaido region

The observation for the two westernmost sites SIOW and MRT2 contrast with each other. These sites are both located near the 1946 Nankaido earthquake source region estimated by Baba and Cummins (2005) (green contour in Figure 5). The displacement rate of SIOW is similar to nearby onshore data from Cape Shiono-misaki and suggests that the interplate region around SIOW has a high-coupling ratio (>75%). On the other hand, the displacement rate at MRT2 is clearly smaller than the onshore data from Cape Muroto-zaki, e.g., 'Muroto-4' (99.99% CL), as shown in Figure 6. Although the

Table 3 Velocities with respect to the Eurasian plate with component variances and covariances

Site name	Velocity (cm/year)		Variance and Covariance ((cm/year)^2)		
	East (V_e)	North (V_n)	V (V_e)	V (V_n)	Cov (V_e, V_n)
TOK1	−3.7	1.3	0.25	0.05	0.02
TOK2	−4.8	1.7	0.27	0.14	0.06
KUM1	−1.7	1.5	0.53	0.52	−0.11
KUM3	−2.1	1.8	0.19	0.20	0.06
SIOW	−3.6	1.5	0.21	0.33	0.03
MRT2	−2.6	0.7	0.23	0.18	−0.01

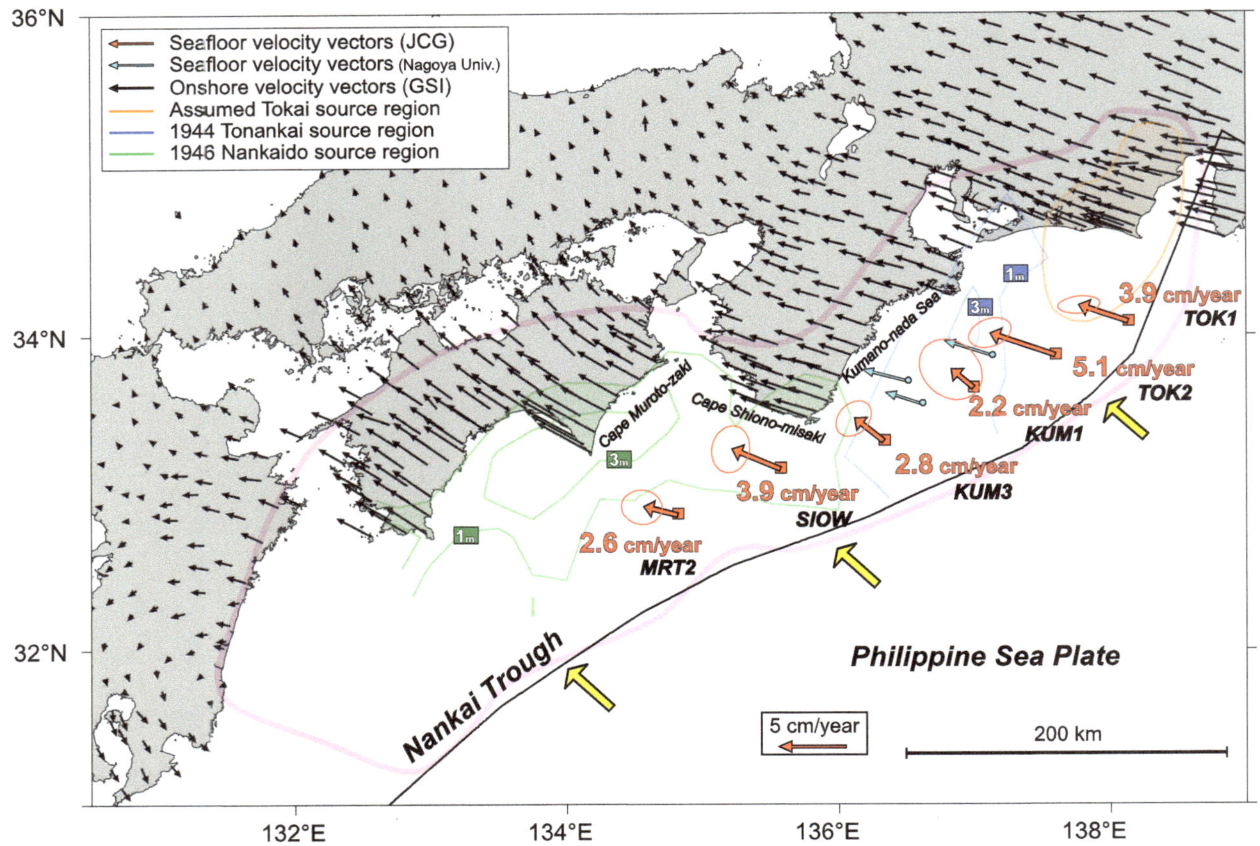

Figure 5 Horizontal velocity vectors detected by seafloor geodetic observations at six seafloor sites. Seafloor velocity vectors are shown by red arrows. Red ellipses indicate the 2-sigma errors. Light blue arrows indicate the seafloor geodetic observations by Nagoya University (Tadokoro et al. 2012). The onshore velocity vectors are calculated for the period from March 2006 to February 2011 using GEONET stations. They are shown as black arrows. Yellow arrows indicate the convergence rate of the Philippine Sea plate under the Eurasian plate (DeMets et al. 1994). The purple region shows the maximum source model of the Nankai Trough great earthquake assumed by the CDMC (2013). The orange region is the source model of the future Tokai earthquake assumed by the CDMC (2013). The blue and green contour lines are coseismic slip distributions associated with the 1944 Tonankai and 1946 Nankaido earthquakes, respectively, obtained by tsunami inversion analyses (Baba and Cummins 2005).

direction of SIOW's velocity vector is similar to those of onshore stations nearby, MRT2 differs in a counterclockwise fashion due to a smaller northward component than at 'Muroto-4' (99.99% CL). The movement at MRT2 is inconsistent with the onshore geodetic observations, suggesting the presence of a weak interplate-coupling region (about 50%) located off Cape Muroto-zaki. This is consistent with the 1946 Nankaido earthquake source region, which has a peak slip region in the deep side.

Relationship with subducting seamount and ridge

In the interplate region beneath TOK1 and TOK2, the Paleo-Zenisu ridge was detected by seismic refraction studies (Kodaira et al. 2003 and Park et al. 2003) (light blue polygon in Figure 7). These and several other studies (e.g., Kodaira et al. 2000 and Duan 2012) interpreted that a subducting seamount increases the normal stress and makes a locally strong coupling region. On the other hand, other seismic surveys (Ranero and von Huene 2000, Bangs et al. 2006, and Mochizuki et al. 2008) and a laboratory study (Dominguez et al. 2000) suggested that interplate coupling is weak on a subducting seamount due to elevated pore pressure. Therefore, the effect of a subducting seamount on an interplate-coupling state is now being debated. Our observations at TOK1 and TOK2 suggest the presence of strong interplate coupling around the Paleo-Zenisu ridge, consistent with the strengthening effect proposed by the former studies.

On the interplate boundary beneath MRT2, the seismic refraction study detected a subducting seamount (Kodaira et al. 2000) (light blue ellipse in Figure 7). The positional relationship between this seamount and the weak interplate-coupling region suggested by our seafloor observations at MRT2 suggests the possibility that the weakening effect proposed by the latter studies is at work.

Therefore, the Paleo-Zenisu ridge and the subducting seamount located off Cape Muroto-zaki have different

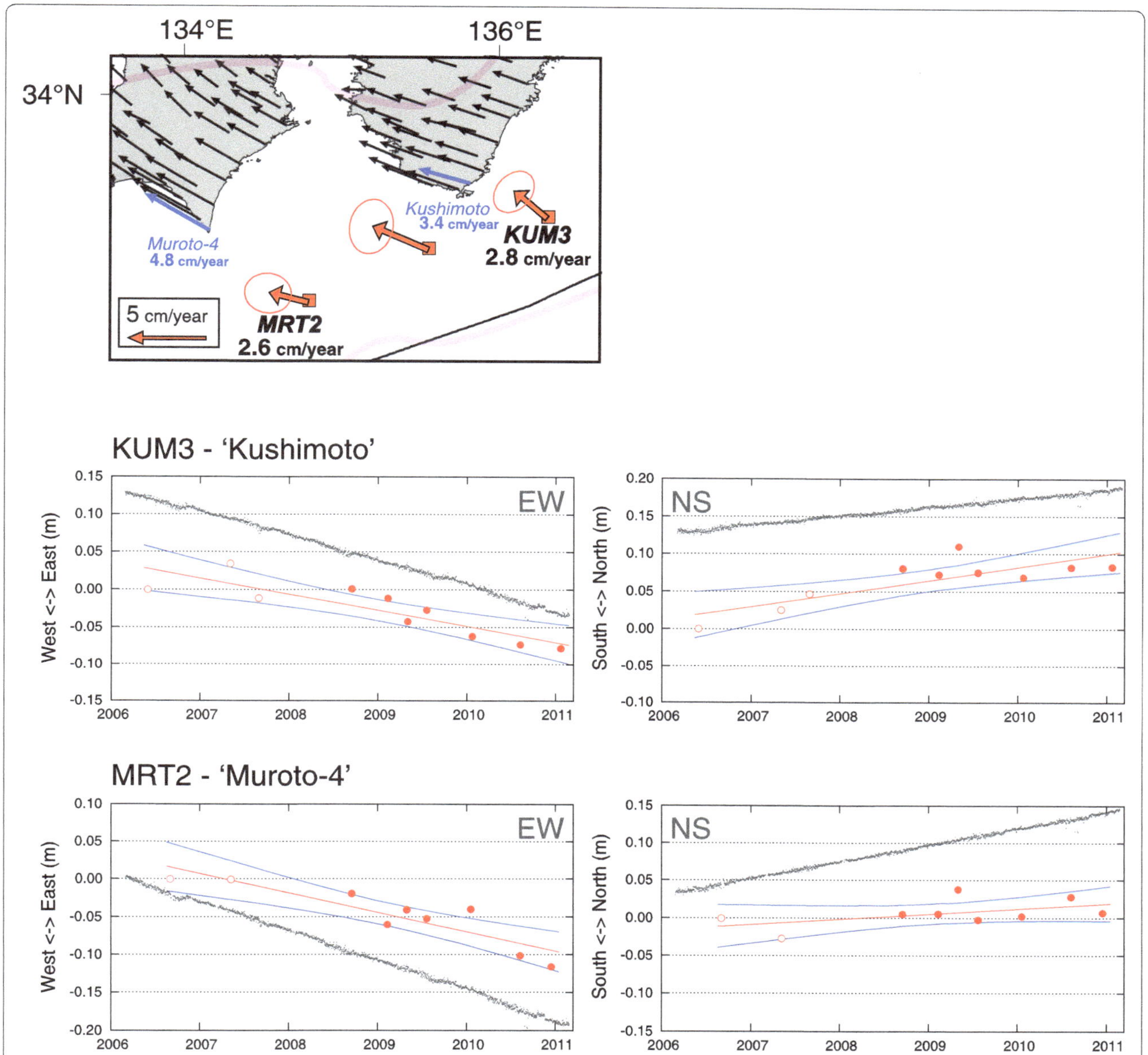

Figure 6 Comparisons of seafloor and onshore time series in east-west and north-south displacements. Time series of sites KUM3 and MRT2 compared with nearby onshore GPS data. Gray dots indicate the time series in the horizontal coordinates of the onshore 'Kushimoto' and 'Muroto-4' GPS station. The position reference is the Eurasian plate.

effects on interplate coupling. This indicates that the effect of a subducting seamount for a plate boundary differs according to various conditions, e.g., the height and width of the seamount, temperature and pressure on the plate boundary, physical properties, and convergence rate. Although the clearest difference between the Paleo-Zenisu ridge and the Muroto-zaki subducting seamount is their breadth in the east-west direction, it is difficult to identify that as the cause of the differing effects on interplate coupling because there are other unknown conditions. Recent numerical simulation studies also offer a similar suggestion (Yang et al. 2012, 2013). They point out that the effect of a subducting seamount varies complicatedly depending on the friction parameters, the depth, and the shape.

Conclusions

We have presented the results of seafloor geodetic observations conducted over about 5 years along the Nankai Trough and found the heterogeneous interplate coupling near the trough axis to have the following features. A strong coupling region is located in the easternmost interplate region around TOK1 above the Paleo-Zenisu ridge that is hypothesized to

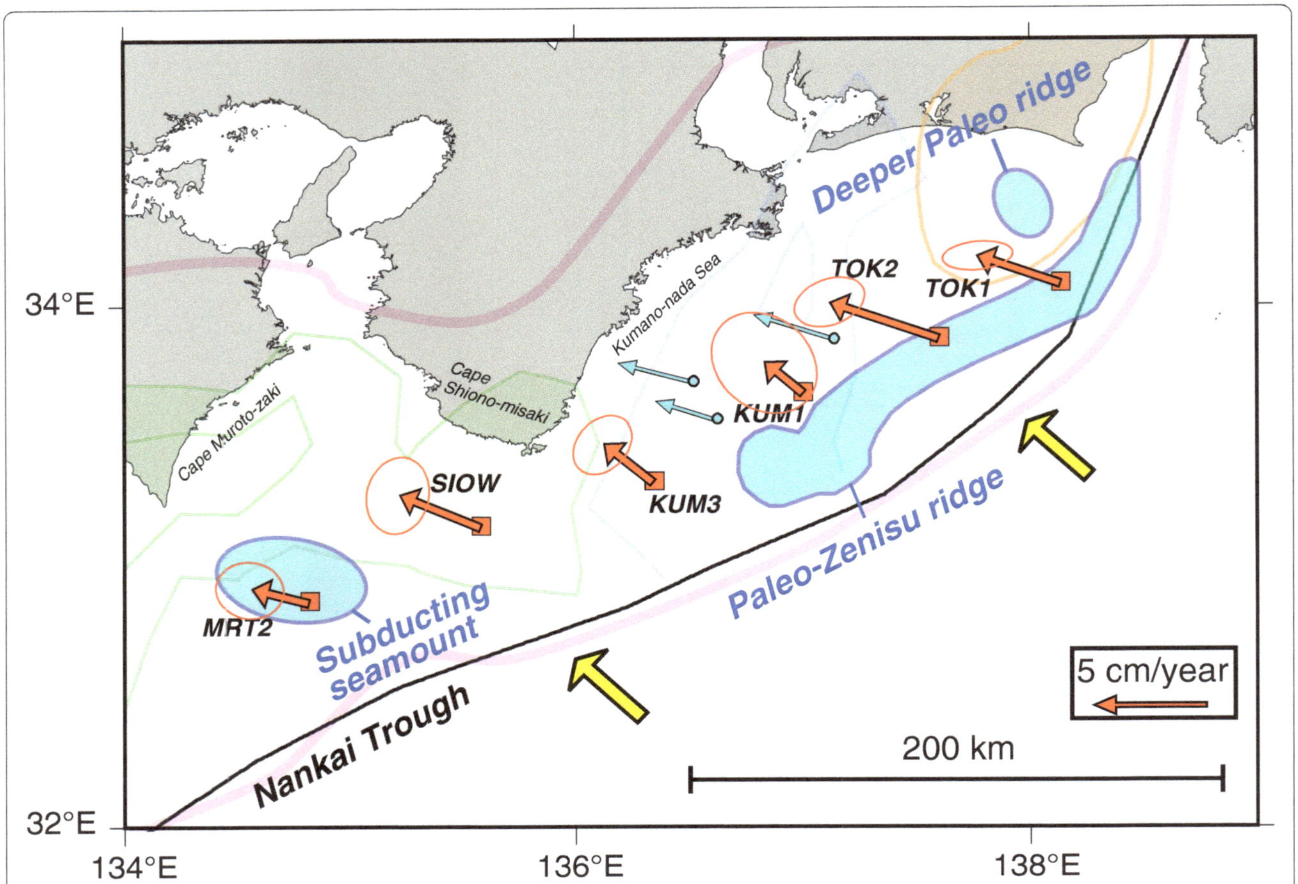

Figure 7 Seafloor velocity vectors and locations of the subducting seamount and ridges. The subducting seamount (Kodaira et al. 2000) and the Paleo-Zenisu and deeper Paleo ridges (Kodaira et al. 2003 and Park et al. 2003) are shown by light blue regions. The light blue and red arrows indicate the seafloor velocity vectors observed by Nagoya University (Tadokoro et al. 2012) and the authors, respectively.

be the source area of the future Tokai earthquake. This strong coupling region is believed extend to the southwest around TOK2. A shallow interplate region around KUM1 and KUM3 does not show a full coupling ratio, which is consistent with the 1944 Tonankai earthquake source region estimated using tsunami data. In the westernmost region, the slow displacement rate observed at MRT2 reflects weak interplate coupling, probably due to the subducting seamount off Cape Muroto-zaki. Our observations above the subducting ridge and seamount indicate that the effect of a subducting seamount on interplate-coupling region differs depending on various conditions.

As above, our seafloor geodetic observations suggested a complication of the interplate coupling on the assumed maximum source model of the Nankai Trough great earthquake. However, the blank offshore region on the west side of MRT2 has still not been revealed. In order to expand the seafloor geodetic observation network to this blank region and interpolate other blank regions, nine new seafloor sites were deployed in addition to the six existing sites after November 2011. These new sites will deliver useful information for discussing the earthquake model and disaster prevention in more detail.

Abbreviations

CDMC: the Central Disaster Management Council of the Japanese government; CL: confidence level; CTD: conductivity temperature depth profiler; GPS: global positioning system; GSI: Geospatial Information Authority of Japan; JHOD: Hydrographic and Oceanographic Department of Japan; XBT: expendable bathy thermograph; XCTD: expendable conductivity temperature depth profiler.

Competing interests

The authors declare that they have no competing interests.

Authors' contributions

YY, TI, MS, and SW proposed the topic and conceived and designed the study. YY, TI, MS, SW, HS, NU, YM, ST, MF, and TY carried out the observations and analyzed the data. YY, TI, MS, SW, HS, NU, YM, ST, MF, TY, MM, and AA helped in interpreting the data and approved the final manuscript.

Acknowledgements

We thank the Geospatial Information Authority of Japan for providing us with the GPS data from the terrestrial transponders for the kinematic GPS analysis and for releasing the daily coordinates of the sites on the GSI's website. We greatly thank

Professor Keiichi Tadokoro for providing us with Nagoya University's seafloor geodetic observations. We also thank Oscar L. Colombo of the NASA Goddard Space Flight Center for providing us with the kinematic GPS software IT (Interferometric Translocation) (Colombo et al. 2000). Many among the staff of the Hydrographic and Oceanographic Department of the Japan Coast Guard, including the crew of the S/Vs Meiyo and Kaiyo, have supported our observations and data analyses. Comments from anonymous reviewers have improved our manuscript. Some figures were produced using the GMT software (Wessel and Smith 1991). Our article is in the science section: 4) solid earth sciences.

Author details

[1]Hydrographic and Oceanographic Department, Japan Coast Guard, 2-5-18 Aomi, Koto-ku, Tokyo 135-0064, Japan. [2]National Research Institute for Earth Science and Disaster Prevention, 3-1 Tennodai, Tsukuba, Ibaraki 305-0006, Japan. [3]Institute of Industrial Science, University of Tokyo, 4-6-1 Komaba, Meguro-ku, Tokyo 153-8505, Japan.

References

Altamimi Z, Collilieux X, Legrand J, Garayt B, Boucher C (2007) ITRF2005: a new release of the International Terrestrial Reference Frame based on time series of station positions and Earth Orientation Parameters. J Geophys Res 112:B09401, doi:10.1029/2007JB004949

Ando M (1975) Source mechanism and tectonic significance of historical earthquakes along the Nankai Trough, Japan. Tectonophysics 27:119–140, doi:10.1016/0040-1951(75)90102-X

Asada A, Yabuki T (2001) Centimeter-level positioning on the seafloor. Proc Jpn Acad Ser B 77:7–12

Baba T, Cummins PR (2005) Contiguous rupture areas of two Nankai Trough earthquakes revealed by high-resolution tsunami waveform inversion. Geophys Res Lett 32:L083005, doi:10.1029/2004GL022320

Bangs NLB, Gulick SPS, Shipley TH (2006) Seamount subduction erosion in the Nankai Trough and its potential impact on the seismogenic zone. Geology 34:701–704, doi:10.1130/G22451.1

Colombo OL, Evans AG, Vigo-Aguiar MI, Ferrandiz JM, Benjamin JJ (2000) Long-baseline (>1000 km), sub-decimeter kinematic positioning of buoys at sea, with potential application to deep sea studies. Proc ION GPS 2000:1476–1484

Central Disaster Management Council of the Japanese Government (2013) http://www.bousai.go.jp/jishin/nankai/taisaku/pdf/1_1.pdf (in Japanese).

DeMets C, Gordon RG, Argus DF, Stein S (1994) Effect of recent revisions to the geomagnetic reversal time scale on estimates of current plate motions. Geophys Res Lett 21:2191–2194, doi:10.1029/94GL02118

Dominguez S, Malavieille J, Lallemand SE (2000) Deformation of accretionary wedges in response to seamount subduction: insights from sandbox experiments. Tectonics 19:182–196, doi:10.1029/1999TC900055

Duan B (2012) Dynamic rupture of the 2011 Mw 9.0 Tohoku-Oki earthquake: roles of a possible subducting seamount. J Geophys Res 117:B05311, doi:10.1029/2011JB009124

Fujita M, Ishikawa T, Mochizuki M, Sato M, Toyama S, Katayama M, Matsumoto Y, Yabuki T, Asada A, Colombo OL (2006) GPS/acoustic seafloor geodetic observation: method of data analysis and its application. Earth Planets Space 58:265–275, doi:10.1186/BF03351923

Ito T, Hashimoto M (2004) Spatiotemporal distribution of interplate coupling in southwest Japan from inversion of geodetic data. J Geophys Res 109:B02315, doi:10.1029/2002JB002358

Kido M, Fujimoto H, Miura S, Osada Y, Tsuka K, Tabei T (2006) Seafloor displacement at Kumano-nada caused by the 2004 off Kii Peninsula earthquakes, detected through repeated GPS/acoustic surveys. Earth Planets Space 58:911–915

Kido M, Osada Y, Fujimoto H, Hino R, Ito Y (2011) Trench-normal variation in observed seafloor displacements associated with the 2011 Tohoku-Oki earthquake. Geophys Res Lett 38:L24303, doi:10.1029/2011GL050057

Kodaira S, Takahashi N, Nakanishi A, Miura S, Kaneda Y (2000) Subducted seamount imaged in the rupture zone of the 1946 Nankaido earthquake. Science 289:104–106, doi:10.1126/science.289.5476.104

Kodaira S, Nakanishi A, Park J-O, Ito A, Tsuru T, Kaneda Y (2003) Cyclic ridge subduction at an inter-plate locked zone off central Japan. Geophys Res Lett 30:6, doi:10.1029/2002GL016595

Loveless JP, Meade BP (2010) Geodetic imaging of plate motions, slip rates, and partitioning of deformation in Japan. J Geophys Res 115:B02410, doi:10.1029/2008JB006248

Matsumoto Y, Fujita M, Ishikawa T, Mochizuki M, Yabuki T, Asada A (2006) Undersea co-seismic crustal movements associated with the 2005 Off Miyagi Prefecture Earthquake detected by GPS/acoustic seafloor geodetic observation. Earth Planets Space 58:1573–1576, doi:10.1186/BF03352663

Matsumoto Y, Fujita M, Ishikawa T (2008) Development of multi-epoch method for determining seafloor station position (in Japanese). Report of Hydrographic and Oceanographic Researches 26:16–22

Mochizuki K, Yamada T, Shinohara M, Yamanaka Y, Kanazawa T (2008) Weak interplate coupling by seamounts and repeating M ~ 7 earthquakes. Science 321:1194–1197, doi:10.1126/science.1160250

Nishimura S, Hashimoto M (2006) A model with rigid rotations and slip deficits for the GPS-derived velocity field in southwest Japan. Tectonophysics 421:187–207, doi:10.1016/j.tecto.2006.04.017

Park J-O, Tsuru GT, Kodaira S, Cummins PR, Kaneda Y (2002) Splay fault branching along the Nankai subduction zone. Science 297:1157–1160, doi:10.1126/science.1074111

Park J-O, Moore GF, Tsuru T, Kodaira S, Kaneda Y (2003) A subducted oceanic ridge influencing the Nankai megathrust earthquake rupture. Earth and Planet Sci Let 217:77–84, doi:10.1016/S0012-821X(03)00553-3

Ranero CR, von Huene R (2000) Subduction erosion along the Middle America convergent margin. Nature 404:748–752, doi:10.1038/35008046

Sagiya T (1999) Interplate coupling in the Tokai district, central Japan, deduced from continuous GPS data. Geophys Res Lett 26:2315–2318, doi:10.1029/1999GL900511

Sagiya T, Thacher W (1999) Coseismic slip resolution along a plate boundary megathrust: the Nankai Trough, southwest Japan. J Geophys Res 104(B1):1111–1129, doi:10.1029/98JB02644

Sagiya T, Miyazaki S, Tada T (2000) Continuous GPS array and present-day crustal deformation of Japan. Pure Appl Geophys 157:2303–2322, doi:10.1007/PL00022507

Sato M, Ishikawa T, Ujihara N, Yoshida S, Fujita M, Mochizuki M, Asada A (2011a) Displacement above the hypocenter of the 2011 Tohoku-oki earthquake. Science 332:1395, doi:10.1126/science.1207401

Sato M, Saito H, Ishikawa T, Matsumoto Y, Fujita M, Mochizuki M, Asada A (2011b) Restoration of interplate locking after the 2005 Off-Miyagi prefecture earthquake, detected by GPS/acoustic seafloor geodetic observation. Geophys Res Lett 38:L01312, doi:10.1029/2010GL045689

Sato M, Fujita M, Matsumoto Y, Saito H, Ishikawa T, Asakura T (2013a) Improvement of GPS/acoustic seafloor positioning precision through controlling the ship's track line. J Geod 87:825–842, doi:10.1007/s00190-013-0649-9

Sato M, Fujita M, Matsumoto Y, Ishikawa T, Saito H, Mochizuki M, Asada A (2013b) Interplate coupling off northeastern Japan before the 2011 Tohoku-oki earthquake, inferred from seafloor geodetic data. J Geophys Res 118:1–10, doi:10.1002/jgrb.50275

Seno T, Stein S, Gripp AE (1993) A model for the motion of the Philippine Sea plate consistent with NUVEL-1 and geological data. J Geophys Res 98(B10):17941–17948

Spiess FN, Chadwell CD, Hildebrand JA, Young LE, Purcell GH Jr, Dragert H (1998) Precise GPS/acoustic positioning of seafloor reference points for tectonic studies. Phys Earth and Planet Inter 108:101–112, doi:10.1016/S0031-9201(98)00089-2

Tadokoro K, Ikuta R, Watanabe T, Ando M, Okuda T, Nagai S, Yasuda K, Sakata T (2012) Interseismic seafloor crustal deformation immediately above the source region of anticipated megathrust earthquake along the Nankai Trough, Japan. Geophys Res Lett 39:L10306, doi:10.1029/2012GL051696

Tanioka Y, Satake K (2001) Detailed coseismic slip distribution of the 1944 Tonankai Earthquake estimated from tsunami waveforms. Geophys Res Lett 28:1075–1078, doi:10.1029/2000GL012284

Watanabe S, Sato M, Fujita M, Ishikawa T, Yokota Y, Ujihara N, Asada A (2014) Evidence of viscoelastic deformation following the 2011 Tohoku-oki earthquake revealed from seafloor geodetic observation. Geophys Res Lett 41:5789–5796, doi:10.1002/2014GL061134

Wessel P, Smith WHF (1991) Free software helps map and display data. Eos Trans AGU 72:441, doi:10.1029/90EO00319

Yang H, Liu Y, Lin J (2012) Effects of subducted seamounts on megathrust earthquake nucleation and rupture propagation. Geophys Res Lett 39:L24302, doi:10.1002/2012GL053892

Yang H, Liu Y, Lin J (2013) Geometrical effects of a subducted seamount on stopping megathrust ruptures. Geophys Res Lett 40:1–6, doi:10.1002/grl.50509

Yasuda K, Tadokoro K, Ikuta R, Watanabe T, Nagai S, Okuda T, Fujii C, Sayanagi K (2014) Interplate locking condition derived from seafloor geodetic data at the northernmost part of the Suruga Trough, Japan. Geophys Res Lett 41:5806–5812, doi:10.1002/2014GL060945

Yoshioka S, Matsuoka Y (2013) Interplate coupling along the Nankai Trough, southwest Japan, inferred from inversion analyses of GPS data: effects of subducting plate geometry and spacing of hypothetical ocean-bottom GPS stations. Tectonophysics 600:165–174, doi:10.1016/j.tecto.2013.01.023

Permissions

All chapters in this book were first published in PEPS, by Springer; hereby published with permission under the Creative Commons Attribution License or equivalent. Every chapter published in this book has been scrutinized by our experts. Their significance has been extensively debated. The topics covered herein carry significant findings which will fuel the growth of the discipline. They may even be implemented as practical applications or may be referred to as a beginning point for another development.

The contributors of this book come from diverse backgrounds, making this book a truly international effort. This book will bring forth new frontiers with its revolutionizing research information and detailed analysis of the nascent developments around the world.

We would like to thank all the contributing authors for lending their expertise to make the book truly unique. They have played a crucial role in the development of this book. Without their invaluable contributions this book wouldn't have been possible. They have made vital efforts to compile up to date information on the varied aspects of this subject to make this book a valuable addition to the collection of many professionals and students.

This book was conceptualized with the vision of imparting up-to-date information and advanced data in this field. To ensure the same, a matchless editorial board was set up. Every individual on the board went through rigorous rounds of assessment to prove their worth. After which they invested a large part of their time researching and compiling the most relevant data for our readers.

The editorial board has been involved in producing this book since its inception. They have spent rigorous hours researching and exploring the diverse topics which have resulted in the successful publishing of this book. They have passed on their knowledge of decades through this book. To expedite this challenging task, the publisher supported the team at every step. A small team of assistant editors was also appointed to further simplify the editing procedure and attain best results for the readers.

Apart from the editorial board, the designing team has also invested a significant amount of their time in understanding the subject and creating the most relevant covers. They scrutinized every image to scout for the most suitable representation of the subject and create an appropriate cover for the book.

The publishing team has been an ardent support to the editorial, designing and production team. Their endless efforts to recruit the best for this project, has resulted in the accomplishment of this book. They are a veteran in the field of academics and their pool of knowledge is as vast as their experience in printing. Their expertise and guidance has proved useful at every step. Their uncompromising quality standards have made this book an exceptional effort. Their encouragement from time to time has been an inspiration for everyone.

The publisher and the editorial board hope that this book will prove to be a valuable piece of knowledge for researchers, students, practitioners and scholars across the globe.

List of Contributors

Elena G Petrova
Faculty of Geography, Lomonosov Moscow State University, GSP-1, Leninskie Gory, Moscow 119991, Russia

Yury V Mironov
Vernadsky State Geological Museum of the Russian Academy of Sciences, Mokhovaya 11, Bldg.11, Moscow 125009, Russia

Hajime Matsushima and Satoshi Ebine
Research Faculty of Agriculture, Hokkaido University, Kita 9, Nishi 9, Kita-ku, Sapporo 060-8589, Japan

Katsunori Furuya
Graduate School of Horticulture, Chiba University, 648, Matsudo, Matsudo-shi, Chiba 271-8510, Japan

Anastasia Petrova
Institute for Oriental Studies of the Russian Academy of Sciences, Rozhdestvenka 12, Moscow 107031, Russia

Norimasa Takayama
Forestry and Forest Products Research Institute, 1 Matsunosato, Tsukuba, Ibaraki 305-8687, Japan

Hirofumi Ueda
School of Design, Sapporo City University, Geijutsu-no-mori 1, Minami-ku, Sapporo 005-0864, Japan

Yoji Aoki
Open University of Japan, Nishikamata 5-12-1, Ota-ku, Tokyo 144-0051, Japan

Roland T Tsunoda
Center for Geospace Studies, SRI International, 333 Ravenswood Ave, Menlo Park, CA 94025, USA

Satoshi Taguchi and Keisuke Hosokawa
Department of Communication Engineering and Informatics, University of Electro-Communications, Chofu, Tokyo 182-8585, Japan
Center for Space Science and Radio Engineering, University of Electro-Communications, Chofu, Tokyo 182-8585, Japan

Satoshi Taguchi
Now at Department of Geophysics, Graduate School of Science, Kyoto University, Kyoto 606-8502, Japan

Yasunobu Ogawa
National Institute of Polar Research, Tachikawa, Tokyo 190-8518, Japan

Virginia Strati, Marica Baldoncini, Fabio Mantovani and Barbara Ricci
Department of Physics and Earth Sciences, University of Ferrara, Via Saragat 1, 44121 Ferrara, Italy

Virginia Strati, Ivan Callegari and Gerti Xhixha
Legnaro National Laboratories, INFN, Viale dell'Università, 2, 35020 Legnaro, Italy

Marica Baldoncini, Fabio Mantovani and Barbara Ricci
Ferrara Section, INFN, Via Saragat 1, 44121 Ferrara, Italy

William F McDonough
Ferrara Section, INFN, Via Saragat 1, 44121 Ferrara, Italy

Syun-Ichi Akasofu
International Arctic Research Center, University of Alaska Fairbanks, Fairbanks, AK, USA

Keiji Horikawa and Jing Zhang
Graduate School of Science and Engineering for Research, University of Toyama, Gofuku 3190, Toyama 930-8555, Japan

Tomohiro Kodaira
Graduate School of Science and Engineering for Education, University of Toyama, Gofuku 3190, Toyama 930-8555, Japan

Masafumi Murayama
Center for Advanced Marine Core Research, Kochi University, B200 Monobe, Nankoku 783-8502, Japan

Daisuke Ishimura
Disaster Science Division, International Research Institute of Disaster Science, Tohoku University, 468-1 Aza-aoba, Aramaki, Aoba-ku, Sendai, Miyagi 980-0845, Japan

Takahiro Miyauchi
Department of Earth Sciences, Graduate School of Science, Chiba University, 1-33 Yayoi-cho, Inage-ku, Chiba, Chiba 263-8522, Japan

Satoshi Kaneshima
Department of Earth and Planetary Science, Kyushu University, Hakozaki, Higashi-ku, 812-8581 Fukuoka, Fukuoka, Japan

Takanori Matsuzawa
National Research Institute for Earth Science and Disaster Prevention, 3-1 Tennodai, 305-0006 Tsukuba, Japan

Charles Le Losq, Bjorn O. Mysen and George D. Cody
Geophysical Laboratory, Carnegie Institution of Washington, 5251 Broad Branch Road NW, Washington, DC 20015, USA

Chi-Wen Chen and Takashi Oguchi
Department of Natural Environmental Studies, Graduate School of Frontier Science, The University of Tokyo, 5-1-5 Kashiwanoha, Kashiwa Chiba 277-8568, Japan

Hitoshi Saito
College of Economics, Kanto Gakuin University, 1-5-1 Mutsuura-higashi, Yokohama, Kanagawa 236-8501, Japan

Hitoshi Saito and Takashi Oguchi
Center for Spatial Information Science, University of Tokyo, 5-1-5 Kashiwanoha, Kashiwa Chiba 277-8568, Japan

Teresa Wong and Viatcheslav S Solomatov
Department of Earth and Planetary Sciences, Washington University, 1 Brookings Drive, 63130-4899 St. Louis, MO, USA

Rachel P. Scudder and Richard W. Murray
Department of Earth & Environment, Boston University, Boston, MA 02215, USA

Julie C. Schindlbeck, Steffen Kutterolf and Folkmar Hauff
GEOMAR Helmholtz Centre for Ocean Research Kiel, 24148 Kiel, Germany

Michael B. Underwood
Department of Earth and Environmental Science, New Mexico Institute of Mining and Technology, Socorro, NM 87801, USA

Samantha Gwizd
Department of Geosciences, Princeton University, Princeton, NJ 08544, USA

Rebecca Lauzon
Division of Earth and Ocean Sciences, Nicholas School of the Environment, Duke University, Durham, NC 27708, USA

Claire C. McKinley and Rachel P. Scudder
Department of Oceanography, Texas A&M University, College Station, TX 77843, USA

Krista M. Soderlund
Institute for Geophysics, John A. & Katherine G. Jackson School of Geosciences, The University of Texas at Austin, J. J. Pickle Research Campus, Building 196 (ROC), 10100 Burnet Road (R2200), Austin, Texas, 78758-4445, USA

Andrey Sheyko
Institut für Geophysik, ETH Zürich, NO H 11.2, Sonneggstrasse 5 8092 Zürich, LA59 Switzerland

Eric M. King
U.S. Global Development Lab, U.S. Agency for International Development, Washington, DC, 20004, USA

Jonathan M. Aurnou
Department of Earth, Planetary, and Space Sciences, University of California, Los Angeles, 595 Charles Young Drive East, Los Angeles, California, 90095-1567, USA

Yousuke Sato, Seiya Nishizawa, Hisashi Yashiro, Yoshiaki Miyamoto, Yoshiyuki Kajikawa and Hirofumi Tomita
RIKEN Advanced Institute for Computational Science, 7-1-26 Minatojima-Minami-machi, Chuo-ku, Kobe, Hyogo 650-0047, Japan

Kazuhiro Miyazaki
Geological Survey of Japan, AIST, Higashi 1-1-1, Tsukuba, 305-8567 Ibaraki, Japan

Yoshinori Takano and Yusuke Yokoyama
Department of Biogeochemistry, Japan Agency for Marine-Earth Science and Technology (JAMSTEC), 2-15 Natsushima, Yokosuka, Kanagawa 237-0061, Japan

Hisaya Kojima, Eriko Takeda, and Manabu Fukui
The Institute of Low Temperature Science, Hokkaido University, N19, W8, Kita-ku, Sapporo, Hokkaido 060-0819, Japan

Yusuke Yokoyama
Atmosphere and Ocean Research Institute, University of Tokyo, 5-1-5 Kashiwanoha, Kashiwa, Chiba 277-8564, Japan

Jun Kawano
Creative Research Institution (CRIS), Hokkaido University, N21 N10, Kita-ku, Sapporo, Hokkaido 001-0021, Japan

Jun Kawano and Takaya Nagai
Faculty of Science, Hokkaido University, N10 W8, Kita-ku, Sapporo, Hokkaido 060-0810, Japan

Hiroshi Sakuma
Environment and Energy Materials Division, National Institute for Materials Science, 1-1 Namiki, Tsukuba, Ibaraki 305-0044, Japan

Yusuke Yokota, Tadashi Ishikawa, Mariko Sato, Shun-ichi Watanabe, Hiroaki Saito, Naoto Ujihara, Yoshihiro Matsumoto, Shin-ichi Toyama, Masayuki Fujita and Tetsuichiro Yabuki
Hydrographic and Oceanographic Department, Japan Coast Guard, 2-5-18 Aomi, Koto-ku, Tokyo 135-0064, Japan

Masashi Mochizuki
National Research Institute for Earth Science and Disaster Prevention, 3-1 Tennodai, Tsukuba, Ibaraki 305-0006, Japan

Akira Asada
Institute of Industrial Science, University of Tokyo, 4-6-1 Komaba, Meguro-ku, Tokyo 153-8505, Japan